Lecture Notes
in Control and Information Sciences 321

Editors: M. Thoma . M. Morari

Wijesuriya P. Dayawansa Anders Lindquist
Yishao Zhou (Eds.)

New Directions and Applications in Control Theory

 Springer

Editors

Dr. Wijesuriya P. Dayawansa
Department of Mathematics
and Statistics
Texas Tech University
79409-1042 Lubbock
USA
E-mail: wijesuriya.dayawansa@ttu.edu

Professor Yishao Zhou
Department of Mathematics
Stockholm University
106 91 Stockholm
Sweden
E-mail: yishao@math.su.se

Professor Anders Lindquist
Department of Mathematics
Royal Institute of Technology
Division Optimization
and Systems Theory
Lindstedtsvägen 25
100 44 Stockholm
Sweden
E-mail: alq@math.kth.se

Library of Congress Control Number: 2005927514

ISSN print edition: 0170-8643
ISSN electronic edition: 1610-7411
ISBN-10   3-540-23953-7 Springer Berlin Heidelberg New York
ISBN-13   978-3-540-23953-6 Springer Berlin Heidelberg New York

Springer is a part of Springer Science+Business Media
springeronline.com
© Springer-Verlag Berlin Heidelberg 2005
Printed in The Netherlands

Typesetting: by the authors and TechBooks using a Springer LATEX macro package

Cover design: *design & production* GmbH, Heidelberg

Printed on acid-free paper   SPIN: 10984413   89/TechBooks   5 4 3 2 1 0

Preface

This volume contains a collection of papers in control theory and applications presented at a conference in honor of Clyde Martin on the occasion of his 60th birthday, held in Lubbock, Texas, November 14-15, 2003.

Clyde Martin has made numerous important contributions dealing with control, stabilization, and observation of linear, nonlinear, and distributed systems. His pioneering work on the analysis of matrix Riccati equations by algebraic-geometric methods has been very important for the development of the area. The Herman-Martin embedding of linear systems led to important breakthroughs not only in linear systems, but eventually also led to deeper understanding of the global observability of nonlinear systems.

Clyde Martin is a good example of a researcher who has a fundamental understanding both of engineering applications and mathematics and has contributed toward lessening the gap between them. His mathematical results have had an impact on control of aircraft, flexible space structures, and robotics. More recently, he has also contributed to areas like biomathematics, medicine and numerical analysis, where he has made important contributions. Recent work has been aimed toward tying together concepts and methods of numerical analysis, control theory and robotics.

This volume is a tribute to Clyde Martin for his many years of leadership in the systems and control community.

Lubbock, Texas
Stockholm
March 2005

Wijesuriya P. Dayawansa
Anders Lindquist
Yishao Zhou

Contents

An Efficient QR Algorithm for a Hessenberg Submatrix of a Unitary Matrix

Gregory S. Ammar[1], William B. Gragg[2], and Chunyang He[3]

[1] Department of Mathematical Sciences, Northern Illinois University, DeKalb, IL 60115 USA
   `ammar@math.niu.edu`
[2] Department of Mathematics, Naval Postgraduate School, Monterey, CA 93943 USA
   `bill_gragg@yahoo.com`
[3] Department of Mathematics and Statistics, University of Missouri–Kansas City, 5100 Rockhill Road, Kansas City, MO 64110–2499 USA
   `chunyang@sbcglobal.net`

Dedicated to Clyde Martin on the occasion of his sixtieth birthday.

Summary. We describe an efficient procedure for implementing the Hessenberg QR algorithm on a class of matrices that we refer to as subunitary matrices. This class includes the set of Szegő-Hessenberg matrices, whose characteristic polynomials are Szegő polynomials, i.e., polynomials orthogonal with respect to a measure on the unit circle in the complex plane. Computing the zeros of Szegő polynomials is important in time series analysis and in the design and implementation of digital filters. For example, these zeros are the poles of autoregressive filters.

1.1 Introduction

A real-valued bounded nondecreasing function $\mu(t)$ on the interval $[-\pi, \pi]$ with infinitely many points of increase defines an inner product according to

$$\langle f, g \rangle = \frac{1}{2\pi} \int_{-\pi}^{\pi} \overline{f(\lambda)} g(\lambda) d\mu(t), \qquad \lambda = \mathrm{e}^{\mathrm{i}t}, \tag{1.1}$$

where i denotes the imaginary unit and the bar denotes complex conjugation. There is then a unique family $\{\psi_k(\lambda)\}_{k=0}^{\infty}$ of monic polynomials such that each $\psi_k(\lambda)$ has degree k and such that

$$\langle \psi_j(\lambda), \psi_k(\lambda) \rangle = \begin{cases} 0 & \text{if } j \neq k, \\ \delta_k^2 > 0 & \text{if } j = k. \end{cases}$$

The polynomials $\{\psi_k\}$ are said to be *orthogonal on the unit circle*, and we call them *Szegő polynomials*.

G.S. Ammar et al.: *An Efficient QR algorithm for a Hessenberg Submatrix of a Unitary Matrix*, LNCIS **321**, 1–14 (2005)
`www.springerlink.com`

The monic Szegő polynomials satisfy the recurrence relation

$$\psi_{k+1}(\lambda) = \lambda\psi_k(\lambda) + \gamma_{k+1}\tilde{\psi}_k(\lambda), \quad k = 0, 1, 2, \ldots, \tag{1.2}$$

where $\psi_0(\lambda) = 1$, and $\tilde{\psi}_k(\lambda) = \lambda^k\bar{\psi}_k(1/\lambda)$ is the polynomial obtained by reversing and conjugating the power basis coefficients of ψ_k. The recurrence coefficients $\gamma_{k+1} \in \mathbb{C}$ are given by

$$\gamma_{k+1} = -\langle 1, \lambda\psi_k(\lambda)\rangle/\delta_k^2, \tag{1.3}$$

and the squared norm of ψ_{k+1} is recursively given by

$$\delta_{k+1}^2 = \delta_k^2(1 - |\gamma_{k+1}|^2),$$

where $\delta_0^2 = \langle 1, 1\rangle$. See, for example, [22, Chapter 11] or [16].

The Szegő recursion coefficients γ_j are known as *Schur parameters*. Since the distribution function $\mu(t)$ has infinitely many points of increase, $|\gamma_j| < 1$ for all j, and the zeros of each $\psi_j(\lambda)$ lie in the open unit disk $|\lambda| < 1$. On the other hand, if $\mu(t)$ has only n points of increase, then $|\gamma_j| < 1$ for $j = 1, 2, \ldots, n-1$ and $|\gamma_n| = 1$. In this case the orthogonal polynomials $\{\psi_j(\lambda)\}_{j=1}^{n-1}$ are defined only up to degree $n-1$, and (1.2) formally defines the monic polynomial ψ_n whose norm $\delta_n = 0$. In this case, the zeros of $\psi_n(\lambda)$ are pairwise distinct, of unit modulus, and equal to $\{e^{it_j}\}$, where $\{t_j\}$ are the points of increase of $\mu(t)$. See [22, Chapter 11] or [16].

Szegő polynomials arise in applications such as signal processing and time series analysis because of their connection with stationary time series. In these applications the Szegő polynomials are sometimes referred to as *backward predictor polynomials* or *Levinson polynomials*, and the Schur parameters are better known as *reflection coefficients* or *partial correlation coefficients*.

The moments associated with $\mu(t)$,

$$\mu_j = \frac{1}{2\pi}\int_{-\pi}^{\pi} e^{-\mathrm{i}jt}d\mu(t), \qquad j = 0, \pm 1, \pm 2, \ldots, \tag{1.4}$$

form a positive definite Toeplitz matrix $M_{n+1} = [\mu_{j-k}]_{j,k=0}^n$. If we use these moments to compute the inner products of polynomials in (1.3) to obtain the Schur parameters, and use these to construct the power basis coefficients of $\psi_k(\lambda)$, $k = 0, 1, \ldots n$, with the Szegő recursion (1.2), the resulting algorithm is known as the *Levinson-Durbin algorithm* for solving the Yule-Walker equations, which is fundamental among fast algorithms for solving Toeplitz systems of equations. In fact, a variety of efficient Toeplitz solvers generate the Szegő polynomials and/or the Schur parameters. See, for example, [12, 3].

The Szegő polynomials $\{\psi_j\}_{j=0}^n$ are determined by the Schur parameters $\{\gamma_j\}_{j=1}^n$, and problems involving the Szegő polynomials can often be re-cast in terms of the Schur parameters. In particular, the Szegő polynomials can be identified as the characteristic polynomials of the leading principal submatrices of an upper Hessenberg matrix H determined by the Schur parameters as follows.

Given n complex parameters $\{\gamma_j\}_{j=1}^n$ with $|\gamma_j| \leq 1$, define the *complementary parameters* $\sigma_j = (1 - |\gamma_j|^2)^{1/2}$. For $j = 1, \ldots, n-1$, define the unitary transformation of order n in the $(j, j+1)$ coordinate plane

$$
G_j(\gamma_j) =
\begin{bmatrix}
I_{j-1} & & & \\
& -\gamma_j & \sigma_j & \\
& \sigma_j & \bar{\gamma}_j & \\
& & & I_{n-j-1}
\end{bmatrix}
\in \mathbb{C}^{n \times n}, \qquad 1 \leq j < n, \qquad (1.5)
$$

where I_k denotes the $k \times k$ identity matrix. Also define a truncated matrix

$$
\tilde{G}_n(\gamma_n) =
\begin{bmatrix}
I_{n-1} & \\
& -\gamma_n
\end{bmatrix}. \qquad (1.6)
$$

Then we call the upper Hessenberg matrix

$$
H_n = G_1(\gamma_1)G_2(\gamma_2)\cdots G_{n-1}(\gamma_{n-1})\tilde{G}_n(\gamma_n) \qquad (1.7)
$$

the *Szegő-Hessenberg matrix* determined by the Schur parameters $\{\gamma_j\}_{j=1}^n$, and we write $H_n = H(\gamma_1, \gamma_2, \ldots \gamma_n)$. Although H_n is mathematically determined by the Schur parameters, we retain the complementary parameters in computational procedures to avoid numerical instability, as σ_j cannot be accurately computed from γ_j when the latter has magnitude close to one. The Szegő-Hessenberg matrix H_n is then represented by *Schur parameter pairs* $\{(\gamma_j, \sigma_j)\}_{j=1}^n$. The last complementary parameter σ_n, which is not needed in H_n, is included for notational convenience.

The leading principal submatrix of order k of H_n is

$$
H_k = H(\gamma_1, \ldots, \gamma_k).
$$

Let $\psi_k(\lambda bda) = \det(\lambda bda I - H_k)$ denote the characteristic polynomial of H_k. It is shown in [12] that these polynomials satisfy the recurrence relation (1.2) with $\psi_0(\lambda) = 1$. Consequently, these characteristic polynomials are the monic Szegő polynomials determined by the Schur parameters $\{\gamma_j\}_{j=1}^n$. See [2] for a short induction proof.

In the special case that $|\gamma_n| = 1$, the Szegő-Hessenberg matrix $H(\gamma_1, \ldots \gamma_n)$ is a *unitary* Hessenberg matrix, and there are a variety of efficient algorithms for computing its eigenvalues. These include the unitary Hessenberg QR algorithm [11], divide-and-conquer methods [14, 15, 5, 6, 17], a matrix pencil approach [7], and a Sturm-sequence type method [8]. The existence of these algorithms stems from the fact that unitary Hessenberg matrices are invariant under unitary similarity transformations, so that the computations can be performed on the Schur parameters that determine the intermediate matrices, rather than on the matrix elements explicitly.

However, the set of Szegő-Hessenberg matrices with $|\gamma_n| < 1$ is not invariant under unitary similarity transformations. In particular, a general Szegő-Hessenberg matrix $H_n = H(\gamma_1, \ldots, \gamma_n)$ satisfies

$$H_n^H H_n = I - \sigma_n^2 e_n e_n^H , \tag{1.8}$$

where the superscript H denotes the conjugate transpose of a matrix and e_n denotes the nth column of I_n. Since $\sigma_n > 0$, this property is preserved by the unitary similarity transformation $Q^H H_n Q$ only if $Q^H e_n$ is proportional to e_n. The development of efficient algorithms for general Szegő-Hessenberg matrices is therefore more complicated than that of unitary Hessenberg matrices.

One approach to efficiently compute the eigenvalues of a Szegő-Hessenberg matrix is a continuation method given in [2]. In order to find the eigenvalues of $H_n = H(\gamma_1, \ldots, \gamma_n)$ with $|\gamma_n| < 1$, we first find the eigenvalues of a unitary Hessenberg matrix $H_n' = H(\gamma_1, \ldots, \gamma_{n-1}, \gamma_n')$, where $|\gamma_n'| = 1$, using any of the established $O(n^2)$ methods. A continuation method is then applied to track the path of each eigenvalue, beginning with those of H_n' on the unit circle, and ending with those of H_n, as the last Schur parameter is varied from γ_n' to γ_n. This results in an $O(n^2)$ algorithm that also lends itself well to parallel computation.

Another approach is presented in [9], in which H_n is viewed as being in a larger class of matrices, called *fellow matrices*, defined as additive rank-one perturbations of unitary Hessenberg matrices. While fellow matrices are not invariant under the QR iteration, each QR iteration adds $O(n)$ additional parameters to the representation of the matrix. One can contain the number of parameters required by using a periodically restarted QR iteration, which leads to an efficient, $O(n^2)$, algorithm for computing the eigenvalues of fellow matrices. See [9] for details.

In this paper we present an efficient implementation of the QR algorithm on another class of matrices that include the Szegő-Hessenberg matrices. In view of (1.8), we consider matrices A that have the property that

$$A^H A = I_n - uu^H \quad \text{with } \|u\| \le 1.$$

We will refer to such a matrix as a *subunitary* matrix. This class of matrices includes the Szegő-Hessenberg matrices, and moreover, is invariant under unitary similarity transformations. Our goal is to present an efficient implementation of the QR algorithm on the set of subunitary Hessenberg matrices.

In [1] it is shown that Szegő-Hessenberg matrices provide an alternative to companion matrices for finding the zeros of a general polynomial from its power basis coefficients. In particular, any polynomial, after a suitable change of variable, can be identified as a Szegő polynomial. Experiments presented in [1] indicate computing the zeros of a polynomial by applying the QR algorithm to an associated Szegő-Hessenberg matrix often yields more accurate results than the traditional use of the QR algorithm on companion matrices. The development of efficient algorithms for Szegő-Hessenberg eigenproblems will therefore have a direct impact on the problem of computing the zeros of a general polynomial.

In Section 1.2 we summarize the mechanics of the bulge chasing procedure for performing one step of the implicit Hessenberg QR algorithm. Some basic properties of subunitary matrices are presented in Section 1.3, where we see that A is a subunitary matrix if and only if it is the leading principal submatrix of a unitary matrix of size one larger. In Section 1.4 we show that a subunitary Hessenberg matrix is represented by approximately $4n$ real parameters, and describe how the QR iteration can be efficiently performed on a subunitary Hessenberg matrix implicitly in terms of the underlying parameters.

1.2 Overview of the Hessenberg QR Algorithm

In finding the eigenvalues of a matrix using the QR algorithm, the matrix is first transformed by a unitary similarity transformation to upper Hessenberg form. The QR algorithm then iteratively generates a sequence of upper Hessenberg matrices by unitary similarity transformations. Implicit implementations of the Hessenberg QR algorithm can be viewed in terms of a *bulge chasing procedure*, which is a general computational procedure for performing a similarity transformation on a Hessenberg matrix to obtain another Hessenberg matrix. This is possible by virtue of the fact that a unitary matrix U such that U^*AU is an upper Hessenberg matrix is essentially determined by its first column. In order to establish notation, we now summarize the mechanics of the bulge chasing procedure that underlies the implicitly shifted QR algorithm. See, for example, [10, 19, 18, 25] for background on bulge-chasing procedures and their application in the implicitly shifted QR algorithm.

Let A be an upper Hessenberg matrix. One step of the Hessenberg QR algorithm with a single shift $\mu \in \mathbb{C}$ applied to A results in a new Hessenberg matrix A', given by

$$A' := RQ + \mu I_n = Q^H A Q$$

where

$$A - \mu I_n =: QR$$

is the QR factorization of the shifted matrix. The transformation Q is essentially determined by its first column, and since A is in upper Hessenberg form, so is Q. Let Q_1 be a unitary transformation in the $(1, 2)$ coordinate plane with the same first column as that of Q, and set $A_0 := A$. The bulge-chasing procedure now proceeds as follows.

The matrix $A_1 := Q_1^H A_0$ is also an upper Hessenberg matrix, and completing the similarity transformation yields the matrix $K_1 := Q_1^H A_0 Q_1$. The matrix K_1 would be a Hessenberg matrix if its $(3, 1)$ element were nonzero. This element is the *bulge*. A unitary transformation Q_2 in the $(2, 3)$ plane is then chosen to to annihilate the bulge in K_1 by left multiplication, so that $Q_2^H K_1 = A_2$ is in Hessenberg form. After multiplying on the right by Q_2 to complete the similarity transformation, the matrix $K_2 = A_2 Q_2$ has a bulge in the (4,2) position. The bulge in K_2 is annihilated by left multiplication by a unitary transformation Q_3 in the $(3,4)$-plane, and the process continues until the bulge is 'chased' diagonally down the matrix until we obtain the Hessenberg matrix $K_{n-1} = Q_{n-1}^H K_{n-2} Q_{n-1}$, which is unitarily similar to the initial Hessenberg matrix A. Finally, a diagonal unitary similarity transformation, equal to the identity matrix except possibly for its (n,n) entry, is performed to make the $(n, n-1)$ entry nonnegative, resulting in $A' = \tilde{Q}_n^H K_{n-1} \tilde{Q}_n^H$. Then $A' = Q^H A Q$, where $Q = Q_1 Q_2 \cdots Q_{n-1} \tilde{Q}_n$, is the result of a single bulge-chasing sweep on the original matrix A.

The pattern of nonzero elements in the intermediate matrices is displayed in Figure 1.1 for $n = 5$.

$$
A_1 = Q_1^H A = \begin{bmatrix} \times & \times & \times & \times & \times \\ \times & \times & \times & \times & \times \\ & \bullet & \times & \times & \times \\ & & \bullet & \times & \times \\ & & & \bullet & \times \end{bmatrix}
\qquad
K_1 = Q_1^H A_0 Q_1 = \begin{bmatrix} \times & \times & \times & \times & \times \\ \times & \times & \times & \times & \times \\ + & \times & \times & \times & \times \\ & & \bullet & \times & \times \\ & & & \bullet & \times \end{bmatrix}
$$

$$
K_2 = Q_2^H K_1 Q_2 = \begin{bmatrix} \times & \times & \times & \times & \times \\ \bullet & \times & \times & \times & \times \\ \oplus & \times & \times & \times & \times \\ & + & \times & \times & \times \\ & & & \bullet & \times \end{bmatrix}
\qquad
K_3 = Q_3^H K_2 Q_3 = \begin{bmatrix} \times & \times & \times & \times & \times \\ \bullet & \times & \times & \times & \times \\ & \bullet & \times & \times & \times \\ & \oplus & \times & \times & \times \\ & & + & \times & \times \end{bmatrix}
$$

$$
K_4 = Q_4^H K_3 Q_4 = \begin{bmatrix} \times & \times & \times & \times & \times \\ \bullet & \times & \times & \times & \times \\ & \bullet & \times & \times & \times \\ & & \bullet & \times & \times \\ & & \oplus & \times & \times \end{bmatrix}
\qquad
A' = \tilde{Q}_5^H K_4 \tilde{Q}_5 = \begin{bmatrix} \times & \times & \times & \times & \times \\ \bullet & \times & \times & \times & \times \\ & \bullet & \times & \times & \times \\ & & \bullet & \times & \times \\ & & & \bullet & \times \end{bmatrix}
$$

Fig. 1.1. Matrix profiles during one bulge chasing sweep. Entries $\times$ represent complex numbers, $\bullet$ represent nonnegative numbers, $+$ represents a fill-in, and $\oplus$ represents a zero element introduced after annihilating a fill-in element

The matrix A' is uniquely determined by A and Q_1 provided that every nonnegative subdiagonal element σ_j' $(j = 1, 2, \ldots, n-1)$ of A' (represented by the $\bullet$ symbols in Figure 1.1) is positive. If the subdiagonal element σ_j' of A' vanishes, then the procedure terminates early, and the eigenproblem for A' deflates into two smaller eigenproblems.

For general Hessenberg matrices A, a bulge-chasing step requires $O(n^2)$ floating-point operations (flops).

1.3 Subunitary Matrices

We will say that $A \in \mathbb{C}^{n \times n}$ is a *subunitary matrix* if $A^H A = I - u u^H$ for some vector u such that $\|u\|_2 \leq 1$. We refer to u as the *departure vector* of the subunitary matrix A, and to $\|u\|_2$ as its *departure norm*.

Many elementary properties of subunitary matrices are easily derived. For example, the proof of the following proposition is immediate.

Proposition 1. *Let A be a subunitary matrix with $\|u\| = \nu$. Then every eigenvalue λbda of A is contained in the annulus $\sqrt{1 - \nu^2} \leq \lambda \leq 1$. Moreover, A has $n - 1$ singular values equal to one, and one singular value equal to $(1 - \nu^2)^{1/2}$.*

The next proposition gives some support to the choice of the appellation *subunitary.*

Proposition 2. *The $n \times n$ matrix A is a subunitary matrix if and only if it is a submatrix of a unitary matrix of order $n + 1$.*

Proof. There is no loss of generality to prove this result for A the leading principal submatrix of a unitary matrix B, since this can be enforced after performing row and column permutations on B. Let B be an $(n+1) \times (n+1)$ matrix, and write B in partitioned form,

$$B = \begin{bmatrix} A & v \\ u^H & \beta \end{bmatrix}, \tag{1.9}$$

where A is an $n \times n$ matrix, $u, v \in \mathbb{C}^n$ are column vectors, and $\beta \in \mathbb{C}$. Then

$$B^H B = \begin{bmatrix} A^H A + u u^H & A^H v + u\beta \\ v^H A + \bar{\beta} u^H & v^H v + |\beta|^2 \end{bmatrix}.$$

Thus, if B is unitary, then $A^H A = I - u u^H$, where $\|u\| \leq 1$.

Conversely, assume that A is a subunitary matrix with departure vector u. If $\|u\| < 1$, then A is nonsingular, and the equation $A^H x = -u$ has a unique solution x. Set $\beta = (1 + \|x\|^2)^{-1/2}$ and $v = \beta x$. Then the matrix B given by (1.9) is unitary. If $\|u\| = 1$, then A is singular with a one-dimensional nullspace spanned by u. Let v be such that $A^H v = 0$ and $\|v\| = 1$, and set $\beta = 0$ to obtain a unitary matrix B whose leading principal submatrix is A. $\square$

Proposition 3. *Let A be a subunitary matrix with departure vector u, and let Q be any unitary matrix. Then:*

1. QA is a subunitary matrix with the same departure vector u.
2. AQ is a subunitary matrix with departure vector $Q^H u$.

It follows immediately that set of subunitary matrices of departure norm ν is invariant under unitary equivalence transformations and unitary similarity transformations. In particular, the set is invariant under the QR iteration. It also follows that for any factorization of the form $A = QX$, where Q is unitary, the matrix X is also subunitary with the same departure vector as that of A. In particular, in the QR factorization of a subunitary matrix A, the upper triangular factor R is a subunitary matrix with the same departure vector u.

Proposition 4. *Let R be a subunitary upper triangular matrix with positive diagonal elements. Then R is uniquely determined by its departure vector u. Moreover, the entries of $R = [\rho_{jk}]_{j,k=1}^n$ are explicitly given by*

$$\rho_{jk} = \begin{cases} \dfrac{-v_j \overline{v}_k}{\kappa_{j-1}\kappa_j} & \text{for } j < k, \\[2mm] \kappa_j/\kappa_{j-1} & \text{for } j = k, \\[2mm] 0 & \text{for } j > k, \end{cases} \tag{1.10}$$

where $u = [v_j]_{j=1}^n$ and

$$\kappa_j^2 = 1 - \sum_{k=1}^{j} |v_k|^2 = \kappa_{j+1}^2 + |v_{j+1}|^2$$

for $j = 0, \ldots, n-1$, with $\kappa_n^2 := 1 - \|u\|_2^2$.

Proof. Since R has full rank, its departure vector u has norm strictly less than one. From $R^H R = I - uu^H$, we see that R is the unique Cholesky factor of the positive definite matrix $I - uu^H$. The formulas (1.10) follow from the Cholesky factorization algorithm. $\square$

If $\|u\| = 1$ and $v_n \neq 0$ (i.e., $\kappa_n = 0$ and $\kappa_{n-1} > 0$), the above formulas for the entries of R remain valid. In this case R is unique up to a unimodular scaling of its last column.

1.4 Efficient QR Iteration on Subunitary Hessenberg Matrices

Let $\mathcal{H}_0$ denote the set of unitary upper Hessenberg matrices with nonnegative subdiagonal elements, and let $\mathcal{H}_1$ denote the set of subunitary upper Hessenberg matrices with nonnegative subdiagonal elements. Then in the QR factorization of $A \in \mathcal{H}_1$,

$$A = HR\,,$$

we have that $H \in \mathcal{H}_0$, and therefore H has a unique Schur parametrization $H = H(\gamma_1, \ldots, \gamma_n)$. This combined with Proposition 4 yields the following result.

Theorem 1. *Any subunitary upper Hessenberg matrix $A \in \mathbb{C}^{n \times n}$ with nonnegative subdiagonal elements and departure vector u with $\|u\| < 1$ is uniquely represented by $4n - 1$ real parameters. In particular, $A = HR$, where $H = H(\gamma_1, \ldots, \gamma_n) \in \mathcal{H}_0$ (with $|\gamma_n| = 1$), and where R is the subunitary upper triangular matrix given in Proposition 4 whose departure vector $u \in \mathbb{C}^n$ is the same as that of A.*

The parametrization of $\mathcal{H}_1$ given in Theorem 1 now allows us to approach the problem of efficiently implementing the QR iteration on $\mathcal{H}_1$. The key is to perform the QR iteration on the parameterized product $A = HR$, keeping the parameterized form of each factor intact during the iteration.

Let $A =: HR$ be the QR factorization of $A \in \mathcal{H}_1$, where

$$H = H(\gamma_1, \ldots, \gamma_n) \in \mathcal{H}_0,$$

and where R is the upper triangular matrix with positive diagonal elements satisfying $R^H R = A^H A = I - uu^H$. In one step of the QR algorithm on A with shift $\tau \in \mathbb{C}$, the QR factorization of the shifted matrix,

$$A - \tau I =: QU,$$

defines the unitary Hessenberg matrix Q that produces the result of the QR step $A' = Q^H A Q$. We seek to compute the QR factorization $A' = H'R' \in \mathcal{H}_1$. The Hessenberg-triangular product representation is maintained through the introduction of a unitary matrix Z such that

$$A' = Q^H A Q = (Q^H H Z)(Z^H R Q) =: H'R'.$$

We will therefore implement the implicitly shifted QR step on $A \in \mathcal{H}_1$ keeping the parameterized factors H and R separate, as in implementations of the QZ algorithm for matrix pencils (see, e.g., [10]).

Let us now consider the individual steps of an implicit QR step on $A = HR \in \mathcal{H}_1$. Let Q_1 denote the initial transformation in the $(1, 2)$ coordinate plane that implicitly defines Q and A'. Write $H = G_1 G_2 \cdots \tilde{G}_n$, where each $G_j = G_j(\gamma_j, \sigma_j)$. In addition to the matrices G_j, throughout the following discussion, Q_j, T_j, Z_j will denote unitary transformations in the $(j, j+1)$-plane (for $j < n$), and $\tilde{Q}_n$, $\tilde{T}_n$, $\tilde{Z}_n$ will denote unitary matrices that differ from the identity matrix only in the (n, n) entry. We consider the special structure of intermediate matrices generated during the QR step on $A = HR$ as described in Section 1.2.

The initial bulge in the $(3,1)$ position of the matrix

$$K_1 = Q_1^H A Q_1 = Q_1^H H R Q_1$$

arises from the bulge in the $(2,1)$ position of RQ_1. Choose a unitary transformation Z_1 to annihilate this bulge, so that $Z_1^H R Q_1 =: R_1$ is upper triangular with positive diagonal entries. Then $K_1 = (Q_1^H H Z_1) R_1$, and the bulge resides in the Hessenberg factor of the product. Since Z_j commutes with G_k whenever $|j - k| > 1$, we have

$$\begin{aligned}
K_1 &= Q_1^H G_1 G_2 \cdots \tilde{G}_n Z_1 R_1 \\
&= (T_1 G_2 Z_1) G_3 G_4 \cdots \tilde{G}_n R_1 \, ,
\end{aligned}$$

where $T_1 := Q_1^H G_1$. The bulge in K_1 now arises from the $(3, 1)$ entry of the matrix $W_1 = T_1 G_2 Z_1$. Note that W_1 differs from the identity matrix only in its 3×3 leading principal submatrix. Also, R_1 is the subunitary upper triangular matrix determined by its departure vector $u_1 = Q_1^H u$.

A unitary transformation Q_2 is now chosen so that $Q_2^H W_1$ is in upper Hessenberg form, and then choose $G_1' = G_1(\gamma_1', \sigma_1')$ so that the (2,1) entry of $G_1''^H Q_2^H W_1$ is annihilated, and note that since W_1 is unitary, this last matrix is a unitary matrix T_2 which differs from the identity matrix only in the $(2,3)$ principal submatrix. In this way, we obtain unitary plane transformations Q_2, G_1', and T_2 such that

$$W_1 = T_1 G_2 Z_1 =: Q_2 G_1' T_2$$

and

$$K_1 = Q_2 G_1' T_2 G_3 G_4 \cdots \tilde{G}_n R_1 \ .$$

The similarity transformation defined by Q_2 on K_1 then yields the matrix K_2 with bulge in the (4,2) position,

$$\begin{aligned}
K_2 = Q_2^H K_1 Q_2 &= G_1' T_2 G_3 G_4 \cdots \tilde{G}_n R_1 Q_2 \\
&= G_1' T_2 G_3 G_4 \cdots \tilde{G}_n Z_2 (Z_2^H R_1 Q_2) \\
&= G_1' (T_2 G_3 Z_2) G_4 \cdots \tilde{G}_n R_2 \\
&= G_1' (Q_3 G_2' T_3) G_4 \cdots \tilde{G}_n R_2 \\
&= Q_3 (G_1' G_2') T_3 G_4 \cdots \tilde{G}_n R_2 \ ,
\end{aligned}$$

where Z_2 is chosen so that $Z_2^H R_1 Q_2^H =: R_2$ is the upper triangular subunitary matrix with departure vector $u_2 = Q_2^H u_1$, and the identification

$$T_2 G_3 Z_2 =: W_2 =: Q_3 G_2' T_3$$

is made as above, using the matrix W_2 that differs from the identity matrix only in the 3×3 principal submatrix from rows and columns $(2,3,4)$. Now $K_3 = Q_3^H K_2 Q_3$ has a bulge in the $(5,3)$ position, and the process continues until we obtain the upper Hessenberg matrix

$$K_{n-1} = \tilde{Q}_n G_1' G_2' \cdots G_{n-1}' \tilde{T}_n R_{n-1}$$

and

$$\begin{aligned}
A' = \tilde{Q}_n^H K_{n-1} \tilde{Q}_n &= G_1' G_2' \cdots G_{n-1}' (\tilde{T}_n \tilde{Q}_n)(\tilde{Q}_n^H R_{n-1} \tilde{Q}_n) \\
&= G_1' G_2' \cdots G_{n-1}' \tilde{G}_n' R_n \\
&= H(\gamma_1', \ldots, \gamma_n') R_n \\
&= H' R' \ .
\end{aligned}$$

Thus, the transition from $A = HR$ to $A' = H'R'$ can be achieved by keeping the Hessenberg factors and upper triangular factors separate. Moreover, individual operations can be performed on 2×2 and 3×3 matrices. This leads to the following algorithm for performing one QR step on a subunitary Hessenberg matrix using $O(n)$ flops.

Algorithm 1.

Input: Schur parameter pairs $\{\gamma_j, \sigma_j\}_{j=1}^n$ of $H = H(\gamma_1, \ldots, \gamma_n) \in \mathcal{H}_0$, departure vector $u = [v_j] \in \mathbb{C}^n$ of a subunitary upper triangular matrix R, with $\|u\| = \nu < 1$, and initial unitary transformation Q_1 in the $(1, 2)$ coordinate plane.

Output: Schur parameter pairs $\{\gamma_j', \sigma_j'\}_{j=1}^n$ of $H' = H(\gamma_1', \ldots, \gamma_n') \in \mathcal{H}_0$, departure vector $u' \in \mathbb{C}^n$ of the subunitary upper triangular matrix R', with $\|u'\| = \nu$, such that $A' = H'R'$ is the QR factorization of the subunitary upper Hessenberg matrix $A' = Q^H A Q = Q^H H Z Z^H R Q$ that results from one bulge chasing sweep applied to the initial subunitary Hessenberg matrix $A = HR$ with initial transformation Q_1.

Set $\kappa_n := (1 - \|u\|_2^2)^{1/2}$.

Set $\kappa_j := (\kappa_{j+1}^2 + |v_j|^2)^{1/2}$, $j = n - 1, n - 2, \ldots, 0$.

Form the 2×2 matrix $T_1 = Q_1^H G_1$, where $G_1 = G_1(\gamma_1, \sigma_1)$.

For $j = 1, 2, \ldots n - 1$

> Form the 2×2 matrix $X = \begin{bmatrix} \rho_{jj} & \rho_{j,j+1} \\ 0 & \rho_{j+1,j+1} \end{bmatrix}$ according to (1.10).

> Overwrite $XQ_j =: X =: \begin{bmatrix} \xi_{11} & \xi_{12} \\ \xi_{21} & \xi_{22} \end{bmatrix}$.

> Set $\rho_{jj}' = (|\xi_{11}|^2 + |\xi_{21}|^2)^{1/2}$, $\mu_j := -\xi_{11}/\rho_{jj}'$, $\kappa_j = \xi_{21}/\rho_{jj}'$.

> Update $\begin{bmatrix} v_j \\ v_{j+1} \end{bmatrix} := Q_j^H \begin{bmatrix} v_j \\ v_{j+1} \end{bmatrix}$ and $\kappa_j := \rho_{jj}' \kappa_{j-1}$.

> Form $Z_j = G_j(\mu_j, \kappa_j)$ and $G_{j+1} = G_{j+1}(\gamma_{j+1}, \sigma_{j+1})$.

> Form the 3×3 matrix $W = T_j G_{j+1} Z_j$.

> Set $\sigma_j' = (|\omega_{21}|^2 + \omega_{31}^2)^{1/2}$, $\alpha_{j+1} = -\omega_{21}/\sigma_j'$, $\beta_{j+1} = \omega_{31}/\sigma_j'$.

> Form $Q_{j+1} = \begin{bmatrix} -\alpha_{j+1} & \beta_{j+1} \\ \beta_{j+1} & \bar{\alpha}_{j+1} \end{bmatrix} = G_{j+1}(\alpha_{j+1}, \beta_{j+1})$.

> Overwrite $Q_{j+1}^H W =: W$. Set $\gamma_j' = -\omega_{11}$.

> Form $G_j' = G_j(\gamma_j', \sigma_j')$, and overwrite $(G_j')^H W =: W$.

> Set $T_{j+1} = \begin{bmatrix} \omega_{22} & \omega_{23} \\ \omega_{32} & \omega_{33} \end{bmatrix}$.

End (for j).

Form the 2×2 matrix $W = T_{n-1} \tilde{G}_n Z_{n-1}$.

Set $\sigma_{n-1}' = |\omega_{21}|$, $\alpha_n = -\omega_{21}/\sigma_j'$.

Form $\tilde{Q}_n = \begin{bmatrix} 1 & 0 \\ 0 & -\alpha_n \end{bmatrix}$.

Overwrite $\tilde{Q}_n^H W =: W$, and set $\gamma_j' = -\omega_{11}$.

12   G.S. Ammar et al.

Form $G'_{n-1} = G_{n-1}(\gamma'_{n-1}, \sigma'_{n-1})$, and overwrite $(G'_{n-1})^H W =: W$.

Set $\tilde{T}_n = \begin{bmatrix} 1 & 0 \\ 0 & \omega_{22} \end{bmatrix}$.

Update $v_n = -\bar{\alpha}_n v_n$.

Set $\tilde{T}_n \tilde{Q}_n =: \tilde{G}'_n =: \begin{bmatrix} 1 & 0 \\ 0 & -\gamma'_n \end{bmatrix}$.

Set $u' := u$.

end algorithm.

1.5 Concluding Remarks

We considered some basic aspects of subunitary matrices, and showed that the QR algorithm can be implemented on subunitary Hessenberg matrices using $O(n)$ flops per iteration. The resulting subunitary Hessenberg QR (SUHQR) algorithm can be regarded as a generalization of the general idea behind the unitary Hessenberg QR (UHQR) algorithm. In particular, UHQR operates on the Schur parameters of intermediate unitary Hessenberg matrices to implicitly perform unitary similarity transformations on the initial matrix, while SUHQR implicitly performs unitary equivalence transformations on both a unitary Hessenberg matrix and a subunitary upper triangular matrix, using the Schur parameters of the former and the departure vector of the latter.

The algorithm outlined above represents only a beginning for the study of QR iterations on $\mathcal{H}_1$. A next step is to further streamline the algorithm by implementing it directly on the parameters that determine the intermediate matrices, rather than explicitly on the entries of these matrices. This will result in an implementation more closely related to the original implementation of the UHQR algorithm given in [11]. In fact, several different implementations of the UHQR algorithm have been developed [23, 24], and because the SUHQR algorithm is essentially a generalization of UHQR, there will be corresponding variants on the SUHQR algorithm as well.

A numerical stability analysis of our algorithm remains open. We make no claim on this matter here. In [13] it is shown that the original version of the UHQR algorithm is not numerically stable. A source of instability is identified there, and a modified UHQR algorithm is proposed to avoid the instability. Numerical experiments in [13] confirm the improved stability of this modified UHQR algorithm. A detailed error analysis of UHQR algorithms has been performed by Stewart [20, 21], where additional potential instabilities are identified and resolved with provably numerical stable implementations of the UHQR algorithm. Certainly this recent work on stabilizing the UHQR algorithm will be relevant for SUHQR as well.

Acknowledgement

GA began his study of the QR algorithm during his work toward his doctorate under the direction of Clyde Martin. In [4] it is shown that the basic QR algorithm can be interpreted in terms of a linear action on the full flag manifold. This is in analogy with the study of control-theoretic matrix Riccati equations in terms of linear actions on Grassmannians, which was initiated by Hermann and Martin. GA takes this opportunity to thank Clyde Martin for introducing him to the QR algorithm, which continues to provide for interesting and fruitful study.

References

1. G. Ammar, D. Calvetti, W. Gragg, and L. Reichel. Polynomial zerofinders based on Szegő polynomials. *J. Comput. Appl. Math.*, 127:1–16, 2001.
2. G. S. Ammar, D. Calvetti, and L. Reichel. Continuation methods for the computation of zeros of Szegő polynomials. *Lin. Alg. Appl.*, 249:125–155, 1996.
3. G. S. Ammar and W. B. Gragg. The generalized Schur algorithm for the superfast solution of Toeplitz systems. In J. Gilewicz, M. Pindor, and W. Siemaszko, editors, *Rational Approximation and its Applications in Mathematics and Physics*, number 1237 in Lecture Notes in Mathematics, pages 315–330. Springer-Verlag, Berlin, 1987.
4. G. S. Ammar and C. F. Martin. The geometry of matrix eigenvalue methods. *Acta Applicandae Math.*, 5:239–278, 1986.
5. G. S. Ammar, L. Reichel, and D. C. Sorensen. An implementation of a divide and conquer algorithm for the unitary eigenproblem. *ACM Trans. Math. Software*, 18:292–307, 1992.
6. G. S. Ammar, L. Reichel, and D. C. Sorensen. Algorithm 730: An implementation of a divide and conquer algorithm for the unitary eigenproblem. *ACM Trans. Math. Software*, 20:161, 1994. (Mathematical software.).
7. A. Bunse-Gerstner and L. Elsner. Schur parameter pencils for the solution of the unitary eigenproblem. *Lin. Alg. Appl.*, 154–156:741–778, 1991.
8. A. Bunse-Gerstner and C. He. On the Sturm sequence of polynomials for unitary Hessenberg matrices. *SIAM J. Matrix Anal. Appl.*, 16(4):1043–1055, October 1995.
9. D. Calvetti, S.-M. Kim, and L. Reichel. The restarted QR-algorithm for eigenvalue computation of structured matrices. *J. Comput. Appl. Math.*, 149:415–422, 2002.
10. G. H. Golub and C. F. Van Loan. *Matrix Computations*. Johns Hopkins University Press, third edition, 1996.
11. W. B. Gragg. The QR algorithm for unitary Hessenberg matrices. *J. Comput. Appl. Math.*, 16:1–8, 1986.
12. W. B. Gragg. Positive definite Toeplitz matrices, the Arnoldi process for isometric operators, and Gaussian quadrature on the unit circle. *J. Comput. Appl. Math.*, 46:183–198, 1993. Originally published (in Russian) in: E.S. Nikolaev, Ed., *Numerical Methods in Linear Algebra*, Moscow Univ. Press, 1982, pp. 16–32.

13. W. B. Gragg. Stabilization of the uhqr algorithm. In Z. Chen, Y. Li, C. Micchelli, and Y. Xu, editors, *Advances in Computational Mathematics: Proceedings of the Guangzhou International Symposium*, pages 139–154, New York, 1999. Marcel Dekker.

14. W. B. Gragg and L. Reichel. A divide and conquer method for the unitary eigenproblem. In M. T. Heath, editor, *Hypercube Multiprocessors 1987*, Philadelphia, 1987. SIAM.

15. W. B. Gragg and L. Reichel. A divide and conquer method for unitary and orthogonal eigenproblems. *Numer. Math.*, 57:695–718, 1990.

16. U. Grenander and G. Szegő. *Toeplitz Forms and Their Applications*. Chelsea, New York, second edition, 1984.

17. M. Gu, R. Guzzo, X.-B. Chi, and X.-Q. Cao. A stable divide and conquer algorithm for the unitary eigenproblem. *SIAM J. Matrix Anal. Appl.*, 25(2):385–404, 2003.

18. B. N. Parlett. *The Symmetric Eigenvalue Problem*. Society for Industrial and Applied Mathematics, Philadelphia, 1998.

19. G. W. Stewart. *Introduction to Matrix Computations*. Academic Press, New York, 1973.

20. M. Stewart. Stability properties of several variants of the unitary Hessenberg QR algorithm. In V. Olshevsky, editor, *Structured Matrices in Mathematics, Engineering, and Computer Science II (Boulder, CO, 1999)*, volume 281 of *Contemporary Math.*, pages 57–72, Providence, RI, 2001. Amer. Math. Soc.

21. M. Stewart. An error analysis of a unitary Hessenberg QR algorithm. Preprint, 2004.

22. G. Szegő. *Orthogonal Polynomials*. American Math. Soc., Providence, RI, fourth edition, 1975.

23. T.-L. Wang and W. B. Gragg. Convergence of the shifted QR algorithm for unitary Hessenberg matrices. *Math. Comp.*, 71:1473–1496, 2002.

24. T.-L. Wang and W. B. Gragg. Convergence of the unitary QR algorithm with a unimodular Wilkinson shift. *Math. Comp.*, 72:375–385, 2003.

25. D. S. Watkins. *Fundamentals of Matrix Computations*. John Wiley and Sons, New York, 1991.

Quantum SubRiemannian Dynamics

Anthony M. Bloch[1]*, Roger Brockett[2]†, and Alberto G. Rojo[3]

[1] Department of Mathematics, University of Michigan, Ann Arbor, MI 48109
   abloch@math.lsa.umich.edu
[2] Division of Engineering and Applied Sciences, Harvard University, Cambridge,
   MA 02138
   brockett@hrl.harvard.edu
[3] Department of Physics, Oakland University, Rochester, MI 48309
   rojo@oakland.edu

With best wishes to Clyde Martin on the occasion of his 60th birthday.

Summary. The Hamiltonians that arise in optimal control include some that correspond directly to physical systems and some that do not. For those that do, one can often identify an appropriate quantization procedure and thus associate a Schrodinger equation with an optimal control problem. However, optimal control problems that can not be converted to Lagrangian form, such as those that arise when the system is controllable because of a Lie bracket effect, are more difficult and no straight forward method seems to exist. The purpose of this paper is to describe some possibilities for assigning a Schrodinger equation to such systems. We do not attempt to formulate a general theory but focus on selected examples. We show that this approach leads to interesting "quantum" nonholonomic phenomena.

2.1 Introduction

There has been a great deal of interest in recent years in optimal control of underactuated control systems and in quantum control. In this paper we brings these notions together in considering the quantization the optimal dynamics, giving us a generalization of classical constrained variational dynamics to the quantum setting. This work is related to the subRiemannian dynamical systems arising in optimal control theory. (see e.g. [Brockett(1981)] and [Bloch(2003)] and references therein). Since we are quantizing a Hamiltonian the dynamics remains Hamiltonian in the quantum context and is hence is in the spirit of subRiemannian or variational nonholonomic dynamics as opposed to mechanical nonholonomic mechanics as discussed in [Bloch(2003)].

---

*Research partially supported by the National Science Foundation
†Research partially supported by the National Science Foundation

A.M. Bloch et al.: *Quantum SubRiemannian Dynamics*, LNCIS **321**, 15–26 (2005)
www.springerlink.com

We restrict ourselves here to considering some key examples of interest - the Heisenberg integrator and the rolling penny. However, the generalization is clear and we will consider aspects of the general setting in a future publication.

We point out here some interesting quantum phenomena that arise in this setting.

2.2 The Simplest Model

2.2.1 The Optimal Control Point of View

We consider a set of control equations

$$\begin{aligned}
\dot{x} &= u, \\
\dot{y} &= v, \\
\dot{z} &= A_1 u + A_2 v,
\end{aligned} \tag{2.1}$$

where $A_1(x, y)$ and $A_2(x, y)$ are smooth functions of x and y. The variational problem is to minimize

$$\eta = \min \frac{1}{2} \int (u^2 + v^2) dt \tag{2.2}$$

The special case $A_1 = y$ and $A_2 = -x$ has been extensively investigated in the literature.

To find the Euler-Lagrange equations we introduce the Lagrange multiplier λ and a corresponding Lagrangian

$$L\left(x, \dot{x}, y, \dot{y}, z, \dot{z}, \lambda, \dot{\lambda}\right) = \tfrac{1}{2}\left(\dot{x}^2 + \dot{y}^2\right) + \frac{\lambda}{2}\left(\dot{z} - A_1\dot{x} - A_2\dot{y}\right).$$

This gives rise to the second order variational equations involving a constant λ,

$$\begin{aligned}
\ddot{x} &= \frac{\lambda}{2}\left(\frac{\partial A_2}{\partial x} - \frac{\partial A_1}{\partial y}\right)\dot{y}, \\
\ddot{y} &= \frac{\lambda}{2}\left(\frac{\partial A_1}{\partial y} - \frac{\partial A_2}{\partial x}\right)\dot{x}.
\end{aligned} \tag{2.3}$$

With a view toward the physical interpretations to come, we introduce a "vector potential" $\mathbf{A} = (A_1, A_2, 0)$ and further define B_z via

$$(\nabla \times \mathbf{A})_z = \frac{\partial A_2}{\partial x} - \frac{\partial A_1}{\partial y} \equiv B_z.$$

In terms of B_z the Euler–Lagrange equations take the form

$$\ddot{x} = \frac{\lambda}{2} B_z \dot{y},$$

$$\ddot{y} = -\frac{\lambda}{2} B_z \dot{x}, \tag{2.4}$$

We can reduce this to a single vector equation by introducing $\mathbf{v} = (\dot{x}, \dot{y}, 0)$ and $\mathbf{B} = (0, 0, B_z)$,

$$\frac{d\mathbf{v}}{dt} + \mathbf{v} \times \frac{\lambda}{2}\mathbf{B} = 0. \tag{2.5}$$

These later changes of variable have been introduced so that the description can be identified as the motion of a particle confined to two dimensions but moving in a magnetic field directed normal to the plane of motion.

Of course the variational problem can also be analyzed using the Pontryagin maximum principle. To do this, form a Hamiltonian as a function of $(x, y, z, p_1, p_2, p_3, u, v)$

$$H = p_1 u + p_2 v + p_3(A_1 u + A_2 v) - \frac{u^2}{2} - \frac{v^2}{2}. \tag{2.6}$$

The stationarity conditions

$$\frac{\partial H}{\partial u} = 0 \ ; \ \ \frac{\partial H}{\partial v} = 0$$

yield

$$u = p_1 + p_3 A_1,$$

$$v = p_2 + p_3 A_2. \tag{2.7}$$

Hence, when made stationary with respect to u and v, Hamiltonian becomes

$$H = \frac{1}{2}\left\{(p_1 + A_1 p_3)^2 + (p_2 + A_2 p_3)^2\right\}. \tag{2.8}$$

With the appropriate sign conventions this is the Hamiltonian for a particle in a magnetic field.

2.2.2 Quantum Setting

The presence of a charge on the particle q and and a magnetic field with vector potential A requires a simple modification of the Hamiltonian of a particle. This is standard, for example in quantum optics, and is summarized by

$$h(p, x) \mapsto h(p - qA, x)$$

(see, for example [Scully and Zubiary(1997)].) Applying this to the system at hand and adjusting the costate variable p_3 in the appropriate way gives the Hamiltonian

$$H = \frac{1}{2}\left\{(p_1 + A_1 p_3)^2 + (p_2 + A_2 p_3)^2\right\}. \tag{2.9}$$

If A_1 and A_2 do not depend on z, the standard quantization procedure yields the Schrodinger equation

$$i\hbar\frac{\partial\Psi}{\partial t} = -\frac{\hbar^2}{2m}\left\{\frac{\partial^2}{\partial x^2} + \frac{\partial^2}{\partial y^2} + (A_1^2 + A_2^2)\frac{\partial^2}{\partial z^2} + 2A_1\frac{\partial^2}{\partial y\partial z} - 2A_2\frac{\partial^2}{\partial x\partial z}\right\}\Psi$$
$$\equiv H\Psi. \tag{2.10}$$

On the other hand, the idea of quantum diffusion gives an alternative derivation which is more closely dependent on sample path ideas. (For the unconstrained of case see e.g. [Zinn-Justin(1996)].) Start from a discrete model of an otherwise free particle diffusing (moving quantum-mechanically) on a $3D$ cubic lattice of lattice constant δ, and then take the limit $\delta \to 0$. We take the motion in the xy plane as unconstrained, meaning that the particle is allowed to hop horizontally from site to site without restrictions. The equation of motion for $\Psi(x, y, z)$, subject to the constraint

$$dz = A_1 dx + A_2 dy \tag{2.11}$$

is of the form

$$i\hbar\frac{\partial\Psi}{\partial t} = -V\left\{\Psi(x + \delta, y, z + A_1\delta) + \Psi(x - \delta, y, z + A_2\delta)\right.$$
$$+ \Psi(x, y + \delta, z + A_1\delta) + \Psi(x, y - \delta, z - A_2\delta)$$
$$\left. - \Psi(x, y, z)\right\}, \tag{2.12}$$

with $V = \hbar^2/m\delta^2$ playing the role of a hopping constant. Expanding to lowest order in δ we obtain equation (2.10) again.

Now, since H has translational invariance in the z direction, in the special case where $A_1 = y$ and $A_2 = -x$ the eigenfunctions of energy $\epsilon_{k,\nu}$ are of the form

$$\Psi = e^{ikz}\phi_{k,\nu}(x, y), \tag{2.13}$$

with $\phi_{k,\nu}(x, y)$ solutions of a reduced Hamiltonian: $H_k\phi_{k,\nu} = \epsilon_{k,\nu}\phi_{k,\nu}$, with

$$H_k = -\frac{\hbar^2}{2m}\left\{\frac{\partial^2}{\partial x^2} + \frac{\partial^2}{\partial y^2} - k^2(x^2 + y^2) + 2ik\left(x\frac{\partial}{\partial y} - y\frac{\partial}{\partial x}\right)\right\}$$
$$\equiv \frac{1}{2m}\left(\mathbf{p} - q\mathbf{A}\right)^2, \tag{2.14}$$

with

$$\mathbf{A} = \frac{mv_z}{q}(y, -x), \tag{2.15}$$

and

$$v_z = \frac{\hbar k}{m} \tag{2.16}$$

the velocity in the z direction (which is a constant of motion). In other words, the motion in the xy plane corresponds to a particle in a magnetic field B of magnitude

$$B = \frac{mv_z}{2q} \tag{2.17}$$

This is consistent with the classical analysis above.

Note that the spectrum of H_k is equally spaced (as in the harmonic oscillator) infinitely degenerate Landau levels. This is very interesting but well known so we omit the analysis here (see e.g. [Basdevant and Dalibard(2002)]. We note also that the controlling this system is expected to be difficult just as for the harmonic oscillator.

Finally, we point out that there is a connection with classical Wiener diffusion. Replacing the controls u and v by independent white noise terms give rise to the Itô equations

$$dx = dw_1$$

$$dy = dw_2$$

$$dz = A_1(x,y)dw_1 - A_2 dw_2$$

having the associated Fokker-Planck equation

$$\frac{\partial \rho(t,x,y,z)}{\partial t} = \frac{1}{2}\left(\left(\frac{\partial}{\partial x} + A_1\frac{\partial}{\partial z}\right)^2 + \left(\frac{\partial}{\partial y} + A_2\frac{\partial}{\partial z}\right)^2\right)\rho(t,x,y,z)$$

diffusing in three dimensions.

Note that the alternative choice of constraint given by

$$\dot{z} = \frac{x\dot{y}}{2} \tag{2.18}$$

corresponds to the same physics with a change in gauge, this time the vector potential being

$$\mathbf{A} = \frac{mv_z}{2q}(0,-x). \tag{2.19}$$

2.3 The Knife Edge

Here we consider the quantum problem corresponding to a knife edge constraint.

2.3.1 Classical Setting

We consider first the classical optimal control problem:

$$\min \frac{1}{2} \int (u^2 + v^2)dt \tag{2.20}$$

subject to the equations

$$\dot{x} = u \cos \theta,$$
$$\dot{y} = u \sin \theta,$$
$$\dot{\theta} = v, \tag{2.21}$$

which enforces the constraint $\dot{x} \sin \theta - \dot{y} \cos \theta = 0$.

This corresponds physically to a blade moving the $x - y$ plane at an angle θ to the x-axis.

Note that a fiber bundle picture for this system may be seen by considering it in the form

$$\dot{x} = u \cos \theta,$$
$$\dot{\theta} = v,$$
$$\dot{y} = \dot{x} \tan \theta, \tag{2.22}$$

so the bundle if over the base spanned by the x, θ directions and the fiber is in the y direction. Note that the connection does have a singularity as written.

This problem may also be solved via the maximum principle. We form the Hamiltonian

$$H = p_1 u \cos \theta + p_2 u \sin \theta + p_3 v - \frac{u^2}{2} - \frac{v^2}{2}. \tag{2.23}$$

The optimality conditions

$$\frac{\partial H}{\partial u} = 0 = \frac{\partial H}{\partial v}$$

yield

$$u = p_1 \cos \theta + p_2 \sin \theta,$$
$$v = p_3. \tag{2.24}$$

Hence the optimal Hamiltonian becomes

$$H = \frac{1}{2} \left\{ (p_1 \cos \theta + p_2 \sin \theta)^2 + (p_3)^2 \right\}. \tag{2.25}$$

2.3.2 Heuristic Derivation of the Quantum Hamiltonian

We now turn to the quantum situation.

First we present a heuristic derivation of the Hamiltonian.

Consider the unconstrained motion first.

$$H_0 = -\frac{\hbar^2}{2I}\frac{\partial^2}{\partial\theta^2} - \frac{\hbar^2}{2m}\left(\frac{\partial^2}{\partial x^2} + \frac{\partial^2}{\partial y^2}\right) \tag{2.26}$$

$$\equiv -\frac{\hbar^2}{2I}\frac{\partial^2}{\partial\theta^2} - \frac{\hbar^2}{2m}\left(\cos\theta\frac{\partial}{\partial x} + \sin\theta\frac{\partial}{\partial y}\right)^2$$

$$-\frac{\hbar^2}{2m}\left(\sin\theta\frac{\partial}{\partial x} - \cos\theta\frac{\partial}{\partial y}\right)^2, \tag{2.27}$$

where we have decomposed the translational term in "directions" perpendicular and parallel to the instantaneous angular variable. Now write the translational component as

$$-\frac{\hbar^2}{2m}\left(\cos\theta\frac{\partial}{\partial x} + \sin\theta\frac{\partial}{\partial y}\right)^2 - \frac{\hbar^2}{2m_\perp}\left(\sin\theta\frac{\partial}{\partial x} - \cos\theta\frac{\partial}{\partial y}\right)^2, \tag{2.28}$$

where we introduced a mass $m_\perp \neq m$ for the motion perpendicular to the instantaneous angular position. Taking the limit $m_\perp \to \infty$ we obtain

$$H = -\frac{\hbar^2}{2I}\frac{\partial^2}{\partial\theta^2} - \frac{\hbar^2}{2m}\left(\cos\theta\frac{\partial}{\partial x} + \sin\theta\frac{\partial}{\partial y}\right)^2. \tag{2.29}$$

Since, implicitly, the two masses depend on the instantaneous value of the angular variable, the constraint is dynamical.

This is consistent with the Hamiltonian given by the maximum principle above from equation 2.25.

2.3.3 Non-classical Solutions

First we show that the constraints are obeyed. Using $\dot{O} = (i/\hbar)[H, O]$,

$$\dot{x}_{\rm op} = -i\frac{\hbar}{m}\cos\theta\left(\cos\theta\frac{\partial}{\partial x} + \sin\theta\frac{\partial}{\partial y}\right),$$

$$\dot{y}_{\rm op} = -i\frac{\hbar}{m}\sin\theta\left(\cos\theta\frac{\partial}{\partial x} + \sin\theta\frac{\partial}{\partial y}\right), \tag{2.30}$$

which implies that the following "constraint" is satisfied

$$\dot{x}_{\rm op}\sin\theta - \dot{y}_{\rm op}\cos\theta = 0. \tag{2.31}$$

Thus the "constraint operator" is identically zero and in particular the expected value of the constraint in satisfied.

It is important to note the satisfaction of the constraint is consistent with the Heisenberg uncertainty principle. The intuitive idea is as follows. For a particle of mass m moving in the x-direction for example the uncertainty principle is expressed by the relation $[x, p_x] = i\hbar$. In the setting here we are considering effectively the limiting case of m and hence p_x going to infinity. Hence there is no contradiction in assuming x is zero. Similarly the constraint operator may be identically zero without contradicting the uncertainty principle.

The translational invariance of H means that $\mathbf{k} \equiv k(\cos\theta_{\mathbf{k}}, \sin\theta_{\mathbf{k}})$ is a good quantum number:

$$\psi_{\mathbf{k},\nu}(\mathbf{x}, \theta) = e^{i\mathbf{k}\cdot\mathbf{x}} u_{\mathbf{k}}^{\nu}(\theta), \tag{2.32}$$

with $u_{\mathbf{k}}^{\nu}(\theta)$ satisfying the Mathieu equation:

$$\left(-\frac{\hbar^2}{2I}\frac{\partial^2}{\partial\theta^2} + \frac{\hbar^2 k^2}{2m}\cos^2(\theta - \theta_{\mathbf{k}}) \right) u_{\mathbf{k}}^{\nu}(\theta) = E_{\mathbf{k}}^{\nu} u_{\mathbf{k}}^{\nu}(\theta). \tag{2.33}$$

Note that $\theta_{\mathbf{k}}$ parametrizes infinite degenerate solutions that are orthogonal because of the plane wave prefactor in (2.32).

Also, note that for $I \to \infty$ the solutions of (2.33) are

$$u_{\mathbf{k}}^{\nu}(\theta) = \delta(\theta - \theta_\nu), \tag{2.34}$$

with energies

$$E_{\mathbf{k}}^{\nu} = \frac{\hbar^2 k^2}{2m}\cos^2(\theta_\nu - \theta_{\mathbf{k}}). \tag{2.35}$$

This solution corresponds to the knife edge at a fixed angle θ_ν, and the center of mass described by a plane wave with a kinetic energy given by the projection of the wave vector in the directional parallel to the knife edge of the penny. Since one can think of the motion perpendicular to the (fixed) plane of the knife edge as having infinite mass, we see why it does not contribute to the energy.

For finite and small I an interesting family of non classical solutions appear. Take $\theta_{\mathbf{k}} = 0$. Since k is large we expand around the minimum of the potential $(\theta = \pi/2)$ and the Mathieu equation corresponds to the Harmonic oscillator:

$$\left(-\frac{\hbar^2}{2I}\frac{\partial^2}{\partial\theta^2} + \frac{\hbar^2 k^2}{2m}\theta^2 \right) u_{\mathbf{k}}^{\nu}(\theta) = E_{\mathbf{k}}^{\nu} u_{\mathbf{k}}^{\nu}(\theta), \tag{2.36}$$

with energies

$$E_{\mathbf{k}}^{\nu} = \left(\nu + \frac{1}{2}\right)\frac{\hbar^2}{\sqrt{mI}}|k|, \quad \nu = 0, 1, \cdots \tag{2.37}$$

We remark that these energies that formally resemble photons with a velocity given by

$$c = \frac{\hbar}{\sqrt{mI}}. \tag{2.38}$$

In particular the energy levels are equally spaced and linear in k. (Recall that photon energies levels are given by $E = (n + 1/2)\hbar\omega = (n + 1/2)\hbar ck$.)

The meaning of this solution is that the wave function has non-zero extent transverse to the knife edge direction due to the zero point energy of the angular variable.

2.3.4 Path Integral Treatment

We now consider a path integral treatment of this problem.

$$\mathcal{U}(\mathbf{x}_f\theta_f t_f, \mathbf{x}_i\theta_i t_i) = \langle \mathbf{x}_f\theta_f | e^{-\frac{i}{\hbar}\hat{H}(t_f - t_i)} | \mathbf{x}_i\theta_i \rangle$$

$$= \langle \mathbf{x}_f\theta_f | \left(e^{-i\frac{\epsilon}{\hbar}\hat{H}} \right)^M | \mathbf{x}_i\theta_i \rangle, \tag{2.39}$$

with

$$\epsilon = \frac{t_f - t_i}{M}. \tag{2.40}$$

Denote intermediate times t_n as

$$t_n = t_i + (n - 1)\epsilon, \tag{2.41}$$

and insert $M - 1$ closure relations in (2.39):

$$\mathcal{U} = \int \prod_{k=1}^{M-1} d\mathbf{x}_k d\theta_k \langle \mathbf{x}_f\theta_f | e^{-i\frac{\epsilon}{\hbar}\hat{H}} | \mathbf{x}_{M-1}\theta_{M-1} \rangle \langle \mathbf{x}_{M-1}\theta_{M-1} | e^{-i\frac{\epsilon}{\hbar}\hat{H}} | \mathbf{x}_{M-2}\theta_{M-2} \rangle$$

$$\times \langle \mathbf{x}_{M-2}\theta_{M-2} | \cdots e^{-i\frac{\epsilon}{\hbar}\hat{H}} | \mathbf{x}_1\theta_1 \rangle \langle \mathbf{x}_1\theta_1 | e^{-i\frac{\epsilon}{\hbar}\hat{H}} | \mathbf{x}_i\theta_i \rangle. \tag{2.42}$$

We use the usual approximation for the infinitesimal evolution operator

$$\langle \mathbf{x}_n\theta_n | e^{-i\frac{\epsilon}{\hbar}\hat{H}} | \mathbf{x}_{n-1}\theta_{n-1} \rangle$$

$$= \int d\mathbf{p} \sum_{r=-\infty}^{\infty} \langle \mathbf{x}_n\theta_n | \mathbf{p}, r \rangle \langle \mathbf{p}, r | e^{-i\frac{\epsilon}{\hbar}\hat{H}} | \mathbf{x}_{n-1}\theta_{n-1} \rangle$$

$$= \left(\frac{1}{2\pi\hbar} \right)^3 \int d\mathbf{p} \sum_{r=-\infty}^{\infty} e^{i\left[\frac{\mathbf{p}}{\hbar} \cdot (\mathbf{x}_n - \mathbf{x}_{n-1}) + r(\theta_n - \theta_{n-1}) \right]}$$

$$\times \exp -i\frac{\epsilon}{\hbar} \left[\frac{\hbar^2 r^2}{2I} + \frac{1}{2m} (p_x \cos\theta_{n-1} + p_y \sin\theta_{n-1})^2 \right]. \tag{2.43}$$

Now define

$$p_\| = p_x \cos\theta_{n-1} + p_y \sin\theta_{n-1},$$
$$p_\perp = p_y \cos\theta_{n-1}, -p_x \sin\theta_{n-1}$$
$$\Delta_x = x_n - x_{n-1}, \qquad \Delta_y = y_n - y_{n-1}, \tag{2.44}$$
$$\Delta_\| = \Delta_x \cos\theta_{n-1} + \Delta_y \sin\theta_{n-1},$$
$$\Delta_\perp = \Delta_y \cos\theta_{n-1} - \Delta_x \sin\theta_{n-1}.$$

$$\tag{2.45}$$

With these definitions the integrals over $d\mathbf{p}$ have the form

$$\left(\frac{1}{2\pi\hbar}\right)^2 \int dp_\| dp_\perp \, e^{i\frac{p_\perp}{\hbar}\Delta_\perp} e^{i\left[\frac{p_\|}{\hbar}\Delta_\| - \frac{\epsilon}{\hbar}\frac{1}{2m}p_\|^2\right]} \tag{2.46}$$

$$= \delta(\Delta_\perp) \left(\frac{m}{2\pi}i\epsilon\hbar\right)^{1/2} e^{\frac{i}{\hbar}\frac{m}{2}\left(\frac{\Delta_\|}{\epsilon}\right)^2}$$

$$\equiv \delta(\Delta_\perp) \left(\frac{m}{2\pi}i\epsilon\hbar\right)^{1/2} e^{\frac{i}{\hbar}\frac{m}{2}\frac{\Delta_\|^2+\Delta_\perp^2}{\epsilon^2}}$$

$$= \delta(\Delta_\perp) \left(\frac{m}{2\pi}i\epsilon\hbar\right)^{1/2} e^{\frac{i}{\hbar}\frac{m}{2}\left(\frac{\mathbf{x}_n-\mathbf{x}_{n-1}}{\epsilon}\right)^2}$$

$$\sum_{r=-\infty}^{\infty} e^{-i\frac{\epsilon}{\hbar}\left[\frac{\hbar^2 r^2}{2I}+r(\theta_n-\theta_{n-1})\right]} \tag{2.47}$$

$$= \left(\frac{i\hbar\epsilon}{2\pi I}\right)^{1/2} \int_{-\infty}^{\infty} dx e^{-i\frac{\epsilon}{\hbar}\frac{\hbar^2 x^2}{2I}} \underbrace{\sum_{r=-\infty}^{\infty} e^{ir\left[(\theta_n-\theta_{n-1})-2x\frac{\epsilon}{\hbar}\frac{\hbar^2}{2I}\right]}}_{\delta\left((\theta_n-\theta_{n-1})-2x\frac{\epsilon}{\hbar}\frac{\hbar^2}{2I}\right)}$$

$$= \left(\frac{i\hbar\epsilon}{2\pi I}\right)^{1/2} e^{-i\frac{\epsilon}{\hbar}\frac{I}{2}\left(\frac{\theta_n-\theta_{n-1}}{\epsilon}\right)^2}.$$

Collecting the above results we arrive at

$$\mathcal{U}(\mathbf{x}_f\theta_f t_f, \mathbf{x}_i\theta_i t_i)$$
$$= \int \mathcal{D}[\mathbf{x}(t),\theta(t)]\delta\left[\dot{y}(t)\cos\theta(t) - \dot{x}(t)\sin\theta(t)\right] e^{-\frac{i}{\hbar}\int_i^f \mathcal{L}_0(\mathbf{x}(t),\theta(t))} \tag{2.48}$$

with $\mathcal{L}_0$ the free action.

2.4 The Rolling Penny

The rolling penny may be analyzed in similar fashion.

Here we have Hamiltonian

$$H = \frac{1}{2}\left(m\dot{x}^2 + m\dot{y}^2 + I\dot{\theta}^2 + J\dot{\phi}^2\right)$$

with constraints

$$\dot{x} = R\cos\theta\dot{\phi}$$
$$\dot{y} = R\sin\theta\dot{\phi}. \tag{2.49}$$

This is similar to knife edge but with the the penny now rolling with roll angle ϕ at an angle θ to the x-axis.

Note that we cannot just substitute the constraints in the Hamiltonian and then quantize for this gives the trivial Hamiltonian

$$H = \frac{1}{2}((I + mR^2)\dot{\phi}^2 + J\dot{\theta}^2).$$

(This is equivalent to decomposing into constraint and orthogonal to constraint directions the letting the orthogonal mass go to zero.

The correct subRiemannian problem is:

$$\min \frac{1}{2}\int(u^2 + v^2)dt \tag{2.50}$$

subject to

$$\dot{\phi} = u$$
$$\dot{\theta} = v$$
$$\dot{x} = R\cos\theta\, u,$$
$$\dot{y} = R\sin\theta\, u, \tag{2.51}$$

Note that this is not dynamical.

This problem may also be solved via the maximum principle. We form the Hamiltonian

$$H = p_1 u + p_2 v + p_3 R\cos\theta u + p_4 R\sin\theta u - \frac{u^2}{2} - \frac{v^2}{2}. \tag{2.52}$$

The optimality conditions

$$\frac{\partial H}{\partial u} = 0 = \frac{\partial H}{\partial v}$$

yield

$$u = p_1 + p_3 R\cos\theta + p_4 R\sin\theta,$$
$$v = p_2. \tag{2.53}$$

Hence the optimal Hamiltonian becomes

$$H = \frac{1}{2}\left\{(p_1 + p_3 R\cos\theta + p_4 R\sin\theta)^2 + (p_2)^2\right\}. \tag{2.54}$$

Hence the quantum Hamiltonian in this case is

$$H = -\frac{\hbar^2}{2}\frac{\partial^2}{\partial\theta^2} - \frac{\hbar^2}{2}\left(\frac{\partial}{\partial\phi} + R\cos\theta\frac{\partial}{\partial x} + R\sin\theta\frac{\partial}{\partial y}\right)^2. \tag{2.55}$$

If we multiply by p_1 here by a small parameter which we let tend to zero we recover the knife edge dynamics. Hence the dynamics here is essentially similar to that of the knife edge.

We note also that rolling penny quantum dynamics may also be derived via a quantum mechanical diffusion argument as we did for the Heisenberg system. Finally, we note that equation (2.55) is translationally invariant as well as invariant under rotations in ϕ. This means that the solutions are identical to those of Hamiltonian (2.33), with the addition of a "flux quantum" through the "ring" described by the periodic variable θ.

2.5 Conclusion

In this paper we have analyzed the quantum versions of some key systems in subRiemannian dynamics. In a future paper we intend to discuss the general setting for such systems as well as related quantum nonholonomic dynamics.

References

[Basdevant and Dalibard(2002)] Basdevant, J.-L. and J. Dalibard [2002] Quantum Mechanics, Springer, New York.

[Bloch(2003)] Bloch, A.M., with J. Baillieul, P.E. Crouch and J.E. Marsden [2003] *Nonholonomic Mechanics and Control*, Springer Verla.

[Bloch and Crouch(1993)] Bloch, A. M. and P. E. Crouch [1993], Nonholonomic and vakonomic control systems on Riemannian manifolds, *Fields Institute Communications* **1**, 25–52.

[Brockett(1981)] Brockett, R. W. [1981], Control theory and singular Riemannian geometry, in *New Directions in Applied Mathematics* (P. J. Hilton and G. S. Young, eds.), Springer-Verlag, 11–27.

[Bloch, Krishnaprasad, Marsden, and Murray(1996)] Bloch, A.M., P.S. Krishnaprasad, J.E. Marsden, and R. Murray [1996], Nonholonomic mechanical systems with symmetry, *Arch. Rat. Mech. An.*, **136**, 21–99.

[Marsden and Ratiu(1999)] Marsden, J. E. and T. S. Ratiu [1999], *Introduction to Mechanics and Symmetry*, Springer-Verlag, Texts in Applied Mathematics, **17**; First Edition 1994, Second Edition, 1999.

[Scully and Zubiary(1997)] Scully, M. O. and M. S. Zubiary [1997], *Quantum Optics*, (Chapter 5). Cambridge University Press.

[Zinn-Justin(1996)] Zinn-Justin, J. [1996], *Quantum Field Theory and Critical Phenomena*, Oxford University Press.

A Homotopy Continuation Solution of the Covariance Extension Equation*

Christopher I. Byrnes[1], Giovanna Fanizza[2], and Anders Lindquist[2]

[1] Washington University, St. Louis, MO, USA
`chrisbyrnes@seas.wustl.edu`
[2] Royal institute of Technology, Stockholm, Sweden
`alq@math.kth.se`

Dedicated to Clyde Martin on the occation of his 60th birthday

Algebraic geometry plays an important role in the theory of linear systems for (at least) three reasons. First, the Laplace transform turns expressions about linear differential systems into expressions involving rational functions. In addition, many of the transformations studied in linear systems theory, like changes of coordinates or feedback, turn out to be the action of algebraic groups on algebraic varieties. Finally, when we study linear quadratic problems in optimization and estimation, all roads eventually lead either to the Riccati equation or to spectral factorization.

Clyde Martin was a pioneer in applying algebraic geometry to linear systems in all three of these theaters. Perhaps the work which is closest to the results we discuss in this paper was his joint study, with Bob Hermann, of the matrix Riccati equation as a flow on a Grassmannian.

In this paper we study the steady state form of a discrete-time matrix Riccati-type equation, connected to the rational covariance extension problem and to the partial stochastic realization problem. This equation, however, is nonstandard in that it lacks the usual kind of definiteness properties which underlie the solvability of the standard Riccati equation. Nonetheless, we prove the existence and uniqueness of a positive semidefinite solution. We also show that this equation has the proper geometric attributes to be solvable by homotopy continuation methods, which we illustrate in several examples.

3.1 The Covariance Extension Equation

We will consider real sequences

$$c = (c_0, c_1, \ldots, c_n) \tag{3.1}$$

---

*This research was supported in part by grants from AFOSR, NSF, VR, the Göran Gustafsson Foundation, and Southwestern Bell.

C.I. Byrnes et al.: *A Homotopy Continuation Solution of the Covariance Extension Equation,*
LNCIS **321**, 27–42 (2005)
`www.springerlink.com`

that are positive in the sense that

$$
T_n = \begin{bmatrix} c_0 & c_1 & \cdots & c_n \\ c_1 & c_0 & \cdots & c_{n-1} \\ \vdots & \vdots & \ddots & \vdots \\ c_n & c_{n-1} & \cdots & c_0 \end{bmatrix} > 0,
$$

and we consider Schur polynomials

$$
\sigma(z) = z^n + \sigma_1 z^{n-1} + \cdots + \sigma_n; \tag{3.2}
$$

i.e, polynomials with all their roots in the open unit disc. For simplicity, we normalize by taking $c_0 = 1$. Motivated by the partial stochastic realization problem and the rational covariance extension problem, which we will briefly review in Section 3, we form the following n vectors and $n \times n$ matrix:

$$
\sigma = \begin{bmatrix} \sigma_1 \\ \sigma_2 \\ \vdots \\ \sigma_n \end{bmatrix}, \quad h = \begin{bmatrix} 1 \\ 0 \\ \vdots \\ 0 \end{bmatrix}, \quad \text{and} \quad \Gamma = \begin{bmatrix} -\sigma_1 & 1\,0 & \cdots & 0 \\ -\sigma_2 & 0\,1 & \cdots & 0 \\ \vdots & \vdots\,\vdots & \ddots & \vdots \\ -\sigma_{n-1} & 0\,0 & \cdots & 1 \\ -\sigma_n & 0\,0 & \cdots & 0 \end{bmatrix}. \tag{3.3}
$$

Defining $u_1, u_2, \ldots, u_n$ via

$$
\frac{z^n}{z^n + c_1 z^{n-1} + \cdots + c_n} = 1 - u_1 z^{-1} - u_2 z^{-2} - u_3 z^{-3} - \ldots \tag{3.4}
$$

we also form

$$
u = \begin{bmatrix} u_1 \\ u_2 \\ \vdots \\ u_n \end{bmatrix} \qquad U = \begin{bmatrix} 0 & & & \\ u_1 & 0 & & \\ u_2 & u_1 & & \\ \vdots & \vdots & \ddots & \\ u_{n-1} & u_{n-2} & \cdots & u_1 \; 0 \end{bmatrix}. \tag{3.5}
$$

We shall also need the function $g : \mathbb{R}^{n \times n} \to \mathbb{R}^n$ defined by

$$
g(P) = u + U\sigma + U\Gamma Ph. \tag{3.6}
$$

From these quantities, in [4], we formed the Riccati-like matrix equation

$$
P = \Gamma(P - Phh'P)\Gamma' + g(P)g(P)', \tag{3.7}
$$

which we sought to solve in the space of positive semidefinite matrices satisfying the additional constraint

$$
h'Ph < 1, \tag{3.8}
$$

where $'$ denotes transposition. We refer to this equation as the covariance extension equation (CEE).

To this end, define the semialgebraic sets

$$X = \{(c, \sigma) \mid T_n > 0, \ \sigma(z) \text{ is a Schur polynomial}\}$$

and

$$Y = \{P \in \mathbb{R}^{n \times n} \mid P \geq 0, \ h'Ph < 1\}.$$

On $X \times Y$ we define the rational map

$$F(c, \sigma, P) = P - \Gamma(P - Phh'P)\Gamma' - g(P)g(P)'$$

Of course its zero locus

$$Z = F^{-1}(0) \subset X \times Y$$

is the solution set to the covariance extension equations. We are interested in the projection map restricted to Z

$$\pi_X(c, \sigma, P) = (c, \sigma).$$

For example, to say that π_X is surjective is to say that there is always a solution to CEE, and to say that π_X is injective is to say that solutions are unique. One of the main results of this paper is the following, which, in particular, implies that CEE has a unique solution $P \in Y$ for each $(c, \sigma) \in X$ [4, Theorem 2.1].

Theorem 1. *The solution set Z is a smooth semialgebraic manifold of dimention $2n$. Moreover, π_X is a diffeomorphism between Z and X.*

In particular the map π_X is smooth with no branch points and every smooth curve in X lifts to a curve in Z. These observations imply that the homotopy continuation method will apply to solving the covariance extension equation [1].

3.2 Proof of Theorem 1

The rational covariance extension problem is to find polynomials

$$a(z) = z^n + a_1 z^{n-1} + \cdots + a_n \tag{3.9a}$$

$$b(z) = z^n + b_1 z^{n-1} + \cdots + b_n \tag{3.9b}$$

satisfying the interpolation condition

$$\frac{b(z)}{a(z)} = \frac{1}{2} + c_1 z^{-1} + \cdots + c_n z^{-n} + O(z^{-n-1}) \tag{3.10}$$

and the positivity condition

$$\frac{1}{2}\left[a(z)b(z^{-1}) + b(z)a(z^{-1})\right] = \rho^2\sigma(z)\sigma(z^{-1}) \qquad (3.11)$$

for some positive real number ρ. Given (3.9), let a and b be the n-vectors $a := (a_1, a_2, \ldots, a_n)$ and $b := (b_1, b_2, \ldots, b_n)$, respectively.

In [4] we proved:

Theorem 2. *There is a one-to-one correspondence between symmetric solutions P of the covariance extension equation (3.7) such that $h'Ph < 1$ and pairs of monic polynomials (3.9a)-(3.9b) satisfying the interpolation condition (3.10) and the positivity condition (3.11). Under this correspondence*

$$a = (I - U)(\Gamma Ph + \sigma) - u, \qquad (3.12a)$$

$$b = (I + U)(\Gamma Ph + \sigma) + u, \qquad (3.12b)$$

$$\rho = (1 - h'Ph)^{1/2}, \qquad (3.12c)$$

and P is the unique solution of the Lyapunov equation

$$P = JPJ' - \frac{1}{2}(ab' + ba') + \rho^2\sigma\sigma', \qquad (3.13)$$

where

$$J = \begin{bmatrix} 0 & 1 & 0 & \cdots & 0 \\ 0 & 0 & 1 & \cdots & 0 \\ \vdots & \vdots & \vdots & \ddots & \vdots \\ 0 & 0 & 0 & \cdots & 1 \\ 0 & 0 & 0 & \cdots & 0 \end{bmatrix} \qquad (3.14)$$

is the upward shift matrix. Moreover the following conditions are equivalent

1. $P \geq 0$
2. $a(z)$ is a Schur polynomial
3. $b(z)$ is a Schur polynomial

and, if they are fulfilled,

$$\deg v(z) = \operatorname{rk} P. \qquad (3.15)$$

We can now prove Theorem 1. Let $\mathcal{P}_n$ be the space of pairs (a, b) whose quotient is positive real. Clearly, the mapping

$$f : \mathcal{P}_n \to X,$$

sending (a, b) to the corresponding (c, σ), is smooth. Our main result in [5] asserts that f is actually a diffeomorphism. In particular, for each positive sequence (3.1) and each monic Schur polynomial (3.2), there is a unique pair of polynomials, (a, b), satisfying (3.10) and (3.11), and consequently (a, b) solves the rational covariance extension problem corresponding to (c, σ). Moreover, by Theorem 2, there is a unique corresponding solution to the covariance extension equation, which is positive semi-definite.

Since J is nilpotent, the Lyapunov equation (3.13) has a unique solution, P, for each right hand side of equation (3.13). Moreover, the right hand side is a smooth function on X and, using elementary methods from Lyapunov theory, we conclude that P is also smooth as a function on X. As the graph in $X \times Y$ of a smooth mapping defined on X, Z is a smooth manifold of dimension $2n = \dim X$. Moreover, this mapping has the smooth mapping π_X as its inverse. Therefore, π_X is a diffeomorphism.

Remark 1. Our proof, together with the results in [5], shows more. Namely, that Z is an analytic manifold and that π_X is an analytic diffeomorphism with an analytic inverse.

3.3 Rational covariance extension and the CEE

As described above, given a positive sequence (3.1), the rational covariance extension problem – or the covariance extension problem with degree constraint – amounts to finding a pair (a, b) of Schur polynomials (3.9a)-(3.9b) satisfying the interpolation condition

$$\frac{b(z)}{a(z)} = \frac{1}{2} + c_1 z^{-1} + \cdots + c_n z^{-n} + O(z^{-n-1}) \qquad (3.16)$$

and the positivity condition

$$\frac{1}{2}\left[a(z)b(z^{-1}) + b(z)a(z^{-1})\right] > 0 \quad \text{on } \mathbb{T}. \qquad (3.17)$$

Then there is a Schur polynomial (3.2) such that

$$\frac{1}{2}\left[a(z)b(z^{-1}) + b(z)a(z^{-1})\right] = \rho^2 \sigma(z)\sigma(z^{-1}) \qquad (3.18)$$

for some positive normalizing coefficient ρ, and

$$\mathrm{Re}\left\{\frac{b(e^{i\theta})}{a(e^{i\theta})}\right\} = \left|\rho\frac{\sigma(e^{i\theta})}{a(e^{i\theta})}\right|^2. \qquad (3.19)$$

Georgiou [9, 10] raised the question whether there exists a solution for each choice of σ and answered this question in the affirmative. He also conjectured that this assignment is unique. This conjecture was proven in [5] in a more general context of well-posedness.

In [4] we showed that, for any $(c, \sigma) \in X$, CEE has a unique solution $P \in Y$ and that the unique solution corresponding to σ to the rational covariance extension problem is given by

$$a = (I - U)(\Gamma Ph + \sigma) - u, \tag{3.20a}$$

$$b = (I + U)(\Gamma Ph + \sigma) + u. \tag{3.20b}$$

Clearly the interpolation condition (3.16) can be written

$$b = 2c + (2C_n - I)a, \tag{3.21}$$

where

$$c = \begin{bmatrix} c_1 \\ c_2 \\ \vdots \\ c_n \end{bmatrix} \qquad C_n = \begin{bmatrix} 1 & & & & \\ c_1 & 1 & & & \\ c_2 & c_1 & 1 & & \\ \vdots & \vdots & \vdots & \ddots & \\ c_{n-1} & c_{n-2} & c_{n-3} & \cdots & 1 \end{bmatrix}.$$

Using the fact that $C_n u = c$ and $C_n(I - U) = I$, it was shown in [4] that (3.21) can be written

$$a = \frac{1}{2}(I - U)(a + b) - u. \tag{3.22}$$

For a fixed $(c, \sigma) \in X$, let $H : Y \to \mathbb{R}^{n \times n}$ be the map sending P to $F(c, \sigma, P)$, and let

$$dH(P; Q) := \lim_{t \to 0} \frac{H(P + tQ) - H(P)}{t}$$

be the derivative in the direction $Q = Q'$. A key property needed in the homotopy continuation solution of the CEE is the fact that this derivative is full rank.

Proposition 1. *Given $(c, \sigma) \in X$, let $P \in Y$ be the corresponding solution of CEE. Then, if $dH(P; Q) = 0$, $Q = 0$.*

Proof. Suppose that $dH(P; Q) = 0$ for some Q. Then

$$H(P) + \lambda dH(P; Q) = 0$$

for any $\lambda \in \mathbb{R}$. Since

$$dH(P; Q) = Q - \Gamma Q \Gamma' + \Gamma Phh'Q\Gamma' + \Gamma Qhh'P\Gamma' - g(P)h'Q\Gamma'U' - U\Gamma Qhg(P)',$$

this can be written

$$H(P_\lambda) = \lambda^2 R(Q), \tag{3.23}$$

where $P_\lambda := P + \lambda Q$ and

$$R(Q) := 2\Gamma Qhh'Q\Gamma' - 2U\Gamma Qhh'Q\Gamma'U'.$$

Proceeding as in the proof of Lemma 4.6 in [4], (3.23) can be written

$$P_\lambda = JP_\lambda J' - \frac{1}{2}(a_\lambda b'_\lambda + b_\lambda a'_\lambda) + \rho_\lambda^2 \sigma\sigma' - \lambda^2 R(Q), \qquad (3.24)$$

where

$$a_\lambda = (I - U)(\Gamma P_\lambda h + \sigma) - u, \qquad (3.25a)$$

$$b_\lambda = (I + U)(\Gamma P_\lambda h + \sigma) + u, \qquad (3.25b)$$

$$\rho_\lambda = (1 - h' P_\lambda h)^{1/2}. \qquad (3.25c)$$

Observe that

$$a_\lambda = \frac{1}{2}(I - U)(a_\lambda + b_\lambda) - u, \qquad (3.26)$$

and hence (a_λ, b_λ) satisfies the interpolation condition (3.22), or, equivalently, (3.16), for all $\lambda \in \mathbb{R}$.

Multipying (3.24) by $z^{j-i} = z^{n-i} z^{-(n-j)}$ and summing over all $i, j = 1, 2, \ldots, n$, we obtain

$$\frac{1}{2}\left[a_\lambda(z)b_\lambda(z^{-1}) + b_\lambda(z)a_\lambda(z^{-1})\right] = \rho_\lambda^2 \sigma(z)\sigma(z^{-1}) - \lambda^2 \sum_{i=1}\sum_{j=1} R_{ij}(Q)z^{j-i}$$

$$(3.27)$$

again along the calculations of the proof of Lemma 4.6 in [4]. Since $\sigma(z)\sigma(z^{-1}) > 0$ on $\mathbb{T}$,

$$\rho_\lambda^2 \sigma(z)\sigma(z^{-1}) - \lambda^2 \sum_{i=1}\sum_{j=1} R_{ij}(Q)z^{j-i} > 0 \quad \text{on } \mathbb{T}$$

for $|\lambda|$ sufficiently small. Then there is a Schur polynomial σ_λ and a positive constant $\hat{\rho}_\lambda$ such that

$$\hat{\rho}_\lambda^2 \sigma_\lambda(z)\sigma_\lambda(z^{-1}) = \rho_\lambda^2 \sigma(z)\sigma(z^{-1}) - \lambda^2 \sum_{i=1}\sum_{j=1} R_{ij}(Q)z^{j-i}.$$

Therefore,

$$\frac{1}{2}\left[a_\lambda(z)b_\lambda(z^{-1}) + b_\lambda(z)a_\lambda(z^{-1})\right] = \hat{\rho}_\lambda^2 \sigma_\lambda(z)\sigma_\lambda(z^{-1}) \qquad (3.28)$$

for $|\lambda|$ sufficiently small.

Now recall that $a_0 = a$ and $b_0 = b$ are Schur polynomials and that the Schur region is open in $\mathbb{R}^n$. Hence there is an $\varepsilon > 0$ such that $a_\varepsilon(z), a_{-\varepsilon}(z), b_\varepsilon(z)$ and $b_{-\varepsilon}(z)$ are also Schur polynomials and (3.28) holds for $\lambda = \pm\varepsilon$.

Consequently, $(a_\varepsilon, b_\varepsilon)$ and $(a_{-\varepsilon}, b_{-\varepsilon})$ both satisfy the interpolation condition (3.16) and the positivity condition (3.18) corresponding to the same $\sigma := \sigma_\varepsilon = \sigma_{-\varepsilon}$. Therefore, since the solution to the rational covariance extension problem corresponding to σ is unique, we must have $a_\varepsilon = a_{-\varepsilon}$ and $b_\varepsilon = b_{-\varepsilon}$, and hence in view of (3.24), $P_\varepsilon = P_{-\varepsilon}$; i.e, $Q = 0$, as claimed.

3.4 Reformulation of the Covariance Extension Equation

Solving the covariance extension equation (3.7) amounts to solving $\frac{1}{2}n(n-1)$ nonlinear scalar equations, which number grows rapidly with increasing n. As in the theory of fast filtering algorithms [11, 12], we may replace these equations by a system of only n equations. In fact, setting

$$p = Ph \tag{3.29}$$

the covariance extension equation can be written

$$P - \Gamma P \Gamma' = -\Gamma p p' \Gamma' + (u + U\sigma + U\Gamma p)(u + U\sigma + U\Gamma p)' \tag{3.30}$$

If we could first determine p, P could be obtained from (3.30), regarded as a Lyapunov equation. We proceed to doing precisely this.

It follows from Theorem 2 that (3.30) may also be written

$$P = JPJ' - \frac{1}{2}(ab' + ba') + \rho^2 \sigma\sigma', \tag{3.31}$$

with a, b and ρ given by (3.12). Multiplying (3.31) by $z^{j-i} = z^{n-i}z^{-(n-j)}$ and summing over all $i, j = 1, 2, \ldots, n$, we obtain precisely (3.11), which in matrix form becomes

$$S(a) \begin{bmatrix} 1 \\ b \end{bmatrix} = 2\rho^2 \begin{bmatrix} d \\ \sigma_n \end{bmatrix} \tag{3.32}$$

or, symmetrically,

$$S(b) \begin{bmatrix} 1 \\ a \end{bmatrix} = 2\rho^2 \begin{bmatrix} d \\ \sigma_n \end{bmatrix}, \tag{3.33}$$

where

$$S(a) = \begin{bmatrix} 1 & \ldots & a_{n-2} & a_{n-1} & a_n \\ a_1 & \ldots & a_{n-1} & a_n & \\ a_2 & \ldots & a_n & & \\ \vdots & \reflectbox{$\ddots$} & & & \\ a_n & & & & \end{bmatrix} + \begin{bmatrix} 1 & a_1 & a_2 & \ldots & a_n \\ & 1 & a_1 & \ldots & a_{n-1} \\ & & 1 & \ldots & a_{n-2} \\ & & & \ddots & \vdots \\ & & & & 1 \end{bmatrix} \tag{3.34}$$

and

$$d = \begin{bmatrix} 1 + \sigma_1^2 + \sigma_2^2 + \cdots + \sigma_n^2 \\ \sigma_1 + \sigma_1\sigma_2 + \sigma_{n-1}\sigma_n \\ \sigma_2 + \sigma_1\sigma_3 + \sigma_{n-2}\sigma_n \\ \vdots \\ \sigma_{n-1} + \sigma_1\sigma_n \end{bmatrix}. \tag{3.35}$$

Inserting (3.12) and (3.29) in (3.32) yields

$$S(a(p)) \begin{bmatrix} 1 \\ b(p) \end{bmatrix} = 2(1 - h'p) \begin{bmatrix} d \\ \sigma_n \end{bmatrix}, \tag{3.36}$$

where

$$a(p) = (I - U)(\Gamma p + \sigma) - u, \tag{3.37a}$$

$$b(p) = (I + U)(\Gamma p + \sigma) + u \tag{3.37b}$$

are functions of p. More precisely, (3.36) are $n+1$ equations in the n unknown p. However, from (3.12) we have

$$\frac{1}{2}(a_n + b_n) = \rho^2 \sigma_n,$$

which is precisely the last equation in (3.32). Hence (3.36) is redundant and can be deleted to yield

$$ES(a(p)) \begin{bmatrix} 1 \\ b(p) \end{bmatrix} = 2(1 - h'p)d, \tag{3.38}$$

where E is the $n \times (n + 1)$ matrix

$$E = \begin{bmatrix} I_n & 0 \end{bmatrix}. \tag{3.39}$$

These n equations in n unkowns $p_1, p_2, \ldots, p_n$ clearly has a unique solution $\hat{p}$, for CEE has one.

3.5 Homotopy Continuation

Suppose that $(c, \sigma) \in X$. To solve the corresponding covariance extension equation

$$P = \Gamma(P - Phh'P)\Gamma' + g(P)g(P)' \tag{3.40}$$

for its unique solution $\hat{P}$, we first observe that the solution is particularly simple if $c = c_0 = 0$. Then $u = 0$, $U = 0$ and (3.40) reduces to

$$P = \Gamma(P - Phh'P)\Gamma' \tag{3.41}$$

having the unique solution $P = 0$ in Y. Consider the deformation

$$c(\nu) = \nu c, \quad \nu \in [0, 1].$$

Clearly, $(c(\nu), \sigma) \in X$, and consequently the equation

$$H(P, \nu) := P - \Gamma(P - Phh'P)\Gamma' - g(P, \nu)g(P, \nu)' = 0, \tag{3.42}$$

where

$$g(P, \nu) = u(\nu) + U(\nu)\sigma + U(\nu)\Gamma Ph$$

with

$$u(\nu) = \begin{bmatrix} 1 & & & & \\ \nu c_1 & 1 & & & \\ \nu c_2 & \nu c_1 & 1 & & \\ \vdots & \vdots & \vdots & \ddots & \\ \nu c_{n-1} & \nu c_{n-2} & \nu c_{n-3} & \cdots & 1 \end{bmatrix}^{-1} \begin{bmatrix} \nu c_1 \\ \nu c_2 \\ \vdots \\ \nu c_n \end{bmatrix}$$

and

$$U(\nu) = \begin{bmatrix} 0 & & & \\ u_1(\nu) & 0 & & \\ u_2(\nu) & u_1(\nu) & & \\ \vdots & \vdots & \ddots & \\ u_{n-1}(\nu) & u_{n-2}(\nu) & \cdots & u_1(\nu) & 0 \end{bmatrix},$$

has a unique solution $\hat{P}(\nu)$ in Y.

The function $H : Y \times [0,1] \to \mathbb{R}^{n \times n}$ is a homotopy between (3.40) and (3.41). In view of Theorem 1, the trajectory $\{\hat{P}(\nu)\}_{\nu=0}^1$ is continuously differentiable and has no turning points or bifurcations. Consequently, homotopy continuation can be used to obtain a computational procedure. However, the corresponding ODE will be of dimension $O(n^2)$. Therefore, it is better to work with the reduced equation (3.38), which yields an ODE of order n.

To this end, setting

$$V := \{p \in \mathbb{R}^n \mid p = Ph, P \in Y\},$$

consider instead the homotopy $G : V \times [0,1] \to \mathbb{R}^n$ defined by

$$G(p,\nu) := ES(a(p)) \begin{bmatrix} 1 \\ b(p) \end{bmatrix} - 2(1 - h'p)d,$$

where $a(p)$ and $b(p)$ are given by (3.37). *A fortiori* the corresponding trajectory $\{\hat{p}(\nu)\}_{\nu=0}^1$ is continuously differentiable and has no turning points or bifurcations. Differentiating

$$G(p,\nu) = 0$$

with respect to ν yields

$$ES(a) \begin{bmatrix} 0 \\ \dot{b} \end{bmatrix} + ES(b) \begin{bmatrix} 0 \\ \dot{a} \end{bmatrix} + 2h'\dot{p}d = 0,$$

where dot denotes derivative and

$$\dot{a} = (I - U)\Gamma\dot{p} - \dot{U}(\Gamma p + \sigma) - \dot{u}, \tag{3.43a}$$

$$\dot{b} = (I + U)\Gamma\dot{p} + \dot{U}(\Gamma p + \sigma) + \dot{u}, \tag{3.43b}$$

or, which is the same,

$$ES\left(\frac{a+b}{2}\right)\begin{bmatrix}0\\ \Gamma\dot{p}\end{bmatrix} - ES\left(\frac{b-a}{2}\right)\begin{bmatrix}0\\ U\Gamma\dot{p}\end{bmatrix} + dh'\dot{p} =$$

$$= ES\left(\frac{b-a}{2}\right)\begin{bmatrix}0\\ \dot{U}\Gamma p + \dot{U}\sigma + \dot{u}\end{bmatrix}.$$

In view of (3.37), this may be written

$$\left[\hat{S}(\Gamma p + \sigma) - \hat{S}(U\Gamma p + U\sigma + u) + dh'\right]\dot{p} = \hat{S}(U\Gamma p + U\sigma + u)(\dot{U}\Gamma p + \dot{U}\sigma + \dot{u}),$$

where $\hat{S}(a)$ is the $n \times n$ matrix obtained by deleting the first column and the last row in (3.34). Hence we have proven the following theorem.

Theorem 3. *The differential equation*

$$\dot{p} = \left[\hat{S}(\Gamma p + \sigma) - \hat{S}(U(\nu)\Gamma p + U(\nu)\sigma + u(\nu)) + dh'\right]^{-1} \times$$

$$\hat{S}(U(\nu)\Gamma p + U(\nu)\sigma + u(\nu))(\dot{U}(\nu)\Gamma p + \dot{U}(\nu)\sigma + \dot{u}(\nu)),$$

$$p(0) = 0$$

has a unique solution $\{\hat{p}(\nu); 0 \le \nu \le 1\}$. Moreover, the unique solution of the Lyapunov equation

$$P - \Gamma P \Gamma' = -\Gamma\hat{p}(1)\hat{p}(1)'\Gamma' + (u + U\sigma + U\Gamma\hat{p}(1))(u + U\sigma + U\Gamma\hat{p}(1))',$$

where $U = U(1)$ and $u = u(1)$, is also the unique solution of the covariance extension equation (3.7).

The differential equation can be solved by methods akin to those in [3].

3.6 Simulations

We illustrate the method described above by two examples, in which we use covariance data generated in the following way. Pass white noise through a given stable filter

$$\text{white noise} \xrightarrow{\ w\ } \boxed{\ w(z)\ } \xrightarrow{\ y\ }$$

with a rational transfer function

$$w(z) = \frac{\hat{\sigma}(z)}{\hat{a}(z)}$$

of degree $\hat{n}$, where $\hat{\sigma}(z)$ is a (monic) Schur polynomial. This generates a time series

$$y_0, y_1, y_2, y_3, \ldots, y_N, \tag{3.44}$$

from which a covariance sequence is computed via the biased estimator

$$\hat{c}_k = \frac{1}{N} \sum_{t=k+1}^{N} y_t y_{t-k}, \tag{3.45}$$

which actually provides a sequences with positive Toepliz matrices. By setting $c_k := \hat{c}_k/\hat{c}_0$ we obtained a normalized covariance sequence

$$1, c_1, c_2, \ldots, c_n, \quad n \geq \hat{n}. \tag{3.46}$$

Example 1: Detecting the positive degree

Given a transfer function $w(z)$ of degree $\hat{n} = 2$ with zeros at $0.37e^{\pm i}$ and poles at $0.82e^{\pm 1.32i}$, estimate the covariance sequence (3.46) for $n = 2, 3, 4, 5$ and 6. Given these covariance sequences, we apply the algorithm of this paper to compute the $n \times n$ matrix P, using the zero polynomial $\sigma(z) = z^{n-\hat{n}}\hat{\sigma}(z)$, thus keeping the trigonometric polynomial $|\sigma(e^{i\theta})|^2$ constant. For each value of n, 100 Monte Carlo simulations are performed, and the average of the singular values of P are computed and shown in Table 1.

Table 3.1. Singular values of solution P of the CEE

n=2	n=3	n=4	n=5	n=6
.42867	.42892	.42922	.42967	.43004
.25322	.25368	.25388	.25407	.25433
	$3.0409 \cdot 10^{-6}$	$2.5042 \cdot 10^{-5}$	$2.1045 \cdot 10^{-4}$	$4.3479 \cdot 10^{-4}$
		$2.6563 \cdot 10^{-7}$	$1.6027 \cdot 10^{-6}$	$1.0086 \cdot 10^{-4}$
			$3.6024 \cdot 10^{-7}$	$9.1628 \cdot 10^{-7}$
				$1.8882 \cdot 10^{-7}$

For each $n > 2$, the first two singular values are considerably larger than the others. Indeed, for all practical purposes, the singular values below the line in Table 1 are zero. Therefore, as the dimension of P increases, its rank remains close to 2. This is to say that the positive degree [4] of the covariance sequence (3.46) is approximately 2 for all n. In Fig. 3.2 the spectral density for $n = 2$ is plotted together with those obtained by taking $n > 2$, showing no major difference.

Next, for $n = 4$, we compute the solution of the CEE with

$$\sigma(z) = \hat{\sigma}(z)(z - 0.6e^{1.78i})(z - 0.6e^{-1.78i}).$$

As expected, the rank of the 4×4 matrix solution P of the CEE, is approximately 2, and, as seen in Fig. 3.3, $a(z)$ has roots that are very close to cancelling the zeros $0.6e^{\pm 1.78i}$ of $\sigma(z)$.

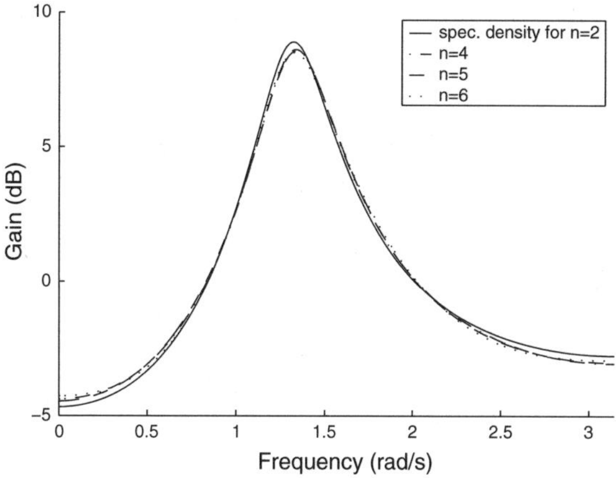

Fig. 3.1. The given spectral density $(n = 2)$ and the estimated one for $n = 4, 5, 6$.

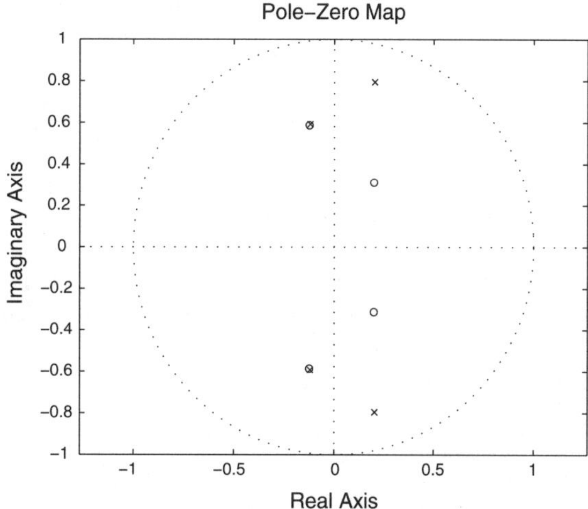

Fig. 3.2. The spectral zeros (o) and the corresponding poles (x) for $n = 4$.

Example 2: Model reduction

Next, given a transfer function $w(z)$ of degree 10 with zeros

$$0.99e^{\pm 1.78i}, \ 0.6e^{\pm 0.44i}, \ 0.55e^{\pm 2i}, \ 0.98e^{\pm i}, \ 0.97e^{\pm 2.7i}$$

and poles

$$0.8e^{\pm 2.6i}, \ 0.74e^{\pm 0.23i}, \ 0.8e^{\pm 2.09i}, \ 0.82e^{\pm 1.32i}, \ 0.77e^{\pm 0.83i}$$

as in Fig. 3.3, we generate data (3.44) and a corresponding covariance sequence (3.46). Clearly, there is no zero-pole cancellation.

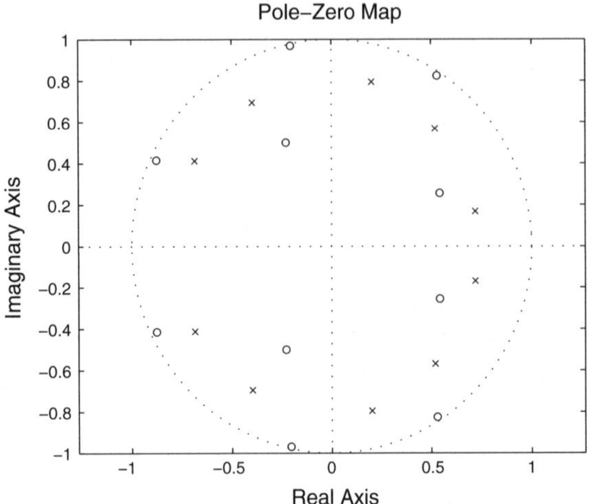

Fig. 3.3. Zeros (o) and the corresponding poles (x) of $w(z)$.

Nevertheless, the rank of the 10×10 matrix solution P of CEE is close to 6. In fact, its singular values are equal to

$$1.1911 \; 0.1079 \; 0.0693 \; 0.0627 \; 0.0578 \; 0.0434$$
$$0.0018 \; 0.0012 \; 0.0009 \; 0.0008$$

The last four singular values are quite small, establishing an approximate rank of 6. The estimated spectral density ($n = 10$) is depicted in Fig. 3.4 together with the theoretical spectral density.

Clearly six zeros are dominant, namely

$$0.98e^{\pm i}, \; 0.99e^{\pm 1.78i}, \; 0.97e^{\pm 2.7i},$$

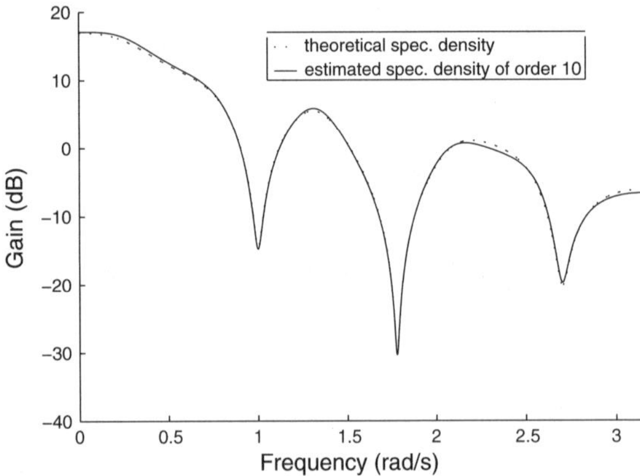

Fig. 3.4. $n = 10$ estimate of spectral density together with the true spectral density.

and these can be determined from the estimated spectral density in Fig. 3.4. Therefore applying our algorithm to the reduced covariance sequence 1, $c_1, \ldots, c_6$ using the six dominant zeros to form $\sigma(z)$, we obtain a 6×6 matrix solution P of CEE and a corresponding reduced order system with poles and zeros as in Fig. 3.5. Comparing with Fig. 3.3, we see that the poles are located in quite different locations. Nevertheless, the corresponding reduced-order spectral estimate, depicted in Fig. 3.6, is quite accurate.

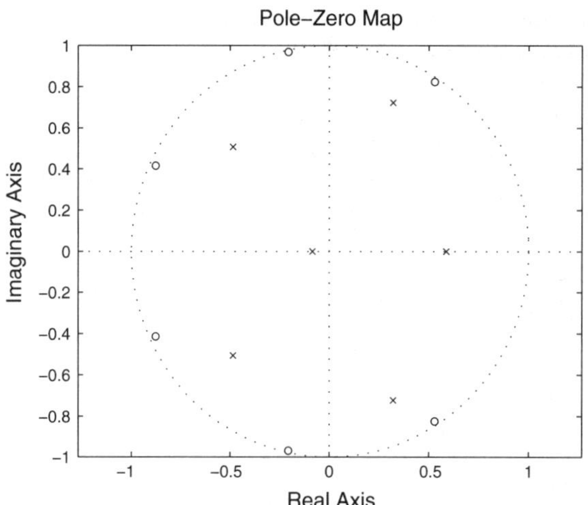

Fig. 3.5. Zeros (o) and poles (x) of the reduced-order system.

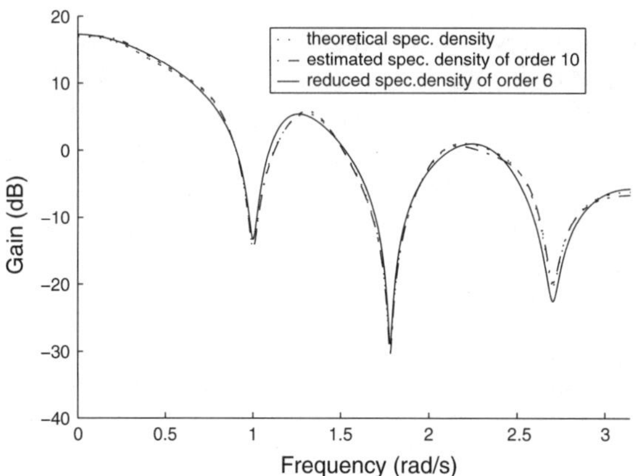

Fig. 3.6. Reduced-order estimate of spectral density ($n = 6$) together with that of $n = 10$ and the true spectral density.

References

1. J.C. Alexander, *The Topological Theory of an Embedding Method*, H. Wacker ed., Academic Press, N.Y., 1978.
2. Eugene L. Allgower and Kurt Georg, *Numerical Continuation Method, An Introduction*, Springer-Verlag, 1990.
3. A. Blomqvist, G.Fanizza and R. Nagamune , *Computation of bounded degree Nevanlinna-Pick interpolants by solving nonlinear equation.* The proceedings of the 42nd IEEE Conference of Decision and Control. (2003) 4511-4516.
4. C. I. Byrnes and A. Lindquist, *On the partial stochastic realization problem,* IEEE Trans. Automatic Control **AC-42** (August, 1997), 1049–1069.
5. C. I. Byrnes, A. Lindquist, S. V. Gusev, and A. S. Matveev, A complete parameterization of all positive rational extensions of a covariance sequence, *IEEE Trans. Automat. Control,* **40** (1995), 1841–1857.
6. C. I. Byrnes, S.V. Gusev, and A. Lindquist, *From finite covariance windows to modeling filters: A convex optimization approach,* SIAM Review **43** (Dec. 2001), 645–675.
7. C. I. Byrnes, P.Enqvist and A. Lindquist, *Cepstral coefficients, covariance lags and pole-zero models for finite data strings,* IEEE Trans. Signal Processing **SP-50** (April, 2001), 677–693.
8. P.Enqvist, *A homotopy approach to rational covariance extension with degree constraint,* Int.J.Applied Mathematics and Computer Science **SP-11** (2001), 1173–1201.
9. T. T. Georgiou, *Partial realization of covariance sequences,* CMST, Univ. Florida, Gainesville, 1983.
10. T. T. Georgiou, *Realization of power spectra from partial covariance sequences,* IEEE Transactions Acoustics, Speech and Signal Processing **ASSP-35** (1987), 438–449.
11. A. Lindquist, *Some reduced-order non-Riccati equations for linear least-squares estimation: the stationary, single-output case,* Int. J. Control **24** (1976), 821–842.
12. A. Lindquist, *On Fredholm integral equations, Toeplitz equations and Kalman-Bucy filtering,* Applied mathematics and optimization **1** (1975), 355–373.

4

Set-point Boundary Control for a Viscous Burgers Equation

C.I. Byrnes[1], D.S. Gilliam[2], A. Isidori[3], and V.I. Shubov[4]

[1] Systems Science and Mathematics, Washington University, St. Louis, MO 63130
`chrisbyrnes@seas.wustl.edu`
[2] Mathematics and Statistics, Texas Tech University, Lubbock, TX 79409
`gilliam@math.ttu.edu` Name and Address of your Institute
`name@email.address`
[3] Dipart. di Info. e Sistemistica, Università di Roma "La Sapienza", 00184 Rome, ITALY
`isidori@zach.wustl.edu`
[4] Mathematics and Statistics, Texas Tech University, Lubbock, TX 79409
`vshubov@math.ttu.edu`

Summary. This paper is concerned with problems of set point regulation of a boundary controlled viscous Burgers' equation. The methodology for obtaining the desired controls for solving these problems is based on the recent nonequilibrium theorem which has been applied successfully for finite dimensional nonlinear systems. In particular a dynamic control law is obtained, in each case, from outputs of the associated augmented zero dynamics. Our proofs rely on the authors earlier work in which there is established a generalized root locus methodology via convergence of trajectories for the boundary controlled viscous Burgers' equation.

4.1 Introduction

Problems of output regulation, such as tracking and disturbance rejection, are of extreme importance in practical applications of control theory. Such problems have lead to a rich history in the systems and control literature and have been studied from many different points of view using a wide variety of mathematical tools. In the present work we study problems of set point regulation of a boundary controlled viscous Burgers' equation. In our main result we consider a pair of Neumann boundary controls and measured outputs at the end points of the spatial domain. In our second example we seek a controller that will produce a multivariable control that will drive the two outputs to a pair of prescribed constant values. In our second example we introduce a single Neumann boundary control through the right end of the spatial interval and choose and choose as output the value of the state at the same endpoint. The problem here is to find a controller that will drive the

C.I. Byrnes et al.: *Set-point Boundary Control for a Viscous Burgers Equation*, LNCIS **321**, 43–60 (2005)
`www.springerlink.com`

output to a prescribed constant value – setpoint control. We note that the case of a single set point can also be obtained from the first problem but in our development it would require two controls. The point of our second example is that this problem is particularly easy to solve with only a single control.

The controls for solving these problems are constructed by incorporating two design principles. The first is based on a classical design methodology using a *Proportional Error* feedback control for stabilization. The second is derived from the methodology recently developed for nonequilibrium output regulation by C.I. Byrnes and A. Isidori [5]. Their methodology has been applied successfully for finite dimensional nonlinear systems. These methods are particularly appealing in the case in which a well defined zero dynamics is available. In the present our boundary control systems provides a very simple notion of a zero dynamics. Indeed, this methodology has been exploited in our earlier work [3] in which we established a nonlinear enhancement of classical root locus constructs by proving convergence of trajectories for the closed loop system (with proportional error feedback) to trajectories of the associated zero dynamics in the high gain limit. Additionally, in the work [3] we proved semiglobal stabilization using our convergence of trajectories results and the fact that the zero dynamics was *minimum phase*, i.e., globally exponentially stable. These same results form the core of our proofs in the present work and for that reason we shall refer to estimates from [3] rather than try to repeat the many lengthy technical arguments here. The important feature used here is that the convergence of trajectories in [3] is established, not only in, $L^2(0,1)$ but also in $H^1(0,1)$.

As we have already mentioned, we consider two somewhat different set point control problems. In the first problem we consider a pair of boundary controls and pair of set points M_0 and M_1 – Multi-Input Multi-Output (MIMO). In the second problem we employ a single boundary control in order to achieve a single set point value M – Single Input Single Output (SISO). We note that in the first problem when $M_0 = M_1 = M$, the first problem provides the same conclusion as obtained in the second problem but using two controls instead of one. The results are also established somewhat differently.

We also note that for both problems we provide two different compensators. Each compensator consists on a *Proportional Error* term plus either a dynamic or static additional part. The utility of the dynamic controller is that using it does not require any a priori knowledge of the steady state response of the augmented zero dynamics. On the other hand use of the additional static controller greatly simplifies implementation of the compensator.

4.2 Statement of the Main Problem

In this paper we consider a plant modeled by a Neumann boundary controlled viscous Burgers' equation on spatial interval $0 \leq x \leq 1$.

$$z_t(x,t) = \epsilon z_{xx}(x,t) - z(x,t)z_x(x,t) \tag{4.1}$$
$$z_x(0,t) = u_0(t) \tag{4.2}$$
$$z_x(1,t) = u_1(t) \tag{4.3}$$
$$z(x,0) = \varphi(x). \tag{4.4}$$

The co-located measured outputs are taken to be

$$y_j(t) = z(j,t), \quad j = 0,1. \tag{4.5}$$

For this system our objective is to find controls u_0, u_1 so that for for any real numbers M_0, M_1, the errors defined by

$$e_j(t) = z(j,t) - M_j, \quad j = 0,1 \tag{4.6}$$

satisfy

$$e_j(t) \xrightarrow{t \to \infty} 0, \quad j = 0,1. \tag{4.7}$$

4.3 Dynamic Compensator

Following classical design paradigms we seek a compensator as a sum of two parts. The first part provides stability and second ensures tracking the desired set points. To this end we seek a stabilizing proportional error feedback plus additional boundary control terms which will be derived from the plants zero dynamics. As we have already mentioned in the introduction, this approach is based on a nonequilibrium output regulation developed by C.I. Byrnes and A. Isidori [5] combined with a nonlinear enhancement of classical root-locus design methods as developed for Burgers' equation in [3].

Based on classical design methods we look for controls as a sum of two parts: the first part is a *Proportional Error* feedback for stabilization; the second part is a dynamic control designed to ensure tracking of the nonzero set-points. Thus we seek controls in the form

$$u_j(t) = (-1)^j k_j e_j(t) + \Gamma_j(t), \quad k_j > 0, \quad j = 0,1. \tag{4.8}$$

Here the terms $(-1)^j k_j e_j(t)$ provide stabilizing Proportional Error feedback controls. For the design of the additional controls $\Gamma_j(t)$ we first define the *Zero Dynamics* which is the system obtained from (4.1)-(4.3) by replacing the equations for the boundary conditions by constraining the errors $e_j(t)$ to zero. Thus we obtain the system

$$\xi_t(x,t) = \epsilon \xi_{xx}(x,t) - \xi(x,t)\xi_x(x,t) \tag{4.9}$$
$$\xi(0,t) = M_0 \tag{4.10}$$
$$\xi(1,t) = M_1 \tag{4.11}$$
$$\xi(x,0) = \psi(x) \tag{4.12}$$

where $\psi \in L^2(0,1)$.

Remark 1. An informal calculation that motivates our definition of the zero dynamics can be derived from the system obtained by introducing the controls (4.8) into the boundary terms in (4.2), (4.3). These control laws produce the non-homogeneous boundary conditions

$$z_x(0,t) - k_0(z(0,t) - M_0) = \Gamma_0(t), \quad z_x(1,t) + k_1(z(1,t) - M_1) = \Gamma_1(t)$$

which, in turn, provides a closed loop system

$$z_t(x,t) = \epsilon z_{xx}(x,t) - z(x,t)z_x(x,t) \tag{4.13}$$
$$z_x(0,t) + k_0(z(0,t) - M_0) = \Gamma_0(t) \tag{4.14}$$
$$z_x(1,t) - k_1(z(1,t) - M_1) = \Gamma_1(t) \tag{4.15}$$
$$z(x,0) = \varphi(x)$$

Now if we formally divide the equations defining the boundary conditions in (4.14), (4.15) by the gain parameters k_j we have

$$\frac{z_x(j,t)}{k_j} - (-1)^j(z(j,t) - M_j) = \frac{\Gamma_j(t)}{k_j}.$$

Standard root locus methodology would suggest passing to the high gain limits $k_j \to \infty$ from which one obtains the boundary conditions (4.10), (4.11). For the Burgers' system considered here this procedure was made rigorous in [3].

The main result of this work is that the additional controls $\Gamma_j(t)$ can be easily obtained from outputs of the zero dynamics as

$$\Gamma_j(t) = \xi_x(j,t), \quad j = 0,1 \tag{4.16}$$

where ξ is the solution of the zero dynamics problem (4.9)-(4.12).

In this case the composite closed loop system, consisting of the plant (4.1)-(4.4) and the zero dynamics (here considered as a controller) (4.9)-(4.12) coupled through the controls defined by (4.8) and (4.16), is

$$z_t - \epsilon z_{xx} + zz_x = 0 \tag{4.17}$$
$$z_x(0,t) - k_0(z(0,t) - w_0(t)) = \xi_x(0,t) \tag{4.18}$$
$$z_x(1,t) + k_1(z(1,t) - w_1(t)) = \xi_x(1,t) \tag{4.19}$$
$$z(x,0) = \varphi(x). \tag{4.20}$$
$$\xi_t - \epsilon\xi_{xx} + \xi\xi_x = 0 \tag{4.21}$$
$$\xi(0,t) = M_0 \tag{4.22}$$
$$\xi(1,t) = M_1 \tag{4.23}$$
$$\xi(x,0) = \psi(x) \tag{4.24}$$

We can now state the main result of this work.

Theorem 1. *For any $R > 0$ there exists $K(R) > 0$ such that the following statement holds for the closed loop system (4.17) – (4.23). If the initial function in (4.20) and (4.23) satisfy*

$$\varphi, \psi \in L^2(0, 1), \quad \|\varphi\|, \|\psi\| \leq R \tag{4.25}$$

and the gain parameters

$$k_0, k_1 \geq K(R) \tag{4.26}$$

we have

$$e_j(t) \xrightarrow{t \to \infty} 0, \quad j = 0, 1. \tag{4.27}$$

4.4 Outline of Proof of Main Result

A straightforward calculation shows that the difference $\eta = z - \xi$ between the functions $z(x, t)$ and $\xi(x, t)$ satisfy the following equation, initial and boundary conditions

$$\eta_t - \epsilon \eta_{xx} + \eta \eta_x + (\eta \xi)_x = 0 \tag{4.28}$$
$$\eta_x(0, t) - k_0 \eta(0, t) = 0 \tag{4.29}$$
$$\eta_x(1, t) + k_1 \eta(1, t) = 0 \tag{4.30}$$
$$\eta(x, 0) = \eta_0(x) = \varphi(x) - \psi(x). \tag{4.31}$$

To prove Theorem 1 and hence show that our feedback law solves the regulator problem stated in the introduction we will prove the following statement.

Proposition 1. *If η is the solution of the problem (4.28)-(4.31) then the error*

$$e_j(t) = \eta(j, t) \xrightarrow{t \to \infty} 0, \quad j = 0, 1 \tag{4.32}$$

for all $k_0, k_1 \geq K(R)$.

This proposition is an immediate corollary of the following theorem, which provides extra information about the behavior of $\eta(x, t)$.

Theorem 2. *For $R > 0$ there exists $K(R) > 0$ such that the following statement holds. If the initial function in (4.31) satisfies*

$$\eta_0 \in L^2(0, 1), \quad |\eta_0\| \leq R \tag{4.33}$$

and the gain parameters

$$k_0, k_1 \geq K(R) \tag{4.34}$$

then the solution η of the initial-boundary value problem (4.28)-(4.31) satisfies the estimates:

$$\|\eta(t)\| \leq C_0(R)e^{-\alpha t}, \quad t \geq 0; \tag{4.35}$$

$$\|\eta(t)\|_{H^1(0,1)} \leq C_1(R,t_0)e^{-\alpha t}, \quad t \geq t_0 > 0. \tag{4.36}$$

Here α is a positive constant and C_0 and C_1 are positive continuous functions of their arguments, and

$$\lim_{t_0 \to 0} C_1(R,t_0) = \infty.$$

If, in addition, $\eta_0 \in H^1(0,1)$ and $\|\eta_0\|_{H^1(0,1)} \leq R$ then (4.36) holds for $t \geq 0$ and $C_1 = C_1(R)$.

Proposition 1 follows from Theorem 2 because of the Sobolev embedding: $H^1(0,1) \subset C[0,1]$. This embedding implies that

$$\|\eta(\cdot,t)\|_{C[0,1]} = \max_{x \in [0,1]} |\eta(x,t)| \leq M\|\eta(t)\|_{H^1(0,1)}, \tag{4.37}$$

where $M > 0$. Comparing (4.37) with (4.36) we see that (4.32) holds if $k_0, k_1 \geq K(\|\eta_0\|)$.

The proof of Theorem 2 follows along the same lines as the proof of Theorem 5.1 from the authors' paper [3]. In fact, the aformentioned theorem gives the desired result in the particular case when $\xi = 0$. To obtain the proof of Theorem 2 we have to repeat all the steps of the proof of Theorem 5.1 in [3] plus on each step we must handle an additional linear term $(\xi\eta)_x$ which is weaker than the first three terms in equation (4.28). This can be done by an application of standard techniques (see, e.g., [3], [10]).

Theorem 2 can be obtained as a corollary of the following results.

Theorem 3. *(a) Let $T > 0$ and consider the problem (4.28)-(4.31) on the time interval $[0,T]$. Assume $\eta_0 \in L^2(0,1)$. Then the problem (4.28)-(4.31) has a unique weak solution $\eta^K(x,t)$ which belongs to the space*

$$\mathcal{M}_T = C([0,T], L^2(0,1)) \cap L^2(([0,T], H^1(0,1))$$

with the norm

$$\|\eta\|_{\mathcal{M}_T} = \max_{0 \leq t \leq T} \|\eta(t)\| + \left(\int_0^T \|\eta(t)\|_{H^1(0,1)} \, dt\right)^{1/2}.$$

(b) Moreover, for $t > 0$ the weak solution is $C^{2,1}$-smooth and, therefroe, satisfies (4.28) in the classical sense. More precisely, $\eta^K \in H^{4,2}([0,1] \times [t_0,T])$ for any $t_0 > 0$ (we refer to [10] Section 2 for definitions of $C^{2,1}$ and $H^{4,2}$ spaces.)

Theorem 4. *The statement of Theorem 2 holds for the solution of equation (4.28) with initial conditions (4.31) and the Dirichlet boundary conditions*

$$\eta(0,t) = \eta(1,t) = 0. \tag{4.38}$$

Theorem 5. *(a) Let η^K and η be the solutions of the problems (4.28)-(4.31) and (4.28), (4.38), (4.31) respectively. Then*

$$\lim_{k_0,k_1\to\infty} \|\eta^K(t) - \eta(t)\| = 0 \quad \text{for any } t > 0. \tag{4.39}$$

(b) Moreover,

$$\lim_{k_0,k_1\to\infty} \|\eta^K(t) - \eta(t)\|_{H^1(0,1)} = 0 \quad \text{for any } t > 0, \tag{4.40}$$

and the convergence is uniform on any interval $[t_0, T]$ with $t_0 > 0$.

Theorem 6. *Consider the problem (4.28)-(4.31)*

(a) Assume that $\eta_0 \in L^2(0,1)$. There exists sufficiently small $\rho > 0$ such that if

$$\|\eta_0\| \le \rho \tag{4.41}$$

then the solution η^K satisfies the inequality

$$\|\eta^K(t)\| \le \|\eta_0\| e^{-\alpha t}, \quad \alpha > 0, \quad t \ge 0, \tag{4.42}$$

i.e., $\eta = 0$ is a exponentially stable equilibrium in the $L^2(0,1)$-norm.
(b) Assume that $\eta_0 \in H^1(0,1)$. There exists $\rho_1 > 0$ such that if

$$\|\eta_0\| \le \rho_1 \tag{4.43}$$

then η^K satisfies

$$\|\eta^K(t)\|_{H^1(0,1)} \le C\left(\|\eta_0\|_{H^1(0,1)}\right) e^{-\alpha t}, \quad \alpha > 0, \quad t > 0. \tag{4.44}$$

Here $C(\tau) > 0$ is a continuous function of $\tau \in [0, \infty)$.

Theorem 7. *The estimates (4.42), (4.44) hold for the solution η of the problem (4.28), (4.38), (4.31) without the assumption that the norms of η are small, i.e.,*

$$\|\eta(t)\| \le \|\eta_0\| e^{-\alpha t}, \quad \alpha > 0, \quad t \ge 0, \quad \text{for every } \eta_0 \in L^2(0,1). \tag{4.45}$$

$$\|\eta^K(t)\|_{H^1(0,1)} \le \mathcal{D}\left(\|\eta\|_{H^1(0,1)}\right) e^{-\alpha t}, \quad \alpha > 0, \quad t > 0, \text{ for every } \eta_0 \in H^1(0,1) \tag{4.46}$$

where $\mathcal{D}(\tau) > 0$ is a continuous function of $\tau \in [0, \infty)$.

Remark 2. Explicit estimates from below for α in (4.42), (4.44), (4.45), (4.46) can be given. But these estimates are immaterial for us.

Theorem 2 is a direct corollary of Theorems 3 - 7. Its derivation from Theorem 2.6 (in [3]) is similar (but easier) than the derivation of Theorem 5.1 from Theorems 2.1, 2.2, 2.3, 2.4 and Propositions 5.1, 5.2 in [3].

4.5 Static Controllers

As we have seen from our work in Sections 4.3 and 4.4 our multi-dimensional set-point control problem can be solved using a pair of dynamic controls consisting of a (finite dimensional) proportional error feedback plus a finite dimensional output of an infinite dimensional controller. In the case of set-point control it would be natural to expect that we should be able to carry out this control objective with constant values for the gains Γ_j. This is indeed the case as we will show in this section. The reason why this works derives from as analysis of the *steady-state* behavior of the zero dynamics system (4.9)-(4.12).

For the zero dynamics system (4.9)-(4.12) a complete description of the solution for all $M_j \in \mathbb{R}$ can be obtained using the Hopf-Cole transformation. The main results of such an analysis can be found in the paper [2]. However, in order to obtain explicit information concerning the structure of stationary solutions (which are not contained in [2]) we present here a detailed discussion in the Appendix A. The Hopf-Cole transformation

$$\xi(x,t) = -2\epsilon \frac{v_x(x,t)}{v(x,t))} \tag{4.47}$$

transforms the zero dynamics into the following problem for the heat equation:

$$v_t(x,t) = \epsilon v_{xx}(x,t) \tag{4.48}$$

$$v_x(0,t) + \widetilde{M_0}v(0,t) = 0 \tag{4.49}$$

$$v_x(1,t) + \widetilde{M_1}v(1,t) = 0 \tag{4.50}$$

$$\widetilde{M_0} = M_0/(2\epsilon), \quad \widetilde{M_1} = M_1/(2\epsilon) \tag{4.51}$$

$$v(x,0) = \exp\left(-1/2\int_0^x \psi(s)\,ds\right). \tag{4.52}$$

An analysis of (4.48)-(4.52) (see Appendix A) shows that for the problem (4.9)-(4.12) has a single global asymptotically stable equilibrium which we denote by $\widehat{\xi}(x)$, i.e. the solution of

$$\widehat{\xi}''(x) - \widehat{\xi}(x)\widehat{\xi}'(x) = 0$$
$$\widehat{\xi}(0) = M_0$$
$$\widehat{\xi}(1) = M_1.$$

We refer to Appendix A for an explicit formula for $\widehat{\xi}(x)$ depending on the values of M_0 and M_1.

It is also easy to verify that the convergence of the time dependent solution to the corresponding stationary solution takes place in the following sense.

Proposition 2. *For every M_0, $M_1 \in \mathbb{R}$ and every initial condition $\psi \in L^2(0,1)$, the solution $\xi(x,t)$ of the zero dynamics problem satisfies*

$$\|\xi(\cdot,t) - \widehat{\xi}(\cdot)\|_{H^1(0,1)} \xrightarrow{t\to\infty} 0.$$

As an immediate consequence of this result and the the results of Section 4.3 we have the following result.

Theorem 8. *For any M_0, $M_1 \in \mathbb{R}$ and $\psi \in L^2(0,1)$ let $\widehat{\xi}$ be the stationary solution determined in Proposition 2 and define the controls $u_j(t)$ for our problem (4.1)-(4.4) by*

$$u_j(t) = (-1)^j k_j e_j(t) + \widehat{\xi}'(j), \quad j = 0,1. \tag{4.53}$$

Then the following results hold. For $R > 0$ there exists $K(R) > 0$ such that the following statement holds for the closed loop system obtained from (4.1) – (4.4) with feedback (4.53). If the initial function in φ and ψ satisfy

$$\varphi, \psi \in L^2(0,1), \quad \|\varphi\|, \|\psi\| \le R \tag{4.54}$$

and the gain parameters

$$k_0, k_1 \ge K(R) \tag{4.55}$$

we have

$$e_j(t) \xrightarrow{t\to\infty} 0, \quad j = 0,1. \tag{4.56}$$

In the special case $M_0 = M_1 = M$ the stationary solution is given by $\widehat{\xi}(x) = M$ so the controls (4.53) take the particularly simple form of proportional error controls

$$u_j(t) = (-1)^j k_j e_j(t) \quad j = 0,1 \tag{4.57}$$

since $\widehat{\xi}'(j) = 0$.

4.6 The Single-Input Single-Output Case

As we have just seen in the last section, the case of a single set point, i.e. $M_0 = M_1 = M$ is very special. To make this point more clear we note in this section that in this case there are two interesting results which deserve attention. First, for a single set point one needs only one boundary control. Secondly, in this case the entire state of the closed loop system converges to the the set point, i.e., for every $x_1 \in [0,1]$ we have $z(x_1,t) \to M$ as $t \to \infty$.

To see why this is the case, we replace the system (4.1)-(4.6) by a corresponding system with a single input and a single output

$$z_t(x,t) = \epsilon z_{xx}(x,t) - z(x,t)z_x(x,t) \tag{4.58}$$

$$z_x(0,t) = 0 \tag{4.59}$$

$$z_x(1,t) = u(t) \tag{4.60}$$

$$z(x,0) = \varphi(x). \tag{4.61}$$

52   C.I. Byrnes et al.

In this case we consider a single co-located, measured output given by

$$y(t) = z(1, t) \tag{4.62}$$

and we seek a control $u(t)$ so that for M from a prescribed bounded set of real numbers, the error satisfies

$$e(t) = z(1, t) - M \xrightarrow{t \to \infty} 0. \tag{4.63}$$

We choose a proportional error feedback

$$u(t) = -ke(t), \quad k > 0. \tag{4.64}$$

With this we have the following result.

Theorem 9. *For $R > 0$ there exists $K(R) > 0$ such that the following statement holds for the closed loop system (4.58) – (4.61) with control (4.64). If the initial function $\varphi \in L^2(0, 1)$ in (4.61) satisfies*

$$\|\varphi\| \leq R \tag{4.65}$$

and the gain parameter

$$k \geq K(R) \tag{4.66}$$

(where, as before, $K(R)$ is a suitable large constant depending only on R) we have

$$e(t) \xrightarrow{t \to \infty} 0. \tag{4.67}$$

Actually, much more is true. For any $x_1 \in [0, 1]$ we have

$$z(x_1, t) \xrightarrow{t \to \infty} M. \tag{4.68}$$

Remark 3.   1. The proof of this result follows exactly as does the results in Section 4.3, discussed in Section 4.4, from the results in [3]. Indeed, in this case the modifications are even simpler. Note, for example, if we set $\eta(x, t) = z(x, t) - M$ and use the control (4.64), then η is a solution of the system

$$\eta_t(x, t) = \epsilon \eta_{xx}(x, t) - \eta(x, t)\eta_x(x, t) - M\eta_x(x, t)$$
$$\eta_x(0, t) = 0$$
$$\eta_x(1, t) + k\eta(1, t) = 0$$
$$\eta(x, 0) = \varphi(x) - M.$$

In modifying the arguments in Section 4.4 we replace the system (4.28)-(4.31) by the above system.
   2. We could just have easily chosen to set the problem up at $x = 0$ rather than at $x = 1$.

3. Recall that the controls from Section 4.3 with $M_0 = M_1 = M$ were proportional error controls; no additional control terms were required. The main point of this section is that this same result holds with a single proportional error control.

4. We note that there has been other works related to the problem of this section. In particular, in [1] a cubic nonlinear boundary feedback scheme is proposed for the control problem with a single set point M. This feedback law is definitely efficient and successfully solves the problem in the case $M_0 = M_1 = M$, as is clear from the estimates and numerical simulations given in [1]. However, the results of [1] require an additional justification: the global in time existence and uniqueness of the solution must be shown. The authors of [1] appeal to Theorem 7.4, Chapter 5 in [10] which is not applicable for their feedback control (the boundary terms must be at most linear in z_x).

5. Note that in this work we treat only co-located actuators and sensors. The non-co-located case is more interesting, requires somewhat different techniques and is the subject of another forthcoming work.

6. We also note that much more general tracking, as well as, disturbance rejection problems can also be done using this same methodology. These results will also appear in a forthcoming work.

4.7 Numerical Examples

We present two numerical examples. The first is a MIMO example and the second is an SISO example.

Example 1. In this example we have taken as an initial condition $\varphi(x) = 2x(1-x)$ and set $M_0 = 1$, $M_1 = 5$. We have also set $k_0 = k_1 = 5$. The static feedback from the stationary solutions were used for the second part of the controls. Thus we consider the problem

$$z_t - \epsilon z_{xx} + z z_x = 0$$

$$z_x(0,t) - k_0(z(0,t) - M_0) = \widehat{\xi}'(0)$$

$$z_x(1,t) + k_1(z(1,t) - M_1) = \widehat{\xi}'(1)$$

$$z(x,0) = \varphi(x)$$

where, with the values $M_0 = 1$, $M_1 = 5$ the stationary solution (see Appendix A) comes from region II so that

$$\widehat{\xi}(x) = -2\epsilon r_1 \tanh(r_1 x - \theta)$$

where r_1 and θ are determined in Appendix A. Thus we have

$$\widehat{\xi}'(x) = -2\epsilon r_1^2 \operatorname{sech}^2(r_1 x - \theta)$$

and

$$\widehat{\xi}'(0) = -2\epsilon r_1^2 \operatorname{sech}^2(\theta), \quad \widehat{\xi}'(0) = -2\epsilon r_1^2 \operatorname{sech}^2(r_1 - \theta).$$

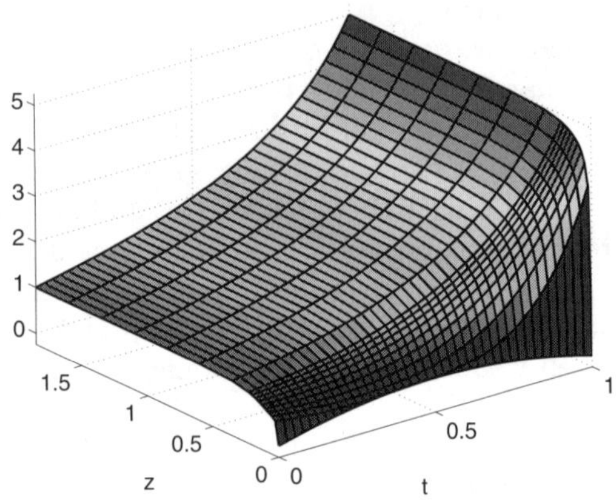

Solution Surface

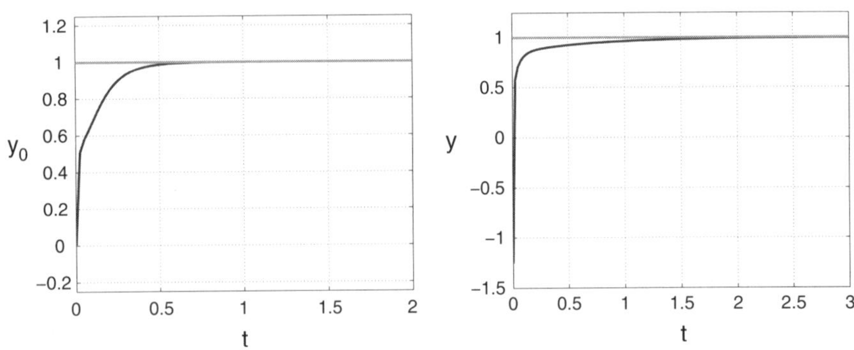

Outputs $y_0(t)$, $y_1(t)$ with $k_0 = k_1 = 5$, $M_0 = 1$, $M_1 = 5$

It is clear from the above figures that we have achieved our goal of set-point tracking for the outputs at $x = 0$ and $x = 1$.

In the next example we consider a single set point M and following the development in Section 4.6 we solve the problem of regulation using the proportional error feedback (4.64).

Example 2. In this example we have taken as an initial condition $\varphi(x) = 10(1/2 - x)^3$ and set $M = 1$ and $k = 10$. Following the development in Section 4.6 we consider the closed loop system

$$z_t - \epsilon z_{xx} + z z_x = 0$$
$$z_x(0, t) = 0$$
$$z_x(1, t) + k(z(1, t) - M) = 0$$
$$z(x, 0) = \varphi(x)$$

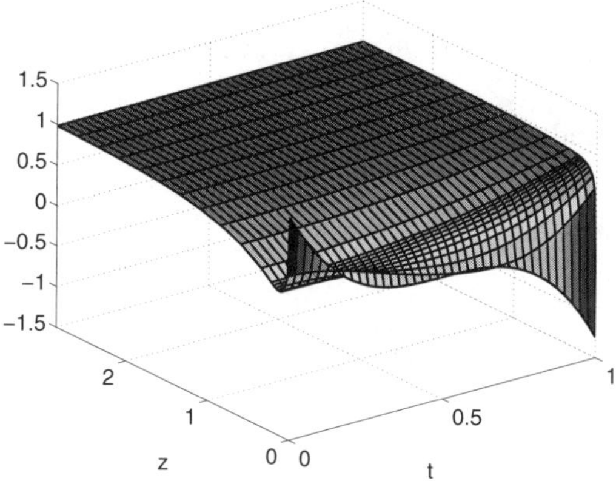

Solution Surface

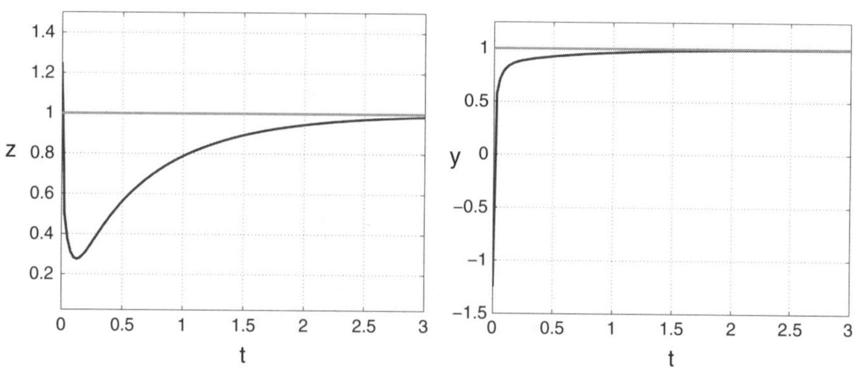

(left) $z(0, t)$, (right) $y(t)$ with $k = 10,\ \ M = 1$

Once again it is clear that we have solved the set-point control problem, but in addition, as our theory predicts, we actually asymptotically obtain the set-point for every $x \in [0, 1]$.

A Appendix

Analysis of the zero dynamics system (4.9)-(4.12) via the Hopf-Cole transformation (4.47) is well known (see, for example, [2]). But, as we have mentioned

56   C.I. Byrnes et al.

in Section 4.5, precise information of the global asymptotically stable equilibrium of the zero dynamics system is necessary for implementation.

The equilibrium solution and its properties is most easily obtained from an analysis of the heat problem (4.48)-(4.52). The solution to this heat problem can be obtained from a generalized Fourier series expansion in terms of the eigenvalues and eigenfunctions of the operator

$$A = \epsilon \frac{d^2}{dx^2} \mathcal{D}(A) = \{\varphi \in H^2(0,1) \ : \ \varphi'(j) - (-1)^j \varphi(j) = 0, \quad j = 0,1\}.$$

The eigenvalues and eigenfunctions are obtained from the Sturm-Liouville problem

$$\epsilon \psi'' = \lambda \psi \tag{4.69}$$

$$\psi'(0) + \widehat{M_0} \psi(0) = 0$$

$$\psi'(1) + \widehat{M_1} \psi(1) = 0$$

where

$$\widetilde{M_0} = \frac{M_0}{2\epsilon}, \quad \widetilde{M_1} = \frac{M_1}{2\epsilon}.$$

This Sturm-Liouville problem has eigenvalues λ_j and associated eigenfunction $\psi_j(x)$. The solution of the heat problem is given by

$$v(x,t) = \sum_{j=1}^{\infty} e^{\lambda_j t} \langle v(\cdot,0), \psi_j \rangle \psi_j(x).$$

Using this we can compute the corresponding solution to the zero dynamics system

$$\xi(x,t) = -2\epsilon \frac{v_x(x,t)}{v(x,t)} = -2\epsilon \frac{\displaystyle\sum_{j=1}^{\infty} e^{\lambda_j t} \langle v(\cdot,0), \psi_j \rangle \psi_j'(x).}{\displaystyle\sum_{j=1}^{\infty} e^{\lambda_j t} \langle v(\cdot,0), \psi_j \rangle \psi_j(x).}$$

The long time behavior of this solution is easily determined as the single stationary solution

$$\widehat{\xi}(x) = -2\epsilon \frac{\psi_1'(x)}{\psi_1(x)} \tag{4.70}$$

where $\psi_1(x)$ is the first eigenfunction (i.e., the one associated with the largest eigenvalue). As is well known this first function has no zeros on $[0,1]$ and therefore $\widehat{\xi} \in L^2(0,1)$. Actually, much more is true but we only need that $\widehat{\xi} \in C^2(0,1)$.

The following analysis provides details concerning the precise representation of global asymptotically stable equilibrium given in (4.70) for the zero dynamics system (4.9)-eqrefzd4 for all values of M_0 and M_1.

As $\widetilde{M}_0$ and $\widetilde{M}_1$ vary over all real numbers there are three possibilities:

$$\lambda_1 < 0, \quad \lambda_1 = 0, \quad \lambda_1 > 0.$$

In each of these cases we consider the exact form of the stationary solution $\widehat{\xi}$ given in (4.47).

1. $\lambda_1 = 0$: In this case (4.69) gives $\psi = ax + b$ which satisfies the corresponding boundary conditions if and only if

$$\widetilde{M}_0 - \widetilde{M}_1 + \widetilde{M}_0\widetilde{M}_1 = 0. \tag{4.71}$$

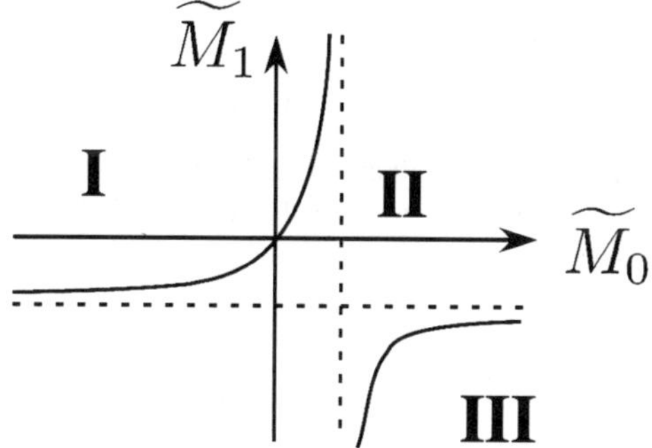

Regions determined by M_0, M_1

This equation divides the $\widetilde{M}_0 \times \widetilde{M}_1$ plane into three disjoint regions. In region I $\lambda_1 < 0$, and in regions II and III $\lambda_1 > 0$. The distinction between regions II and III is that in region II there is exactly one positive eigenvalue λ_1 but in region III there are two positive eigenvalues $\lambda_2 < \lambda_1$.
In this case we have $\psi_1(x) = 1 - \widetilde{M}_0 x$ so

$$\widehat{\xi}(x) = -2\epsilon \frac{-\widetilde{M}_0}{1 - \widetilde{M}_0 x}.$$

Notice that $\widetilde{M}_0 = 1$ and $\widetilde{M}_1 = -1$ are not allowed so as we expect $\widehat{\xi}(x)$ is in $L^2(0,1)$.

2. In region I we have

$$\widetilde{M}_1 > \left(\frac{-\widetilde{M}_0}{\widetilde{M}_0 - 1} \right).$$

In this region, $\widetilde{M}_0$ is typically negative and $\widetilde{M}_1$ is typically positive. In region I the first eigenvalue can be found geometrically from setting $\lambda = -\mu^2$ with $\mu > 0$ and analyzing the intersections of the graphs corresponding to the left and right hand sides of the equation

$$\tan(\mu) = \frac{(\widetilde{M}_1 - \widetilde{M}_0)\mu}{(\mu^2 + \widetilde{M}_0\widetilde{M}_1)}.$$

58   C.I. Byrnes et al.

The eigenfunction ψ corresponding to the first eigenvalue $\lambda_1 = -\mu_1^2$ is given by

$$\psi_1(x) = \mu_1 \cos(\mu_1 x) + \widetilde{M}_0 \sin(\mu_1 x).$$

If we define θ by

$$\cos(\theta) = \frac{\mu_1}{\sqrt{\mu_1^2 + \widetilde{M}_0^2}}, \quad \sin(\theta) = \frac{\widetilde{M}_0}{\sqrt{\mu_1^2 + \widetilde{M}_0^2}}$$

then we can take

$$\psi_1(x) = \cos(\mu_1 x + \theta)$$

and therefore,

$$\widehat{\xi}(x) = 2\epsilon\mu_1 \tan(\mu_1 x + \theta).$$

3. In the regions II and III the first eigenvalue can be found geometrically from setting $\lambda = r^2$ with $r > 0$ and analyzing the intersections of the graphs corresponding to the left and right hand sides of the equation

$$\tanh(r) = \frac{(\widetilde{M}_0 - \widetilde{M}_1)r}{(r^2 - \widetilde{M}_0\widetilde{M}_1)}.$$

The eigenfunction ψ corresponding to the first eigenvalue $\lambda_1 = r_1^2$ is given by

$$\psi_1(x) = r_1 \cosh(r_1 x) - \widetilde{M}_0 \sinh(r_1 x).$$

If we define θ by

$$\cosh(\theta) = \frac{r_1}{\sqrt{r_1^2 + \widetilde{M}_0^2}}, \quad \sinh(\theta) = \frac{\widetilde{M}_0}{\sqrt{r_1^2 + \widetilde{M}_0^2}}$$

then we can take

$$\psi_1(x) = \cosh(r_1 x - \theta)$$

and therefore,

$$\widehat{\xi}(x) = -2\epsilon r_1 \tanh(r_1 x - \theta).$$

In region II the diagonal $\widetilde{M}_0 = \widetilde{M}_1 \equiv \widetilde{M}$ presents a special case that must be treated separately. In this case, with $\tilde{\lambda} = \lambda/\epsilon = r^2 > 0$, the eigenvalue problem takes the form

$$\psi'' = r^2\psi \tag{4.72}$$

$$\psi'(0) + \widetilde{M}\psi(0) = 0$$

$$\psi'(1) + \widetilde{M}\psi(1) = 0.$$

The general solution of (4.72) is

$$\psi(x) = a\frac{\sinh(rx)}{r} + b\cosh(rx).$$

Applying the boundary conditions we find that r must satisfy

$$(r^2 - \widetilde{M}^2)\frac{\sinh(r)}{r} = 0$$

so

$$r = \widetilde{M},$$

which implies

$$\psi(x) = \frac{\sinh(\widetilde{M}x)}{r} - \cosh(\widetilde{M}x).$$

Thus we have

$$\widehat{\xi}(x) = -2\epsilon\frac{\psi'(x)}{\psi(x)} = -2\epsilon\frac{-\widetilde{M}\left(\dfrac{\sinh(\widetilde{M}x)}{r} - \cosh(\widetilde{M}x)\right)}{\left(\dfrac{\sinh(\widetilde{M}x)}{r} - \cosh(\widetilde{M}x)\right)} = M.$$

Therefore, in the case $M_0 = M_1 = M$ the stationary solution is

$$\widehat{\xi}(x) = M. \tag{4.73}$$

Acknowledgement

Research supported in part by Grants from the AFOSR and the NSF.

References

1. A. Balogh and M. Krstic, "Burgers' equation with nonlinear boundary feedback: H^1 stability, well posedness, and simulation," Mathematical Problems in Engineering, vol. 6, pp. 189-200, 2000.
2. C. Boldrighini, L. Triolo, "Absense of turbulence in a unidimensional model of fluid motion," *Meccanica*, (1977), 12, 15-18.
3. C.I. Byrnes, D.S. Gilliam, and V.I. Shubov, "Boundary control, stabilization and zero–pole dynamics for a nonlinear distributed parameter system," *Journal of Robust and Nonlinear Control*, 9, 737-768 (1999).
4. C.I. Byrnes, D.S. Gilliam, and V.I. Shubov "On the global dynamics for a boundary controlled viscous Burgers' equation," *J. Dynam. Control Systems*, 4 no. 4, 457–519 (1998).

5. C. I. Byrnes, A. Isidori, Limit Sets, "Zero Dynamics, and Internal Models in the Problem of Nonlinear Output Regulation," *IEEE Trans. Aut. Control*, V. 48, No. 10, October 2003, p 1712-1723.

6. C. I. Byrnes, A. Isidori, "Nonlinear Internal Models for Output Regulation," submitted to *IEEE Trans. Aut. Control*.

7. C. I. Byrnes, D.S. Gilliam, A. Isidori, V.I. Shubov, Output regulation and flows on complete metric spaces.

8. A. Isidori and C.I. Byrnes. Output regulation of nonlinear systems. *IEEE Trans. Autom. Contr.*, **AC-25**: 131–140, 1990.

9. C. I. Byrnes, A. Isidori, Asymptotic stabilization of minimum phase nonlinear systems, *IEEE Trans. Autom. Control*, **AC-36**: 1122–1137, 1991.

10. Ladyzhenskaya, O.A., V.A. Solonnikov, N.N. Ural'ceva, Linear and Quasilinear Equations of Parabolic Type, Translations of AMS, Vol. 23, 1968.

11. T.I. Zelenyak, M.M. Lavrentiev Jr., M.P. Vishnevskii, "Qualitative Theory of Parabolic Equations, Part 1," VSP, Utrecht, The Netherlands, 1997.

5

Semi-tensor Product of Matrices and its Applications to Dynamic Systems*

Daizhan Cheng

Institute of Systems Science, Chinese Academy of Sciences, Beijing 100080, P.R. China
`dcheng@iss03.iss.ac.cn`

Dedicated to Professor Clyde Martin on the occasion of his 60th birthday

Summary. In this paper we introduce the semi-tensor product of matrices, which is a generalization of conventional matrix product. Three of its applications will be discussed: Numerical solution to Morgan's problem, Estimation of region of attraction of a stable equilibrium, and Non-regular linearization.

5.1 Introduction

A new matrix product, namely the left semi-tensor product, denoted by $\ltimes$, has been proposed recently by the author [4]. It has been successfully applied to several control problems. The paper is a survey on this new matrix product, its basic properties, and its various applications.

Given two matrices $A \in M_{m \times n}$ and $B \in M_{p \times q}$. Only when $n = p$ the (conventional) matrix product is defined. But sometimes we may need a product for $n \neq p$. The following example shows that this extension is natural in multi-linear computation.

Example 1. Let V be an n dimensional vector space, $\{d_1, \cdots, d_n\}$ be a basis of V, and V^* be the dual vector space of V with basis $\{e_1, \cdots, e_n\}$. Let $\omega : \underbrace{V \times \cdots \times V}_{r} \times \underbrace{V^* \times \cdots \times V^*}_{s} \to \mathbb{R}$ be a tensor of type (r, s) (i.e., with covariant order r and contra-variant order s. Denoted as $\omega \in \mathcal{T}_s^r$)[2].

Denoted by

$$\omega_{j_1 \cdots j_s}^{i_1 \cdots i_r} = \omega(d_{i_1}, \cdots, d_{i_r}; e_{j_1}, \cdots, e_{j_s}), \quad i_1, \cdots, j_s = 1, \cdots, n.$$

Then we get a $n^s \times n^r$ matrix, M_ω, called the structure matrix, as

---

*Supported partly by National Natural Science of Foundation of China.

D. Cheng: *Semi-tensor Product of Matrices and its Applications to Dynamic Systems*, LNCIS **321**, 61–79 (2005)
`www.springerlink.com`

$$M_\omega = \begin{pmatrix} \omega^{1\cdots11}_{1\cdots11} & \omega^{1\cdots12}_{1\cdots11} & \cdots & \omega^{n\cdots nn}_{1\cdots11} \\ \omega^{1\cdots11}_{1\cdots12} & \omega^{1\cdots12}_{1\cdots12} & \cdots & \omega^{n\cdots nn}_{1\cdots12} \\ \vdots & & & \\ \omega^{1\cdots11}_{n\cdots nn} & \omega^{1\cdots12}_{n\cdots nn} & \cdots & \omega^{n\cdots nn}_{n\cdots nn} \end{pmatrix}. \tag{5.1}$$

Here the rows and the columns of the matrix are arranged by indexes $(i_1, \cdots, i_r)$ and $(j_1, \cdots, j_2)$ respectively in alphabetic order.

Let $\sigma_i = (\sigma^i_1, \cdots, \sigma^i_n)$, $i = 1, \cdots, s$ (precisely, $\sigma_i = \sum_{k=1}^n \sigma^i_k e_k$) and $X_j = (x^j_1, \cdots, x^j_n)^T$, $j = 1, \cdots, r$ be sets of co-vectors and vectors respectively. Then it is easy to see that

$$\omega(\sigma_1, \cdots, \sigma_s; X_1, \cdots, X_r) = (\sigma_1 \otimes \cdots \otimes \sigma_s) M_\omega (X_1 \otimes \cdots \otimes X_r). \tag{5.2}$$

It we use the semi-tensor product (refer to the next section), it is easy to see that

$$\omega(\sigma_1, \cdots, \sigma_s; X_1, \cdots, X_r) = (\sigma_s \ltimes \cdots \ltimes \sigma_1) M_\omega (X_1 \ltimes \cdots \ltimes X_r). \tag{5.3}$$

Now what is the advantage of (5.3) over (5.2)? Since the semi-tensor product is a generalization of the conventional matrix product, (5.3) can also be expressed as

$$\omega(\sigma_1, \cdots, \sigma_s; X_1, \cdots, X_r) = \sigma_s \ltimes \cdots \ltimes \sigma_1 \ltimes M_\omega \ltimes X_1 \ltimes \cdots \ltimes X_r. \tag{5.4}$$

Given a fixed vector X, a mapping $id_X : \mathcal{T}^r_s \mapsto \mathcal{T}^{r-1}_s$ is defined as:

$$id_X(\omega)(\sigma_1, \cdots, \sigma_s; X_1, \cdots, X_{r-1}) = \omega(\sigma_1, \cdots, \sigma_s; X, X_1, \cdots, X_{r-1}).$$

This mapping is important in physics. Now it is easy to see that the structure matrix of $\tilde{\omega} := id_x(\omega)$ is

$$M_{\tilde{\omega}} = M_o \ltimes X. \tag{5.5}$$

This is a typical example of the semi-tensor product. $\square$

Next example is also interesting. It shows that the conventional matrix product is "incomplete", while the semi-tensor product, as an generalization of the prior, reminds this incompleteness.

To the end of this section we give an example to show that as an generalization of the conventional matrix product the semi-tensor product is more convenient.

Example 2. Consider $X, Y, Z, W \in \mathbb{R}^n$. We define

$$E := (XY^T)(ZW^T) \in M_{n \times n}. \tag{5.6}$$

By associativity of the product and the fact that $Y^T Z$ is a scalar, we have that

$$E = X(Y^T Z)W^T = (Y^T Z)XW^T = Y^T ZXW^T = Y^T(ZX)W^T.$$

Now we meet a problem: legal procedure leads to an illegal expression, ZX! If we consider all the conventional matrix product as a special case of left semi-tensor product, then it is easy to see that

$$Y^T \ltimes (Z \ltimes X) \ltimes W^T = E.$$

$\square$

The paper is organized as follows: Section 2 gives the definitions and some basic properties of the semi-tensor product. Section 3 considers the Morgan's problem be using new product. Section 4 gives the estimation of region of attraction of a stable equilibrium. Section 5 considers non-regular linearization on nonlinear systems. Some related works are indicated in Section 6.

5.2 Semi-tensor Product

Definition 1. *1. Let X be a row vector of dimension np, and Y be a column vector with dimension p. Then we split X into p equal-size blocks as $X^1, \cdots, X^p$, which are $1 \times n$ rows. Define the left semi-tensor product, denoted by $\ltimes$, as*

$$\begin{cases} X \ltimes Y = \sum\limits_{i=1}^{p} X^i y_i \in \mathbb{R}^n, \\ Y^T \ltimes X^T = \sum\limits_{i=1}^{p} y_i (X^i)^T \in \mathbb{R}^n. \end{cases} \tag{5.7}$$

2. Let $A \in M_{m \times n}$ and $B \in M_{p \times q}$. If either n is a factor of p, say $nt = p$ and denote it as $A \prec_t B$, or p is a factor of n, say $n = pt$ and denote it as $A \succ_t B$, then we define the left semi-tensor product of A and B , denoted by $C = A \ltimes B$, as the following: C consists of $m \times q$ blocks as $C = (C^{ij})$ and each block is

$$C^{ij} = A^i \ltimes B_j, \quad i = 1, \cdots, m, \quad j = 1, \cdots, q,$$

where A^i is i-th row of A and B_j is the j-th column of B.

Example 3. 1. Let $X = (x_1 \ x_2 \ x_3 \ x_4) \in \mathbb{R}^4$ and $Y = \begin{pmatrix} y_1 \\ y_2 \end{pmatrix} \in \mathbb{R}^2$. Then

$$X \ltimes Y = (x_1 \ x_2) \cdot y_1 + (x_3 \ x_4) \cdot y_2 = (x_1 y_1 + x_3 y_2 \quad x_2 y_1 + x_4 y_2).$$

2. Let

$$A = \begin{pmatrix} a_{11} \ a_{12} \\ a_{21} \ a_{22} \end{pmatrix}, \quad B = \begin{pmatrix} b_{11} \ b_{12} \ b_{13} \\ b_{21} \ b_{22} \ b_{23} \\ b_{31} \ b_{32} \ b_{33} \\ b_{41} \ b_{42} \ b_{43} \end{pmatrix}.$$

Then

$$A \ltimes B = \begin{pmatrix} a_{11}b_{11} + a_{12}b_{31} & a_{11}b_{12} + a_{12}b_{32} & a_{11}b_{13} + a_{12}b_{33} \\ a_{11}b_{21} + a_{12}b_{41} & a_{11}b_{22} + a_{12}b_{42} & a_{11}b_{23} + a_{12}b_{43} \\ a_{21}b_{11} + a_{22}b_{31} & a_{21}b_{12} + a_{22}b_{32} & a_{21}b_{13} + a_{22}b_{33} \\ a_{21}b_{21} + a_{22}b_{41} & a_{21}b_{22} + a_{22}b_{42} & a_{21}b_{23} + a_{22}b_{43} \end{pmatrix}.$$

$\square$

Remark 1. Note that when $n = p$ the left semi-tensor product coincides with the conventional matrix product. Therefore, the left semi-tensor product is only a generalization of the conventional product. For convenience, we may omit the product symbol $\ltimes$.

Some fundamental properties of the left semi-tensor product are collected in the following:

Proposition 1 ([4]).
The left semi-tensor product satisfies (as long as the related products are well defined)

1. (Distributive rule)

$$\begin{aligned} A \ltimes (\alpha B + \beta C) &= \alpha A \ltimes B + \beta A \ltimes C; \\ (\alpha B + \beta C) \ltimes A &= \alpha B \ltimes A + \beta C \ltimes A, \quad \alpha, \beta \in \mathbb{R}. \end{aligned} \tag{5.8}$$

2. (Associative rule)

$$\begin{aligned} A \ltimes (B \ltimes C) &= (A \ltimes B) \ltimes C; \\ (B \ltimes C) \ltimes A &= B \ltimes (C \ltimes A). \end{aligned} \tag{5.9}$$

Proposition 2. *Let $A \in M_{p \times q}$ and $B \in M_{m \times n}$. If $q = km$, then*

$$A \ltimes B = A(B \otimes I_k); \tag{5.10}$$

If $kq = m$, then

$$A \ltimes B = (A \otimes I_k)B. \tag{5.11}$$

Proposition 3. *1. Assume A and B are of proper dimensions such that $A \ltimes B$ is well defined. Then*

$$(A \ltimes B)^T = B^T \ltimes A^T. \tag{5.12}$$

2. In addition assume both A and B are invertible, then

$$(A \ltimes B)^{-1} = B^{-1} \ltimes A^{-1}. \tag{5.13}$$

Proposition 4. *Assume $A \in M_{m \times n}$ is given.*
1. Let $Z \in \mathbb{R}^t$ be a row vector. Then

$$A \ltimes Z = Z \ltimes (I_t \otimes A); \tag{5.14}$$

2. Let $Z \in \mathbb{R}^t$ be a column vector. Then

$$Z \ltimes A = (I_t \otimes A) \ltimes Z. \tag{5.15}$$

Note that when $\xi \in \mathbb{R}^n$ is a column (or a row), then $\underbrace{\xi \ltimes \cdots \ltimes \xi}_{k}$ is well defined, and denoted as $\xi^k := \underbrace{\xi \ltimes \cdots \ltimes \xi}_{k}$.

Next, we define a *swap matrix,* which is also called a permutation matrix in [17]. Many properties can be found in [4]. The swap matrix, $W_{[m,n]}$ is an $mn \times mn$ matrix constructed in the following way: label its columns by $(11, 12, \cdots, 1n, \cdots, m1, m2, \cdots, mn)$ and its rows by $(11, 21, \cdots, m1, \cdots, 1n, 2n, \cdots, mn)$. Then its element in the position $((I, J), (i, j))$ is assigned as

$$w_{(IJ),(ij)} = \delta_{i,j}^{I,J} = \begin{cases} 1, & I = i \text{ and } J = j, \\ 0, & \text{otherwise.} \end{cases} \tag{5.16}$$

When $m = n$ we simply denote by $W_{[n]}$ for $W_{[n,n]}$.

Example 4. Let $m = 2$ and $n = 3$, the swap matrix $W_{[2,3]}$ is constructed as

$$
W_{[2,3]} = \begin{array}{c} \\ \end{array}
\begin{pmatrix}
1 & 0 & 0 & 0 & 0 & 0 \\
0 & 0 & 0 & 1 & 0 & 0 \\
0 & 1 & 0 & 0 & 0 & 0 \\
0 & 0 & 0 & 0 & 1 & 0 \\
0 & 0 & 1 & 0 & 0 & 0 \\
0 & 0 & 0 & 0 & 0 & 1
\end{pmatrix}
\begin{array}{l} (11) \\ (21) \\ (12) \\ (22) \\ (13) \\ (23) \end{array} .
$$

with columns labelled $(11)\ (12)\ (13)\ (21)\ (22)\ (23)$.

$\square$

Let $A = (a_{ij}) \in M_{m \times n}$. The row stacking form of A, denoted by $V_r(A)$, is

$$V_r(A) = (a_{11}\ a_{12}\ \cdots\ a_{1n}\ \cdots\ a_{m1}\ a_{m2}\ \cdots\ a_{mn})^T.$$

The column stacking form of A, denoted by $V_c(A)$, is

$$V_c(A) = (a_{11}\ a_{21}\ \cdots\ a_{m1}\ \cdots\ a_{1n}\ a_{2n}\ \cdots\ a_{mn})^T.$$

As a consequence of the definition, the following "swap" property is useful in the sequel.

Proposition 5. *1. Let $X \in \mathbb{R}^m$ and $Y \in \mathbb{R}^n$ be two columns. Then*

$$W_{[m,n]} \ltimes X \ltimes Y = Y \ltimes X, \quad W_{[n,m]} \ltimes Y \ltimes X = X \ltimes Y. \tag{5.17}$$

2. Let $A \in M_{m \times n}$. Then

$$W_{[m,n]}V_r(A) = V_c(A); \quad W_{[n,m]}V_c(A) = V_r(A). \tag{5.18}$$

Proposition 6. *Let $A \in M_{m \times n}$ and $B \in M_{p \times q}$. Then*

$$W_{[m,p]}(A \otimes B)W_{[q,n]} = B \otimes A. \tag{5.19}$$

Swap matrix can be constructed in the following way: Denote the i-th canonical basic element in $\mathbb{R}^n$ by δ_i^n. That is, δ_i^n is the i-th column of I_n. Then it is easy to prove that

$$W_{[m,n]} = \left(\delta_1^n \ltimes \delta_1^m \cdots \delta_n^n \ltimes \delta_1^m \cdots \delta_1^n \ltimes \delta_m^m \cdots \delta_n^n \ltimes \delta_m^m\right). \tag{5.20}$$

The following factorization formula is useful for simplifying swap matrix computation.

Proposition 7.

$$W_{[pq,r]} = \left(W_{[p,r]} \otimes I_q\right)\left(I_p \otimes W_{[q,r]}\right). \tag{5.21}$$

For our later results, we need also to introduce the tensor expression of polynomials. Let $x = (x_1, \cdots, x_n)^T \in \mathbb{R}^n$. Then a k-th degree homogeneous polynomial can be expressed as $F \ltimes x^k$, where $x^k := \underbrace{x \ltimes \cdots \ltimes x}_{k}$, and the coefficient vector F is a $1 \times n^k$ row. Briefly, $Fx^k := F \ltimes x^k$.

Proposition 8. *Let Fx^m and Gx^n be m-th and n-th homogeneous polynomials respectively. Then the product is*

$$(Fx^m)(Gx^n) = F \ltimes G \ltimes x^{m+n}. \tag{5.22}$$

Next, we consider the differential of a matrix with differentiable function entries.

Definition 2. *Let $H = (h_{ij}(x))$ be a $p \times q$ matrix with the entries $h_{ij}(x)$ as smooth functions of $x \in \mathbb{R}^n$. Then the differential of H is defined as a $p \times nq$ matrix obtained by replacing each element h_{ij} by its differential $dh_{ij} = \left(\frac{\partial h_{ij}(x)}{\partial x_1}, \cdots, \frac{\partial h_{ij}(x)}{\partial x_n}\right)$.*

Our goal is to apply it to polynomials. We construct an $n^{k+1} \times n^{k+1}$ matrix Φ_k as

$$\Phi_k = \sum_{s=0}^{k} I_{n^s} \otimes W_{[n^{k-s},n]}. \tag{5.23}$$

Then we have the following differential form of x^k, which is fundamental in the further approach.

Proposition 9.

$$D(x^{k+1}) = \Phi_k \ltimes x^k. \tag{5.24}$$

The higher degree differential is defined inductively as:

$$D^{k+1} A(x) = D\left(D^k A(x)\right), \quad k \geq 1.$$

Using the above expression, the Taylor expression of a vector field $f(x) \in V(\mathbb{R}^n)$ is expressed as

$$f(x) = f(x_0) + \sum_{i=1}^{\infty} \frac{1}{i!} D^i f(x_0)(x - x_0)^i, \tag{5.25}$$

which has exactly the same form as the Taylor expansion of one variable case. A similar early work can be found in [15], where the tensor product is used. Unlike semi-tensor product, the associativity between conventional and tensor products fails. It can be seen here and later that most of the formula deductions are impossible without the semi-tensor product.

We may omit the symbol $\ltimes$ and consider all the matrix products are left semi-tensor product. When the dimensions meet the requirement of the conventional matrix product, it coincides with the conventional one.

5.3 Morgan's Problem

This section considers the solvability of Morgan's problem by using semi-tensor product. Consider a linear system

$$\begin{cases} \dot{x} = Ax + Bu, & x \in R^n, \ u \in R^m, \\ y = Cx, & y \in R^p. \end{cases} \tag{5.26}$$

Morgan's problem [18] asked that when we can find a control $u = Kx + Hv$, with $v \in R^p$ such that the closed-loop system is decoupled, i.e., each v_i controls y_i and doesn't affect y_j, $j \neq i$. Rigorously speaking, the transfer matrix is non-singular and diagonal. Denote by $\rho_1, \cdots, \rho_p$ the relative degrees of the outputs. Then when $m = p$ the answer to Morgan's problem is

Theorem 1 ([14]). *When $m = p$, the Morgan's problem is solvable iff the decoupling matrix*

$$D = \begin{pmatrix} c_1 A^{\rho_1 - 1} B \\ \vdots \\ c_p A^{\rho_p - 1} B \end{pmatrix} \tag{5.27}$$

is non-singular.

As $m > p$, the problem has been discussed over 35 years. It was claimed to be solved several times. But then some counter-examples were discovered. So far, it is still an open problem. The problem was partly solved in [22] etc.

From Theorem 3.1, the following is obvious:

Lemma 1. *The Morgan's problem is solvable, iff there exist $K \in M_{m \times n}$, $H \in M_{m \times p}$, and $1 \leq \rho_i \leq n$, $i = 1, \cdots, p$, such that*

$$c_i(A + BK)^{t_i} BH = 0, \quad t_i = 0, \cdots, \rho_i - 2, \quad i = 1, \cdots, p. \tag{5.28}$$

and

$$D = \begin{pmatrix} c_1(A + BK)^{\rho_1 - 1}BH \\ \vdots \\ c_p(A + BK)^{\rho_p - 1}BH \end{pmatrix} \tag{5.29}$$

is non-singular.

Let

$$W(K) = \begin{pmatrix} c_1 B \\ c_1(A + BK)B \\ \vdots \\ c_1(A + BK)^{\rho_1 - 2}B \\ \vdots \\ c_p(A + BK)^{\rho_p - 2}B \end{pmatrix}, \quad T(K) = \begin{pmatrix} c_1(A + BK)^{\rho_1 - 1}B \\ \vdots \\ c_p(A + BK)^{\rho_p - 1}B \end{pmatrix}.$$

Then (5.28) becomes

$$W(K)H = 0, \tag{5.30}$$

and (5.29) becomes

$$D = T(K)H. \tag{5.31}$$

Since $1 \leq \rho_i \leq n$; $i = 1, \cdots, p$, we may consider the solvability of Morgan's problem for fixed ρ_i; $i = 1, \cdots, p$, because there are only finite cases (n^p) have to be verified. For statement ease, in the rest of this chapter we consider the solvability of the Morgan's problem under a set of fixed ρ_i unless elsewhere stated.

Lemma 2. *The Morgan's problem is solvable, iff there exists a $K \in M_{m \times n}$ such that (i)*

$$\mathrm{Im}(T^T(K)) \cap \mathrm{Im}(W^T(K)) = 0, \tag{5.32}$$

(ii) $T(K)$ has full row rank.

The following theorem is an immediate consequence of the above lemma.

Theorem 2. *The Morgan's problem is solvable for fixed ρ_i, iff there exists a $K_0 \in M_{m \times n}$ such that*

$$rank \begin{pmatrix} T(K_0) \\ W(K_0) \end{pmatrix} = p + rank(W(K_0)). \tag{5.33}$$

Definition 3. *Let $A(K)$ be a matrix with the entries $a_{ij}(K)$ as polynomials of the entries of K, where $K \in M_{m \times n}$. The essential rank of $A(K)$, denoted by $rank_e(A(K))$ is defined as*

$$rank_e(A(K)) = \max_{K \in M_{m \times n}} rank(A(K)).$$

Now (under fixed ρ_i) denote

$$rank_e(T(K)) = t, \quad rank_e(W(K)) = s, \quad rank_e \begin{pmatrix} T(K) \\ W(K) \end{pmatrix} = q.$$

Since the essential rank is easily computable, the following corollary is convenient for some cases.

Corollary 1. *The Morgan's problem is solvable, if $q = p + s$.*

Using semi-tensor product, a numerical method can be used to verify the conditions of Theorem 3.4.

To begin with we have to calculate $T(K)$ and $W(K)$. Denote $Z = V_r(K) \in R^{mn}$, we first express $T(K)$ and $W(K)$ as the polynomial matrices of Z. The product of two matrices can be expressed by a semi-tensor product of matrices:

Lemma 3. *Let a matrix $A \in M_{n \times m}$ be given.*
(i) $x \in R^n$ is a row, then

$$xA = V_r^T(A) \ltimes x^T. \tag{5.34}$$

(ii) If $Y \in M_{p \times n}$, then

$$YA = (I_p \otimes V_r^T(A)) \ltimes V_r(Y). \tag{5.35}$$

Now we expand $(A + BK)^t$ in the following way:

$$(A + BK)^t = \sum_{i=0}^{2^t - 1} P_i(A, BK),$$

where P_i is a product of t elements, which are either A or BK. It is constructed in the following way: Convert i into a binary form, and replace each "0" by "A" and "1" by "BK".

70   D. Cheng

For instance, if $i = 0 = \underbrace{0 \cdots 0}_{t}$ then $P_0(A, BK) = A \cdots A = A^t$; if $i = 2 = \underbrace{0 \cdots 1\,0}_{t}$ then $P_2(A, BK) = A \cdots ABKA = A^{t-2}BKA$.

Collecting terms with same number of "K", we can easily get the following expression:

$$c_k(A + BK)^t B = \sum_{i=0}^{t} \sum_{j=1}^{T_i} S_0^{ij} K S_1^{ij} K \cdots S_{t-1}^{ij} K S_t^{ij}, \quad k = 1, \cdots, p, \quad (5.36)$$

where $T_i = \binom{i}{t}$. Using Lemma 3.7 and Proposition 2.7 , (5.36) can be expressed as

$$
\begin{aligned}
c_k(A &+ BK)^t B \\
&= \sum_{i=0}^{t} \sum_{j=1}^{T_i} S_0^{ij} \theta(I_m \otimes V_r^T(S_1^{ij}))\theta Z\theta \cdots \theta(V_r^T(I_m \otimes S_t^{ij}))\theta Z \\
&= \sum_{i=0}^{t} \sum_{j=1}^{T_i} S_0^{ij} \theta(I_m \otimes V_r^T(S_1^{ij}))\theta(I_{m^2 n} \otimes V_r^T(S_2^{ij})) \\
&\quad \theta \cdots \theta(I_{m^t n^{t-1}} \otimes V_r^T(S_t^{ij}))\theta Z^t, \quad k = 1, \cdots, p.
\end{aligned}
\quad (5.37)
$$

Using (5.29), we can easily express $W(K)$ and $T(K)$ in the standard polynomial form as:

$$W(K) = W_0 + W_1\theta Z + \cdots + W_{l-1}\theta Z^{l-1} \in M_{d \times m}, \quad d = \sum_{i=1}^{p} \rho_i - p \quad (5.38)$$

$$T(K) = T_0 + T_1\theta Z + \cdots + T_l\theta Z^l \in M_{p \times m}.$$

where $l = \max\{\rho_i - 1 | i = 1, \cdots, p\}$.

Let W^s be the set of s rows of $W(K)$, then the size of W^s is

$$|W^s| = \frac{d!}{s!(d-s)!}.$$

Now the Morgan's problem can be formulated as the follows.

Proposition 10. *The Morgan's problem is solvable, iff there exists $1 \le s \le m - p + 1$ such that*

$$R(Z) := \sum_{L \in W^s} \det(L(Z)L^T(Z)) = 0, \quad (5.39)$$

and

$$J(Z) := \sum_{L \in W^{s-1}} \det\left(\binom{T(Z)}{L(Z)} (T^T(Z)\, L^T(Z)) \right) > 0 \quad (5.40)$$

have a solution Z.

We refer to [3] for details.

5.4 Stability Region for a Nonlinear System

Consider a smooth nonlinear system of the form

$$\dot{x} = f(x), \quad x \in \mathbb{R}^n, \tag{5.41}$$

where $f(x)$ is an analytic vector field.

Let x_e be an equilibrium point of (5.41). The stable and unstable sub-manifolds of x_e are defined respectively as

$$\begin{aligned} W^s(x_e) &= \{p \in \mathbb{R}^n \,|\lim_{t \to \infty} x(t,p) \to x_e\} \\ W^u(x_e) &= \{p \in \mathbb{R}^n \,|\lim_{t \to -\infty} x(t,p) \to x_e\}. \end{aligned} \tag{5.42}$$

Suppose x_s is a stable equilibrium point of (5.41). The region of attraction of x_s is defined as

$$A(x_s) = \left\{p \in \mathbb{R}^n \,\middle|\, \lim_{t \to \infty} x(t,p) \to x_s\right\} \tag{5.43}$$

The boundary of the region of attraction is denoted by $\partial A(x_s)$.

An equilibrium point x_e is *hyperbolic* if the Jacobian matrix of f at x_e, denoted by $J_f(x_e)$, has no eigenvalues with zero real part. A hyperbolic equilibrium point is said to be of type-k if $J_f(x_e)$ has k positive real part eigenvalues.

The problem of determining the region of attraction of power systems has been discussed intensively by many researches. Particularly, [23] and [13] proved that for a stable equilibrium point x_s the stability boundary is composed of the stability sub-manifolds of equilibrium points on the boundary of the region of attraction under the assumptions that

(i) the equilibrium points on the stability boundary $\partial A(x_s)$ are hyperbolic;

(ii) the stable and unstable sub-manifolds of the equilibrium points on the stability boundary $\partial A(x_s)$ satisfy the transversality condition;

(iii) every trajectory on the stability boundary $\partial A(x_s)$ approaches one of the equilibrium points as $t \to \infty$.

It is well known that the stability boundary is of dimension $n-1$ [11, 12]. Therefore, the stability boundary is composed of the closure of stability sub-manifolds of type-1 equilibrium points on the boundary. Calculating the stability sub-manifold of type-1 becomes an important task. Several approximations have been proposed [21], [19].

In this section we use the semi-tensor product to calculate the stability sub-manifold. Without loss of generality, we can assume the type-1 equilibrium point we concern is $x_u = 0$.

Using Taylor series expansion, we express the vector field f in (5.41) by

$$f(x) = \sum_{i=1}^{\infty} F_i x^i = Jx + F_2 x^2 + \cdots. \tag{5.44}$$

Where $F_1 = J = J_f(0)$, and $F_i = \frac{1}{i!} D^i f(0)$ are known $n \times n^i$ matrix.

We use A^{-T} for the inverse of A^T, which is the transpose of A.

72   D. Cheng

Lemma 4. *Let A be a hyperbolic matrix. Denote by V_s and V_u the stable and unstable sub-spaces of A respectively and by U_s and U_u the stable and unstable sub-spaces of A^{-T} respectively. Then*

$$V_s^\perp = U_u, \quad V_u^\perp = U_s. \tag{5.45}$$

The following is an immediate consequence of the above lemma.

Corollary 2. *Let A be a type-1 matrix and assume its only unstable eigenvalue is μ. Let η be the eigenvector of A^T with respect to eigenvalue μ. Then η is perpendicular to the stable subspace of A.*

The following is a set of necessary and sufficient conditions for the stable submanifold of a type-1 equilibrium point to satisfy.

Theorem 3. *Assume $x_u = 0$ is a type-1 equilibrium point of system (5.41).*

$$W^s(e_u) = \{x \mid h(x) = 0\}. \tag{5.46}$$

Then $h(x)$ is uniquely determined by the following necessary and sufficient conditions (5.47)-(5.49).

$$h(0) = 0, \tag{5.47}$$

$$h(x) = \eta^T x + 0(\|x\|^2), \tag{5.48}$$

$$L_f h(x) = \mu h(x), \tag{5.49}$$

where $L_f h(x)$ is the Lie derivative of $h(x)$ with respect to f; η is an eigenvector of $J_f^T(0)$ with respect to its only positive eigenvalue μ.

For computational ease, we denote the Taylor expansion of $h(x)$ as

$$h(x) = H_1 x + H_2 x^2 + H_3 x^3 + \cdots = H_1 x + \frac{1}{2} x^T \Psi x + H_3 x^3 + \cdots . \tag{5.50}$$

Note that in the above we use two forms for the quadratic term: The semi-tensor product form, $H_2 x^2$, and the quadratic form as $\frac{1}{2} x^T \Psi x$, where $\Psi = Hess(h(0))$ is the Hessian matrix of $h(x)$ at $x = 0$, and $H_2 = V_c^T(\frac{1}{2}\Psi)$, which is the column stacking form of the matrix $\frac{1}{2}\Psi$.

Lemma 5. *The quadratic form in the stable sub-manifold equation (5.50) satisfies*

$$\Psi(\frac{\mu}{2} I - J) + (\frac{\mu}{2} I - J^T)\Psi = \sum_{i=1}^n \eta_i Hess(f_i(0)), \tag{5.51}$$

where μ and η are defined as in Corollary 4.2 for $A = F_1 = J := J_f(0)$, $Hess(f_i)$ is the Hessian matrix of the i-th component f_i of f.

Lemma 6. *Equation (5.51) has unique symmetric solution.*

Denote by V_c^{-1} the inverse mapping of V_c. That is, it recovers a (square) matrix, A, from its column stacking form, $V_c(A)$.

Summarizing Lemmas 4.1, 4.4, 4.5, we have the following quadratic form approximation of the stable sub-manifold.

Theorem 4. *The stable sub-manifold of x_u, expressed as $h(x) = 0$, can be expressed as*

$$h(x) = H_1 x + \frac{1}{2} x^T \Psi x + 0(\|x\|^3), \tag{5.52}$$

where

$$\begin{cases} H_1 = \eta^T \\ \Psi = V_c^{-1} \left\{ \left[\left(\frac{\mu}{2} I_n - J^T \right) \otimes I_n + I_n \otimes \left(\frac{\mu}{2} I_n - J^T \right) \right]^{-1} V_c \left(\sum_{i=1}^n \eta_i Hess(f_i(0)) \right) \right\}, \end{cases}$$

where μ and η are defined as in Corollary 4.2 for $J = F_1$, $Hess(f_i)$ is the Hessian matrix of the i-th component f_i of f.

Next, we consider the whole Taylor expansion of the stable sub-manifold. In later calculation of the differentials we have to calculate Φ_k. For this purpose we need the following:

Proposition 11.

$$W_{[n^s,n]} = \prod_{i=0}^{s-1} \left(I_{n^i} \otimes W_{[n,n]} \otimes I_{n^{s-i-1}} \right) \tag{5.53}$$

Next, we will solve H_k from (5.47)-(5.49). The problem is x^k is a redundant basis of the k-th degree homogeneous polynomials, so from (5.47)-(5.49) we can not get unique solution. To overcome this obstacle we consider the natural basis of the k-th homogeneous polynomials. Let $S \in Z_+^n$. The natural basis is defined as

$$B_n^k = \{ x^S \mid S \in Z_+^n, \ |S| = k \}.$$

Now we arrange the elements in B_n^k in alphabetic order. That is, for $S^1 = (s_1^1, \cdots, s_n^1)$ and $S^2 = (s_1^2, \cdots, s_n^2)$, we sign the order as $x^{S^1} \prec x^{S^2}$ if there exists a t, $1 \leq t \leq n-1$, such that

$$s_1^1 = s_1^2, \ \cdots, \ s_t^1 = s_t^2, \ s_{t+1}^1 > s_{t+2}^2.$$

Then we arrange the elements in B_n^k as a column and denote it as $x_{(k)}$. We have

74   D. Cheng

Proposition 12. *The size of the basis B_n^k is*

$$|B_n^k| := d = \frac{(n+k-1)!}{k!(n-1)!}; \quad k \geq 0, \quad n \geq 1. \tag{5.54}$$

Two matrices $T_N(n,k) \in M_{n^k \times d}$ and $T_B(n,k) \in M_{d \times n^k}$ can be defined to convert one basis to the other, that is,

$$x^k = T_N(n,k)x_{(k)}, \quad x_{(k)} = T_B(n,k)x^k,$$

and

$$T_B(n,k)T_N(n,k) = I_d.$$

Now, instead of solving H_k, we may try to solve G_k, where

$$H_k x^k = G_k x_{(k)}.$$

A straightforward computation shows the following:

Proposition 13. *The symmetric coefficient set, H_k is unique. Moreover,*

$$H_k = G_k T_B(n,k), \quad G_k = H_k T_N(n,k). \tag{5.55}$$

Now we consider the higher degree terms of the equation $h(x)$ of the stable sub-manifold. Denote

$$f(x) = F_1 x + F_2 x^2 + \cdots;$$

$$h(x) = H_1 x + H_2 x^2 + \cdots.$$

Note that $F_1 = J_f(0) = J$, $H_1 = \eta^T$, and H_2 is uniquely determined by (5.52).

Proposition 14. *The coefficients, H_k, $k \geq 2$, of $h(x)$ satisfy the following equation.*

$$\left[\sum_{i=1}^{k} H_i \Phi_{i-1}((I_{n^{i-1}} \otimes F_{k-i+1}) - \mu H_k \right] x^k = 0, \quad k \geq 2, \tag{5.56}$$

The following result, which is a summary of the above arguments, is generically applicable.

Theorem 5. *Assume the matrices*

$$C_k := \mu I_d - T_B(n,k)\Phi_{k-1}(I_{n^{k-1}} \otimes F_1)T_N(n,k), \quad k \geq 3 \tag{5.57}$$

are non-singular, then

$$G_k = \left[\sum_{i=1}^{k-1} G_i T_B(n,i)\Phi_{i-1}(I_{n^{i-1}} \otimes F_{k-i+1}) \right] T_N(n,k)C_k^{-1} \tag{5.58}$$

We refer to [6], [7] and the references there for more details.

5.5 Non-regular Linearization

Consider an affine nonlinear system

$$\dot{x} = f(x) + \sum_{i=1}^{m} g_i(x)u_i, \quad f(0) = 0, \quad x \in \mathbb{R}^n, u \in \mathbb{R}^m. \tag{5.59}$$

The vector fields $f(x)$ and $g_i(x)$ etc. are assumed to be analytic (C^ω) vector fields to assure the convergence of the Taylor expansion of the vector fields etc.

The static state feedback linearization is stated as follows: Find a state feedback control

$$u = \alpha(x) + \beta(x)v, \tag{5.60}$$

and a diffeomorphism $z = \Phi(x)$ such that in coordinate frame z the closed-loop system can be expressed as a completely controllable linear system.

When $\beta(x)$ is an $m \times m$ non-singular matrix, the problem is called a regular linearization problem. When $\beta(x)$ is an $m \times k$ matrix with $k < m$, it is called the non-regular linearization. When $k = 1$, it is called single-input linearization. Recently, non-regular linearization problem has been investigated in [20], [15].

This section considers the non-regular linearization via normal form approximation and the tool of semi-tensor product.

Lemma 7. *System (5.59) is non-regular state feedback linearizable, iff it is single-input linearizable.*

Lemma 8 ([20]). *Let $A = J_f(0)$ be the Jacobian matrix of f at the origin, $B = g(0)$. If system (5.1) is linearizable, then (A, B) is completely controllable.*

Formal reduction of nonlinear dynamical systems to linear systems can be traced back to a long time ago. The following result, which was the fundamental result in Poincare's dissertation, is essential for our approach.

Theorem 6 (Poincare's Theorem [1]). *Consider a C^ω dynamic system*

$$\dot{x} = Ax + f_2(x) + f_3(x) + \cdots, \quad x \in \mathbb{R}^n, \tag{5.61}$$

where $f_i(x)$, $i \geq 2$ are i-th degree homogeneous vector fields. If A is non-resonant, there exists a formal change of coordinates

$$x = y + h(y), \tag{5.62}$$

where $h(y)$ corresponds to the sum of possibly infinite homogeneous vector polynomials $h_m(y)$, $m \geq 2$, that is

$$h(y) = h_2(y) + h_3(y) + \cdots, \tag{5.63}$$

such that system (5.61) can be expressed as

$$\dot{y} = Ay.$$

We recall that for a matrix A, let $\lambda = (\lambda_1, \cdots, \lambda_n)$ be its eigenvalues. A is a resonant matrix if there exists $m = (m_1 \cdots, m_n) \in Z_+^n$, and $|m| \geq 2$, i.e., $m_i \geq 0$ and $\sum_{i=1}^n m_i \geq 2$, such that for some s,

$$\lambda_s = \langle m, \lambda \rangle.$$

The Poincare's result has been extended to SISO control systems [16].

Proposition 15. *Let $\lambda = (\lambda_1, \cdots, \lambda_n)$ be the eigenvalues of a given Hurwitz matrix A. A is non-resonant if*

$$\max\{|Re(\lambda_i)| \mid \lambda_i \in \sigma(A)\} \leq 2 \min\{|Re(\lambda_i)| \mid \lambda_i \in \sigma(A)\} \tag{5.64}$$

A constant vector, $b = (b_1, \cdots, b_n)^T \in \mathbb{R}^n$, is said to be of non-zero component if none of its components is zero, i.e., $b_i \neq 0$, $i = 1, \cdots, n$.

Proposition 16. *A linear control system*

$$\dot{x} = Ax + \sum_{i=1}^m b_i u_i := Ax + Bu, \quad x \in \mathbb{R}^n, \ u \in \mathbb{R}^m \tag{5.65}$$

is completely controllable, iff there exists two matrices F, G such that the closed-loop system

$$\dot{x} = (A + BF)x + BGv/$$

can be converted, by a linear coordinate change, into the following form

$$\dot{z} = Az + bv := \begin{pmatrix} d_1 & 0 & \cdots & 0 \\ 0 & d_2 & \cdots & 0 \\ & & \ddots & \\ 0 & 0 & \cdots & d_n \end{pmatrix} z + \begin{pmatrix} b_1 \\ b_2 \\ \vdots \\ b_n \end{pmatrix} v, \tag{5.66}$$

where d_i, $i = 1, \cdots, n$ are distinct and b is of non-zero component.

From the above proposition we can call (5.66) the reduced single-input feedback A-diagonal (RSIFAD) canonical form.

The following corollary is an immediate consequence.

Proposition 17. *Consider a system*

$$\dot{x} = Ax + \sum_{i=1}^m g_i(x)u_i, \tag{5.67}$$

where A is a diagonal matrix with distinct diagonal elements d_i. If there is a constant vector of non-zero component $b \in Span\{g_i\}$, then the system is single-input linearizable.

From now on, we assume that the linearized matrix A satisfies the following assumption:

A1. A is a diagonal matrix with distinct diagonal elements d_i and is non-resonant.

We are ready to consider the non-regular linearization problem. To begin with, using Taylor series expression on $f(x)$ with the form of semi-tensor product, we express system

$$\dot{x} = f(x) \tag{5.68}$$

as

$$\dot{x} = Ax + F_2 x^2 + F_3 x^3 + \cdots, \tag{5.69}$$

where F_k are $n \times n^k$ constant matrices, and x^k are as stated above. Then assuming

$$\mathrm{ad}_{Ax}\, \eta_k = F_k x^k,$$

we can easily obtain that

$$\eta_k = \left(\Gamma_k^n \odot F_k \right) x^k, \quad x \in \mathbb{R}^n. \tag{5.70}$$

Here $\odot$ is the Hadamard product of matrices [24].

Now we are ready to present our main result:

Theorem 7. *Assume A satisfies A1. Then system (5.68) can be transformed into a linear form*

$$\dot{z} = Az, \tag{5.71}$$

by the following coordinate transformation:

$$z = x - \sum_{i=2}^{\infty} E_i x^i, \tag{5.72}$$

where E_i are determined recursively as

$$\begin{aligned} E_2 &= \Gamma_2 \odot F_2 \\ E_s &= \Gamma_s \odot \left(F_s - \sum_{i=2}^{s-1} E_i \Phi_{i-1}(I_{n^{i-1}} \otimes F_{s+1-i}) \right), \quad s \geq 3. \end{aligned} \tag{5.73}$$

The advantage of this Taylor series expression is that it doesn't require recursive computation of the intermediate forms of the system under transferring coordinates z_i, $i = 1, 2, 3, \cdots$.

Now consider the linearization of system (5.59). Denote $A = \frac{\partial f}{\partial x}|_0$, $B = g(0)$, and assume (A, B) is a completely controllable pair. Then we can find feedback K and a linear coordinate transformation T, such that $\tilde{A} = T^{-1}(A + BK)T$ satisfies assumption A1. For the sake of simplicity, we call the above transformation an *NR-type transformation*.

Using the notations and algorithm proposed above the following result is immediate.

Theorem 8. *System (5.59) is single-input linearizable, iff there exist an NR-type transformation and a constant vector b of non-zero component such that*

$$b \in Span \left\{ \left(I - \sum_{i=2}^{\infty} E_i \Phi_{i-1} x^{i-1} \right) g_j \mid j = 1, \cdots, m \right\}. \tag{5.74}$$

We refer to [5] for details.

5.6 Conclusion

This paper has introduced the semi-tensor product of matrices, which is a generalization of conventional matrix product. Three of its applications were introduced: Numerical solution to Morgan's problem, Estimation of region of attraction of a stable equilibrium, and Non-regular linearization. The method has also been used to investigate the linear symmetry of nonlinear systems [8], to some algebraic and differential geometric problems [9], and to some physical problems [10]. The right semi-tensor product of matrices and both products for two matrices with arbitrary dimensions can be found in [10]. The author is confident that the semi-tensor product of matrices is a new useful tool.

References

1. V.I. Arnold, *Geometrical Methods on the Theory of Ordinary Differential Equations*, Mew York, Springer-Verlag, 177-188, 1983.
2. W.M. Boothby, *An Introduction to Differentiable Manifolds and Riemannian Geometry*, 2nd ed., Academic Press, 1986.
3. D. Cheng, Semi-tensor product of matrices and its application to Morgan's problem, Science in China, Series F, Vol. 44, No.3 195-212, 2001.
4. D. Cheng, *Matrix and Polynomial Approach to Dynamic Control Systems*, Science Press, Beijing, 2002.
5. D. Cheng, X. Hu, Y. Wang, Non-regular feedback linearization of nonlinear systems via a normal form algorithm, *Automatica*, Vol. 40, No. 3, 439-447, 2004.
6. D. Cheng, J. Ma, Q. Lu, S. Mei, Quadratic Form of Stable Sub-manifold for Power Systems, *Int. J. Rob. Nonl. Contr.*, No. 14, 773-788, 2004.
7. D. Cheng, J. Ma, Calculation of Stability Region, *Proc. 42nd IEEE CDC'03*, Maui, 5615-5620, 2003.
8. D. Cheng, G. Yang, Z. Xi, Nonlinear systems possing linear symmetry, *Int. J. Rob. Nonl.*, (submitted).
9. D. Cheng, L. Zhang, On semi-tensor product of matrices and its applications, *ACTA Math. Appl. Sinica*, English Series, Springer-Verlag, Vol. 19, No. 2, 219-228, 2003.
10. D. Cheng, Y. Dong, Semi-tensor product of matrices and its some applications to physics. *Methods and Applications of Analysis*, Vol. 10, No. 4, pp 565-588, 2004.

11. H.D. Chiang, M. Hirsch and F. Wu, Stability regions of nonlinear autonomous dynamical systems, *IEEE Trans. Automat. Contr.*, Vol. 33, No. 1, pp 16-27, 1988.

12. H.D. Chiang, F.F. Wu, Foundations of the potential energy boundary surface method for power system transient stability analysis, *IEEE Trans. Circ. Sys.*, Vol. 35, No. 6, pp 712-728, 1988.

13. H.D. Chiang, Chia-Chi Chu, Theoretical foundation of the BCU method for direct stability analysis of network-reduction power system models with small transfer conductances, *IEEE Trans. Circ. Sys.*, Vol. 42, No. 5, 252-265, 1995.

14. P.L. Falb, W.A. Wolovich, Decoupling and synthesis of multivariable control systems, *IEEE Trans. Aut. Contr.*, Vol. 12, No. 6, 651-668, 1967.

15. S.S. Ge, Z. Sun, T.H. Lee, Nonregular feedback linearization for a class of second-order nonlinear systems, *Automatica*, Vol. 37, 1819-1824, 2001.

16. A.J. Krener, S. Karahan, and M.Hubbard, pproximate normal forms of nonlinear systems, *Proc. 27th Conf. Dec. Contr.*, 1223-1229, 1988.

17. J.R. Magnus, H. Neudecker, *Matrix Differential Calculus with Applications in Statistics and Econometrics*, Revised Ed., John Wiley & Sons, 1999.

18. B.S. Morgan, The synthesis of linear multivariable systems by state feedback, *JACC*, Vol. 64, 468-472, 1968.

19. S. Saha, A.A. Fouad, W.H. Kliemamm, V. Vittal, Stability boundary approximation of a power system using the real normal form of vector fields, *IEEE Trans, Power Sys.*, Vol. 12, No. 2, 797-802, 1997.

20. Z. Sun, X. Xia, On nonregular feedback linearization, *Automatica*, Vol. 33, No. 7, 1339-1344, 1997.

21. V. Venkatasubramanian, and W. Ji, Numerical approximation of $(n-1)$-dimensional stable manifolds in large systems such as the power system, *Automatica*, Vol. 33, No. 10, 1877–1883, 1997.

22. W.M. Wonham, *Linear Multo-variable Control, A Geometric Approach*, Lecture Notes in Economics and Mathematical Systems, Soringer-Verlag, 1974.

23. J.G. Zaborszky, J.G. Huang, B. Zheng and T.C. Leung, On the phase protraits of a class of large nonlinear dynamic systems such as the power systems, *IEEE Trans. Autom. Contr.*, Vol. 33, No. 1, 4-15, 1988.

24. F. Zhang, *Matrix Theory, Basic Results and Techniques*, Springer-Verlag, 1999.

From Empirical Data to Multi-Modal Control Procedures*

Florent Delmotte[1] and Magnus Egerstedt[2]

[1] School of Electrical and Computer Engineering, Georgia Institute of Technology, Atlanta, GA 30332, USA
`florent@ece.gatech.edu`
[2] School of Electrical and Computer Engineering, Georgia Institute of Technology, Atlanta, GA 30332, USA
`magnus@ece.gatech.edu`

Summary. In this paper we study the problem of generating control programs, i.e. strings of symbolic descriptions of control-interrupt pairs (or modes) from input-output data. In particular, we take the point of view that such control programs have an information theoretic content and thus that they can be more or less effectively coded. As a result, we focus our attention on the problem of producing low-complexity programs by recovering the strings that contain the smallest number of distinct modes. An example is provided where the data is obtained by tracking ten roaming ants in a tank.

6.1 Introduction

As the complexity of many control systems increases, due both to the system complexity (e.g. manufacturing systems, [6]) and the complexity of the environment in which the system is embedded (e.g. autonomous robots [1, 17]), multi-modal control has emerged as a useful design tool. The main idea is to define different modes of operation, e.g. with respect to a particular task, operating point, or data source. These modes are then combined according to some discrete switching logic and one attempt to formalize this notion is through the concept of a *Motion Description Language* (MDL) [5, 9, 15, 18].

Each string in a MDL corresponds to a control program that can be operated on by the control system. Slightly different versions of MDLs have been proposed, but they all share the common feature that the individual atoms, concatenated together to form the control program, can be characterized by control-interrupt pairs. In other words, given a dynamical system

---

*This work was supported by NSF through the programs EHS NSF-01-161 (grant # 0207411) and ECS NSF-CAREER award (grant # 0237971).

F. Delmotte and M. Egerstedt: *From Empirical Data to Multi-Modal Control Procedures*, LNCIS
321, 81–95 (2005)
`www.springerlink.com`

$$\dot{x} = f(x, u), \ x \in \mathbb{R}^N, \ u \in U$$
$$y = h(x), \ y \in Y, \tag{6.1}$$

together with a control program $(k_1, \xi_1), \ldots, (k_z, \xi_z)$, where $k_i : Y \to U$ and $\xi_i : Y \to \{0, 1\}$, the system operates on this program as $\dot{x} = f(x, k_1(h(x)))$ until $\xi_1(h(x)) = 1$. At this point the next pair is read and $\dot{x} = f(x, k_2(h(x)))$ until $\xi_2(h(x)) = 1$, and so on[3].

Now, a number of results have been derived for such (and similar) systems, driven by strings of symbolic inputs, i.e. when the control and interrupt sets are finite. For example, in [4], the set of reachable states was characterized, while [10] investigates optimal control aspects of such systems. In [8, 15, 18], the connection between MDLs and robotics was investigated. However, in this paper we continue the development begun in [3], where the control programs are viewed as having an information theoretic content. In other words, they can be coded more or less effectively. Within this context, one can ask questions concerning minimum complexity programs, given a particular control task.

But, in order to effectively code symbols, drawn from a finite alphabet, one must be able to establish a probability distribution over the alphabet. If such a distribution is available then Shannon's celebrated source coding theorem [20] tells us that the minimal expected code length l satisfies

$$\mathcal{H}(\mathcal{A}) \leq l \leq \mathcal{H}(\mathcal{A}) + 1, \tag{6.2}$$

where $\mathcal{A}$ is the alphabet, and where the *entropy* is given by

$$\mathcal{H}(\mathcal{A}) = \sum_{a \in \mathcal{A}} p(a) \log_2 \frac{1}{p(a)}, \tag{6.3}$$

where $p(a)$ is the probability of drawing the symbol a from $\mathcal{A}$. The main problem that we will study is how to produce an empirical probability distribution over the set of modes, given a string of input-output data. Such a probability distribution would be useful when coding control procedures since a more common mode should be coded using fewer bits than an uncommon one. The ability to code control programs effectively has a number of potential applications, from teleoperated robotics, control over communication constrained networks, to minimum attention control.

The outline of this paper is as follows: In Section 6.2 we introduce motion description languages, and in Section 6.3, we define the problem at hand and show how this can be addressed within the MDL framework. In Section 6.4 we show how to find mode sequences that contain the smallest number of distinct modes, followed by a description of how these types of sequences can be modified for further complexity reductions, in Section 6.5. In Section 6.6, we give an example where the control programs are obtained from data generated by 10 roaming ants.

---

[3]Note that the interrupts can also be time-triggered, but this can easily be incorporated by a simple augmentation of the state space.

6.2 Motion Description Languages

The primary objects of study in this paper are so called *motion description languages* (MDLs). Given a finite set, or alphabet, $\mathcal{A}$, by $\mathcal{A}^\star$ we understand the set of all strings of finite length over $\mathcal{A}$, with the binary operation of concatenation defined on $\mathcal{A}^\star$. Relative to this operation, $\mathcal{A}^\star$ is a semigroup, and if we include the empty string in $\mathcal{A}^\star$ it becomes a monoid, i.e. a semigroup with an identity, and a *formal language* is a subset of the free monoid over a finite alphabet. (See for example [12] for an introduction to this subject.)

The concept of a *motion alphabet* has been proposed recently in the literature as a finite set of symbols representing different control actions that, when applied to a specific system, define segments of motion [5, 7, 11, 14, 18]. A MDL is thus given by a set of strings that represent such idealized motions, i.e. a MDL is a subset of the free monoid over a given motion alphabet. Particular choices of MDLs become meaningful only when the language is defined relative to the physical device that is to be controlled, as shown in Equation (6.1).

Now, in order to make matters somewhat concrete, we illustrate the use of MDLs with a navigation example, found in [8]. What makes the control of mobile robots particularly challenging is the fact that the robots operate in unknown, or partially unknown environments. Any attempt to model such a system must take this fact into account. We achieve this by letting the robot make certain observations about the environment, and we let the robot dynamics be given by

$$\dot{x} = u, \; x, u \in \mathbb{R}^2$$
$$y_1 = o_d(x), \; y_2 = c_f(x),$$

where o_d is an odometric (possibly quantized) position estimate of x, and c_f is the (possibly quantized) contact force from the environment. The contact force could either be generated by tactile sensors in contact with the obstacle or by range sensors such as sonars, lasers, or IR-sensors.

Relative to this robot it is now possible to define a MDL for executing motions that drive the robot toward a given goal, located at x_F, when the robot is not in contact with an obstacle. On the other hand, when the robot is in contact with an obstacle, it seems reasonable to follow the contour of that obstacle in a clock-wise or counter clock-wise fashion, as suggested in [13]. We let the MDL be given by the set $\sigma_{GA} \cdot (\sigma_{OA} \cdot \sigma_{GA})^\star$, where GA and OA denote "goal-attraction" and "obstacle-avoidance" respectively, and where $a^\star = \{\emptyset, a, aa, aaa, ...\}$. The individual modes $\sigma_{GA} = (k_{GA}, \xi_{GA})$ and $\sigma_{OA} = (k_{OA}, \xi_{OA})$ are furthermore given by

$$\begin{cases} k_{GA}(y_1, y_2) = \kappa(x_F - y_1) \\ \xi_{GA}(y_1, y_2) = \begin{cases} 0 \text{ if } \langle y_2, x_F - y_1 \rangle \geq 0 \\ 1 \text{ otherwise} \end{cases} \end{cases}$$

$$\begin{cases} k_{OA}(y_1, y_2) = cR(-\pi/2)y_2 \\ \xi_{OA}(y_1, y_2) = \begin{cases} 0 \text{ if } \langle y_2, x_F - y_1 \rangle < 0 \text{ or } \angle(x_F - y_1, y_2) < 0 \\ 1 \text{ otherwise.} \end{cases} \end{cases}$$

The idea here is that the goal is located at x_F, and when the robot is not in contact with an obstacle, x_F is taken as a set-point in a proportional feedback law, provided by the mapping $k_{GA}(y_1, y_2) = \kappa(x_F - y_1)$, with $\kappa > 0$. When the robot is in contact with an obstacle, no set-point is needed, and $k(y_1, y_2)$ is simply given by $cR(-\pi/2)y_2$, where $c > 0$, $R(\theta)$ is a rotation matrix, and the choice of $\theta = -\pi/2$ corresponds to a clockwise negotiation of the obstacle. Note that in this example, the interrupts trigger as new obstacles are encountered, and in the definition of the interrupts, $\angle(\gamma, \delta)$ denotes the angle between the vectors γ and δ. An example of using this multi-modal control sequence is shown in Figure 6.1.

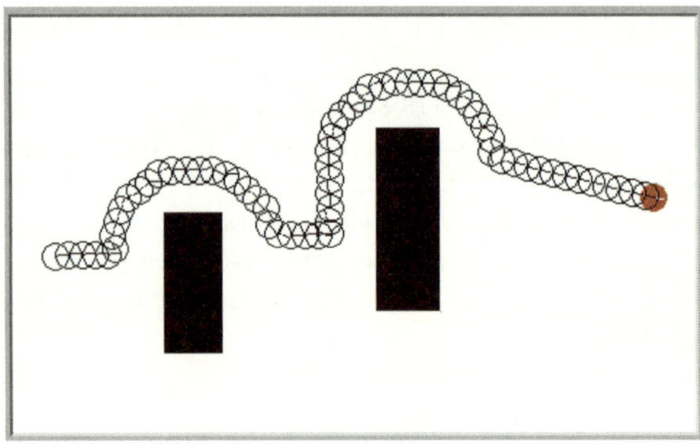

Fig. 6.1. A multi-modal input string is used for negotiating two rectangular obstacles. Depicted is a simulation of a Nomadic Scout in the Nomadic **Nserver** environment.

6.3 Specification Complexity

If we now assume that the input and output spaces (U and Y respectively) in Equation (6.1) are finite, which can be justified by the fact that all physical sensors and actuators have a finite range and resolution, the set of all possible modes $\Sigma_{total} = U^Y \times \{0, 1\}^Y$ is finite as well. We can moreover adopt the point of view that a data point is measured only when the output or input change values, i.e. when a new output or input value is encountered. This corresponds to a so called *Lebesgue sampling*, in the sense of [2]. Under this sampling policy, we can define a mapping $\delta : \mathbb{R}^N \times U \to \mathbb{R}^N$ as $x_{p+1} = \delta(x_p, k(h(x_p)))$, given the control law $k : Y \to U$, with a new time update occurring whenever a new output or input value is encountered. For such a system, given the input string $(k_1, \xi_1), \ldots, (k_z, \xi_z) \in \Sigma^*$ where $\Sigma \subseteq \Sigma_{total}$, the evolution is given by

$$\begin{cases} x(q+1) = \delta(x(q), k_{l(q)}(y(q))), \; y(q) = h(x(q)) \\ l(q+1) = l(q) + \xi_{l(q)}(y(q)). \end{cases} \tag{6.4}$$

Now, given a mode sequence of control-interrupt pairs $\sigma \in \Sigma^*$, we are interested in how many bits we need in order to specify σ. If no probability distribution over Σ is available, this number is given by the *description length*, as defined in [19]:

$$\mathcal{D}(\sigma, \Sigma) = |\sigma| \log_2(card(\Sigma)),$$

where $|\sigma|$ denotes the length of σ, i.e. the total number of modes in the string. This measure gives us the number of bits required for describing the sequence in the "worst" case, i.e. when all the modes in Σ are equally likely. However, if we can establish a probability distribution p over Σ, the use of optimal codes can, in light of Equation (6.2), reduce the number of bits needed, which leads us to the following definition:

Definition (Specification Complexity): *Given a finite alphabet Σ and a probability distribution p over Σ. We say that a word $\sigma \in \Sigma^*$ has specification complexity*

$$\mathcal{S}(\sigma, \Sigma) = |\sigma| \mathcal{H}(\Sigma).$$

An initial attempt at establishing a probability distribution over $\Sigma \subseteq U^Y \times \{0,1\}^Y$ was given in [3]. In that work, the main idea was to recover modes (and hence also the empirical probability distribution) from empirical data. For example, supposing that the mode string $\sigma = \sigma_1 \sigma_2 \sigma_1 \sigma_3$ was obtained, then we can let $\Sigma = \{\sigma_1, \sigma_2, \sigma_3\}$, and the corresponding probabilities become $p(\sigma_1) = 1/2$, $p(\sigma_2) = 1/4$, $p(\sigma_3) = 1/4$. In such a case where we let Σ be built up entirely from the modes in the sequence σ, the empirical specification complexity depends solely on σ:

$$\mathcal{S}^e(\sigma) = |\sigma| \mathcal{H}^e(\sigma) = - \sum_{i=1}^{M(\sigma)} \lambda_i(\sigma) \log_2 \frac{\lambda_i(\sigma)}{|\sigma|}, \tag{6.5}$$

where $M(\sigma)$ is the number of distinct modes in σ, $\lambda_i(\sigma)$ is the number of occurrences of mode σ_i in σ, and where we use superscript e to stress the fact that the probability distribution is obtained from empirical data.

Based on these initial considerations, the main problem, from which this work draws its motivation, is as follows:

Problem (Minimum Specification Complexity): *Given an input-output string $S = (y(1), u(1)), (y(2), u(2)), \ldots, (y(n), u(n)) \in (Y \times U)^n$, find the minimum specification complexity mode string $\sigma \in \Sigma^*_{total}$ that is consistent with the data. In other words, find σ that solves*

$$P(\Sigma_{total}, \boldsymbol{y}, \boldsymbol{u}) : \begin{cases} \min_{\sigma \in \Sigma^*_{total}} \mathcal{S}^e(\sigma) \\ \text{subject to } \forall q \in \{1, \ldots, n\} \\ \begin{cases} \sigma_{l(q)} = (k_{l(q)}, \xi_{l(q)}) \in \Sigma_{total} \\ k_{l(q)}(y(q)) = u(q) \\ \xi_{l(q)}(y(q)) = 0 \Rightarrow l(q+1) = l(q), \end{cases} \end{cases}$$

where the last two constraints ensure consistency of σ with the data S, and where $\boldsymbol{y} = (y(1), \ldots, y(n))$, $\boldsymbol{u} = (u(1), \ldots, u(n))$ give the empirical data string.

Note that this is slightly different than the formulation in Equation (6.4) since we now use $\sigma_{l(q)}$ to denote a particular member in $U^Y \times \{0, 1\}^Y$ instead of the $l(q)$-th element in σ. Unfortunately, this problem turns out to be very hard to address directly. However, the easily established property

$$0 \le \mathcal{H}^e(\sigma) \le \log_2(M(\sigma)), \; \forall \sigma \in \Sigma^*_{total}$$

allows us to focus our efforts on a more tractable problem. Here, the last inequality is reached when all the $M(\sigma)$ distinct modes of σ are equally likely.

As a consequence, we have $\mathcal{S}^e(\sigma) \le |\sigma| \log_2(M(\sigma))$ and thus it seems like a worth-while endeavor, if we want to find low-complexity mode sequences, to try to minimize either the length of the mode sequence $|\sigma|$ or the number of distinct modes $M(\sigma)$. In fact, minimization of $|\sigma|$ was done in [3], while the minimization of $M(\sigma)$ is the main pursuit in this paper:

Problem (Minimum Distinct Modes):
Given an input-output string $S = (y(1), u(1)), (y(2), u(2)), \ldots, (y(n), u(n)) \in (Y \times U)^n$, find σ that solves

$$Q(\Sigma_{total}, \boldsymbol{y}, \boldsymbol{u}) : \begin{cases} \min_{\sigma \in \Sigma^*_{total}} M(\sigma) \\ \text{subject to } \forall q \in \{1, \ldots, n\} \\ \begin{cases} \sigma_{l(q)} = (k_{l(q)}, \xi_{l(q)}) \in \Sigma_{total} \\ k_{l(q)}(y(q)) = u(q) \\ \xi_{l(q)}(y(q)) = 0 \Rightarrow l(q+1) = l(q). \end{cases} \end{cases}$$

6.4 Always Interrupt Sequences

Definition (Always Interrupt Sequence):   *We will refer to any mode string $\sigma = \sigma_1 \ldots \sigma_n \in \Sigma^*_{total}$ such that*

$$\begin{cases} M(\sigma) \triangleq card\{\sigma_i \mid \sigma_i \in \sigma\} \\ \qquad \triangleq card\{l(q) \mid q = 1, \ldots, n\} \\ \qquad = \max_{y \in Y}(card\{u \mid (y, u) \in S) \\ \xi_{l(q)}(y(q)) = 1, \qquad q = 1, \ldots, n \end{cases}$$

$$(6.6)$$

$$(6.7)$$

as an Always Interrupt Sequence (AIS)[4]

Here, Equation (6.6) means that the total number of distinct modes $M(\sigma)$ used in the AIS is equal to the maximum number of different input values u associated with an output value y in the sense that (u,y) appears as an input-output pair in the data string. One direct consequence of Equation (6.7) is that the length of an AIS is equal to the length n of the input-output string it is consistent with.

Existence: *Given an input-output string $S \in (Y \times U)^n$ there always exist an Always Interrupt Sequence consistent with the data.*

Proof: The consistency of a mode string with the data S is ensured by the two conditions :

$$\forall q \in \{1,\ldots,n\}, \begin{cases} k_{l(q)}(y(q)) = u(q) \\ \xi_{l(q)}(y(q)) = 0 \Rightarrow l(q+1) = l(q) \end{cases}$$

Let M denote $\max_{y \in Y}(card\{u \mid (y,u) \in S\})$. For every $y \in Y$, there exist $m \leq M$ distinct values of u such that $(y,u) \in S$. One possible way to construct an AIS consistent with the data is to associate one distinct mode from the M available modes with each of the different values of u, i.e.

$$\forall(i,j) \text{ such that } y(i) = y(j),\ u(i) \neq u(j) \Rightarrow l(i) \neq l(j)$$

By doing so for every value of y encountered in S, we ensure that

$$\forall(i,j), \begin{cases} l(i) = l(j) \\ y(i) = y(j) \end{cases} \Rightarrow u(i) = u(j)$$

so that the first condition is met. The second condition is always met since by definition of the AIS, $\xi_{l(q)}(y(q)) = 1$, $q = 1,\ldots,n$.

Hence we have constructed an AIS that is consistent with the data S. $\Box$ One important fact should be noted here. In the proof, we proposed one particular AIS but there are many different ways to construct an AIS consistent with the data.

Example. Given the following input-output string

y	0 0 1 2 2 0 1 1 0 1 2 2 1 2 1 0 2 0 2 1
u	4 2 1 2 3 0 3 3 1 1 4 4 0 2 3 4 4 0 1 0

We have $Y = \{0,1,2\}$ and:

---

[4]Note that given a finite set C, by $card(C)$ we understand the number of different elements in C, e.g. $card(\{c_1, c_2, c_1, c_3\}) = 3$.

$$card\{u \mid (0, u) \in S\} = card\{4, 2, 0, 1, 4, 0\} = 4$$
$$card\{u \mid (1, u) \in S\} = card\{1, 3, 3, 1, 0, 3, 0\} = 3$$
$$card\{u \mid (2, u) \in S\} = card\{2, 3, 4, 4, 2, 4, 1\} = 4$$

so that an AIS will use $M = \max\{4, 3, 4\} = 4$ modes here.

As seen in the previous proof for existence, one way to build an AIS is to, for each $y \in Y$, establish an injective mapping between U and U^Y. For example, we can use:

mode	$y = 0$	$y = 1$	$y = 2$
1	$k_1(0) = 4$	$k_1(1) = 1$	$k_1(2) = 2$
2	$k_2(0) = 2$	$k_2(1) = 3$	$k_2(2) = 3$
3	$k_3(0) = 0$	$k_3(1) = 0$	$k_3(2) = 4$
4	$k_4(0) = 1$		$k_4(2) = 1$

Thus we get the following l-string:

y	0	0	1	2	2	0	1	1	0	1	2	2	1	2	1	0	2	0	2	1
u	4	2	1	2	3	0	3	3	1	1	4	4	0	2	3	4	4	0	1	0
l	1	2	1	1	2	3	2	2	4	1	3	3	3	1	2	1	3	3	4	3

and the corresponding mode sequence is $\sigma = \sigma_1 \sigma_2 \sigma_1 \sigma_1 \sigma_2 \sigma_3 \sigma_2 \sigma_2 \sigma_4 \sigma_1 \sigma_3 \sigma_3 \sigma_3 \sigma_1 \sigma_2 \sigma_1 \sigma_3 \sigma_3 \sigma_4 \sigma_3$.

Theorem: *Any mode string consistent with a given input-output string S is such that its number of modes is greater than or equal to*

$$M = \max_{y \in Y}(card\{u(q) \mid (y, u(q)) \in S, \ q \in \{1, \ldots, n\}\}).$$

Proof: Suppose that there exists a mode string σ consistent with the data using only $m < M$ modes. Consider the value of $y \in Y$ such that $card\{u(q) \mid (y, u(q)) \in S, \ q \in \{1, \ldots, n\}\} = M$ and label it y_M. In other words, there exist M different values of $u \in U$ such that $(y_M, u) \in S$. As $m < M$ there must exist two couples $(y_M, u(i))$ and $(y_M, u(j))$ in S with $u(i) \neq u(j)$ that are associated with the same mode, say $\sigma_x = \sigma_{l(i)} = \sigma_{l(j)}$. As the mode string is supposed to be consistent, we can write $k_x(y_M) = u(i)$ for the first couple and $k_x(y_M) = u(j)$ for the second one. But as $u(i) \neq u(j)$ we have a contradiction.

Consequently, any mode string consistent with a given input-output string S must use at least M modes. $\square$

Corollary. *Any AIS consistent with the data is a solution to the problem* $Q(\Sigma_{total}, \mathbf{y}, \mathbf{u})$.

Proof: To be consistent with the data, a mode string must use at least M modes. An AIS consistent with the data uses exactly M modes. Thus it solves $Q(\Sigma_{total}, \mathbf{y}, \mathbf{u})$. $\square$

Theorem. *Given an input-output sequence $S \in (Y \times U)^n$ with a minimum number of distinct modes M, the number of possible AIS is bounded above by M^{n-M}.*

Proof: First, let us consider $y_M \in Y$ such that $card\{u(q) \mid (y_M, u(q)) \in S, \ q \in \{1,\ldots,n\}\} = M$ and $S_M = \{(y_M, u(q)) \in S, \ q \in \{1,\ldots,n\}\}$. To be consistent with the data, each distinct input-output pair in S_M must correspond to a different mode. There are $P_{S_M} = M!$ ways in which this can be achieved.

Now consider the other values of y. S_M contains at least M pairs $(y_M, u(q))$ so that we now have to look at the contribution of at most $n - M$ other pairs $(y(q), u(q))$ in S. Let S_m denote the corresponding set. Each element in S_m can potentially be associated with up to M modes so that S_m can add $P_{S_m} \leq M^{n-M}$ to the total number of possibilities in a multiplicative fashion.

The total number of modes can thus be bounded by: $P = \frac{1}{M!} P_{S_M} P_{S_m} \leq M^{n-M}$, where the division by $M!$ avoids counting sequences that differ from one another by permutations of the mode indexes.

Note that this bound can be reached when S_M contains exactly M elements and S_m contains $n - M$ elements that all have a distinct value for y. $\square$ A conclusion to draw from this is that there is a large number of AIS and one question would be to pick the one that minimizes $\mathcal{S}^e(\sigma)$. However, as will be seen in the next section, there are potentially better ways of obtaining low complexity programs by abandoning the AIS structure.

6.5 Modified Mode Sequences

Here we introduce a method that reduces the length of a given AIS. The idea is to modify the interrupt function ξ of each mode and make it be equal to zero (i.e. no mode change) whenever possible. Ideally, a sequence like $\sigma = \sigma_1 \sigma_1 \sigma_2 \sigma_2 \sigma_2 \sigma_1 \sigma_1 \sigma_1 \sigma_2 \sigma_2$ could then be reduced to $\sigma' = \sigma_1 \sigma_2 \sigma_1 \sigma_2$. This method, if plausible, would not add any new modes. The resulting sequence would still use exactly M distinct modes and would thus be another solution to $Q(\Sigma_{total}, \mathbf{y}, \mathbf{u})$. But as the ξ functions are modified, the resulting mode sequence is no longer an AIS. In this matter, the resulting sequence will be referred to as a *Sometimes Interrupt Sequence* (SIS).

Algorithm (Sometimes Interrupt Sequence): *Given an AIS $\sigma = \sigma_{l(1)}, \sigma_{l(2)}, \ldots, \sigma_{l(n)}$ consistent with an input-output sequence $S = (y(1), u(1)), \ldots, (y(n), u(n))$, we construct the associated SIS by :*

1. *keeping $\xi_x(y(q)) = 1$ for all mode σ_x whenever $\exists q \in \{1,\ldots,n\}$ such that $l(q) = x$ and $l(q+1) \neq x$,*
2. *changing all the other values of ξ to zero, for all modes.*

Theorem. *The SIS derived from an AIS consistent with the data is consistent with the data.*

Proof: We recall here again the two conditions for consistency :

$$\forall q \in \{1,\ldots,n\}, \quad \begin{cases} k_{l(q)}(y(q)) = u(q) \\ \xi_{l(q)}(y(q)) = 0 \Rightarrow l(q+1) = l(q). \end{cases}$$

The modifications of the AIS mode sequence only concern the ξ functions, i.e. the interrupts. Thus, we just have to prove that the modified sequence does not violate the second consistency condition.

Suppose we have a case where $\xi_{l(q)}(y(q)) = 0$ and $l(q+1) \neq l(q)$. This is impossible as it contradicts the first step in the construction of the SIS. Thus the second condition for consistency is always met and the SIS derived from an AIS which is consistent with the data is consistent with the data as well. $\square$

Example. Consider the following input-output string and the given AIS mode sequence σ (or equivalently the l string) which is consistent with this data.

y	1	0	2	0	2	2	1	2	2	0	1	2
u	0	1	1	0	0	0	1	1	0	1	1	1
l	$1 \to 2$		$2 \to 1$		1		$1 \to 2$		$2 \to 1 \to 2$		2	2

Now construct the associated SIS :

1. The arrows in the table show us whenever $l(q) \neq l(q+1)$, i.e. whenever we need to keep $\xi_{l(q)}(y(q)) = 1$. Here, we need to keep $\xi_1(1) = 1$, $\xi_1(2) = 1$ and $\xi_2(2) = 1$.
2. So we can set $\xi_1(0) = 0$, $\xi_2(0) = 0$ and $\xi_2(1) = 0$.

Consequently, the mode switches happening at $q = 2, 4, 7, 10$ and 11 have been suppressed and the above mode string has been reduced from $\sigma = \sigma_1\sigma_2\sigma_2\sigma_1\sigma_1\sigma_1\sigma_2\sigma_2\sigma_2\sigma_1\sigma_2\sigma_2\sigma_2$ with length $N = 12$ to $\sigma = \sigma_1\sigma_2\sigma_1\sigma_1\sigma_2\sigma_1\sigma_2$ with length 7.

It can be easily shown that the action of removing one element from a mode string σ strictly reduces its specification complexity $\mathcal{S}^e(\sigma)$. The SIS is thus a mode sequence with lower complexity than the AIS it is derived from.

6.6 What Are the Ants Doing?

In this section we consider an example where ten ants *(Aphaenogaster cockerelli)* are placed in a tank with a camera mounted on top, as seen in Figure 6.2. A 52 second movie is shot from which the Cartesian coordinates, x and y, and the orientation, θ, of every ant is calculated every 33ms using a vision-based tracking software. This experimental setup is provided by Tucker Balch and Frank Dellaert at the Georgia Institute of Technology Borg Lab[5] [16].

---

[5]http://borg.cc.gatech.edu

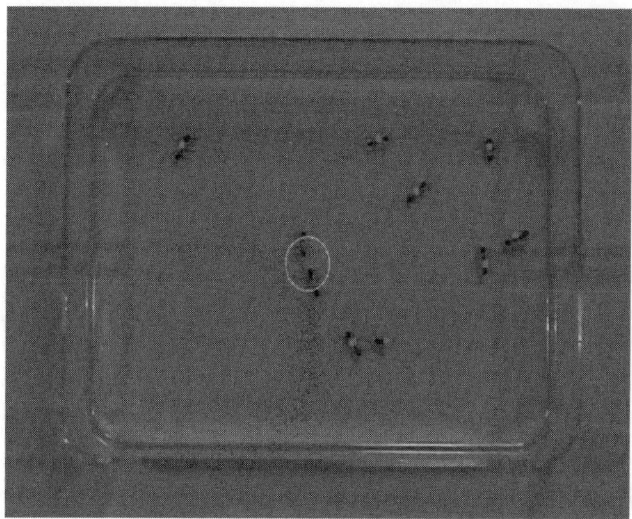

Fig. 6.2. Ten ants are moving around in a tank. The circle around two ants means that they are "docking", or exchanging information.

From this experimental data, an input-output string is constructed for each ant as follows: At each sample time k, the input $u(k)$ is given by $(u_1(k), u_2(k))$ where $u_1(k)$ is the quantized angular velocity and $u_2(k)$ is the quantized translational velocity of the ant at time k. Moreover, the output $y(k)$ is given by $(y_1(k), y_2(k), y_3(k))$ where $y_1(k)$ is the quantized angle to the closest obstacle, $y_2(k)$ is the quantized distance to the closest obstacle, and $y_3(k)$ is the quantized angle to the closest goal. Here, an *obstacle* is either a point on the tank wall or an *already visited* ant within the visual scope of the ant, and a *goal* is an ant that has not been visited recently. Figure 6.3 gives a good illustration of these notions of visual scope, goals and obstacles.

In this example, we choose to quantize $u_1(k), u_2(k), y_1(k), y_2(k)$ and $y_3(k)$ using 8 possible values for each. Thus $u(k)$ and $y(k)$ can respectively take 64 and 512 different values. For each ant, a mode sequence σ_1 with the shortest length , in the sense of [3], and the SIS associated with a particular choice of AIS σ_2 have been computed from the input-output string of length $n = 106$. Results including string length, number of distinct modes, entropy and specification complexity of these two sequences for each of the ten ants are given in Table 6.1.

In Table 6.1, results marked with a star are optimal. For σ_1, it is the length $|\sigma|$ that is minimized and for σ_2, it is the number of distinct modes $M(\sigma)$. It should however be noted that the length of σ_2 is in fact less than $n = 106$ as the mode sequence is not an AIS but a SIS.

The minimum length sequence σ_1 has been constructed using the dynamic programming algorithm given in [3], in which every element of the mode

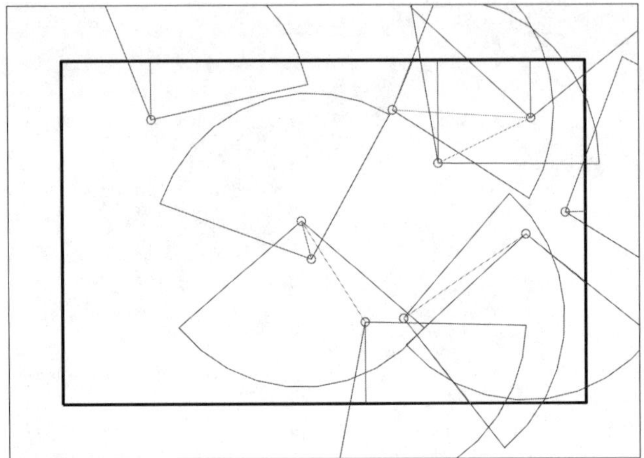

Fig. 6.3. This figure shows the conical visual scope as well as the closest obstacles (dotted) and goals (dashed) for each individual ant.

Table 6.1.

ant#	$|\sigma|$		$M(\sigma)$		$\mathcal{H}^e(\sigma)$		$\mathcal{S}^e(\sigma)$	
	σ_1	σ_2	σ_1	σ_2	σ_1	σ_2	σ_1	σ_2
1	21^*	57	21	5^*	4.4	1.4	92	82
2	34^*	66	34	5^*	5.1	1.5	172	99
3	25^*	68	25	6^*	4.6	2.0	116	139
4	33^*	64	33	6^*	5.0	1.8	166	116
5	20^*	65	20	6^*	4.3	1.9	86	121
6	26^*	73	26	6^*	4.7	1.8	122	133
7	33^*	71	33	6^*	5.0	2.0	166	145
8	19^*	74	19	7^*	4.2	2.2	80	166
9	25^*	71	25	10^*	4.6	2.4	116	169
10	23^*	60	23	4^*	4.5	1.7	104	102

sequence is a new mode. Consequently, $|\sigma_1| = M(\sigma_1)$. Moreover, the entropy of σ_1 is exactly equal to $\log_2(|\sigma_1|)$ as every mode is used only once in the sequence. The entropy of σ_2 is always smaller because the number of distinct modes is minimized and the modes are not equally recurrent in σ_2.

Finally, the specification complexity is smaller with σ_1 for five of the ten ants, and smaller with σ_2 for the five others. On the average, there is a little advantage for σ_2, with a total of 1152 bits compared to 1220 bits for σ_1.

An efficient way to ensure a low complexity coding would be to estimate both sequences for each ant and pick the one with lowest specification complexity. In our example, the total number of bits needed to encode the ten mode sequences using this coding strategy is 1064 bits.

It should be noted, however, that even though we have been able to recover mode strings, these strings can not be directly used as executable control programs without some modifications. Since the input-output string is generated from empirical data, measurement errors will undoubtedly be possible. Moreover, the dynamic system on which the control program is to be run (e.g. we have implemented mode strings obtained from the ant data on mobile robots) may not correspond exactly to the system that generated the data. Hence, a given input string might not result in the same output string on the original system and on the system on which the mode sequence is run.

For example, consider the case where we recovered the mode (k, ξ) and where the available empirical data only allows us to define the domain of k and ξ as a proper subset of the total output space Y, denoted here by Y_k or Y_ξ.[6] But, while executing this mode, it is conceivable that a measurement $y \notin Y_k$ is encountered, at which point some choices must be made. We here outline some possible ways in which this situation may be remedied:

- If $y_p \in Y_k$ and $\xi(y_p) = 0$, but the next measurement $y_{p+1} \notin Y_k$, we can replace $k(y_{p+1})$ with $k(y_p) \in U$ as well as let $\xi(y_{p+1}) = 0$. As would be expected, this approach sometimes produces undesirable behaviors, such as robots moving around indefinitely in a circular motion.
- If $y_p \notin Y_k$, but $y_p \in Y_{\tilde{k}}$ for some other mode pair $(\tilde{k}, \tilde{\xi})$ in the recovered mode sequence, we can let $k(y_p)$ be given by the most recurrent input symbol $\tilde{u} \in U$ such that $\tilde{k}(y_p) = \tilde{u}$. This method works as long as y_p belongs to the domain of at least one mode in the sequence. If this is not the case, additional choices must be made.
- If y_p does not belong to the domain of any of the modes in the sequence, we can introduce a norm on Y, and pick $\tilde{y}$ instead of y_p, where $\tilde{y}$ minimizes $\|y_p - \tilde{y}\|_Y$ subject to the constraint that $\tilde{y}$ belongs to the domain for at least one mode in the sequence.

Note that all of these choices are heuristic in the sense that there is no fundamental reason for choosing one over the other. Rather they should be thought of as tools for going from recovered mode strings to executable control programs. However, more research is needed on this topic.

6.7 Conclusions

In this paper, we present a numerically tractable solution to the problem of recovering modes from empirical data. Given a string of input-output pairs, the string with the smallest number of distinct modes that is consistent with the data is characterized algorithmically through the notion of an Always Interrupt Sequence. This has implications for how to generate multi-modal

---

[6]From the construction of the modes, these two subsets are always identical, i.e. $Y_k = Y_\xi$.

control laws by observing real systems, but also for the way the control programs should be coded. The algorithms for obtaining AIS and slightly modified derivatives of such strings can be thought of as providing a description of what modes are useful for solving a particular task, from which an empirical probability distribution over the set of modes can be obtained. This probability distribution can be put to work when coding the control programs, since a more common mode should be coded using fewer bits than an uncommon one. This work has thus a number of potential applications from teleoperated robotics, control over communication constrained networks, to minimum attention control.

Acknowledgments

The authors are grateful to Tucker Balch and Frank Dellaert for providing access to their bio-tracking data.

References

1. R.C. Arkin. *Behavior Based Robotics*. The MIT Press, Cambridge, MA, 1998.
2. K.J. Åström and B.M. Bernhardsson. Comparison of Riemann and Lebesgue Sampling for First Order Stochastic Systems. In *IEEE Conference on Decision and Control*, pp. 2011–2016, Las Vegas, NV, Dec. 2002.
3. A. Austin and M. Egerstedt. Mode Reconstruction for Source Coding and Multi-Modal Control. *Hybrid Systems: Computation and Control*, Springer-Verlag, Prague, The Czech Republic, Apr. 2003.
4. A. Bicchi, A. Marigo, and B. Piccoli. Encoding Steering Control with Symbols. *IEEE Conference on Decision and Control*, Maui, Hawaii, Dec. 2003.
5. R.W. Brockett. On the Computer Control of Movement. In the *Proceedings of the 1988 IEEE Conference on Robotics and Automation*, pp. 534–540, New York, April 1988.
6. C.G. Cassandras and S. Lafortune. *Introduction to Discrete Event Systems*. Kluwer Academic Publishers, Norwell, MA, 1999.
7. M. Egerstedt. Some Complexity Aspects of the Control of Mobile Robots. *American Control Conference*, Anchorage, Alaska, May, 2002.
8. M. Egerstedt. Motion Description Languages for Multi-Modal Control in Robotics. In *Control Problems in Robotics, Springer Tracts in Advanced Robotics* , (A. Bicchi, H. Cristensen and D. Prattichizzo Eds.), Springer-Verlag, pp. 75-90, Las Vegas, NV, Dec. 2002.
9. M. Egerstedt and R.W. Brockett. Feedback Can Reduce the Specification Complexity of Motor Programs. *IEEE Transactions on Automatic Control*, Vol. 48, No. 2, pp. 213–223, Feb. 2003.
10. E. Frazzoli. Explicit Solutions for Optimal Maneuver-Based Motion Planning. *IEEE Conference on Decision and Control*, Maui, Hawaii, Dec. 2003.
11. T.A. Henzinger. Masaccio: A Formal Model for Embedded Components. *Proceedings of the First IFIP International Conference on Theoretical Computer Science*, Lecture Notes in Computer Science 1872, Springer-Verlag, 2000.

12. J.E. Hopcroft, R. Motwani, and J.D. Ullman. *Introduction to Automata Theory, Languages, and Computation, 2nd Ed.*, Addison-Wesley, New York, 2001.
13. J.E. Hopcroft and G. Wilfong. Motion of Objects in Contact. *The International Journal of Robotics Research*, Vol. 4, No. 4, pp. 32–46, 1986.
14. D. Hristu and S. Andersson. Directed Graphs and Motion Description Languages for Robot Navigation and Control. *Proceedings of the IEEE Conference on Robotics and Automation*, May. 2002.
15. D. Hristu-Varsakelis, M. Egerstedt, and P.S. Krishnaprasad. On The Structural Complexity of the Motion Description Language MDLe. *IEEE Conference on Decision and Control*, Maui, Hawaii, Dec. 2003.
16. Z. Khan, T. Balch and F. Dellaert. *An MCMC-based Particle Filter for Tracking Multiple Interacting Targets*, Technical Report number GIT-GVU-03-35 October 2003
17. D. Kortenkamp, R.P. Bonasso, and R. Murphy, Eds. *Artificial Intelligence and Mobile Robots*. The MIT Press, Cambridge, MA, 1998.
18. V. Manikonda, P.S. Krishnaprasad, and J. Hendler. Languages, Behaviors, Hybrid Architectures and Motion Control. In *Mathematical Control Theory*, Eds. Willems and Baillieul, pp. 199–226, Springer-Verlag, 1998.
19. J. Rissanen. *Stochastic Complexity in Statistical Inquiry*, World Scientific Series in Computer Science, Vol. 15, River Edge, NJ, 1989.
20. C. E. Shannon. Mathematical Theory of Communication. *Bell Syst. Tech. J.*, Vol. 27, pp. 379-423, July and pp. 623-656, October, 1948.

7

Durable Goods and the Distribution of Wealth

Joseph W. Gruber and Robert F. Martin[*]

Division of International Finance Board of Governors – Federal Reserve System, USA

Summary. We study the role an illiquid durable consumption good plays in determining the level of precautionary savings and the distribution of wealth in a standard Aiyagari model. Transaction costs induce an inaction region over the durable good and increase the volatility of nondurable consumption. We find the level of precautionary savings is much higher (over 10 percent of wealth) and that the dispersion of wealth is slightly greater than in single asset models (Gini index of .71 for financial assets). The model is parameterized and calibrated to PSID and SCF data.

Keywords: Precautionary Savings, Wealth Distribution, Durable Goods, Prices

7.1 Introduction

In recent years, beginning with Aiyagari (1994), economists have attempted to quantify the importance of precautionary saving and to explain the distributional pattern of asset holding through models with ex-ante identical agents subject to uninsurable idiosyncratic risk. The standard model allows the agent one asset choice and abstracts from the stock of durable consumption goods. In that durable assets differ from financial assets by yielding utility directly and by being relatively less liquid, it seems plausible that a model that incorporates durable goods would yield substantially different results than one that precludes non-financial wealth.

We consider an Aiyagari-style model modified to allow for consumption of a durable good subject to transaction costs, which we take to be representative of housing. The consideration of housing, as both an illiquid asset and a consumption flow, allows us to study the interaction between the durable

---

[*]Contact: robert.f.martin@frb.gov. The views in this paper are solely the responsibility of the authors and should not be interpreted as reflecting the views of the Board of Governors of the Federal Reserve System.

J.W. Gruber and R.F. Martin: *Durable Goods and the Distribution of Wealth*, LNCIS **321**, 97–105 (2005)
www.springerlink.com   © Springer-Verlag Berlin Heidelberg 2005

good, transaction costs, and the precautionary motive for asset accumulation. We also consider the role of housing as collateral in the determination of precautionary saving. Households are free to borrow against their stock of housing, constrained only by their ability to repay with certainty. Households have their labor income and the net-of transaction costs value of the house available to pay off debt.

Transaction costs induce an inaction region over stocks of the durable good. The high frequency volatility of durable consumption declines. At low frequency, changes in the durable good are large, as households make large adjustments at the boundary of the inaction region. Durable illiquidity increases the volatility of other consumption. The decrease in the high frequency volatility of durable consumption decreases the precautionary impulse for saving, while the increase in low frequency durable volatility increases the precautionary motive. The additional volatility of non-durable consumption introduces a "committed expenditure risk" and increases the precautionary motive for saving. Thus, transactions costs have an ambiguous effect on precautionary saving, with the dominance of the offsetting effects dependent on the relative importance of the durable good in the utility function. As a result, for a broad range of parameter values, we find that transactions costs increase the level of precautionary saving over a model with a liquid durable good.[2]

We also consider the impact of transaction costs on the distribution of wealth. Standard Aiyagari-type models have a difficult time replicating the skewed distribution of wealth in the United States, largely because wealthy agents do not have a strong precautionary motive to save. The precautionary motive is concentrated among wealth-poor agents, with a flattening effect on the distribution of wealth. Not only are fluctuations to productivity (labor income) a smaller proportion of income for the very wealthy, but for households with a sufficient accumulation of wealth, the value function is essentially linear. As a result, relatively poor agents save a much higher proportion of their income, implying that poor agents accumulate assets on average much faster than wealthy agents decreasing the dispersion of the wealth distribution.

Transaction costs increase the motive for precautionary saving across the wealth spectrum, as wealthy households increase their durable stock essentially in proportion to increases in wealth. We find that transaction costs have only a modest impact on the degree of wealth dispersion, as measured by the Gini index, as the associated increase in savings is close to linear in wealth.

---

[2]Throughout the paper, we define the level of precautionary savings to be the increase in the stock of liquid assets over the stock of assets that would exist if there was no income risk. Our interest is the impact of transactions costs on precautionary saving. As such, the proper benchmark is a no-risk durable goods model without transactions costs. Also note that, in the absence of risk, the durable good plays no role in determining the equilibrium interest rate and hence no role in determining the aggregate capital stock. The Euler equation for the bond holds and is independent of the (constant) holdings of the durable good.

We parameterize the model to U.S. data. Our income process is estimated from the Panel Survey of Income Dynamics. Importantly we are also able to distinguish between labor income and business income in this data. We are able to estimate both the relative variance of the income process and the transition matrix directly from this data. Using the panel frees us from having to estimate the transition probabilities. Finally, we test the model by comparing the distribution statistics in several dimensions with the Survey of Consumer Finances.

7.2 The Model Economy

We consider an environment with a continuum of infinitely-lived agents. We normalize the total mass of agents to unity. Each agent experiences idiosyncratic labor productivity shocks. The shocks are structured such that there is no aggregate uncertainty. Agents store value in productive capital which they rent, period by period, to an aggregate technology and a durable good which provides a utility flow to the agent. The durable good is subject transaction costs. The durable good is specific to the household and can not be traded or rented without first converting it to productive capital. We restrict ourselves to steady states.

7.2.1 Endowment

Each agent is endowed with a stochastic sequence of labor endowments, $\{e_t\}_{t=0}^{\infty}$. Each agent's labor endowment follows a Markov process with transition function Q. $Q(a,c)$ gives the probability of moving to a state c if the current state is a. Each agent is endowed with an initial capital stock and an initial stock of the durable good. Since we study only steady states, the initial endowment of assets is not critical to the analysis.

7.2.2 Preferences

Households derive utility from consumption of a nondurable good and the service flow of a durable good. Let c denote consumption of the nondurable good, and h the stock of the durable good. We write the per period utility as $u(c,h)$, and total utility as $\sum_{t=0}^{\infty} \beta^t u(c_t, h_t)$.

7.2.3 Technology

Aggregate output, Y, is produced according to an aggregate constant returns to scale production function that takes as inputs capital, K, and efficient units of labor, L, $Y = F(K, L)$. The labor input is the aggregation of all household's efficiency units of labor. Capital is the aggregation of household saving. We have the following restrictions on Y, and K: 1) Y is a strictly concave function of aggregate capital. 2) There exists a maximum sustainable aggregate capital stock denoted $K_{\max}$. In order to bound preferences from below later on, we assume a lower bound for capital denoted $K_{\min}$. As a result, $K \in [K_{\min}, K_{\max}]$. Capital is assumed to depreciate at rate δ_k.

7.2.4 The Durable Good

In order to change the stock of the durable good the household must incur a destruction expense, λ_s, which is proportional to the current stock of the durable and a construction expense, λ_b, which is proportional to the new level of the durable. Once the destruction costs are incurred, durable capital may be transformed to productive capital, consumed, or returned to the stock of durables. The stock of durable capital depreciates at rate δ_h not necessarily equal to δ_k.

Durable capital accumulation lives on the interval $h \in \left[0, \frac{\max_K (F(K,L)-K)}{\max(\lambda_s+\lambda_b,\delta_h)}\right]$. If transaction costs and δ_h are all zero, h is unbounded above. We restrict attention to cases where either transaction costs or the depreciation rate is strictly positive. In addition, we assume a lower bound for durable capital denoted $h_{\min}$.

7.2.5 Market Arrangements

There are no state contingent markets for the household specific shock, e. Households holds assets $a \in [\underline{a}, \infty)$ that pay an interest rate r. We assume that households are restricted by a lower bound on their asset holdings $\underline{a}$. The lower bound arises endogenously as the quantity that ensures the household is capable of repaying its debt in all states of the world, defined as the natural limit. Because the durable can be transformed, albeit at a cost, into productive capital the agent is allowed to borrow against the durable stock. In addition, agents will be able to borrow against the present value of their labor income. In order to demonstrate the interaction between the borrowing constraint and the precautionary motive, in many of the simulations below, we will tighten the borrowing constraints of the agents. The absence of state-contingent markets and the presence of a lower bound on asset accumulation are sufficient to produce a non-trivial distribution over assets.

Households rent capital and efficient labor units to the aggregate technology each period for which they receive interest r and wage w, determined by their respective marginal products in the aggregate production function.

7.2.6 The Household's Problem

The individual household's state variables are its shock, stock of the durable good, and assets, $\{e, h, a\}$. The problem that the household solves is

$$v(e,h,a) = \max_{c,\,a',\,h'} \left\{ u(c,h') + \beta \sum_{e'} \pi_{e,e'} v(e',h',a') \right\}$$

$$\text{s.t. } a' = ew + (1+r)a + D(h',h) - c$$

$$D(h',h) = \left\{ \begin{array}{c} 0 \text{ if } h' = (1-\delta_h)\,h \\ (1-\lambda_s)\,(1-\delta_h)\,h - (1+\lambda_b)h' \text{ o/w} \end{array} \right\}$$

$$a' \in A = [\underline{a},\infty), \ h' \in H = [0,\infty), \ c \in C = [0,\infty)$$

where r and w are the return on assets and the rental rate for efficiency units of labor. $D(h',h)$ takes choices of h' given h and converts them to current period assets. A prime denotes next period's value and c is current consumption.

Under the conditions set forth above and given the equilibrium conditions set forth below for r and w, the household's problem has a solution which we denote $a' = g^a\,(e,h,a)$, $h' = g^h(e,h,a)$, and $c = g^c\,(e,h,a)$ with an upper and lower bound on asset holdings, $\{a,\ \bar{a}\}$, and on durable holdings, $\{h,\ \bar{h}\}$, such that $a \geq g^a\,(e,h,a) \geq \bar{a}$ and $h \geq g^h\,(e,h,a) \geq \bar{h}$ for all $s \in S$. Where $\{s,S\}$ are the compact notation $s = \{e,h,a\}$ and $S = \{ExHxA\}$. With respect to assets the strict concavity of the aggregate production function is sufficient.

7.2.7 Equilibrium

We are now in a position to define a steady state equilibrium. We need only add the condition that marginal productivities yield factor prices as functions of a measure of households, μ. A steady state equilibrium for this economy is a set of function of the household problem $\{v,g^a,g^h,g^c\}$, and a μ, such that: (i), Factor inputs are obtained by aggregating over households: $K = \int_S a d\mu$, and $L = \int_S e d\mu$; (ii), factor prices are factor marginal productivities, $r = F_1(K,L) - \delta_k$, and $w = F_2(K,L)$; (iii), given μ, K, and L, the functions $\{v,g^a,g^h,g^c\}$ solve the household's decision problem described earlier; (iv), the goods market clears: $\int_S \left[g^c\,(s) + g^a\,(s) + g^h\,(s)\right] d\mu = F(K,L) + (1-\delta)\,K + \int_S D(g^h\,(s),h)d\mu$, and (v), the measure of households is stationary: $\mu\,(b) = \int_S P\,(s,b)\,d\mu$, for all $b \subset B$.

7.3 Parameterization and Calibration

7.3.1 Endowment

We choose to approximate the labor endowment process from the PSID using a four state Markov process. We calculate the incomes process to be the same (in relative terms) as the median of each quartile. We scale the incomes so that the aggregate labor supply is approximately 1. The income states, transition probabilities, and other model parameters are reported in Tables 1A-1C.

Table 1A. Model Parameters

β	σ	δ_k	θ	α
.95	4	.1	.36	0.7

Table 1B. Income States

.3198	.7144	1.0885	1.789

Table 1C. Transition Matrix

.65	.22	.08	.05
.23	.55	.18	.05
.09	.17	.56	.18
.05	.04	.15	.76

The transition probabilities are computed as the average probability of transition between quartiles each year. We take the average over the years 1990-1994. We do this for the years 1990-1991, 1991-1992, 1992-1993, 1993-1994 then average the outcomes. Using the quartiles will prevent us from achieving the tails of the distribution.

7.3.2 Preferences

We choose the following per period utility function:

$$u(c,h)) = \frac{\left(c^{\alpha} h^{1-\alpha}\right)^{1-\sigma}}{1-\sigma}, \qquad \alpha \in [0,1], \ \ \sigma > 0$$

The parameter α determines the share of durables in consumption and σ is the intertemporal elasticity of substitution. For $\alpha = 1$, the model is identical to one with durable goods.

7.3.3 Technology

We choose a standard Cobb-Douglas production function with capital and labor as inputs,

$$Y = K^{\theta} L^{1-\theta}$$

7.3.4 The Durable Good

The transaction cost structure which we feel best captures the structure of U.S. housing markets is one which incorporates both fixed and proportional transaction costs. Hence, the household pays a fee to sell their existing stock of housing and then pays a second fee to purchase the new stock. The fee on the sale of the existing stock of housing is a fixed cost because the household can not affect the size of the transaction in their current choice set. The fee on the purchase of new housing acts as an asymmetric proportional transaction cost. The transaction cost depends on the size of the transaction but is also decreasing in the size of the purchase. We make both of these costs dependent on the size of the relevant transaction, e.g. selling costs are proportional to the existing stock of the durable good.

7.4 Results

We test the model by comparing the distribution statistics to wealth statistics taken from the 2001 Survey of Consumer Finances (SCF). The SCF comprises detailed wealth and asset holding information for 4442 respondents.

The distribution of assets across quintiles is depicted in Table 3. Wealth is considerably more evenly distributed in the sub-sample, both because the high end of the distribution has been curtailed by the elimination of owners of business wealth, and the low end shortened by the consideration of homeowners. Financial wealth is less evenly distributed on account of the large number of homeowners with mortgages and net negative financial positions. Housing is much more evenly distributed in the total sample than is total wealth, largely because of compression in the high end of the distribution. Table 3 also reports the percentage of total wealth held as housing equity for each quintile of the total wealth distribution, considering only those households with positive total wealth. In the total sample the importance of housing equity peaks in the third quintile, while in our home-owning sub-sample the importance of housing equity is uniformly decreasing.

Table 2. SCF Data[3]

		Quintile					Top	
	Gini	First	Second	Thrid	Fourth	Fifth	90–95%	95–99%
Total Wlth								
Total Sample	.81	−.11	1.25	4.48	11.79	82.6	12.3	25.28
Sub-Sample	.63	1.19	4.67	9.83	19.02	65.3	13.38	18.96
Asset Wlth								
Total Sample	1.18	−14.48	−2.00	0.28	6.06	110.1	16.28	32.51
Sub-Sample	2.66	−71.50	−32.13	−11.0	10.57	204.1	37.15	69.07
House Wlth								
Total Sample	.64	0	1.52	12.24	22.6	63.7	13.9	18.44
Sub-Sample	.41	4.93	10.79	15.26	22.12	46.91	11.36	12.94

---

[3]Sub-sample excludes households reporting business and other property wealth as well as households with a head of household younger than 25 or older than 55

Table 3 gives the information for three different levels of the symmetric construction and deconstruction costs. We can see that the precautionary motive is increasing in the transaction costs. The stock of liquid assets increases between 7.8 percent and 17.5 percent over the model with no income risk and the same average labor supply. Increasing the transaction costs from 1 percent to 5 percent increases the stock of capital almost 9 percent. Notice, the distribution over capital is much more disperse than the distribution over holdings of the durable good, an important empirical fact. Changes in transaction costs have a marginal effect on the dispersion in the distribution.[4] The Gini index for wealth increases a mere 1.6 percent as we move from 1 to 5 percent transaction costs. The dispersion over housing is increased more but still only moves 2.7 percent.

Table 3.

		Quintile					Top	
	Gini	First	Second	Third	Fourth	Fifth	90−95%	95−99%
House Value								
$\lambda = .01$	.2356	.0800	.1493	.2102	.2497	.3108	.0794	.0684
$\lambda = .03$	.2461	.0765	.1469	.2284	.2304	.3178	.0817	.0722
$\lambda = .05$	.2529	.0764	.1428	.2059	.2538	.3214	.0827	.0724
Asset Wealth								
$\lambda = .01$	.5963	.0052	.0343	.1070	.2356	.5999	.1595	.1687
$\lambda = .03$	.5903	.0071	.0407	.0399	.3091	.6032	.1615	.1733
$\lambda = .05$	.5998	.0070	.0409	.0994	.2301	.6226	.1676	.1730

Aggregate Stocks					
	K	H	H/TW	r	Prec Sav
$\lambda = .01$	4.12	10.52	.71	1.045	7.8%
$\lambda = .03$	4.21	10.25	.71	1.044	10.2%
$\lambda = .05$	4.49	10.19	.71	1.042	17.5%

7.5 Conclusion

We have studied the role illiquid housing markets play in determining the amount of precautionary savings and the wealth distribution in heterogeneous agents model economies with idiosyncratic uncertainty. Incorporating illiquid

---

[4]Because financial wealth is much more disperse than housing wealth. We can always increase the Gini index of total wealth by increasing the share of nondurable consumption in the utility function. The Gini index for total wealth is the weighted sum of the Gini index for housing and the Gini index for financial assets. Increasing the share, however, alters the ratio of housing to total wealth.

housing into a model with income risk increases the need for precautionary savings at all levels of wealth, and increases the level of precautionary savings substantially over the model without income risk. The level of precautionary saving is increasing in the size of the transaction costs and is decreasing in the household's ability to borrow against their durable stock. We find a larger effect of uncertainty on precautionary saving than the typical model. In a pure Aiyagari-style model, the level of precautionary savings is about 3 % of the aggregate capital stock. Our model raises the aggregate capital stock an additional 7-17% over that found in a model without risk.

Previous studies have examined models with a single type of capital. The relevant comparison between our model and those studies is in the Gini index for financial assets. In this dimension, we do quite well. Using an income process with much smaller volatility (in comparison to Castaneda et al) and a mean (aggregate labor supply of 1) comparable to Aiyagari's, we achieve a Gini index for financial assets on the order of .7. This is very close to that observed in the data once business wealth is excluded from the sample (see Table 3). Our Gini index for total wealth is only slightly higher than that found for financial assets in earlier work. The low Gini index for total wealth is a function of the very equal distribution of housing wealth in our model.

References

1. Aiyagari, S. Rao (1994) Uninsured Idiosyncratic Risk and Aggregate Savings Quarterly Journal of Economics 109, 659-684.
2. Castaneda, Ana, Javier Diaz-Gimenez, and Jose-Victor Rios-Rull (2003) Accounting for the U.S. Earnings and Wealth Inequality. Forthcoming Journal of Political Economy
3. Carroll, Chris (1992). The Buffer-Stock Theory of Saving: Some Macroeconomic Evidence. Brookings Papers on Economic Activity 2, 61-135.
4. Diaz, Antonia, Josep Pijoan-Mas, and Jose-Victor Rios-Rull (2003) Precautionary Savings and Wealth Distribution under Habit Formation Preferences. Forthcoming Journal of Monetary Economics.
5. Hubbard, G., J. Skinner, and S. Zeldes (1994). The Importance of Precautionary Motives in Explaining Individual and Aggregate Savings. Carnegie-Rochester Conference Series on Public Policy 40, 59-125.
6. Meghir, C. and G. Weber. (1996). Intertemporal Nonseparability or Borrowing Restrictions? A Disaggregate Analysis Using a U.S. Consumption Panel. Econometrica 64(5), 1151-1181.
7. Rios-Rull, Jose-Victor (1998). Computing Equilibria in Models with Heterogenous Agents. In Ramon Marimon and Andrew Scott (Eds.), Computational Methods for the Study of Dynamic Economies, Chapter 9. Oxford University Press.
8. Rodriguez, Santiago, Javier Diaz-Gimenez, Vincenzo Quadrini, and Jose-Victor Rios-Rull (2002). Updated acts on the U.S. Distributions of Earnings, Income, and Wealth. Federal Reserve Bank of Minneapolis Quarterly Review Summer 2002, Vol 26, No. 3, 2-35.

Endomorphisms of Hopf Algebras and a Little Bit of Control

Michiel Hazewinkel

CWI POBox 94079, 1090GB Amsterdam, The Netherlands
`mich@cwi.nl`

Summary. Firstly this paper summarizes some of the the relations between control theory and coalgebra and Hopf algebra theory, particularly the fact that the cofree coalgebra over a finite rank free module consists precisely of the realizable power series in noncommuting variables. The second half of the paper concerns generalizations of the Hopf algebra of permutations which in turn generalizes the Hopf algebras of noncommutative symmetric functions and the Hopf algebra of quasisymmetric functions.

Dedication. To Clyde F Martin on the occasion of his 60-th birthday and to the world of control which somehow manages to sneak in almost everywhere.

8.1 Introduction

It is relatively well known that system and control theory, mostly the part that is known as realization theory, and Hopf algebras have something to do with one another. Two independent standard references in this direction are [13, 15]. See also [14].

As a matter of fact the cofree coalgebra over a module perfectly captures on the one hand the idea of recursiveness for power series (in several noncommuting variables) and also the idea of realizability. For a precise statement see section 2 below.

This, however, is not the real subject of this talk and merely serves to point out that there are fundamental links between the worlds of Hopf algebras and that of control. A number of other relevant references here are [2, 3, 7, 10, 11, 12, 14, 19]. The most important Hopf algebra in all these papers is *Shuffle*, the Hopf algebra of all words in a given alphabeth as basis, the shuffle product as multiplication and cut as comultiplication. *Shuffle* will be described in detail below as it is also a basic ingredient in the construction of the Hopf algebras of endomorphisms which form the main topic of this talk.

M. Hazewinkel: *Endomorphisms of Hopf Algebras and a Little Bit of Control*, LNCIS **321**, 107–122 (2005)
`www.springerlink.com`

The real aim of this talk is to describe a number of very beautiful Hopf algebras in the fond and not unreasonable hope that one day they will find applications in systems and control and elsewhere.

Before that I take time out to discuss the notion of multivariable recursiveness and its relations with control (realization theory type) ideas on the one hand and free coalgebras on the other hand.

8.2 Recursiveness

Let $\{X_a : a \in I\}$ be a finite or infinite set of indeterminates indexed by a set I. Let I^* denote the monoid (under concatenation) of all words over the set I (including the empty word (which serves as the unit element)). An element of I^* is usually written in these pages as

$$\alpha = [a_1, ..., a_n] \quad \text{where } a_i \in I$$

The length of such a word is m, $\lg(\alpha) = m$. A power series in the X_a, $a \in I$ over a commutative ring with unit element k is simply a formal expression

$$f = \sum_{\alpha \in I^*} f_\alpha X_\alpha \tag{8.1}$$

where the coefficients f_α are in k and X_α is short for $X_{a_1}, X_{a_2}, ..., X_{a_n}$. The degree n component of f is

$$f^n = \sum_{\lg(\alpha)=n} f_\alpha X_\alpha \tag{8.2}$$

Equivalently, a power series is a function $I^* \to k$, $\alpha \mapsto f(\alpha) = f_\alpha$.

In the case of one variable, everyone knows it means for a power series to be recursive. There are obvious generalization to the cases of more variables. For instance:

f is left polynomially recursive if there is a finite set of monomials X_{λ_j}, $j = 1, ..., l$, with corresponding coefficients c_j such that for each large enough n

$$f^n = \sum_{j=1}^{l} c_j X_{\lambda_j} f^{n-\lg(\lambda_j)} \tag{8.3}$$

and there is similar notion of right polynomial recursiveness. These notions, however, turn out not to be general enough.

A power series f over k, see (8.1), is left recursive if there is a finite set of monomials X_{λ_i}, $i = 1, .., l$ and for some fixed $s > \max\{\lg(\lambda_i), i = 1, ..., l\}$, there are coefficients $c_{\gamma,i} \in k$, for each $i \in \{1, .., l\}$ and word $\gamma \in I^*$ of length s, such that for $n \geq s$ for each $\alpha \in I^*$ of length n

$$f(\alpha) = \sum_{i=1}^{l} c_{\alpha\,\mathrm{pre}(s),i} f(\lambda_i \alpha_{\mathrm{suf}}) \tag{8.4}$$

where if β, γ are two words over J, $\beta\gamma$ is the concatenation of them and where for a word α of length $\geq s$, $\alpha_{\mathrm{pre}(s)}$ is the prefix of α of length s and α_{suf} is the corresponding suffix (or tail), so that $\alpha = \alpha_{\mathrm{pre}(s)}\alpha_{\mathrm{suf}}$.

There is of course an entirely analogous notion of right recursive. Note that these notions of recursiveness exactly capture the idea of recursiveness in the sense that a coefficient $f(\alpha)$ for $\lg(\alpha)$ large enough is (from the left or the right) a linear combination of coefficients for words of lesser length in a uniform manner (same coefficients).

A power series (8.1) is left Schützenberger recursive if and only if there is a finite nonempty set of words S closed under taking prefixes (so that in any case the empty word is in S) such that for all words $\beta \in T = SX \setminus S$ there are coefficients $a_{\beta\alpha}$, $\alpha \in S, \beta \in T$ such that for all words γ

$$f_{\beta\gamma} = \sum_{\alpha \in S} a_{\beta,\alpha} f_{\alpha\gamma} \tag{8.5}$$

At first sight this does not look all that recursive. For one thing the word $\alpha\gamma$ occurring on the right hand side of (8.5) may very well have longer length than the word $\beta\gamma$. However, for each word ω let α be the longest prefix of ω that is in S, and write $\omega = \alpha\omega'$, then (8.5) is recursive with respect to the length of ω'. Again, there is a corresponding notion of right Schützenberger recursiveness.

Given a power series (8.1) consider the infinity by infinity matrix with rows and columns indexed by the monoid of words I^* whose entry at location (α, β) is the coefficient $f_{\alpha\beta}$. This matrix is called the Hankel matrix of f.

Let $A\langle X_\alpha : a \in I \rangle$ be the free associative algebra over k in the indeterminates X_a. The monomials X_α, $\alpha \in I^*$ form a basis (as a k-module) for $k\langle X_a : a \in I \rangle$. A power series f as in (8.1) is the same thing as a k-module morphism $k\langle X_a : a \in I \rangle \to k$, viz the one that assigns to X_α the coefficient f_α. A power series f is called representative if there is a finite number n and power series g_i, h_i, $i = 1, ..., n$ such that for each $\alpha, \beta \in I^*$

$$f_{\alpha\beta} = \sum_{i=1}^{n} g_{i\alpha} h_{i\beta} \tag{8.6}$$

The origin of the terminology is obvioius. If ρ is a representation of the monoid I^* in the $n \times n$ matrices over k then the entries of ρ as functions on I^* have property (8.6).

Finally, to end this string of definitions, the power series f is called realizable, called recognizable in the automata theory world, if there is a number N and $N \times N$ matrices $\rho(X_a)$, $a \in I$, an $N \times 1$ matrix b (column vector), an $1 \times N$ matrix c (row vector) such that for all words $\alpha = [a_1, ..., a_m]$

$$f_{\boldsymbol{\alpha}} = c\boldsymbol{\rho}(X_{a_1}) \cdots \boldsymbol{\rho}(X_{a_n})b \tag{8.7}$$

A fundamental theorem is now:

Theorem 1. *Let I be finite and let k be a Noetherian integral domain. Then the following seven properties for a power series f are equivalent*

  (i) f is left recursive
 (ii) f is right recursive
(iii) f is left Schützenberger recursive
(iv) f is right Schützenberger recursive
  (v) $H(f)$ has finite rank
 (vi) f is representative
(vii) f is realizable.

The proof is a mixture of ideas and constructions of Schützenberger, see [5, 22], Fliess, [9], and Rouchaleou, see [20]. See [16] for details.

Let $FPS(\mathcal{X})$ be the k-module of all formal power series in the set of indeterminates $\mathcal{X} = \{X_a : a \in I\}$. The polynomials in $FPS(\mathcal{X})$ are those power series for which only finitely many coefficients are nonzero[1].

By definition the submodule of rational power series, $FPS_{rat}(\mathcal{X})$, is the smallest submodule that contains the polynomials and is closed under multiplication and taking inverses when possible (which is exactly when the constant term is an invertible element of k). The famous Kleene-Schützenberger theorem now says:

Theorem 2. *A power series in a finite number of variables over a Noetherian integral domain is rational if and only if it is realizable.*

Again, see [16] for a proof.

Now consider the following decomposition structure

$$X_{\boldsymbol{\alpha}} \mapsto \boldsymbol{\mu}(X_{\boldsymbol{\alpha}}) = \sum_{\beta\gamma=\alpha} X_\beta \otimes X_\gamma$$

or more precisely

$$\boldsymbol{\mu}(f) = \sum_{\alpha,\beta\gamma=\alpha} f_{\boldsymbol{\alpha}} X_\beta \otimes X_\beta \tag{8.8}$$

As a rule it is not possible to write $\boldsymbol{\mu}(f)$ as a *finite* $\sum_j g_j \otimes h_j$ but it can happen; for instance this always happens for polynomials. As a matter of fact, practically by definition, $\boldsymbol{\mu}(f)$ can be written as such a finite sum iff f

---

[1] Warning. In case there is an infinity of indeterminates there is another possible definition of polynomials, namely those power series f for which its n-th degree component is zero for n large enough. This one is also appropriate in some contexts and is for instance used when discussing symmetric functions (in an infinity of indeterminates, which is the proper context for symmetric functions).

is representative. And then the equivalence "representative" $\Leftrightarrow$ "realizable" immediately gives that $\mu(f)$ can be written as a finite sum (with possibly different g_j, h_j)

$$\mu(f) = \sum_j g_j \otimes h_j, \quad g_j, h_j \in \mathrm{FPS}_{\mathrm{rat}}(\mathcal{X})$$

(where thereoms 2 and 1 are used to identify $FPS_{rat}(\mathcal{X}) = FPS_{representative}(\mathcal{X}) = FPS_{realizable}(\mathcal{X})$). Thus the decomposition structure induces a morphism of modules (a comultiplication)

$$\mu : FPS_{\mathrm{rat}}(\mathcal{X}) \to FPS_{rat}(\mathcal{X}) \otimes FPS_{rat}(\mathcal{X}) \tag{8.9}$$

There is also an augmentation or counit

$$\varepsilon : FPS_{rat}(\mathcal{X}) \to k, \quad f \mapsto f^0 \tag{8.10}$$

and (8.9), (8.10) combine to define a coalgebra structure on $FPS_{rat}(\mathcal{X})$ in the sense of the following definition.

Definition 1. *A coalgebra over k is a k-module C together with two k-module morphisms $\mu : C \to C \otimes_k C$ (called comultiplication) and $\varepsilon : C \to A$ (called counit) such that*

$$\begin{aligned}
(\mathrm{id} \otimes \mu)\mu &= (\mu \otimes \mathrm{id})\mu \, (coassiciativity) \\
(\varepsilon \otimes \mathrm{id})\mu &= \mathrm{id}, (\mathrm{id} \otimes \varepsilon)\mu = \mathrm{id} (\text{ counit property})
\end{aligned} \tag{8.11}$$

where all tensor products are over k and the canonical identifications $C \otimes_k k = C = k \otimes_k C$ are used.

This notion is dual to that of an algebra as is seen by writing out the definition of an algebra in diagram form and reversing all arrows. See any standard text on Hopf algebras, e.g. [1, 8, 18, 23].

The coalgebra $FPS_{rat}(\mathcal{X})$ comes with a natural morphism of k-modules

$$\pi : FPS_{rat}(\mathcal{X}) \to \prod_{a \in I} kX_a, \quad f \mapsto f^1$$

making it the cofree coalgebra over the module $\prod_{a \in I} kX_a$ in the sense of the following universal property[2]

For every coalgebra C over k and morphism of k-modules $: \varphi : C \to \prod_{a \in I} kX_a$ there is a unique morphism of coalgebras $\tilde{\varphi} : C \to FPS_{rat}(X)$ such that $\pi\tilde{\varphi} = \varphi$.

Thus $FPS_{rat}(\mathcal{X})$ is a kind of universal receptor for coalgebra structures, which goes some way to explaining its dominant role in control and automata theory.

---

[2]When I is finite this module is simply the free module on the finite basis formed by the $X\tilde{\mathrm{O}}$s. When I is infinite this is not the case.

For k a field and M the free module over k with basis $\{X_a : a \in I\}$ the theorem that $\hat{T}M_{\mathrm{repr}}$ is the cofree coalgebra over M is known as the Block-Leroux theorem, [6]. Here, if I is finite, $\hat{T}M_{\mathrm{repr}} = FPS_{repr}(\mathcal{X})$; but if I is infinite $\hat{T}M$ is a submodule (much smaller) of $FPS_{repr}(\mathcal{X})$.

There are several multiplication structures on $\mathrm{FPS}_{\mathrm{rat}}(\mathcal{X})$ that make it into a Hopf algebra (see below for this notion). For instance the shuffle product and the Hadamard product. For control the shuffle product seems to be the most relevant. [3]

For an infinity of indeterminates there are various analogues of theorems 1 and 2. All kinds of finiteness conditions come into play, not only vis à vis the definitions of recursive but also with respect to what kind of power series f are considered. See [16].

8.3 The Idea of a Hopf Algebra

Before stating the formal definition, let me try my hand at giving some sort of informal and intuitive idea of what a Hopf algebra is.

Intuitively:

Algebra = something with a composition.
Coalgebra = something with a decomposition.
Hopf algebra = something with both a composition and a decompostion in a compatible way.

The most important compatibility condition is:

Comultiplication respects multiplications (and vice versa).

This is often called the Hopf property.
It does not say that the comultiplication undoes the multiplication (or vice versa) but rather that they are sort of orthogonal to each other. Think of the comultiplication as giving all ways of composing an object into a left part and a right part and of multiplication as vertical composition. Then, in words, the condition says that all left parts of the decomposition of a product are all products of left parts of the two factors and the corresponding right parts are the products of the corresponding right parts of the decompositions of the factors involved. In pictures (stacking things vertically is composition; cutting things into left and right parts is decomposition):

---

[3]The results reported on in this section shed additional light on the relations between control theory and Hopf algebras. It is certainly significant that the realizable power series precisely form a cofree coalgebra with respect to the alphabeth concerned. This does not mean that there is now a nice coherent well organized picture of all these matters. In fact, organizing what is known from the references cited, and filling in what is missing, is a nice potential PhD level research topic.

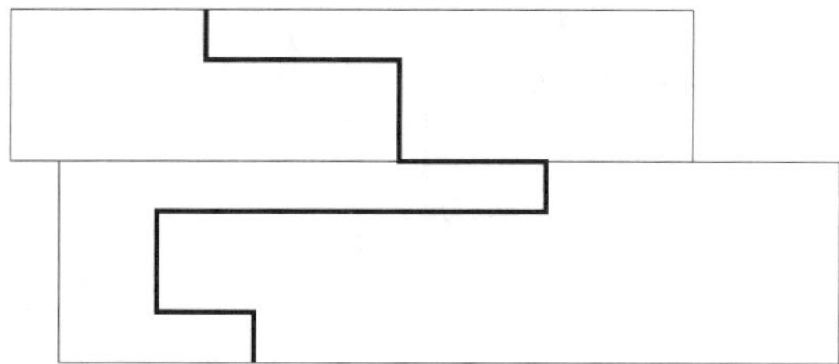

Here the top box is object 1 and the bottom one is object 2. The two boxes together are the cmposition of object 1 and object 2 (in that order). The first three segments of the jagged heavy line give a decomposition of object 1 into a left part L1 and a right part R1 and the last five segments of that heavy jagged line give a decomposition into a left part L2 and a right part R2 of object 2. The complete heavy broken line gives the corresponding decomposition of the product of object 1 and 2 into a left part that is the product of L1 and L2 and a right part that is the product of R1 and R2.

In fact there is a most important and beautiful bialgebra (which is in fact a Hopf algebra) that looks almost exactly like this. Consider stacks of rows of unit boxes.

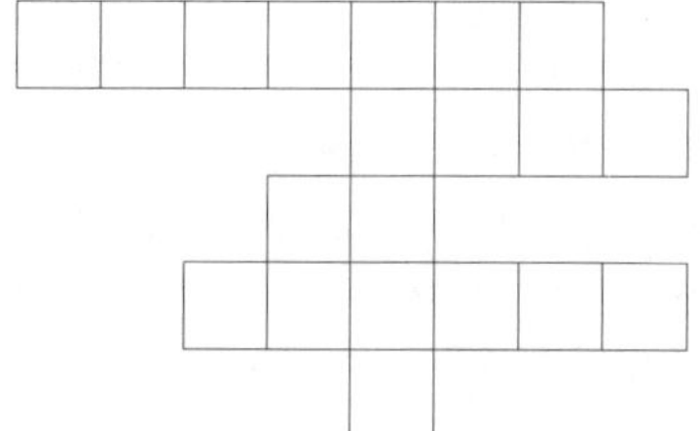

Here two such stacks are considered equivalent if they have the same number of boxes in each layer. Thus it only matters how many boxes there are in each layer. Empty layers are ignored. The case depicted is hence given by the word $[7, 4, 2, 6, 1]$ over the positive integers N. The empty word is permitted and corresponds to any stack of empty layers. The possible decompostions of a stack are obtained by cutting each layer into two parts. Two of these, for the example at hand, are indicated below.

These two correspond to the decompositions $[2, 1, 1] \otimes [5, 4, 1, 5, 1]$ and $[6, 3, 2, 3, 1] \otimes [1, 1, 3]$. A convenient way of encoding this algebraically is to consider the free associative algebra $\mathbf{Z}(Z)$ over the integers in the indeterminates $Z_1, Z_2, Z_3, \ldots$. A basis (as a free Abelian group) for this algebra is given by 1 and all (noncommutative) monomials $Z_{i_1}, Z_{i_2}, \ldots, Z_{i_m}$. This monomial encodes the stack with layers of $i_1, i_2, \ldots, i_m$ boxes. Thus the example above corresponds to the monomial $Z_7 Z_4 Z_2 Z_6 Z_1$. The comultiplication on $\mathbf{Z}(Z)$ is given by the algebra homomorphism determined by

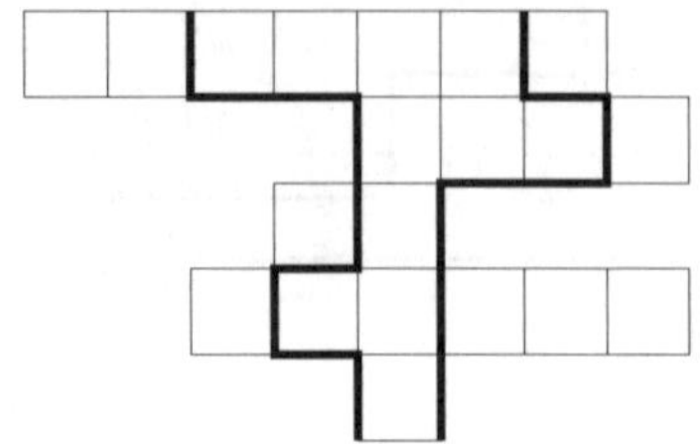

$$\boldsymbol{\mu}(Z_n) = \sum_{i+j=n} Z_i \otimes Z_j \text{ where } Z_0 = 1 \text{ and } i,j \in \mathbf{N} \cup \{0\} = \{0, 1., 2, 3, ...\}$$

(8.12)

This, together with the counit

$$\varepsilon(Z_n) = 0, n \geq 1 \tag{8.13}$$

defines the bialgebra (Hopf algebra) *NSymm* of noncommutative symmetric functions.

To conclude this section, here are the formal defintions of bialgebras and Hopf algebras.

Definition 2 (Bialgebra). *A bialgebra over k is a k-module H equiped with four module morphisms m : $H \otimes H \rightarrow H$ (multiplication, composition), e : $k \rightarrow H$ (unit element), $\boldsymbol{\mu}$: $H \rightarrow H \otimes H$ (comultiplication, decomposition), $\varepsilon : H \rightarrow k$ (counit) such that*

(i) (H, m, e) is an (associative) algebra (with unit element $e(1)$)
(ii) $(H, \boldsymbol{\mu}, \varepsilon)$ is a (coassociative) coalgebra (with counit, see (8.11) above)
(iii) $\boldsymbol{\mu}$ and ε are algebra morphisms

or, equivalently (iii)' m and e are coalgebra morphisms.

Definition 3 (Hopf algebra). *A Hopf algebra over k is a bialgebra H with one extra morphism $\boldsymbol{\iota} : H \rightarrow H$, called antipode, such that*

$$m(\boldsymbol{\iota} \otimes \mathrm{id})\boldsymbol{\mu} = \mathrm{eboldsymbol}\varepsilon, \mathrm{m}(\mathrm{id} \otimes \boldsymbol{\iota})\boldsymbol{\mu} = \mathrm{e}\varepsilon$$

The morphism $\boldsymbol{\mu}$is something very much like the inverse in group theory.

8.4 The Shuffle Hopf Algebra

Consider the monoid N^* all words over the alphabeth $\mathbb{N}$ of positive integers (including the empty word) and let *Shuffle* be the free **Z**-module (Abelian group) with basis $\mathbb{N}^*$. The shuffle product of two words $\boldsymbol{\alpha} = [a_1, a_2, ..., a_m]$ and $\boldsymbol{\beta} = [b_1, b_2, ..., b_n]$ is defined as follows. Take a sofar empty word with $n+m$ slots. Choose m of the available $n+m$ slots and place in it the natural numbers from $\boldsymbol{\alpha}$ in their original order; place the entries from $\boldsymbol{\beta}$ in their original order in the remaining n slots. The product of the two words $\boldsymbol{\alpha}$ and $\boldsymbol{\beta}$ is the sum (with multiplicities) of all words that can be so obtained. So, for instance

$$[a,b] \times_{\mathrm{sh}} [c,d] = [a,b,c,d] + [a,c,b,d] + [a,c,d,b]$$
$$+ [c,a,b,d] + [c,a,d,b] + [c,d,a,b]$$
$$[1] \times_{\mathrm{sh}} [1,1,1] = 4[1,1,1,1]$$

This defines a commutative associative multiplication on *Shuffle* for which the empty word is a unit element. Moreover with cut as a comultiplication

$$\boldsymbol{\mu}([a_1, a_2, ..., a_m]) = \sum_{i=0}^{m} [a_1, ..., a_i] \otimes [a_{i+1}, ..., a_m]$$

counit $\varepsilon([\,]) = 1$, $\varepsilon(\boldsymbol{\alpha}) = 0$ if $\lg(\boldsymbol{\alpha}) \geq 1$, where the length of a word $\boldsymbol{\alpha} = [a_1, ..., a_m]$ is $\lg(\boldsymbol{\alpha}) = m$, and antipode

$$\iota([a_1, a_2, ..., a_m] = (-1)m[a_m, a_{m-1}, ..., a_1]$$

Shuffle becomes a Hopf algebra. It is not difficult to verify this directly. Some remarks re *Shuffle*

-   Over a field of characteristic zero *Shuffle* is free. Over the integers this is definitely not the case
-   There is a natural quantum version, *qShuffle* which interpolates between concatenation and *Shuffle* as regards its algebra structure.

As to both remarks I do not know what they could imply for control.

8.5 The Hopf Algebra of Permutations of Reutenauer, Malvenuto, Poirier (the MPR Hopf Algebra)

As an Abelian group MPR it is the free Abelian group with as basis all permutations on n letters, all$n$

$$MPR = \oplus_{n=0}^{\infty} \mathbf{Z}S_n,$$
$$\mathbf{Z}S_0 = \mathbf{Z}, \mathbf{Z}S_1 = \mathbf{Z}[1], \mathbf{Z}S_2 = \mathbf{Z}[1,2] \oplus \mathbf{Z}[2,1], ... \tag{8.14}$$

Permutations are written as special words

$$\sigma = \begin{pmatrix} 1 & 2 & \cdots & m \\ a_1 & a_2 & \cdots & a_m \end{pmatrix} = [a_1, a_2, ..., a_m]$$

Thus a word over the natural numbers is a permutation if and only if it is injective (no letter occurs more than once) and surjective (there are no gaps, meaning that if $\mathrm{supp}(\boldsymbol{\alpha}) = \{a_1, ..., a_m\}$ is the support of $\boldsymbol{\alpha}$ then $\mathrm{supp}(\boldsymbol{\alpha})$ is an initial interval $\{1, ..., m'\}$ for some $m' \in \mathbf{N}$).

Multiplication example

$$[2,1] \times_{MPR} [1,2] = [2,1,3,4] + [2,3,1,4] + [2,3,4,1] + [3,2,1,4] +$$
$$[3,2,4,1] + [3,4,2,1]$$

Comultiplication example

$$\boldsymbol{\mu}_{MPR}([2,4,3,1] = [] \otimes [2,4,3,1] + [1] \otimes [3,2,1] + [1,2] \otimes [2,1] +$$
$$[1,3,2] \otimes [1] + [2,4,3,1] \otimes []$$

The rules behind these examples are as follows. Given a word $\boldsymbol{\alpha} = [a_1, a_2, ..., a_m]$ without repeated letters let $\mathrm{st}(\boldsymbol{\alpha}) = [b_1, ..., b_m]$ be the permutation word on m letters defined by $b_i < b_j \Leftrightarrow a_i < a_j$. For instance $\mathrm{st}([6,8,3,5]) = [3,4,1,2]$. Then the comultiplication on MPR is given by

$$\boldsymbol{\mu}_{MPR}([a_1, ..., a_m]) = \sum_{j=1}^{m} \mathrm{st}([a_1, ..., a_j]) \otimes \mathrm{st}([a_{j+1}, ..., a_m]) \tag{8.15}$$

The multiplication is easier to describe

$$m_{MPR}([a_1, ..., a_m] \otimes [b_1, ..., b_n]) = [a_1, ..., a_m] \times_{sh} [m + b_1, ..., m + b_n] \tag{8.16}$$

There is also an antipode. Using the grading, see (8.14), it is easy to prove that an antipode exists; writing down an explicit formula is a different matter, but see [4].

Once the formulas have been guessed it is not really difficult to prove that MPR is indeed a Hopf algebra. The question is more, where do these formulas come from and are there natural generalizations. The next sections give answers to these questions.

8.6 The Word Hopf Algebra, WHA

As an Abelian group WHA is the free Abelian group on all words over the natural numbers (including the emtpy word (which serves as the identity). An example of the multiplication formula is

$$[2,3,2] \times_{WHA} [4,1] = [2,3,2,7,4] + [2,3,7,2,4] + [2,7,3,2,4] + [7,2,3,2,4]$$
$$+ [2,3,7,4,2] + [2,7,3,4,2] + [7,2,3,4,2] + [2,7,4,3,2]$$
$$+ [7,2,4,3,2] + [7,4,2,3,2]$$
$$= [2,3,2] \times_{sh} [7,4]$$

So the multiplication is again a shifted shuffle product with the shift equal to $ht(\boldsymbol{\alpha}) = \max\{a_1, ..., a_m\}$. The comultiplication is more mysterious. Here is an example

$$\boldsymbol{\mu}_{WHA}([3,3,2,2,2,5,9,6]) = [\ ] \otimes [3,3,2,2,2,5,9,6] + [1,1] \otimes [2,2,2,4,8,5]+$$
$$+ [3,3,2,2,2] \otimes [2,6,3] + [3,3,2,2,2,5] \otimes [4,1]$$
$$+ [3,3,2,2,2,5,8] \otimes [1] + [3,3,2,2,2,5,9,6] \otimes [\ ]$$

and I do not think it is easy to guess what could be the general recipe behind this formula.

8.7 Convolution

The multiplication parts of MPR and WHA are relatively easy to understand. Let H be a Hopf algebra and let φ, ψ, be two endomorphisms of H (as Abelian groups, or vector spaces, of k-modules, as the case may be). Then a new endomorphism is obtained by convolution[4] as follows

$$H \xrightarrow{\mu} H \otimes H \xrightarrow{\varphi \otimes \psi} H \otimes H \xrightarrow{m} H \tag{8.17}$$

Now take $H = $ Shuffle and interpret a permutation $\sigma \in S_m$ as the endomorphism of *Shuffle* that takes any word $\boldsymbol{\alpha} = [a_1, ..., a_m]$ of length m into $[a_{\sigma(1)}, ..., a_{\sigma(m)}]$ and is zero on all other words. Thus, if $[s_1, ..., s_m]$ is the word describing the permutation a word $\boldsymbol{\alpha} = [a_1, ..., a_m]$ gets taken into $[a_{s_1}, ..., a_{s_m}]$. Then it is easy to see that convolution exactly gives the multiplication on MPR.

This also works for WHA for several different interpretations of words as endomorphisms of *Shuffle*.

For instance there is the very natural the interpretation that says that a word $\boldsymbol{\alpha} = [a_1, ..., a_m]$ acts on words from *Shuffle* as follows.

Let $ht(\boldsymbol{\alpha}) = \max\{a_i\}$. Then $\boldsymbol{\alpha}$ acts on words of length unequal to $ht(\boldsymbol{\alpha})$ as zero and a word $[u_1, ..., u_{ht(\boldsymbol{\alpha})}]$ is taken to $[u_{a_1}, u_{a_2}, ..., u_{a_m}]$.

There are other interpretations that one can think of, but for the multiplication this one works, as do some others (meaning that they induce the same multiplications. As it will turn out the interpretation just given for letting arbitrary words act as endomorphisms of *Shuffle* is not the right one.

8.8 Coconvolution

Everything in the Hopf algebra world tends to have something like a dual version. So let's try coconvolution. Given an endomorphim φ of a Hopf algebra H look at

$$H \otimes H \xrightarrow{m} H \xrightarrow{\varphi} H \xrightarrow{\mu} H \otimes H$$

---

[4]In the sace of certain Hopf algebras coming from groups this can indeed be seen to be classical convolution, see [18].

This give an element of $End(H \otimes H)$ which (for infinite dimensional things like *Shuffle*) is quite a different (much larger) object than $End(H) \otimes End(H)$.[5] Still there might be a suitable projection or something. Let's try an example

$$[u1, u2] \times_{\mathrm{sh}} [v_1, v_2, v_3] = [u_1, u_2, v_1, v_2, v_3] + [u_1, v_1, u_2, v_2, v_3]$$
$$+ \cdots + [v_1, v_2, v_3, u_1, u_2]$$

which under the permutation $\sigma = [3, 1, 4, 5, 2]$ goes to

$$[v_1, u_1, v_2, v_3, u_2] + [u_2, u_1, v_2, v_3, v_1] + \cdots + [v_3, v_1, u_1, u_2, v_2]$$

Apply cut to this and look for terms with support $\{u_1, u_2\} \otimes \{v_1, v_2, v_3\}$. There is just one such, viz

$$[u_2, u_1] \otimes [v_2, v_3, v_1] = [2, 1] \otimes [2, 3, 1]([u_1, u_2] \otimes [v_1, v_2, v_3])$$

which fits with the comultiplication on MPR. Indeed the (length $= 2$) tensor (length $= 3$) component of $\mu([3, 1, 4, 5, 2])$ is

$$\mathrm{st}([3, 1]) \otimes \mathrm{st}([4, 5, 2]) = [2, 1] \otimes [2, 3, 1]$$

And, indeed, quite generally, this procedure produces formula (8.15). It is, however, a quite dodgy procedure. There is no guarantee whatever that the resulting comultiplication will indeed yield a Hopf algebra. In this case it does. In other cases it does not, as in the example discussed in the next section.

8.9 Generalization to Words (Not Just Permutations)?

There is a perfectly nice way, already used, to make any word, not just permutations, act on *Shuffle*.

Let $ht(\alpha) = \max\{a_i\}, \alpha = [a_1, a_2, ..., a_m]$. Then acts on words of the form $[u_1, ..., u_{ht(\alpha)}]$ by taking such a word to $[u_{a_1}, ..., u_{a_m}]$.

Does this also give a nice Hopf algebra. Perhaps WHA from above.

Answer: NO, the resulting comultiplication does not satisfy the Hopf property with respect to convolution.

---

[5]When H is a free finite rank module over k, $End(H) = H \otimes H^*$ and $End(H \otimes H) = End(H) \otimes End(H)$. Then coconvolution defines a comultiplication on $End(H) = H \otimes H^*$. Moreover, convolution and coconvolution combine to define a Hopf algebra structure on $End(H)$. The resulting Hopf algebra is in fact the tensor product of the Hopf algebra H and its dual H^*.

This Hopf algebra comes with a second multiplication (composition of endomorphisms) and a second comultiplication (cocomposition of morphisms). There are also a unit and counit for this, viz

evaluation : $H \otimes H^* \to k$ and its dual coevaluation. The second multiplication (comultiplication) may or may not be distributive over the first one. But in any case these special Hopf algebras have a richer structure than the standard ones. As far as I know nothing much has been done in the way of systematic investigation of these Hopf algebras of endomorphisms.

8.10 The Right Generalization to Words

Here is a sufficiently complicated example

$$[9,10,9,5,3,13,12,3] \leftrightarrow \begin{pmatrix} [u_1,u_1,u_1,u_2,u_2,u_3,u_3,u_3,u_3,u_4,u_5,u_5,u_6] \\ [u_3,u_4,u_3,u_2,u_1,u_6,u_5,u_1] \end{pmatrix}$$

$$(8.18)$$

This works as follows. First look at the alphabeth of letters occurring in the left hand side. The smallest integer occurring is 3. This skips 1,2 and together with 3 accounts for 3 u_1's; the next one is 5, skipping 4, given rise to 2 u_2's; the next one is 9, skipping 6,7,8, giving rise to 4 u_3's; then comes 10, nothing skipped so 1 u_4; then 12, skipping 11, so 2 u_5's; and finally 13, nothing skipped, so 1 u_6 . This describes the first row of the right hand side of the correspondence. The second row is obtained by writing down the 9-th, 10-th, 9-th, 5-th, 3-rd, 13-th, 12-th, 3-rd item from the top row.

It is clear how to go from the right hand side fo (8.18) to the left hand side; the top row determines the support of the left hand side and then the bottom row determines the word itself.

The meaning is that a word like the left hand side acts on *Shuffle* as zero on all basis elements except those of the form like the right hand side of (8.18) (for the case at hand) and acts as indicated on the right hand side on basis elements of the shape indicated.

It is somewhat important that we are really dealing with recipes for endomorphisms rather than with concrete endomorphisms. This means that a word acts on all words of one specified shape in a uniform way.

Now apply the same dodgy procedure of section 8 above and the result is a Hopf algebra WHA. Proving that it is actually a Hopf algebra is a bit of a mess though. Fortunately there is a further generalization for which the proof is much easier and which contains this one as a sub Hopf algebra.

8.11 The Double Word Hopf Algebra, $2WHA$

The most general, so far, of these Hopf algebras is the double word Hopf algebra, $2WHA$. It is defined as follows (over the integers or any other base ring). A basis for the underlying free module is formed by pairs of words in indeterminates $x_1, x_2, \ldots$

$$q = \begin{pmatrix} \gamma \\ \gamma' \end{pmatrix}, \quad \mathrm{supp}(\gamma) = \mathrm{supp}(\gamma')$$

$$(8.19)$$

These are symbolic substitutions, not elements of e.g. *Shuffle*. Only the pattern matters, not the actual symbols used. For instance

$$\begin{pmatrix} [x_1,x_1,x_2] \\ [x_1,x_2,x_1] \end{pmatrix} = \begin{pmatrix} [y_1,y_1,z_2] \\ [y_1,z_2,y_1] \end{pmatrix}$$

The multiplication is defined as follows: Let

$$p = \begin{pmatrix} \beta \\ \beta' \end{pmatrix}, \quad \mathrm{supp}(\beta) = \mathrm{supp}(\beta')$$

and

$$q = \begin{pmatrix} \gamma \\ \gamma' \end{pmatrix}, \quad \mathrm{supp}(\gamma) = \mathrm{supp}(\gamma')$$

be two symbolic word transformations (recipes for transforming words). First, if necessary rewrite them so that there are no common symbols between the two:

$$\mathrm{supp}(\beta) \cap \mathrm{supp}(\gamma) = \emptyset.$$

Then the product is

$$m(p \otimes q) = \begin{pmatrix} \beta * \gamma \\ \beta' \times_{sh} \gamma' \end{pmatrix} \tag{8.20}$$

Now let's define the comultiplication: A good cut of a symbolic word β is a cut $\beta = \beta' * \beta''$ such that

$$\mathrm{supp}(\beta') \cap \mathrm{supp}(\beta'') = \emptyset$$

The comultiplication is now

$$\mu(q) = \sum_{\substack{\text{good cuts,} \\ \gamma' * \gamma'' = \gamma'}} \begin{pmatrix} q^{-1}(\gamma'') \\ \gamma'' \end{pmatrix} \otimes \begin{pmatrix} q^{-1}(\gamma''') \\ \gamma''' \end{pmatrix} \tag{8.21}$$

Here $q^{-1}(\gamma'')$ is that subword of γ that is obtained by removing all letters that are not in γ''. This is a Hopf algebra[6] that vastly generalizes MPR and it has a lot of similar extra structure (and much more besides). First of all there are obvious inclusions

$$PMR \subset WHA \subset 2WHA$$

There is a (nondefinite) nondegenerate inner product:

$$\left(\begin{pmatrix} \beta \\ \beta' \end{pmatrix}, \begin{pmatrix} \gamma \\ \gamma' \end{pmatrix} \right) = \delta_{\beta\gamma'}\delta_{\beta'\gamma} \tag{8.22}$$

The Hopf algebra $2WHA$ is autodual with respect to this inner product, and so is its sub Hopf algebra MPR.

There are a second multiplication (composition of endomorphisms) and a second comultiplication (cocomposition of endomorphisms) with unit and counit. There is room for all kinds of endomorphisms and automorphisms

---

[6]The proof is striaghtforward and that immediately also gives profs that MPR and WHA are Hopf algebras.

(Frobenius-like, Verschiebung-like). Unlike in the case of MPR. There quite a few natural sub Hopf algebras. For instance the ones composed of surjective pairs of words, of injective pairs of words, or the one for which the supports of γ and γ' are equal including multiplicites, or combinations of these restrictions. And there are a large number of natural Hopf algebra morphisms between all these; both inclusions and projections coming from various standardization maps, see [21]. ...... All this comes from looking at endomorphism recipes for *Shuffle* and its dual Hopf algebra, *Lie Hopf.* There are other Hopf algebras, such as the dual pair $NSymm$, $QSymm$, for which this endomorphism recipe idea can be carried out (including $2WHA$ itself). This remains totally uninvestigated at this time.

See [17] for much more detail on Hopf algebras of endomorphisms of Hopf algebras and proofs.

References

1. Eiichi Abe, *Hopf algebras*, Cambridge Univ. Press, 1977, 284pp.
2. A A Agrachev, R V Gamkrelidze, *Chronological algebras and nonstationary vectorfields*, J Soviet Math **17**:1(1979), 1650-1675.
3. A A Agrachev, R V Gamkrelidze, *Volterra series and permutation groups*, J. of the Mathematical Sciences **71**:3(1994), 2409-2433. Translated from: Itogi Nauki i Tekhniki, Seriya Sovremennye Problemy Matematiki, Noveishie Dostizheniya, Vol. 39 (1991), 3-40.
4. Marcelo Aguiar, Frank Sottile, *Structure of the Malvenuto-Reutenauer Hopf algebra of permutations*, Preprint, Dept. Math. Texas A&M University, Dept Math, University of Massachusetts at Amherst, 2002.
5. Jean Berstel, *Séries rationelles*. In: Jean Berstel (ed.), Séries formelles en variables non commutatives et applications, LITP Paris, 1978, 5-22.
6. Richard E. Block, Pierre Leroux, *Generalized dual coalgebras of algebras with applications to cofree coalgebras*, J. pure and appl. Algebra **36**(1985), 15 -21.
7. Kuo-Tsai Chen, *Iterated path integrals*, Bull. Amer. Math. Soc. **82**:5(1977), 831-879.
8. Sorin Dascalescu, Constantin Nastasescu, Serban Raianu, *Hopf algebras. An introduction*, Marcel Dekker, 2001.
9. Michel Fliess, Matrices de Hankel, J. Math. Pures et Appl. 53(1974), 197-224. Erratum: ibid. **54** (1975), 481.
10. Michel Fliess, *Fonctionelles causales non linéaires et indéterminées non commutatives*, Bull. Soc. Math. France **109**(1981), 3-40.
11. Robert Grossman, Richard G Larson, *Hopf-algebraic structure of families of trees*, J of Algebra **126**(1989), 184 - 210.
12. Robert Grossman, Richar G Larson, *Hopf -algebraic structure of combinatorial objects and differential operators*, Israel J. Math, **72**(1990), 109-117.
13. Robert Grossman, Richard G Larson, T*he realization of input-output maps using bialgebras*, Forum Math. 4(1992), 109 - 121.
14. Robert Grossman, Richard G Larson, *An algebraic approach to hybrid systems*, Theoretical Computer Science **138**(1995), 101 - 112.

15. Luzius Grunenfelder, *Algebraic aspects of control systems and realizations*, J. of Algebra **165**:3(1994), 446-464.
16. Michiel Hazewinkel, *Cofree coalgebras and multivariable recursiveness*, J. pure and appied Algebra **183**(2003), 61-103.
17. Michiel Hazewinkel, *Hopf algebras of endomorphisms of Hopf algebras*, Preprint, CWI, 2004. In preparation.
18. Michiel Hazewinkel, Nadia Gubareni, Volodymir Kirichenko, *Algebras, rings, and modules*. Volume 2, KAP, 2005, to appear.
19. M Kawski, *Chronological algebras: combinatorics and control*, J. Mathematical Sciences **103**:6(2001), 725-744.
20. Y Rouchaleou, B F Wyman, R E Kalman, *Algebraic structure of linear dynamical systems. III: Realization theory over a commutative ring*, Proc. National Acad. Sci (USA) **69**(1972), 5404-5406.
21. C Schensted,*Longest increasing and decreasing subsequences*, Canadian J. Math. **13**(1961), 179-191.
22. M P Schützenberger, *On the definition of a family of automata*, Information and control **4**(1961), 215-270.
23. Moss E. Sweedler, *Hopf algebras*, Benjamin, 1969, 336pp.

Conditioned Invariant Subspaces and the Geometry of Nilpotent Matrices

Uwe Helmke[1] and Jochen Trumpf[2]

[1] Mathematisches Institut, Universität Würzburg, Am Hubland, 97074 Würzburg, Germany
`helmke@mathematik.uni-wuerzburg.de`

[2] Department of Information Engineering, The Australian National University, Canberra ACT 0200, Australia, and National ICT Australia Ltd.[†]
`Jochen.Trumpf@anu.edu.au`

Dedicated to Clyde F. Martin, on occasion of his 60th birthday

Summary. The focus of this work is on certain geometric aspects of the classification problems for invariant and conditioned invariant subspaces. In this paper, we make an attempt to illustrate the interplay between geometry and control, by focussing on the connections between partial state observers, spaces of invariant and conditioned invariant subspaces, and nilpotent matrices.

9.1 Introduction

Invariant subspaces of linear operators have long played an important role in pure and applied mathematics, including areas such as e.g. operator theory and linear algebra [7],[8],[11], as well as algebraic groups, representation theory and singularity theory [4],[25],[27],[28]. Their role in control theory in connection with matrix Riccati equations and linear optimal control is now well-understood and has been the subject of extensive research during the past decades; see e.g. [22],[16]. We also mention the important connection to geometric control theory and the fundamental concept of conditioned and controlled invariant subspaces [31]. In fact, the so-called "quaker lemma" completely characterizes conditioned invariant subspaces solely in terms of invariant subspaces through the concept of a friend.

[†]National ICT Australia Ltd. is funded by the Australian Government's Department of Communications, Information Technology and the Arts and the Australian Research Council through Backing Australia's Ability and the ICT Centre of Excellence Program.

U. Helmke and J. Trumpf: *Conditioned Invariant Subspaces and the Geometry of Nilpotent Matrices*, LNCIS **321**, 123–163 (2005)
`www.springerlink.com`

The focus of this work is on certain geometric aspects of the classification problems for invariant and conditioned invariant subspaces. The investigation of the geometry of the algebraic variety $\mathrm{Inv}_k(A)$ of k-dimensional A-invariant subspaces of a vector space V goes back to the early work of Steinberg [27]. Motivated by applications to singularity theory, Steinberg raised the problem of analyzing the geometry of $\mathrm{Inv}_k(A)$ and derived important results. He, as well as Springer [25], showed that the geometry of $\mathrm{Inv}_k(A)$ could be used to construct resolutions of singularities for the set of nilpotent matrices. It also led to a new geometric construction of Weyl group representations [26]. Subsequent work by Spaltenstein [24] and others established basic geometric facts about $\mathrm{Inv}_k(A)$, such as the enumeration of irreducible components via Young diagrams, or the computation of topological invariants, such as Borel-Moore homology groups and intersection homologies [3],[13].

Control theory has provided a different and new entry point to this circle of ideas, as was first realized by Shayman, and Hermann and Martin [22],[16],[18]. In fact, the projective variety $\mathrm{Inv}_k(A)$ can be interpeted as a compactification for the solution set of the algebraic Riccati equation and this link deepened considerably the further understanding of the Riccati equation. In [21],[22], Shayman studied the geometric properties of the solution set of the algebraic Riccati equations, by connecting it to the geometry of $\mathrm{Inv}_k(A)$ and the Grassmann manifold. Interesting applications of the Grassmannian approach to numerical linear algebra appeared in [1] and a whole circle of ideas, centering around nilpotent matrices, representations of the symmetric group, Schubert cycles and the classification of state feedback orbits, has been masterfully presented in [12]. Already in the late 70s, the link between invariant subspaces and geometric control objects, such as conditioned invariant subspaces, was well understood. A driving force for their analysis has been their ubiquitous role in e.g. spectral factorization, linear quadratic control, H_∞ and game theory, as well as observer theory, filtering and estimation. However, it is only until recently, that first attempts have been made towards a better understanding of the geometry of the set of conditioned invariant subspaces $\mathrm{Inv}_k(C, A)$; see [9],[10],[19],[6],[17]. The recent Ph.D. thesis [29] contains a comprehensive summary. Nevertheless, it is fair to say that our current knowledge on $\mathrm{Inv}_k(C, A)$ remains limited, with several basic questions unsolved. For example, it is not known, whether or not $\mathrm{Inv}_k(C, A)$ is homotopy equivalent to $\mathrm{Inv}_k(A)$, or if $\mathrm{Inv}_k(C, A)$ is a manifold. Generally speaking, the interactions between linear systems theory and geometry or physics, despite first steps [14],[23], have not been explored to the depth that they deserve and remain a challenging task for future research. We are convinced, that conditioned invariant subspaces are bound to play an important role here.

In this paper, we make an attempt to illustrate the interplay between geometry and control, by focussing on the connections between partial state observers, spaces of invariant and conditioned invariant subspaces, and nilpotent matrices. The two crucial players in our story are on the one hand the set of pairs (A, V) of linear operators and invariant flags, and on the other hand the set of pairs (J, V), of friends J and conditioned invariant flags V, for a given observable pair (C,A) in dual Brunovsky form. We prove that, despite the actual and potential singularities of $\mathrm{Inv}_k(A)$ and $\mathrm{Inv}_k(C, A)$, respectively, these sets of pairs are actually smooth manifolds, being closely related to classical geometric objects such as the cotangent bundle of the flag manifold. After having introduced these spaces and established their manifold property, we then consider desingularizations of the set of nilpotent matrices. Here we make contact with symplectic geometry and the moment map. State feedback transformations enable us to construct suitable transversal slices to nilpotent similarity orbits. Our construction of these Brunovsky slices extends that of Steinberg [28]. The intersections of the similarity orbits with the variety of nilpotent orbits exhibit interesting singularities, including Kleinian singularities of complex surfaces. We show, at least in a generic situation, that the proposed desingularization via conditioned invariant subspaces restricts to a simultaneous desingularization for all the intersection varieties. We regard this as one of our most surprising results, i.e., that conditioned invariant subspaces provide a natural desingularization for Kleinian singularities. Unfortunately, we cannot really give a deeper explanation of this fact, as this would require to extend the parametrization of conditioned invariant subspaces to systems defined on arbitrary semi-simple Lie groups. The link between linear systems and geometry is also visible at the proof level. In fact, in order to construct a miniversal deformation of the variety of nilpotent matrices via Brunovsky slices, we have to extend the well–known fact (commonly referred to as the *Hermann-Martin Lemma*) that state feedback transformations define a submersion on the state feedback group.

From what has been said above, it becomes evident that the pioneering contributions of Clyde F. Martin to systems theory had a great impact on a wide range of research areas. The influential character of his work on other researchers is also visible in this paper at crucial steps. In fact, Clyde has always been a source of inspiration, and a friend. It is a great pleasure to dedicate this paper to him.

9.2 Smoothness Criteria for Vector Bundles

A well-known consequence of the implicit function theorem is that the fibres $f^{-1}(y)$ of a smooth map $f : M \longrightarrow N$ are smooth manifolds, provided the rank of the differential $Df(x)$ is constant on a neighborhood of $f^{-1}(y)$. The situation becomes more subtle if the rank of $Df(x)$ is assumed to be constant only *on* the fibre, but is allowed to vary outside of it. Then we cannot conclude that $f^{-1}(y)$ is smooth and more refined techniques than the implicit function theorem are needed for. In fact, this is exactly the situation that arises when one tries to prove that certain families of conditioned invariant subspaces are smooth manifolds. One technique that can be applied in such a situation where the implicit function theorem fails is by realizing the space as an abstract vector bundle X over a base manifold, suitably embedded into a smooth vector bundle. Then one can try to employ group action arguments to prove that X is indeed a smooth vector bundle. This therefore requires an easily applicable criterion, when a vector subbundle qualifies as a smooth vector bundle. In this section, we will derive a sufficient condition for a (topological) vector bundle to be a smooth vector bundle. The result also provides a sufficient condition in certain situations for the pre-image of a smooth submanifold being a smooth submanifold again. Second, quotients of smooth vector bundles with respect to free and proper Lie group actions are shown to be smooth vector bundles. Finally, we give a sufficient criterion for a projection map from a subset of a product manifold onto one of the factors being a smooth vector bundle. The criterion involves a Lie group that operates on both factors.

We begin with some standard terminology and notations. Throughout the paper let $\mathbb{F}$ be the field $\mathbb{R}$ or $\mathbb{C}$. Recall, that a *Lie group action* of a Lie group G on a manifold M is a smooth map

$$\sigma : G \times M \longrightarrow M, \ (g, x) \mapsto \sigma(g, x) = g \cdot x$$

such that $e \cdot x = x, g \cdot (h \cdot x) = gh \cdot x$ holds for all $g, h \in G, x \in M$. The *orbit space* of σ is defined as the quotient space $M/ \sim_\sigma$ for the associated equivalence relation on M, where

$$m \sim_\sigma m' \text{ if and only if there is } g \in G \text{ with } m' = g \cdot m.$$

Thus the points of M/G are the equivalence classes

$$[m]_{\sim_\sigma} := \{m' \mid m \sim_\sigma m'\}.$$

The orbit space M/G carries a canonical topology, the quotient topology, which is defined as the finest topology for which the *natural projection*

$$\pi : M \longrightarrow M/G, \ m \mapsto [m]_{\sim_\sigma}$$

is continous. In order to study geometric properties of the orbit space M/G one has to consider the *graph map* $\tilde{\sigma}$ of the action. This is defined as

$$\tilde{\sigma} : G \times M \longrightarrow M \times M, \ (g, x) \mapsto (x, g \cdot x).$$

Therefore the image of $\tilde{\sigma}$ is nothing else but the relation $\sim_\sigma$ seen as a subspace of $M \times M$. Under certain circumstances the orbit space M/G is a manifold again. The following necessary and sufficient condition can be found in [5, Theorem 16.10.3].

Proposition 1. *There is a unique manifold structure on M/G such that the natural projection π is a submersion if and only if the image of the graph map $\mathrm{Im}\,\tilde{\sigma}$ is a closed submanifold of $M \times M$.*

Recall that a group action σ is called *free* if the stabilizer subgroups

$$G_x := \{g \in G \mid g \cdot x = x\}$$

are trivial, i.e., $G_x = \{e\}$ for all $x \in M$. Moreover, the action is called *proper* if the graph map is proper, i.e., the inverse image $\tilde{\sigma}^{-1}(K)$ of any compact subset $K \subset M \times M$ is a compact subset of M. Suppose, σ is a free Lie group action of G on M. Then it is easily seen that $\mathrm{Im}\,\tilde{\sigma}$ is a closed submanifold of $M \times M$ if and only if $\tilde{\sigma}$ is a closed map, i.e., maps closed sets to closed sets. Thus using the above quotient manifold criterion we arrive at the following well-known manifold criterion for an orbit space.

Theorem 1. *Let σ be a free and proper Lie group action of G on M. Then the orbit space M/G is a smooth manifold of dimension $\dim M - \dim G$. Moreover, the quotient map $\pi : M \longrightarrow M/G$ is a principal fibre bundle with structure group G.*

After these basic facts we turn to the discussion of vector bundles. The following definition is standard, but repeated here for convenience. Let X and B be Hausdorff spaces and let

$$f : X \longrightarrow B$$

be a continuous surjection. Let $p \in \mathbb{N}$ be fixed. For each point $x \in B$, assume there exists an open neighborhood U and a homeomorphism

$$\phi_U : U \times \mathbb{F}^p \longrightarrow f^{-1}(U)$$

such that

$$f(\phi_U(x,y)) = x$$

holds for all $x \in U$ and all $y \in \mathbb{F}^p$. Then ϕ_U is called a *local trivialization* of f. For each pair ϕ_U and ϕ_V of local trivializations and each point $x \in U \cap V$ suppose in addition that there exists a map $\theta_{V,U,x} \in \mathrm{GL}_p(\mathbb{F})$ such that

$$\phi_V^{-1} \circ \phi_U(x,y) = (x, \theta_{V,U,x}(y))$$

for all $y \in \mathbb{F}^p$, i.e., the induced change of coordinates function on $\mathbb{F}^p$ is linear. Then f is called a *topological vector bundle* with *fiber* $\mathbb{F}^p$. If X and B are smooth manifolds, f is a smooth map, and each ϕ_U is a diffeomorphism, then the bundle is called *smooth*.

The following result constitutes our proposed extension of the implicit function theorem for fibres that are given by topological vector bundles.

Theorem 2. *Let N and M be smooth manifolds, let $X \subset N$ be a topological subspace, let $B \subset M$ be a q-dimensional smooth submanifold and let $f : X \longrightarrow B$ be a topological vector bundle with fiber $\mathbb{F}^p$ such that f is the restriction of a smooth map*

$$F : U_X \longrightarrow M$$

on an open neighborhood U_X of X in N. Suppose, that each local trivialization $\phi_U : U \times \mathbb{F}^p \longrightarrow f^{-1}(U)$ of f is smooth as a map into N and such that $\phi_U^{-1} : f^{-1}(U) \longrightarrow U \times \mathbb{F}^p$ is the restriction of a smooth map

$$\Phi_{U,\mathrm{inv}} :\ U_{f^{-1}(U)} \longrightarrow M \times \mathbb{F}^p,$$

where $U_{f^{-1}(U)}$ is an open neighborhood of $f^{-1}(U)$ in N. Then X is a $(q+p)$-dimensional smooth submanifold of N and f is a smooth vector bundle.

Proof. Given $x_0 \in X$, there exists an open neighborhood U_1 of $f(x_0)$ in B and a local trivialization $\phi_{U_1} : U_1 \times \mathbb{F}^p \longrightarrow f^{-1}(U_1)$ of f. Furthermore, there exists an open neighborhood U_2 of $f(x_0)$ in B and a local coordinate chart $\varphi_{U_2} : U_2 \longrightarrow \varphi_{U_2}(U_2) \subset \mathbb{F}^q$ of B around $f(x_0)$. Set $U := U_1 \cap U_2$. Then U is open in B and $\phi_U := \phi_{U_1}|_{U \times \mathbb{F}^p} : U \times \mathbb{F}^p \longrightarrow f^{-1}(U)$ is a local trivialization of f. Furthermore, $\varphi_U := \varphi_{U_2}|_U : U \longrightarrow \varphi_U(U) = \varphi_{U_2}(U)$ is a local coordinate chart of B around $f(x_0)$. Define a local coordinate chart $\psi_{f^{-1}(U)} : f^{-1}(U) \longrightarrow \varphi_U(U) \times \mathbb{F}^p$ of X around x_0 by

$$x \mapsto (\varphi_U \circ \mathrm{pr}_1 \circ \phi_U^{-1}(x), \mathrm{pr}_2 \circ \phi_U^{-1}(x)).$$

Here pr_1 and pr_2 denote the projections on the first and second factor of $U \times \mathbb{F}^p$, respectively. Note that $f^{-1}(U)$ is open in X, since f is continuous. Note further that $\varphi_U(U) \times \mathbb{F}^p$ is open in $\mathbb{F}^q \times \mathbb{F}^p$, since φ_U is a local coordinate chart of B. Since ϕ_U^{-1} and φ_U are both bijective, so is $\psi_{f^{-1}(U)}$. Furthermore, $\psi_{f^{-1}(U)}$ is continuous as a concatenation of continuous maps, and hence it is a homeomorphism.

Now let $\psi_{f^{-1}(U)}$ and $\psi_{f^{-1}(V)}$ be two such local coordinate charts of X with $f^{-1}(U) \cap f^{-1}(V) \neq \emptyset$. Then

$$\psi_{f^{-1}(V)} \circ \psi_{f^{-1}(U)}^{-1} :\ \varphi_U(U \cap V) \times \mathbb{F}^p \longrightarrow \varphi_V(U \cap V) \times \mathbb{F}^p,$$

which is given by

$$
\begin{aligned}
(z,y) \mapsto\ & (\varphi_V \circ \mathrm{pr}_1 \circ \phi_V^{-1} \circ \phi_U(\varphi_U^{-1}(z), y), \mathrm{pr}_2 \circ \phi_V^{-1} \circ \phi_U(\varphi_U^{-1}(z), y)) = \\
& (\varphi_V \circ \varphi_U^{-1}(z), \theta_{V,U,\varphi_U^{-1}(z)}(y)),
\end{aligned}
$$

is a diffeomorphism, since $\varphi_V \circ \varphi_U^{-1}$ is a diffeomorphism and $\theta_{V,U,\varphi_U^{-1}(z)} \in \mathrm{GL}_p(\mathbb{F})$. It follows that X is a $(p+q)$-dimensional smooth manifold. Since the local coordinate charts of X are continuous, the preimage of any open set in $\mathbb{F}^q \times \mathbb{F}^p$ under any chart is open in X. These preimages form a basis of the topology τ induced on X by its differentiable structure, and thus τ coincides with the subspace topology on X, that is induced by the topology on N. Thus X is a submanifold of N.

Since f and the inverse maps ϕ_U^{-1} of all local trivializations ϕ_U of f are restrictions of smooth maps defined on open subsets of N, they are smooth themselves. Since each ϕ_U is also smooth, it follows that f is a smooth vector bundle. $\square$

The following theorem shows that quotients of smooth vector bundles with respect to free and proper Lie group actions are again smooth vector bundles provided the natural compatibility condition (9.1) holds.

Theorem 3. *Let $f : X \longrightarrow B$ be a smooth vector bundle with fiber $\mathbb{F}^p$ and let*

$$\sigma_X : G \times X \longrightarrow X$$

and

$$\sigma_B : G \times B \longrightarrow B$$

be free and proper actions of the Lie group G on X and B, respectively. For every local trivialization ϕ_U of f, let U consist of full G-orbits (i.e., $x \in U$ implies $\sigma_B(g, x) \in U$ for all $g \in G$) and let

$$\phi_U(\sigma_B(g, x), y) = \sigma_X(g, \phi_U(x, y)) \tag{9.1}$$

for all $g \in G$, $x \in U$ and $y \in \mathbb{F}^p$. Then

$$\bar{f} : X/G \longrightarrow B/G,$$
$$[x]_{\sim_{\sigma_X}} \mapsto [f(x)]_{\sim_{\sigma_B}}$$

is a smooth vector bundle with fiber $\mathbb{F}^p$.

Proof. Let $x \in X$ and $g \in G$ be arbitrary. Then $f(x) \in B$ and hence there exists a neighborhood U of $f(x)$ and a local trivialization ϕ_U such that $x = \phi_U(z, y)$ for appropriate $z \in U$ and $y \in \mathbb{F}^p$. But then (9.1) implies $f(\sigma_X(g, x)) = f(\sigma_X(g, \phi_U(z, y))) = f(\phi_U(\sigma_B(g, z), y)) = \sigma_B(g, z)$. Taking $g = e$ yields $f(x) = z$. But this means

$$f \circ \sigma_X(g, x) = \sigma_B(g, f(x)) \tag{9.2}$$

for all $g \in G$ and $x \in X$.

From (9.2) it follows that $\bar{f}$ is well defined. By Theorem 1, the spaces X/G and B/G are both smooth manifolds. Now consider the following commutative diagram:

$$
\begin{array}{ccc}
X & \xrightarrow{\ f\ } & B \\
{\scriptstyle \pi_X}\downarrow & \searrow & \downarrow{\scriptstyle \pi_B} \\
X/G & \xrightarrow{\ \bar{f}\ } & B/G
\end{array}
$$

Obviously, the map $\pi_B \circ f$ is smooth and hence $\bar{f}$ is smooth by the universal property of quotients ([5, Proposition 16.10.4]).

For every local trivialization ϕ_U of f define a local trivialization of $\bar{f}$ by

$$\bar{\phi}_U : \pi_B(U) \times \mathbb{F}^p \longrightarrow \pi_X(f^{-1}(U)),$$
$$([x]_{\sim_{\sigma_B}}, y) \mapsto [\phi_U(x,y)]_{\sim_{\sigma_X}}.$$

From (9.1) it follows that $\bar{\phi}_U$ is well defined. Since π_B is an open map, it follows that $\pi_B(U)$ is open in B/G. Since π_B is surjective, the sets $\pi_B(U)$ cover B/G. As before, the commutative diagram

$$
\begin{array}{ccc}
U \times \mathbb{F}^p & \xrightarrow{\phi_U} & f^{-1}(U) \\
{\scriptstyle \pi_B \times \mathrm{id}} \downarrow & \searrow & \downarrow {\scriptstyle \pi_X} \\
\pi_B(U) \times \mathbb{F}^p & \xrightarrow{\bar{\phi}_U} & \pi_X(f^{-1}(U))
\end{array}
$$

implies that $\bar{\phi}_U$ is smooth. Since π_X and ϕ_U are both surjective, so is $\pi_X \circ \phi_U$, and hence $\bar{\phi}_U$ is surjective. To see that $\bar{\phi}_U$ is also injective consider $x, x' \in U$ and $y, y' \in \mathbb{F}^p$ with $[\phi_U(x,y)]_{\sim_{\sigma_X}} = [\phi_U(x',y')]_{\sim_{\sigma_X}}$. Then there exists $g \in G$ with $\phi_U(x,y) = \sigma_X(g, \phi_U(x',y')) = \phi_U(\sigma_B(g,x'),y')$, i.e., $x = \sigma_B(g,x')$ and $y = y'$, since ϕ_U is injective. It follows that $([x]_{\sim_{\sigma_B}}, y) = ([x']_{\sim_{\sigma_B}}, y')$ and $\bar{\phi}_U$ is injective. Now the commutative diagram

$$
\begin{array}{ccc}
f^{-1}(U) & \xrightarrow{\phi_U^{-1}} & U \times \mathbb{F}^p \\
{\scriptstyle \pi_X} \downarrow & \searrow & \downarrow {\scriptstyle \pi_B \times \mathrm{id}} \\
\pi_X(f^{-1}(U)) & \xrightarrow{\bar{\phi}_U^{-1}} & \pi_B(U) \times \mathbb{F}^p
\end{array}
$$

implies that $\bar{\phi}_U^{-1}$ is smooth, hence $\bar{\phi}_U$ is a diffeomorphism.

Now let $x \in U$ and $y \in \mathbb{F}^p$. Then

$$\bar{f}(\bar{\phi}_U([x]_{\sim_{\sigma_B}}, y)) = \bar{f}([\phi_U(x,y)]_{\sim_{\sigma_X}})$$
$$= [f(\phi_U(x,y))]_{\sim_{\sigma_B}}$$
$$= [x]_{\sim_{\sigma_B}}.$$

If ϕ_V is another local trivialization of f, $x \in U \cap V$ and $y \in \mathbb{F}^p$, then

$$\bar{\phi}_V([\mathrm{pr}_1 \circ \phi_V^{-1} \circ \phi_U(x,y)]_{\sim_{\sigma_B}}, \mathrm{pr}_2 \circ \phi_V^{-1} \circ \phi_U(x,y)) =$$
$$[\phi_V(\mathrm{pr}_1 \circ \phi_V^{-1} \circ \phi_U(x,y), \mathrm{pr}_2 \circ \phi_V^{-1} \circ \phi_U(x,y))]_{\sim_{\sigma_X}} =$$
$$[\phi_U(x,y)]_{\sim_{\sigma_X}}$$

implies

$$\bar{\phi}_V^{-1} \circ \bar{\phi}_U([x]_{\sim_{\sigma_B}}, y) = \bar{\phi}_V^{-1}([\phi_U(x,y)]_{\sim_{\sigma_X}})$$
$$= ([\text{pr}_1 \circ \phi_V^{-1} \circ \phi_U(x,y)]_{\sim_{\sigma_B}}, \text{pr}_2 \circ \phi_V^{-1} \circ \phi_U(x,y))$$
$$= ([x]_{\sim_{\sigma_B}}, \theta_{V,U,x}(y)).$$

Let furthermore $g \in G$ be arbitrary, then successive use of (9.1) implies

$$\theta_{V,U,\sigma_B(g,x)}(y) = \text{pr}_2 \circ \phi_V^{-1} \circ \phi_U(\sigma_B(g,x), y)$$
$$= \text{pr}_2 \circ \phi_V^{-1} \circ \sigma_X(g, \phi_U(x,y))$$
$$= \text{pr}_2 \circ \phi_V^{-1} \circ \sigma_X(g, \phi_V(\phi_V^{-1} \circ \phi_U(x,y)))$$
$$= \text{pr}_2 \circ \phi_V^{-1} \circ \sigma_X(g, \phi_V(x, \theta_{V,U,x}(y)))$$
$$= \text{pr}_2 \circ \phi_V^{-1} \circ \phi_V(\sigma_B(g,x), \theta_{V,U,x}(y))$$
$$= \theta_{V,U,x}(y).$$

Hence

$$\theta_{V,U,[x]_{\sim_{\sigma_B}}} := \theta_{V,U,x}$$

is well defined and

$$\bar{\phi}_V^{-1} \circ \bar{\phi}_U([x]_{\sim_{\sigma_B}}, y) = ([x]_{\sim_{\sigma_B}}, \theta_{V,U,[x]_{\sim_{\sigma_B}}}(y)).$$

It follows that $\bar{f}$ is a smooth vector bundle with fiber $\mathbb{F}^p$. $\quad\square$

We now present a variant of Theorem 2, to be able to treat product space situations.

Theorem 4. *Let E be a vector space over $\mathbb{F}$ and B be a q-dimensional smooth manifold. Assume that the Lie group G operates smoothly and linearly on E via*

$$\sigma_E : \ G \times E \longrightarrow E, \quad (g,v) \mapsto g \cdot v$$

and smoothly and transitively on B via

$$\sigma_B : \ G \times B \longrightarrow B, \quad (g,b) \mapsto g \cdot b.$$

Let

$$\sigma_{E \times B} : \ G \times (E \times B) \longrightarrow E \times B, \quad (g,(v,b)) \mapsto g \cdot (v,b) := (g \cdot v, g \cdot b)$$

denote the induced action of G on the product manifold $E \times B$ and let $X \subset E \times B$ be a topological subspace which is closed under the action of G, i.e., $x \in X$ implies $g \cdot x \in X$ for all $g \in G$. Suppose, that the continuous map

$$f : \ X \longrightarrow B, \quad (v,b) \mapsto b$$

132   U. Helmke and J. Trumpf

is surjective. Let $b_0 \in B$ and let

$$E_0 := \{v \in E \mid (v, b_0) \in X\}$$

be a p-dimensional vector subspace of E. Assume further, there exists a submanifold $S \subset G$ and an open neighborhood U of $b_0 \in B$ such that

$$\sigma_{b_0} : S \longrightarrow U, \quad s \mapsto s \cdot b_0 = \sigma_B(g, b_0)$$

is a diffeomorphism. Then X is a $(q + p)$-dimensional smooth submanifold of $E \times B$ and f is a smooth vector bundle with fiber $\mathbb{F}^p$.

Proof. We want to apply Theorem 2. Note that for all $g \in G$ and $x \in X$

$$f(g \cdot x) = g \cdot f(x) \tag{9.3}$$

The vector space E_0 is isomorphic to $f^{-1}(b_0) = E_0 \times \{b_0\}$, where $\{b_0\}$ is considered as a trivial vector space. Let

$$\varphi : \mathbb{F}^p \longrightarrow E_0$$

be a vector space isomorphism, then

$$h : \mathbb{F}^p \longrightarrow f^{-1}(b_0), \quad y \mapsto (\varphi(y), b_0)$$

is also a vector space isomorphism which is clearly smooth as a map into E. For every $g \in G$, the map

$$\sigma_g : B \longrightarrow B, \quad b \mapsto g \cdot b$$

is a homeomorphism, and thus $U_g := g \cdot U$ is open for every $g \in G$. Since G operates transitively on B, for every $b \in B$ there exists $g \in G$ such that $b = g \cdot b_0$, hence the open sets U_g, $g \in G$ cover B. For each $g \in G$, define the continuous map

$$\phi_g : U_g \times \mathbb{F}^p \longrightarrow f^{-1}(U_g), \quad (b, y) \mapsto g \cdot \sigma_{b_0}^{-1}(g^{-1} \cdot b) \cdot h(y),$$

then for all $b \in U_g$ and $y \in \mathbb{F}^p$, (9.3) and

$$g \cdot \sigma_{b_0}^{-1}(g^{-1} \cdot b) \cdot b_0 = g \cdot \sigma_{b_0}\left(\sigma_{b_0}^{-1}(g^{-1} \cdot b)\right)$$
$$= b \tag{9.4}$$

imply that

$$f(\phi_g(b, y)) = f\left(g \cdot \sigma_{b_0}^{-1}(g^{-1} \cdot b) \cdot h(y)\right)$$
$$= g \cdot \sigma_{b_0}^{-1}(g^{-1} \cdot b) \cdot f(h(y))$$
$$= g \cdot \sigma_{b_0}^{-1}(g^{-1} \cdot b) \cdot b_0$$
$$= b,$$

i.e., ϕ_g maps indeed into $f^{-1}(U_g)$. Moreover, (9.4) implies for $b \in U_g$ that $\left[g \cdot \sigma_{b_0}^{-1} \left(g^{-1} \cdot b \right) \right]^{-1} \cdot b = b_0$ and hence $x = (v, b) \in f^{-1}(U_g) \subset X$ implies

$$\left[g \cdot \sigma_{b_0}^{-1} \left(g^{-1} \cdot f(x) \right) \right]^{-1} \cdot x = \left(\left[g \cdot \sigma_{b_0}^{-1} \left(g^{-1} \cdot f(x) \right) \right]^{-1} \cdot v, b_0 \right),$$

which lies in X, since X is closed under the action of G, and hence in $f^{-1}(b_0)$. In particular,

$$\left[g \cdot \sigma_{b_0}^{-1} \left(g^{-1} \cdot b \right) \right]^{-1} \cdot v \in E_0 \tag{9.5}$$

for all $(v, b) \in f^{-1}(U_g)$. But then the continuous map

$$\psi_g : f^{-1}(U_g) \longrightarrow U_g \times \mathbb{F}^p, \quad x \mapsto \left(f(x), h^{-1}\left(\left[g \cdot \sigma_{b_0}^{-1} \left(g^{-1} \cdot f(x) \right) \right]^{-1} \cdot x \right) \right)$$

is well defined and we get for all $b \in U_g$ and $y \in \mathbb{F}^p$

$$\begin{aligned}
\psi_g(\phi_g(b, y)) &= \left(f(\phi_g(b, y)), h^{-1}\left(\left[g \cdot \sigma_{b_0}^{-1} \left(g^{-1} \cdot f(\phi_g(b, y)) \right) \right]^{-1} \cdot \phi_g(b, y) \right) \right) \\
&= \left(b, h^{-1}\left(\left[g \cdot \sigma_{b_0}^{-1} \left(g^{-1} \cdot b \right) \right]^{-1} \cdot g \cdot \sigma_{b_0}^{-1} \left(g^{-1} \cdot b \right) \cdot h(y) \right) \right) \\
&= (b, y).
\end{aligned}$$

Thus for all $x \in f^{-1}(U_g)$

$$\begin{aligned}
\phi_g(\psi_g(x)) &= g \cdot \sigma_{b_0}^{-1} \left(g^{-1} \cdot f(x) \right) \cdot h\left(h^{-1}\left(\left[g \cdot \sigma_{b_0}^{-1} \left(g^{-1} \cdot f(x) \right) \right]^{-1} \cdot x \right) \right) \\
&= x
\end{aligned}$$

and therefore ϕ_g is a homeomorphism with inverse ψ_g and hence a local trivialisation of f. Moreover, ϕ_g is smooth as a map into $E \times B$, as it is a concatenation of smooth maps. Let π denote a smooth projector from E onto E_0, then the smooth map

$$k : E \times B \longrightarrow \mathbb{F}^p, \quad (v, b) \mapsto \varphi^{-1}(\pi(v))$$

restricts to h^{-1} on $f^{-1}(b_0) = E_0 \times \{b_0\}$. Obviously, f is the restriction of the smooth map

$$F : E \times B \longrightarrow B, \quad (v, b) \mapsto b$$

to $X \subset E \times B$. Note that $F^{-1}(U_g)$ is open in $E \times B$ since F is continuous and note further that for $x \in F^{-1}(U_g)$ we have $F(x) \in U_g$ and $g^{-1} \cdot F(x) \in U$. Now $\phi_g^{-1} = \psi_g$ is the restriction of the smooth map

$$\Psi_g : F^{-1}(U_g) \longrightarrow U_g \times \mathbb{F}^p, \quad x \mapsto \left(F(x), k\left(\left[g \cdot \sigma_{b_0}^{-1} \left(g^{-1} \cdot F(x) \right) \right]^{-1} \cdot x \right) \right)$$

to $f^{-1}(U_g)$. Finally, let $g_1, g_2 \in G$ and let ϕ_{g_1} and ϕ_{g_2} be the two associated local trivialisations of f, $b \in U_{g_1} \cap U_{g_2}$, and $y \in \mathbb{F}^p$. From (9.4),

134   U. Helmke and J. Trumpf

$$\phi_{g_1}(b, y) = g_1 \cdot \sigma_{b_0}^{-1}\left(g_1^{-1} \cdot b\right) \cdot (\varphi(y), b_0)$$
$$= \left(g_1 \cdot \sigma_{b_0}^{-1}\left(g_1^{-1} \cdot b\right) \cdot \varphi(y), b\right) \in f^{-1}(U_{g_1}) \cap f^{-1}(U_{g_2})$$

and hence with

$$\alpha(g_1, g_2, b) := \left[g_2 \cdot \sigma_{b_0}^{-1}\left(g_2^{-1} \cdot b\right)\right]^{-1} \cdot g_1 \cdot \sigma_{b_0}^{-1}\left(g_1^{-1} \cdot b\right)$$

$\alpha(g_1, g_2, b) \cdot \varphi(y) \in E_0$ follows from (9.5). By (9.4) then

$$\alpha(g_1, g_2, b) \cdot b_0 = \left[g_2 \cdot \sigma_{b_0}^{-1}\left(g_2^{-1} \cdot b\right)\right]^{-1} \cdot b$$
$$= \left[g_2 \cdot \sigma_{b_0}^{-1}\left(g_2^{-1} \cdot b\right)\right]^{-1} \cdot g_2 \cdot \sigma_{b_0}^{-1}\left(g_2^{-1} \cdot b\right) \cdot b_0$$
$$= b_0.$$

Therefore

$$\phi_{g_2}^{-1}(\phi_{g_1}(b, y)) = \left(f(\phi_{g_1}(b, y)), h^{-1}\left(\left[g_2 \cdot \sigma_{b_0}^{-1}\left(g_2^{-1} \cdot f(\phi_{g_1}(b, y))\right)\right]^{-1} \cdot \phi_{g_1}(b, y)\right)\right)$$
$$= \left(b, h^{-1}\left(\left[g_2 \cdot \sigma_{b_0}^{-1}\left(g_2^{-1} \cdot b\right)\right]^{-1} \cdot g_1 \cdot \sigma_{b_0}^{-1}\left(g_1^{-1} \cdot b\right) \cdot h(y)\right)\right)$$
$$= (b, h^{-1}(\alpha(g_1, g_2, b) \cdot (\varphi(y), b_0)))$$
$$= (b, \varphi^{-1}(\alpha(g_1, g_2, b) \cdot \varphi(y))).$$

Since G acts linearly on E and since φ is a vector space isomorphism, the change of coordinates $\theta_{g_1, g_2, b}(y) := \varphi^{-1}(\alpha(g_1, g_2, b) \cdot \varphi(y))$ on $\mathbb{F}^p$ is a linear map. Hence f is a topological vector bundle and the statement follows from Theorem 2. $\square$

9.3 The Cotangent Bundle of the Flag Manifold

With these results on submanifold criteria being out of the way, we can now introduce our main actors on stage. The first one is the cotangent bundle of a flag manifold and its amazing relation to the geometry of nilpotent matrices. We will explain this connection in detail, using the symplectic nature of the cotangent bundle and by computing an associated momentum map. But first some basic definitions and vocabulary.

Recall, that the Grassmann manifold $\text{Grass}(m, \mathbb{F}^n)$ is defined as the set of m-dimensional $\mathbb{F}$-linear subspaces of $\mathbb{F}^n$. It is a smooth, compact manifold and provides a natural generalization of the familiar projective spaces. Flag manifolds in turn provide a generalization of Grassmannians. To define them consider arbitrary integers $n, r \in \mathbb{N}$. A *flag symbol of type* (n, r) is an r-tupel $a = (a_1, \ldots, a_r)$ of numbers $a_1, \ldots, a_r \in \mathbb{N}$ with $a_1 < \cdots < a_r < n$. The *flag manifold of type* a is the set of partial flags $V_1 \subset \cdots \subset V_r$ of linear subspaces of $\mathbb{F}^n$ with prescribed dimensions $a_1, \cdots a_r$. More precisely,

$$\text{Flag}(a, \mathbb{F}^n) = \{(V_1, \ldots, V_r) \in \prod_{i=1}^{r} \text{Grass}(a_i, \mathbb{F}^n) \mid V_1 \subset \cdots \subset V_r\}.$$

endowed with the differentiable structure inherited from the product of Grassmannians. For convenience of notation we set $a_0 = 0$, $a_{r+1} = n$, $V_0 = \{0\}$ and $V_{r+1} = \mathbb{F}^n$. Furthermore, we define $b_i := a_i - a_{i-1}$, $i = 1, \ldots, r+1$. In the case of *complete flags*, i.e., $a_i = i$ for $i = 0, \cdots, n$, or, equivalently, $b_i = 1$ for $i = 1, \ldots, n$, we use the simplified notation $\mathrm{Flag}(\mathbb{F}^n)$ instead of $\mathrm{Flag}(a, \mathbb{F}^n)$.

We are interested in the set

$$M(a, \mathbb{F}^n) = \{(A, (V_1, \ldots, V_r)) \in \mathfrak{gl}_n(\mathbb{F}) \times \mathrm{Flag}(a, \mathbb{F}^n) \mid AV_i \subset V_i, i = 1, \ldots, r\}$$

of pairs of linear maps and the flags they leave invariant. Here $\mathfrak{gl}_n(\mathbb{F})$ denotes the vector space $\mathbb{F}^{n \times n}$ of $n \times n$-matrices. It is also a Lie algebra with the commutator $[A, B] = AB - BA$ as the Lie bracket operation.

Theorem 5. *$M(a, \mathbb{F}^n)$ is a smooth manifold of dimension n^2 and the projection map*

$$\pi : M(a, \mathbb{F}^n) \longrightarrow \mathrm{Flag}(a, \mathbb{F}^n),$$
$$(A, (V_1, \ldots, V_r)) \mapsto (V_1, \ldots, V_r)$$

is a smooth vector bundle.

Proof. We will apply Theorem 4. Clearly, the Lie group $G = \mathrm{GL}_n(\mathbb{F})$ of invertible $n \times n$ matrices operates linearly on the vector space $E := \mathfrak{gl}_n(\mathbb{F})$ by similarity

$$\sigma_E : \mathrm{GL}_n(\mathbb{F}) \times \mathfrak{gl}_n(\mathbb{F}) \longrightarrow \mathfrak{gl}_n(\mathbb{F}), \quad (T, A) \mapsto TAT^{-1}$$

and transitively on the flag manifold $B := \mathrm{Flag}(a, \mathbb{F}^n)$ by the canonical action

$$\sigma_B : \mathrm{GL}_n(\mathbb{F}) \times \mathrm{Flag}(a, \mathbb{F}^n) \longrightarrow \mathrm{Flag}(a, \mathbb{F}^n),$$
$$(T, \mathsf{V}) = (T, (V_1, \ldots, V_r)) \mapsto T \cdot \mathsf{V} := (TV_1, \ldots, TV_r). \tag{9.6}$$

The topological subspace $X := M(a, \mathbb{F}^n) \subset \mathfrak{gl}_n(\mathbb{F}) \times \mathrm{Flag}(a, \mathbb{F}^n)$ is closed under the induced action on the product space, since for every $A \in \mathfrak{gl}_n(\mathbb{F})$ and every $(V_1, \ldots, V_r) \in \mathrm{Flag}(a, \mathbb{F}^n)$ the inclusion $AV_i \subset V_i$ implies $TAT^{-1}TV_i \subset TV_i$ for all $i = 1, \ldots, r$. Since for every flag there exists a linear map that leaves the flag invariant (e.g. the identity), the map $f := \pi$ is surjective.

We set $b_0 := \mathsf{V}_0 \in \mathrm{Flag}(a, \mathbb{F}^n)$, where V_0 is the *standard flag*

$$\mathsf{V}_0 = (V_1, \ldots, V_r) = \left(\mathrm{colspan}\begin{pmatrix} I_{a_1} \\ 0 \end{pmatrix}, \ldots, \mathrm{colspan}\begin{pmatrix} I_{a_r} \\ 0 \end{pmatrix}\right). \tag{9.7}$$

Then the set E_0 of linear maps that leave V_0 invariant is the vector space

$$E_0 := \left\{ \begin{pmatrix} A_{11} & \cdots & \cdots & A_{1(r+1)} \\ 0 & \ddots & & \vdots \\ \vdots & \ddots & \ddots & \vdots \\ 0 & \cdots & 0 & A_{(r+1)(r+1)} \end{pmatrix} \in \mathfrak{gl}_n(\mathbb{F}) \;\middle|\; \begin{array}{l} A_{ij} \in \mathbb{F}^{b_i \times b_j}, \\ 1 \le i \le j \le r+1 \end{array} \right\}.$$

The open Bruhat cell

$$U = \left\{ \left(\text{colspan} \begin{pmatrix} I_{b_1} & 0 & \cdots & & 0 \\ K_{21} & \ddots & \ddots & & \vdots \\ \vdots & & \ddots & \ddots & 0 \\ \vdots & & & \ddots & I_{b_i} \\ K_{(i+1)1} & \cdots & & \cdots & K_{(i+1)i} \\ \vdots & & & & \vdots \\ K_{(r+1)1} & \cdots & & \cdots & K_{(r+1)i} \end{pmatrix} \right)_{i=1}^{r} \ \middle| \ \begin{matrix} K_{jk} \in \mathbb{F}^{b_j \times b_k}, \\ 1 \leq k < j \leq r+1 \end{matrix} \right\}$$

in $\mathrm{Flag}(a, \mathbb{F}^n)$ is an open neighborhood of $b_0 = \mathsf{V}_0$ which is trivially diffeomorphic to the Lie subgroup

$$S := \left\{ \begin{pmatrix} I_{b_1} & 0 & \cdots & & 0 \\ K_{21} & \ddots & \ddots & & \vdots \\ \vdots & & \ddots & \ddots & 0 \\ K_{(r+1)1} & \cdots & K_{(r+1)r} & I_{b_{r+1}} \end{pmatrix} \in \mathrm{GL}_n(\mathbb{F}) \ \middle| \ \begin{matrix} K_{jk} \in \mathbb{F}^{b_j \times b_k}, \\ 1 \leq k < j \leq r+1 \end{matrix} \right\}$$

of $G = \mathrm{GL}_n(\mathbb{F})$ by

$$\sigma_{\mathsf{V}_0} : S \longrightarrow U, \quad T \mapsto T \cdot \mathsf{V}_0.$$

Now the manifold statement follows from Theorem 4. Concerning the dimension formula we observe that the dimension of the vector bundle $M(a, \mathbb{F}^n)$ equals the dimension of the flag manifold plus the dimension of a fibre. The dimension of a fibre is equal to the dimension of E_0, i.e., to the dimension of the space of block upper triangular matrices. On the other hand, the dimension of the flag manifold is equal to the dimension of S, i.e., to the dimension of the set of strictly lower triangular block matrices. As these dimensions add up to n^2, the result follows. $\square$

We want to see that the bundle $M(a, \mathbb{F}^n)$ of Theorem 5 contains an isomorphic copy of the cotangent bundle $\mathrm{T}^*\mathrm{Flag}(a, \mathbb{F}^n)$. Observe that $\mathrm{GL}_n(\mathbb{F})$ acts transitively on $\mathrm{Flag}(a, \mathbb{F}^n)$, with the stabilizer group $H_n = \mathrm{Stab}(\mathsf{V}_0)$ for the standard flag V_0 of (9.7) given by

$$H_n = \left\{ \begin{pmatrix} A_{11} & \cdots & & \cdots & A_{1(r+1)} \\ 0 & \ddots & & & \vdots \\ \vdots & \ddots & \ddots & & \vdots \\ 0 & \cdots & 0 & A_{(r+1)(r+1)} \end{pmatrix} \in \mathrm{GL}_n(\mathbb{F}) \ \middle| \ \begin{matrix} A_{ij} \in \mathbb{F}^{b_i \times b_j}, \\ 1 \leq i \leq j \leq r+1 \end{matrix} \right\}, \quad (9.8)$$

i.e., by the closed Lie subgroup of $\mathrm{GL}_n(\mathbb{F})$ consisting of all block upper triangular matrices. Let $\mathfrak{gl}_n(\mathbb{F})$ and $\mathfrak{h}_n$ denote the Lie algebras of $\mathrm{GL}_n(\mathbb{F})$ and H_n, respectively. Thus

$$\mathfrak{h}_n = \left\{ \begin{pmatrix} A_{11} & \cdots & \cdots & A_{1(r+1)} \\ 0 & \ddots & & \vdots \\ \vdots & \ddots & \ddots & \vdots \\ 0 & \cdots & 0 & A_{(r+1)(r+1)} \end{pmatrix} \in \mathfrak{gl}_n(\mathbb{F}) \;\middle|\; \begin{matrix} A_{ij} \in \mathbb{F}^{b_i \times b_j}, \\ 1 \le i \le j \le r+1 \end{matrix} \right\}. \qquad (9.9)$$

We endow $\mathfrak{gl}_n(\mathbb{F})$ with the nondegenerate symmetric bilinear form

$$(X, Y) := \operatorname{tr}(XY).$$

Note, that the orthogonal complement of $\mathfrak{gl}_n(\mathbb{F})$ with respect to this trace form is exactly the linear subspace $\mathfrak{u}^+$ of strictly upper triangular matrices

$$\mathfrak{u}^+ = \left\{ \begin{pmatrix} 0 & A_{12} & \cdots & A_{1(r+1)} \\ 0 & \ddots & \ddots & \vdots \\ \vdots & \ddots & \ddots & A_{r(r+1)} \\ 0 & \cdots & 0 & 0 \end{pmatrix} \in \mathfrak{gl}_n(\mathbb{F}) \;\middle|\; \begin{matrix} A_{ij} \in \mathbb{F}^{b_i \times b_j}, \\ 1 \le i < j \le r+1 \end{matrix} \right\}. \qquad (9.10)$$

For any $\mathsf{V} = T \cdot \mathsf{V}_0 \in \operatorname{Flag}(a, \mathbb{F}^n)$, the fibre $\sigma_{\mathsf{V}}^{-1}(T \cdot \mathsf{V}_0)$ of $\sigma_{\mathsf{V}} : \operatorname{GL}_n(\mathbb{F}) \longrightarrow \operatorname{Flag}(a, \mathbb{F}^n), g \mapsto g \cdot \mathsf{V} = gT \cdot \mathsf{V}_0$, is the stabilizer subgroup

$$\operatorname{Stab}(\mathsf{V}) = \operatorname{Stab}(T \cdot \mathsf{V}_0) = Ad(T)H_n = \{ TgT^{-1} \mid g \in H_n \}.$$

Therefore the tangent map at the identity matrix I defines a surjective linear map

$$\mathrm{T}_I\, \sigma_{\mathsf{V}} : \mathfrak{gl}_n(\mathbb{F}) \longrightarrow \mathrm{T}_{\mathsf{V}} \operatorname{Flag}(a, \mathbb{F}^n)$$

that vanishes exactly on the Lie subalgebra

$$Ad(T)\mathfrak{h}_n = \{ TXT^{-1} \mid X \in \mathfrak{h}_n \}.$$

By taking the duals, it follows that the associated dual map

$$\mathrm{T}_I^* \, \sigma_{\mathsf{V}} : \mathrm{T}_{\mathsf{V}}^* \operatorname{Flag}(a, \mathbb{F}^n) \longrightarrow \mathfrak{gl}_n^*(\mathbb{F}), \ \lambda \mapsto \lambda \circ \mathrm{T}_I\, \sigma_{\mathsf{V}},$$

maps the cotangent space $\mathrm{T}_{\mathsf{V}}^* \operatorname{Flag}(a, \mathbb{F}^n)$ isomorphically onto the image set

$$\{ \lambda \in \mathfrak{gl}_n^*(\mathbb{F}) \mid Ad(T)\mathfrak{h}_n \subset \operatorname{Ker} \lambda \}.$$

The trace form on the Lie algebra defines an isomorphism

$$\tau : \mathfrak{gl}_n^*(\mathbb{F}) \longrightarrow \mathfrak{gl}_n(\mathbb{F}), \ \tau(\lambda) = X_\lambda,$$

where $X_\lambda \in \mathfrak{gl}_n(\mathbb{F})$ denotes the uniquely determined element satisfying $(X_\lambda, Y) = \lambda(Y)$ for all $Y \in \mathfrak{gl}_n(\mathbb{F})$. By using this isomorphism of the Lie algebra $\mathfrak{gl}_n(\mathbb{F})$ with its dual space $\mathfrak{gl}_n^*(\mathbb{F})$, then $\{ \lambda \in \mathfrak{gl}_n^*(\mathbb{F}) \mid Ad(T)\mathfrak{h}_n \subset \operatorname{Ker} \lambda \}$ becomes identified with the orthogonal complement

$$\tau(\{ \lambda \in \mathfrak{gl}_n^*(\mathbb{F}) \mid Ad(T)\mathfrak{h}_n \subset \operatorname{Ker} \lambda \}) = (Ad(T)\mathfrak{h}_n)^\perp = Ad(T)(\mathfrak{h}_n^\perp) = Ad(T)\mathfrak{u}^+.$$

Since $\mathfrak{u}^+$ is invariant under similarity transformations by elements of H_n this yields a well-defined smooth map

$$\mu : T^*\mathrm{Flag}(a, \mathbb{F}^n) \longrightarrow \mathfrak{gl}_n(\mathbb{F}), \ (\mathsf{V}, \lambda) \mapsto \tau(\lambda \circ T_I\, \sigma_\mathsf{V}) \qquad (9.11)$$

that maps each cotangent space $T^*_\mathsf{V}\,\mathrm{Flag}(a, \mathbb{F}^n)$ isomorphically onto $Ad(T)\mathfrak{u}^+$. By inspection, the image elements are seen to be exactly those matrices A that satisfy $AV_i \subset V_{i-1}$ for $i = 1, \ldots, r+1$. Note that $AV_i \subset V_{i-1}$ implies $AV_i \subset V_i$, since $V_{i-1} \subset V_i$, where $i = 1, \ldots, r+1$. This shows the following result.

Theorem 6. *The smooth map*

$$\Phi : T^*\mathrm{Flag}(a, \mathbb{F}^n) \longrightarrow \mathfrak{gl}_n(\mathbb{F}) \times \mathrm{Flag}(a, \mathbb{F}^n), \ (\mathsf{V}, \lambda) \mapsto (\tau(\lambda \circ T_I\, \sigma_\mathsf{V}), \mathsf{V}) \ (9.12)$$

maps the cotangent bundle $T^*\mathrm{Flag}(a, \mathbb{F}^n)$ *diffeomorphically onto the subbundle*

$$\{(A, (V_1, \ldots, V_r)) \in \mathfrak{gl}_n(\mathbb{F}) \times \mathrm{Flag}(a, \mathbb{F}^n) \,|\, AV_i \subset V_{i-1}, i = 1, \ldots, r+1\}$$

of $M(a, \mathbb{F}^n)$. *This subbundle of dimension* $2 \dim \mathrm{Flag}(a, \mathbb{F}^n)$ *will be denoted by* $N(a, \mathbb{F}^n)$ *in the sequel. In particular, there is a bundle isomorphism of* $T^*\mathrm{Flag}(a, \mathbb{F}^n)$ *with the homomorphism bundle*

$$\bigoplus_{i=1}^{r} \mathrm{Hom}(V_{i+1}/V_i, V_i).$$

For $i = 1, \ldots, r$ let $V_i^{\perp}$ denote the orthogonal complement of V_i with respect to the Euclidean inner product on $\mathbb{F}^n$. Then we can identify the quotient space V_{i+1}/V_i with $V_i^{\perp} \cap V_{i+1}$ and obtain the bundle isomorphism

$$T^*\mathrm{Flag}(a, \mathbb{F}^n) \simeq \bigoplus_{i=1}^{r} \mathrm{Hom}(V_i^{\perp} \cap V_{i+1}, V_i).$$

In the case of flag length $r = 1$ we recover the well known diffeomorphic descriptions of the cotangent bundle of the Grassmannian.

Corollary 1.

$$T^*\mathrm{Grass}(m, \mathbb{F}^n) \simeq \{(A, \mathsf{V}) \in \mathfrak{gl}_n(\mathbb{F}) \times \mathrm{Grass}(m, \mathbb{F}^n) \,|\, A\mathsf{V} = \{0\}, A\mathbb{F}^n \subset \mathsf{V}\}$$
$$\simeq \mathrm{Hom}(\mathsf{V}^{\perp}, \mathsf{V}).$$

Since the cotangent bundle $T^*\mathrm{Flag}(a, \mathbb{F}^n)$ can be identified with the subbundle $N(a, \mathbb{F}^n)$ of the bundle $M(a, \mathbb{F}^n)$, it makes sense to consider the restriction of the projection map onto the first factor

$$\mathrm{pr}_1|_{T^*\mathrm{Flag}(a, \mathbb{F}^n)} : \ T^*\mathrm{Flag}(a, \mathbb{F}^n) \longrightarrow \mathfrak{gl}_n(\mathbb{F}), \ (A, (V_1, \ldots, V_r)) \mapsto A.$$

The amazing fact now is that the linear operators arising in the image (which is equal to the image of μ) are all nilpotent matrices! This is due to the fact that $AV_i \subset V_{i-1}$ for all $i = 1, \ldots, r+1$ implies $A^{r+1}\mathbb{F}^n = A^{r+1}V_{r+1} \subset V_0 = \{0\}$. Moreover, for the *complete* flag manifold $\mathrm{Flag}(\mathbb{F}^n)$ we conclude that the image of μ coincides with the set $\mathcal{N}_n(\mathbb{F})$ of arbitrary nilpotent $n \times n$-matrices over $\mathbb{F}$. This shows the following equivalent description of the cotangent bundle.

Corollary 2.

$$\mathrm{T}^*\mathrm{Flag}(\mathbb{F}^n) \simeq \{(A, (V_1, \ldots, V_n)) \in \mathcal{N}_n(\mathbb{F}) \times \mathrm{Flag}(\mathbb{F}^n) \mid AV_i \subset V_i, i = 1, \ldots, n\}$$

Of course, all this is immediate by inspection, but in order to gain a better geometric understanding of the connection between cotangent vectors and nilpotent matrices, we make contact with some basic symplectic geometry and Hamiltonian mechanics, specialized to the situation at hand.

Recall, that the cotangent bundle T^*M of an arbitrary smooth manifold M is always a symplectic manifold, implying in particular, that each of the cotangent spaces $\mathrm{T}^*_x M$ carries a canonically defined symplectic form ω (the Liouville form). Now suppose a Lie group G with Lie algebra $\mathfrak{g}$ acts smoothly on M via

$$\sigma : G \times M \longrightarrow M, \ (g, x) \mapsto g \cdot x.$$

Note that each diffeomorphism $\sigma_g : M \longrightarrow M, x \mapsto g \cdot x, g \in G$, lifts (by pull-back) to a diffeomorphism $\sigma_g^* : \mathrm{T}^*M \longrightarrow \mathrm{T}^*M$ on the cotangent bundle. By inspection, these diffeomeorphisms are seen to preserve the Liouville symplectic form on T^*M. Therefore the action σ lifts to a symplectic action on the cotangent bundle

$$\hat{\sigma} : G \times \mathrm{T}^*M \longrightarrow \mathrm{T}^*M, \ (g, (x, \lambda)) \mapsto (g \cdot x, \sigma_g^*(x, \lambda)).$$

The tangent map of the induced smooth map $\sigma_x : G \longrightarrow M, g \mapsto g \cdot x, x \in M$ defines a linear map

$$\mathrm{T}_e\, \sigma_x : \mathfrak{g} \longrightarrow \mathrm{T}_x M$$

that vanishes exactly on $\mathfrak{g}_x$, the Lie algebra of the stabilizer subgroup G_x of x. In this setting the concept of a moment map for the induced group action on T^*M is defined. It gives a map $\mu : \mathrm{T}^*M \longrightarrow \mathfrak{g}^*$ from the cotangent bundle to the dual of the Lie algebra $\mathfrak{g}$. It is simply defined by the dual of the tangent map $\mathrm{T}_e\, \sigma_x : \mathfrak{g} \longrightarrow \mathrm{T}_x M$.

Definition 1. *The* moment map *for the G-action $\hat{\sigma}$ on T^*M is the smooth map*

$$\mu^* : \mathrm{T}^*M \longrightarrow \mathfrak{g}^*, \ (x, \lambda) \mapsto \lambda \circ \mathrm{T}_e\, \sigma_x.$$

Duality provides us with an identification of $\mathfrak{g}$ with $\mathfrak{g}^*$, given a nondegenerate symmetric bilinear form $(\cdot, \cdot)$ on $\mathfrak{g}$. It therefore enables us to redefine the moment map as a map from the cotangent bundle T^*M onto the Lie algebra $\mathfrak{g}$. It should be emphasized that the definition below depends on the above choice of a bilinear form. In contrast, the moment map on the cotangent bundle does not require such choices and is intrinsically defined.

Definition 2. *The* dualized moment map

$$\mu : \mathrm{T}^* M \longrightarrow \mathfrak{g}, \ (x, \lambda) \mapsto \mu_x(\lambda)$$

is defined by the characterizing property

$$(\mu_x(\lambda), \xi) = \lambda(\mathrm{T}_e \, \sigma_x(\xi)) \quad \textit{for all } \xi \in \mathfrak{g}.$$

We want to determine the image of the map

$$\mu_x : \mathrm{T}^*_x M \longrightarrow \mathfrak{g}$$

for a given $x \in M$. Let G_x denote the stabilizer subgroup of x in G with Lie algebra $\mathfrak{g}_x$. Let

$$\mathfrak{m}_x := \mathfrak{g}_x^{\perp} := \{\xi \in \mathfrak{g} \,|\, (\xi, \eta) = 0 \ \forall \eta \in \mathfrak{g}_x\}.$$

Since $\mathrm{T}_e \, \sigma_x$ vanishes exactly on $\mathfrak{g}_x$ we see that the image of μ_x^* is given as

$$\mathrm{Im}(\mu_x^*) = \{\lambda \in \mathfrak{g}^* \,|\, \mathfrak{g}_x \subset \mathrm{Ker}\,\lambda\}$$

and therefore

$$\mathrm{Im}(\mu_x) = \mathfrak{m}_x.$$

We now restrict generality by focussing on homogeneous spaces G/H of a Lie group G by a closed Lie subgroup H. Let $\mathfrak{g}$ and $\mathfrak{h}$ denote their Lie algebras, respectively. Thus we consider the transitive G-action on G/H that is defined by left translation

$$\sigma : G \times G/H \longrightarrow G/H, \ (g, \gamma H) \mapsto g\gamma H.$$

It lifts to an action $\hat{\sigma} : G \times \mathrm{T}^*(G/H) \longrightarrow \mathrm{T}^*(G/H)$ on the cotangent bundle. Fix a nondegenerate bilinear form $(\cdot, \cdot)$ on the Lie algebra $\mathfrak{g}$ that is $Ad(G)$-invariant, i.e.

$$(Ad(g)\xi, Ad(g)\eta) = (\xi, \eta) \quad \text{for all } g \in G, \ \xi, \eta \in \mathfrak{g}.$$

Such a form always exists on, e.g., any semisimple Lie algebra $\mathfrak{g}$ and is given by the *Killing form*

$$(\xi, \eta) = \mathrm{tr}(ad_\xi \circ ad_\eta).$$

Thus the dualized moment map

$$\mu : \mathrm{T}^*(G/H) \longrightarrow \mathfrak{g}$$

is well-defined and has image sets

$$\mathrm{Im}(\mu_{gH}) = \mathfrak{m}_{gH},$$

where

$$\mathfrak{m}_{gH} = (Ad(g)\mathfrak{h})^{\perp}$$

denotes the orthogonal complement of the Lie subalgebra $Ad(g)\mathfrak{h}$ with respect to $(\cdot,\cdot)$. By the $Ad(G)$–invariance of $(\cdot,\cdot)$ the above formula then simplifies to

$$Im(\mu_{gH}) = Ad(g)(\mathfrak{h}^{\perp}).$$

After these generalities let us return to our task of interpreting the projection map pr_1 on the cotangent bundle $\mathrm{T}^*\mathrm{Flag}(a,\mathbb{F}^n)$ as a moment map. Thus we consider the Lie group $G = \mathrm{GL}_n(\mathbb{F})$ with Lie algebra $\mathfrak{gl}_n(\mathbb{F})$. We endow $\mathfrak{gl}_n(\mathbb{F})$ with the $Ad(\mathrm{GL}_n(\mathbb{F}))$–invariant nondegenerate bilinear form

$$(X,Y) := \mathrm{tr}(XY).$$

Choose H_n to denote the parabolic subgroup of $\mathrm{GL}_n(\mathbb{F})$ of all block-upper triangular matrices defined by (9.8), let $\mathfrak{h}_n$ denote its associated Lie algebra (9.9). We have observed already, that the orthogonal complement of $\mathfrak{h}_n$ with respect to the trace form is the linear subspace $\mathfrak{u}^+$ of strictly upper triangular matrices (9.10) satisfying

$$(Ad(T)\mathfrak{h}_n)^{\perp} = Ad(T)(\mathfrak{h}_n^{\perp}) = Ad(T)\mathfrak{u}^+.$$

It follows that at every point $\mathsf{V} = T \cdot \mathsf{V}_0 \in \mathrm{Flag}(a,\mathbb{F}^n) \simeq \mathrm{GL}_n(\mathbb{F})/H_n$ the set of image points of the dualized moment map for the canonical $\mathrm{GL}_n(\mathbb{F})$-action on the cotangent bundle $\mathrm{T}^*(\mathrm{GL}_n(\mathbb{F})/H_n)$ coincides with that of the map μ of (9.11). On the other hand, μ is the first factor in the isomorphism stated in Theorem 6. Thus, under the above identifications, this proves our claim that the projection map

$$\mathrm{pr}_1|_{\mathrm{T}^*\mathrm{Flag}(a,\mathbb{F}^n)} : \ \mathrm{T}^*\mathrm{Flag}(a,\mathbb{F}^n) \longrightarrow \mathfrak{gl}_n(\mathbb{F}), \ (A,(V_1,\ldots,V_r)) \mapsto A$$

coincides with the dualized moment map. We conclude:

Theorem 7. *The set $N(a,\mathbb{F}^n)$ of pairs*

$$\{(A,(V_1,\ldots,V_r)) \in \mathfrak{gl}_n(\mathbb{F}) \times \mathrm{Flag}(a,\mathbb{F}^n) \,|\, AV_i \subset V_{i-1}, i = 1,\ldots,r+1\}$$

carries the structure of a symplectic manifold of dimension $2\dim\mathrm{Flag}(a,\mathbb{F}^n)$ such that the $\mathrm{GL}_n(\mathbb{F})$-similarity action

$$(T,(A,(V_1,\ldots,V_r))) \mapsto (TAT^{-1},(TV_1,\ldots,TV_r))$$

becomes a symplectic action. The moment map for this action is

$$\mathrm{pr}_1|_{\mathrm{T}^*\mathrm{Flag}(a,\mathbb{F}^n)} : \ \mathrm{T}^*\mathrm{Flag}(a,\mathbb{F}^n) \longrightarrow \mathfrak{gl}_n(\mathbb{F}), \ (A,(V_1,\ldots,V_r)) \mapsto A.$$

Its image consists of nilpotent matrices.

It is possible to derive similar explicit formulas for the cotangent bundle of homogeneous spaces that are defined by other classical Lie groups. For the fun of it, we mention one further example, the *Lagrangian Grassmann manifold* $\mathrm{LG}(n)$ of m-dimensional Lagrangian subspaces of $\mathbb{F}^{2n \times 2n}$. Thus $\mathrm{LG}(n)$ is a compact submanifold of the Grassmannian $\mathrm{Grass}(n, \mathbb{F}^{2n})$, consisting of all maximal isotropic subspaces of $\mathbb{F}^{2n \times 2n}$ with respect to the standard symplectic form Ω. Let

$$\mathrm{Sp}_n(\mathbb{F}) := \{T \in \mathrm{GL}_n(\mathbb{F}) \mid T^\top \Omega T = \Omega\}$$

denote the Lie group of symplectic transformations and

$$\mathfrak{Ham}(n) = \{X \in \mathfrak{gl}_{2n}(\mathbb{F}) \mid X^\top \Omega + \Omega X = 0\}$$

the associated Lie algebra of Hamiltonian $2n \times 2n$-matrices. Then $\mathrm{Sp}_n(\mathbb{F})$ acts transitively on $\mathrm{LG}(n)$ and therefore $\mathrm{LG}(n)$ is a homogeneous space. Thus the above method applies and we obtain the following result. We leave the details of the proof to the reader.

Theorem 8. *The cotangent bundle of the Lagrangian Grassmannian is diffeomorphic to*

$$\mathrm{T}^*\mathrm{LG}(n) \simeq \{(A, \mathsf{V}) \in \mathfrak{Ham}_n(\mathbb{F}) \times \mathrm{LG}(n) \mid A\mathsf{V} = \{0\}, A\mathbb{F}^n \subset \mathsf{V}\}$$

Moreover, the $\mathrm{Sp}_n(\mathbb{F})$*-similarity action*

$$(T, (A, \mathsf{V}) \mapsto (TAT^{-1}, T \cdot \mathsf{V})$$

is a symplectic action. The moment map for this action is

$$\mathrm{pr}_1|_{\mathrm{T}^*\mathrm{LG}(n)} : \ \mathrm{T}^*\mathrm{LG}(n) \longrightarrow \mathfrak{Ham}_n(\mathbb{F}), \ (A, \mathsf{V}) \mapsto A.$$

Its image consists of nilpotent Hamiltonian matrices.

9.4 A generalization of the Hermann–Martin Lemma

Lemma 1 and Theorem 9 below are taken from the classical paper [15] by Robert Hermann and Clyde Martin (their theorems 4.1-4.3 are formulated in the dual setup of controllable pairs).

Lemma 1 (Hermann-Martin Lemma). *Let* $(C, A) \in \mathbb{F}^{p \times n} \times \mathbb{F}^{n \times n}$ *and consider the map*

$$g : \mathbb{F}^{n \times n} \times \mathbb{F}^{n \times p} \longrightarrow \mathbb{F}^{n \times n},$$
$$(X, Y) \mapsto [X, A] - YC$$

The following statements are equivalent.

1. (C, A) is observable.

2. *Let $Z \in \mathbb{F}^{n \times n}$ then $[A, Z] = 0$ and $CZ = 0$ implies $Z = 0$.*
3. *The map g is surjective.*

As an immediate consequence we get the following theorem.

Theorem 9 (Hermann-Martin). *Let $(C, A) \in \mathbb{F}^{p \times n} \times \mathbb{F}^{n \times n}$. The map*

$$f : \mathrm{GL}_n(\mathbb{F}) \times \mathbb{F}^{n \times p} \longrightarrow \mathfrak{gl}_n(\mathbb{F}),$$
$$(T, J) \mapsto T(A - JC)T^{-1}$$

is a submersion if and only if (C, A) is observable. Then the Brunovsky orbit

$$\Gamma_{(C,A)} = \mathrm{Im}\, f = \{T(A - JC)T^{-1} \in \mathfrak{gl}_n(\mathbb{F}) \mid T \in \mathrm{GL}_n(\mathbb{F}), J \in \mathbb{F}^{n \times p}\}$$

is an open submanifold of $\mathfrak{gl}_n(\mathbb{F})$.

Note, that the elements of the Brunovsky orbit are completely character-ized by Rosenbrock's theorem [20]. In the sequel we will utilize Theorem 9 to construct various manifold structures that are closely related to the flag man-ifolds $\mathrm{Flag}(a, \mathbb{F}^n)$ and the manifold $M(a, \mathbb{F}^n)$ of Section 9.3. These manifolds will turn out to be related to observers for linear control systems (Section 9.5), as well as to desingularisations of the variety $\mathcal{N}_n(\mathbb{F})$ of nilpotent matrices (Sec-tion 9.6). Fix a pair $(C, A) \in \mathbb{F}^{p \times n} \times \mathbb{F}^{n \times n}$ and recall the following standard definition from geometric control theory [31].

Definition 3. *An $\mathbb{F}$-linear subspace V of $\mathbb{F}^n$ is called (C, A)-invariant (or* conditioned invariant*) if there exists a $J \in \mathbb{F}^{n \times p}$ such that $(A - JC)\mathsf{V} \subset \mathsf{V}$. Such a J is called a* friend *of V. An equivalent condition is $A(\mathsf{V} \cap \mathrm{Ker}\, C) \subset \mathsf{V}$.*

This definition is easily extended to flags of subspaces.

Definition 4. *A flag $\mathsf{V} = (V_1, \ldots, V_r) \in \mathrm{Flag}(a, \mathbb{F}^n)$ is called (C, A)-invariant (or* conditioned invariant*) if its elements have a common friend, i.e., if there exists a $J \in \mathbb{F}^{n \times p}$ such that $(A - JC)V_i \subset V_i$, $i = 1, \ldots, r$. For the sake of brevity we will write $(A - JC)\mathsf{V} \subset \mathsf{V}$.*

Note that with our notation $V_0 = \{0\}$ and $V_{r+1} = \mathbb{F}^n$ a flag $\mathsf{V} = (V_1, \ldots, V_r) \in \mathrm{Flag}(a, \mathbb{F}^n)$ is conditioned invariant if and only if $(A - JC)V_i \subset V_i$, $i = 0, \ldots, r + 1$. Now consider the set

$$\mathrm{InvJ}(a, \mathbb{F}^n) := \{(J, \mathsf{V}) \in \mathbb{F}^{n \times p} \times \mathrm{Flag}(a, \mathbb{F}^n) \mid (A - JC)\mathsf{V} \subset \mathsf{V}\}$$

of conditioned invariant flags and their friends and the set

$$\mathrm{InvTJ}(a, \mathbb{F}^n) := \{(T, J, \mathsf{V}) \in P(a, \mathbb{F}^n) \mid T(A - JC)T^{-1}\mathsf{V} \subset \mathsf{V}\},$$

where $P(a, \mathbb{F}^n)$ denotes the product space $\mathrm{GL}_n(\mathbb{F}) \times \mathbb{F}^{n \times p} \times \mathrm{Flag}(a, \mathbb{F}^n)$. As with the flag manifolds we will write $\mathrm{InvJ}(\mathbb{F}^n)$ and $\mathrm{InvTJ}(\mathbb{F}^n)$ in the case of full flags, respectively.

Theorem 10. *Let $(C, A) \in \mathbb{F}^{p \times n} \times \mathbb{F}^{n \times n}$ be observable. Then $\mathrm{InvJ}(a, \mathbb{F}^n)$ and $\mathrm{InvTJ}(a, \mathbb{F}^n)$ are smooth manifolds of dimensions pn and $n^2 + pn$, respectively.*

Proof. Theorem 9 implies that the map

$$\phi : P(a, \mathbb{F}^n) \longrightarrow \mathfrak{gl}_n(\mathbb{F}) \times \mathrm{Flag}(a, \mathbb{F}^n),$$
$$(T, J, \mathsf{V}) \mapsto (T(A - JC)T^{-1}, \mathsf{V})$$

is a submersion and hence the preimage $\mathrm{InvTJ}(a, \mathbb{F}^n) = \phi^{-1}(M(a, \mathbb{F}^n))$ is a smooth maniflod. Now consider the self-map $\varphi : (T, J, \mathsf{V}) \mapsto (T, J, T \cdot \mathsf{V})$ on the product manifold $P(a, \mathbb{F}^n)$, where the dot denotes the $\mathrm{GL}_n(\mathbb{F})$-action (9.6) on $\mathrm{Flag}(a, \mathbb{F}^n)$. Clearly, φ is a diffeomorphism and hence Theorem 9 implies that the map

$$\psi : P(a, \mathbb{F}^n) \longrightarrow \mathfrak{gl}_n(\mathbb{F}) \times \mathrm{Flag}(a, \mathbb{F}^n),$$
$$(T, J, \mathsf{V}) \mapsto (T(A - JC)T^{-1}, T \cdot \mathsf{V})$$

is a submersion. Hence the preimage

$$\begin{aligned}
\psi^{-1}(M(a, \mathbb{F}^n)) &= \{(T, J, \mathsf{V}) \in P(a, \mathbb{F}^n) \mid T(A - JC)T^{-1}T \cdot \mathsf{V} \subset T \cdot \mathsf{V}\} \\
&= \{(T, J, \mathsf{V}) \in P(a, \mathbb{F}^n) \mid (A - JC)\mathsf{V} \subset \mathsf{V}\} \\
&= \mathrm{GL}_n(\mathbb{F}) \times \mathrm{InvJ}(a, \mathbb{F}^n)
\end{aligned}$$

is a smooth manifold. Now the canonical left action of $\mathrm{GL}_n(\mathbb{F})$ on itself induces the free and proper action

$$\sigma : \mathrm{GL}_n(\mathbb{F}) \times \psi^{-1}(M(a, \mathbb{F}^n)) \longrightarrow \psi^{-1}(M(a, \mathbb{F}^n)),$$
$$(S, (T, J, \mathsf{V})) \mapsto (ST, J, \mathsf{V})$$

and hence by Theorem 1 the orbit space $\psi^{-1}(M(a, \mathbb{F}^n))/\mathrm{GL}_n(\mathbb{F}) = \mathrm{InvJ}(a, \mathbb{F}^n)$ is a smooth manifold. To verify the dimension formula, we focus on $\mathrm{InvJ}(a, \mathbb{F}^n)$. Since ψ is a submersion, the codimension of $\psi^{-1}(M(a, \mathbb{F}^n))$ in $P(a, \mathbb{F}^n)$ equals the codimension of $M(a, \mathbb{F}^n)$ in $\mathfrak{gl}_n(\mathbb{F}) \times \mathrm{Flag}(a, \mathbb{F}^n)$. Thus

$$n^2 + pn + \dim \mathrm{Flag}(a, \mathbb{F}^n) - \dim \psi^{-1}(M(a, \mathbb{F}^n)) =$$
$$n^2 + \dim \mathrm{Flag}(a, \mathbb{F}^n) - \dim M(a, \mathbb{F}^n) = \dim \mathrm{Flag}(a, \mathbb{F}^n).$$

On the other hand, $\dim \psi^{-1}(M(a, \mathbb{F}^n)) = n^2 + \dim \mathrm{InvJ}(a, \mathbb{F}^n)$, and the result follows. $\square$

Other spaces of interest are the following two subsets of $\mathrm{InvJ}(a, \mathbb{F}^n)$ and $\mathrm{InvTJ}(a, \mathbb{F}^n)$, respectively, where all the friends J yield nilpotent maps $A - JC$.

$$\mathrm{NilJ}(a, \mathbb{F}^n) := \{(J, \mathsf{V}) \in \mathbb{F}^{n \times p} \times \mathrm{Flag}(a, \mathbb{F}^n) \mid (A - JC)V_i \subset V_{i-1}, i = 1, \ldots, r+1\}$$

and

$$\text{NilTJ}(a, \mathbb{F}^n) := \{(T, J, \mathsf{V}) \in P(a, \mathbb{F}^n) \mid T(A - JC)T^{-1}V_i \subset V_{i-1}, i = 1, \ldots, r+1\}.$$

For the case of full flags it follows from the discussion in Section 9.3 (cf. Corollary 2) that indeed

$$\text{NilJ}(\mathbb{F}^n) = \{(J, \mathsf{V}) \in \text{InvJ}(\mathbb{F}^n) \mid A - JC \in \mathcal{N}_n(\mathbb{F})\} \quad \text{and}$$
$$\text{NilTJ}(\mathbb{F}^n) = \{(T, J, \mathsf{V}) \in \text{InvTJ}(\mathbb{F}^n) \mid T(A - JC)T^{-1} \in \mathcal{N}_n(\mathbb{F})\}.$$

Theorem 11. *Let* $(C, A) \in \mathbb{F}^{p \times n} \times \mathbb{F}^{n \times n}$ *be observable. Then* $\text{NilJ}(a, \mathbb{F}^n)$ *and* $\text{NilTJ}(a, \mathbb{F}^n)$ *are smooth manifolds of dimensions* $pn - n^2 + 2 \dim \text{Flag}(a, \mathbb{F}^n)$ *and* $pn - 2 \dim \text{Flag}(a, \mathbb{F}^n)$, *respectively.*

Proof. The proof follows along the lines of the proof of Theorem 10 but replacing $M(a, \mathbb{F}^n)$ with the subbundle $N(a, \mathbb{F}^n)$ from Theorem 6 which is diffeomorphic to the cotangent bundle $\text{T}^*\text{Flag}(a, \mathbb{F}^n)$. $\square$

In view of the Hermann-Martin Lemma 1, it is obvious that the map g is surjective if and only if the restricted map

$$g_S : \mathbb{F}^{n \times n} \times S \longrightarrow \mathbb{F}^{n \times n},$$
$$(X, Y) \mapsto [X, A] - YC$$

is surjective, where S is any vector subspace of $\mathbb{F}^{n \times p}$ with $S + L = \mathbb{F}^{n \times p}$, and

$$L := \{Y \in \mathbb{F}^{n \times p} \mid \exists X \in \mathbb{F}^{n \times n} : [X, A] = YC\}.$$

In [30, Lemma 1], an explicit formula for this solution set is given for an arbitrary pair (C, A) in dual Brunovsky canonical form. As an example for a subspace of minimal dimension choose $S^* := L^\perp$, where $\perp$ denotes the orthogonal complement with respect to a given inner product on $\mathbb{F}^{n \times p}$. However, if we want to derive an analogon to Theorem 9 for restricted J-sets S, matters become more complicated, as one needs to prove the simultaneous surjectivity of *all* the maps

$$g_{S,J} : \mathbb{F}^{n \times n} \times S \longrightarrow \mathbb{F}^{n \times n},$$
$$(X, Y) \mapsto [X, A - JC] - YC$$

with $J \in S$. Of course, as is stated in Theorem 9, the choice $S = \mathbb{F}^{n \times p}$ works, but one would like to know whether there is also such an S of smaller or even minimal dimension.

The relevance of this question will become clearer in Section 9.6, where we relate it to the construction of miniversal deformations for nilpotent similarity orbits and subsequently to resolutions of singularities in $\mathcal{N}_n(\mathbb{F})$. Here is our conjectured generalization of the Hermann-Martin Lemma 1.

Conjecture 1. *Denote by $\perp$ the orthogonal complement with respect to the inner product $(Y, Z) := \operatorname{Re}\operatorname{tr}(Y^*Z)$ on $\mathbb{F}^{n\times p}$, where $Y^* := \bar{Y}^\top$ denotes Hermitian transpose. Let $(C, A) \in \mathbb{F}^{p\times n} \times \mathbb{F}^{n\times n}$ be observable. Then for all $J \in S^* := L^\perp$ the map $g_{S^*,J}$ is surjective and hence the map*

$$f_{S^*} : \operatorname{GL}_n(\mathbb{F}) \times S^* \longrightarrow \mathbb{F}^{n\times n},$$
$$(T, J) \mapsto T(A - JC)T^{-1}$$

is a submersion.

In Section 9.6 we will prove this conjecture in some special cases. For the time being let us state some of its consequences which follow along the lines of the proofs of Theorem 10 and Theorem 11.

Theorem 12. *Let $(C, A) \in \mathbb{F}^{p\times n} \times \mathbb{F}^{n\times n}$ be observable and let $P^*(a, \mathbb{F}^n)$ denote the product space $\operatorname{GL}_n(\mathbb{F}) \times S^* \times \operatorname{Flag}(a, \mathbb{F}^n)$. If Conjecture 1 is true then all of the following sets are smooth manifolds.*

$\operatorname{InvJ}^*(a, \mathbb{F}^n) := \{(J, \mathsf{V}) \in S^* \times \operatorname{Flag}(a, \mathbb{F}^n) \,|\, (A - JC)\mathsf{V} \subset \mathsf{V}\}$

$\operatorname{InvTJ}^*(a, \mathbb{F}^n) := \{(T, J, \mathsf{V}) \in P^*(a, \mathbb{F}^n) \,|\, T(A - JC)T^{-1}\mathsf{V} \subset \mathsf{V}\}$

$\operatorname{NilJ}^*(a, \mathbb{F}^n) := \{(J, \mathsf{V}) \in S^* \times \operatorname{Flag}(a, \mathbb{F}^n) \,|\, (A - JC)V_i \subset V_{i-1}, i = 1, .., r+1\}$

$\operatorname{NilTJ}^*(a, \mathbb{F}^n) := \{(T, J, \mathsf{V}) \in P^*(a, \mathbb{F}^n) \,|\, T(A - JC)T^{-1}V_i \subset V_{i-1}, i = 1, .., r+1\}$

Once again we will write $\operatorname{InvJ}^*(\mathbb{F}^n)$, $\operatorname{InvTJ}^*(\mathbb{F}^n)$, $\operatorname{NilJ}^*(\mathbb{F}^n)$ and $\operatorname{NilTJ}^*(\mathbb{F}^n)$ in the case of full flags, respectively. In this case we get

$$\operatorname{NilJ}^*(\mathbb{F}^n) = \{(J, \mathsf{V}) \in \operatorname{InvJ}^*(a, \mathbb{F}^n) \,|\, A - JC \in \mathcal{N}_n(\mathbb{F})\} \quad and$$
$$\operatorname{NilTJ}^*(\mathbb{F}^n) = \{(T, J, \mathsf{V}) \in \operatorname{InvTJ}^*(a, \mathbb{F}^n) \,|\, T(A - JC)T^{-1} \in \mathcal{N}_n(\mathbb{F})\}$$

Note that all the starred manifolds are submanifolds of the corresponding unstarred ones.

9.5 Conditioned Invariant Subspaces and Observers

In the case of flag length one, i.e. for Grassmann manifolds, we give an alternative proof that $\operatorname{InvJ}((k), \mathbb{F}^n)$ is a smooth manifold. This is done by the technique developed in Theorem 2 and Theorem 3, i.e. by forming a smooth vector bundle over a smooth base manifold and subsequent quotient construction. The base manifold is the manifold of tracking observer parameters which is introduced below. This section thus highlights a surprisingly close connection of the previously developed ideas with observer theory.

We consider linear finite-dimensional time-invariant control systems of the following form.

$$\dot{x} = Ax + Bu$$
$$y = Cx, \tag{9.13}$$

where $A \in \mathbb{F}^{n \times n}$, $B \in \mathbb{F}^{n \times m}$ and $C \in \mathbb{F}^{p \times n}$. It is known that, given $(C, A) \in \mathbb{F}^{p \times n} \times \mathbb{F}^{n \times n}$, the set

$$\mathrm{Inv}_k(C, A) = \{ \mathsf{V} \in \mathrm{Grass}(\mathrm{k}, \mathbb{F}^n) \, | \, \exists J \in \mathbb{F}^{n \times p} : (A - JC)\mathsf{V} \subset \mathsf{V} \}$$

of (C, A)-invariant subspaces with prescribed dimension k allows a stratification into smooth manifolds, the so-called *Brunovsky-Kronecker strata* [6]. However, it is still unclear whether $\mathrm{Inv}_k(C, A)$ is a smooth manifold itself. Consider now the set

$$\mathrm{InvJ}_k := \mathrm{InvJ}((k), \mathbb{F}^n) = \{ (J, \mathsf{V}) \in \mathbb{F}^{n \times p} \times \mathrm{Grass}(k, \mathbb{F}^n) \, | \, (A - JC)\mathsf{V} \subset \mathsf{V} \}$$

of (C, A)-invariant subpaces with fixed dimension and their friends. We want to see that InvJ_k is a smooth manifold by relating it to the manifold of tracking observer parameters through the construction of a smooth vector bundle.

Definition 5. *A* tracking observer *for the linear function Vx of the state of system* (9.13), *$V \in \mathbb{F}^{k \times n}$, is a dynamical system*

$$\dot{v} = Kv + Ly + Mu, \tag{9.14}$$

$K \in \mathbb{F}^{k \times k}$, $L \in \mathbb{F}^{k \times p}$ and $M \in \mathbb{F}^{k \times m}$, which is driven by the input u and by the output y of system (9.13) *and has the* tracking property*: For every $x(0) \in \mathbb{F}^n$, every $v(0) \in \mathbb{F}^k$ and every input function $u(.)$*

$$v(0) = Vx(0) \;\Rightarrow\; v(t) = Vx(t) \text{ for all } t \in \mathbb{R}.$$

k is called the order *of the observer.*

Note that the tracking property makes a statement about all trajectories of system (9.13): whatever starting point $x(0)$ and whatever input $u(t)$ is chosen, setting $v(0) := Vx(0)$ must make the observer track the given function.

Theorem 13. *System* (9.14) *is a tracking observer for Vx if and only if*

$$VA - KV = LC$$
$$M = VB. \tag{9.15}$$

In this case the tracking error $e(t) = v(t) - Vx(t)$ is governed by the differential equation $\dot{e} = Ke$.

Proof. Let the system (9.14) satisfy equations (9.15). Set $e(t) = v(t) - Vx(t)$. Then

$$\dot{e} = \dot{v} - V\dot{x}$$
$$= (Kv + Ly + Mu) - V(Ax + Bu)$$
$$= Kv + LCx + Mu - VAx - VBu$$
$$= Kv - KVx + KVx + LCx - VAx + Mu - VBu$$
$$= K(v - Vx) - (VA - KV - LC)x + (M - VB)u$$
$$= Ke,$$

where the last equation follows from (9.15). Now $e(0) = 0$, i.e., $v(0) = Vx(0)$ implies $e(t) = 0$, i.e., $v(t) = Vx(t)$ for all $t \in \mathbb{R}$.

Conversely let (9.14) be a tracking observer for Vx. Again set $e(t) = v(t) - Vx(t)$. Then

$$\dot{e} = Ke - (VA - KV - LC)x + (M - VB)u.$$

Let $x(0)$ and $u(0)$ be given and set $v(0) = Vx(0)$, i.e., $e(0) = 0$. Then $e(t) = 0$ for all $t \in \mathbb{R}$ implies

$$\dot{e}(0) = Ke(0) - (VA - KV - LC)x(0) + (M - VB)u(0)$$
$$= (VA - KV - LC)x(0) + (M - VB)u(0)$$
$$= 0.$$

Since $x(0)$ and $u(0)$ were arbitrary it follows $VA - KV - LC = M - VB = 0$, i.e., equations (9.15).   □

The next theorem provides the link to conditioned invariant subspaces.

Theorem 14. *There exists a tracking observer for the linear function Vx of the state of system* (9.13) *if and only if* $\mathsf{V} = \operatorname{Ker} V$ *is* (C, A)-*invariant.*

Proof. Let the system (9.14) be a tracking observer for Vx. According to Theorem 13 it follows $VA - KV = LC$. Let $x \in \operatorname{Ker} V \cap \operatorname{Ker} C$. Then $VAx = VAx - KVx = LCx = 0$ and $Ax \in \operatorname{Ker} V$. With $\mathsf{V} = \operatorname{Ker} V$ it follows $A(\mathsf{V} \cap \operatorname{Ker} C) \subset \mathsf{V}$ and V is (C, A)-invariant.

Conversely let $V \in \mathbb{F}^{k \times n}$ and let $\mathsf{V} = \operatorname{Ker} V$ be (C, A)-invariant. There exists $J \in \mathbb{F}^{n \times p}$ such that $(A - JC)\mathsf{V} \subset \mathsf{V}$. But then there exists a matrix $K \in \mathbb{F}^{k \times k}$ such that $V(A - JC) = KV$. Setting $L := VJ$ yields $VA - KV = LC$. Define $M := VB$. According to Theorem 13 the system $\dot{v} = Kv + Ly + Mu$ is a tracking observer for Vx.   □

If V is of full row rank k then the spectrum of a corestriction of A to $\operatorname{Ker} V$, i.e., of the map $(A - JC)|_{\mathbb{F}^n / \mathsf{V}}$ where J is a friend of V, is reflected in the matrix K of an appropriate tracking observer for Vx.

Theorem 15. *Let $V \in \mathbb{F}^{k \times n}$ be of full row rank k. For every friend $J \in \mathbb{F}^{n \times p}$ of $\mathsf{V} := \operatorname{Ker} V$ there exists a unique tracking observer for Vx such that K is similar to $(A - JC)|_{\mathbb{F}^n / \mathsf{V}}$. Conversely, for every tracking observer $\dot{v} = Kv + Ly + Mu$ for Vx there exists a friend J of V such that $(A - JC)|_{\mathbb{F}^n / \mathsf{V}}$ is similar to K.*

Proof. Let $(A - JC)\mathsf{V} \subset \mathsf{V}$ then there exists a matrix $K \in \mathbb{F}^{k \times k}$ such that $V(A - JC) = KV$, i.e. such that the following diagram commutes. Since V has full row rank, K is uniquely determined.

$$
\begin{array}{ccc}
\mathbb{F}^n & \xrightarrow{\ A - JC\ } & \mathbb{F}^n \\
{\scriptstyle V}\downarrow & & \downarrow{\scriptstyle V} \\
\mathbb{F}^k & \xrightarrow{\ \ K\ \ } & \mathbb{F}^k
\end{array}
$$

This induces a quotient diagram with the induced map $\bar{V}$ an isomorphism.

$$
\begin{array}{ccc}
\mathbb{F}^n/\mathsf{V} & \xrightarrow{\ (A - JC)_{\mathbb{F}^n/\mathsf{V}}\ } & \mathbb{F}^n/\mathsf{V} \\
{\scriptstyle \bar{V}}\downarrow & & \downarrow{\scriptstyle \bar{V}} \\
\mathbb{F}^k & \xrightarrow{\ \ K\ \ } & \mathbb{F}^k
\end{array}
\qquad (9.16)
$$

But then K is similar to $(A - JC)|_{\mathbb{F}^n/\mathsf{V}}$. Define $L := VJ$ then the first diagram yields $VA - LC = KV$. Define $M := VB$. It follows by Theorem 13 that $\dot{v} = Kv + Ly + Mu$ is a tracking observer for Vx.

Conversely let $\dot{v} = Kv + Ly + Mu$ be a tracking observer for Vx. It follows by Theorem 13 that $VA - KV = LC$. Since V is surjective there exists $J \in \mathbb{F}^{n \times p}$ such that $L = VJ$. But then $V(A - JC) = KV$ and hence $(A - JC)\mathsf{V} \subset \mathsf{V}$, i.e., J is a friend of V. Furthermore, Diagram (9.16) yields that $(A - JC)|_{\mathbb{F}^n/\mathsf{V}}$ is similar to K. $\square$

If the system (9.13) is observable then the connection between (C, A)-invariant subspaces and tracking observers can be made very precise using the following manifold structure. First a technical lemma.

Lemma 2. *Let* $A \in \mathbb{F}^{n \times n}$, $X \in \mathbb{F}^{n \times k}$ *and* $B \in \mathbb{F}^{k \times k}$. *Then* $AX - XB = 0$ *implies* $A^i X - XB^i = 0$ *for all* $i \in \mathbb{N}$. *In particular, let* $A, X \in \mathbb{F}^{n \times n}$ *then* $[A, X] = AX - XA = 0$ *implies* $[A^i, X] = A^i X - XA^i = 0$ *for all* $i \in \mathbb{N}$.

Proof. The proof is by induction. Assume $A^i X - XB^i = 0$ then $A^i X = XB^i$ and $A^{i+1}X - XB^{i+1} = AA^i X - XBB^i = AXB^i - XBB^i = (AX - XB)B^i = 0$ where the last equality follows from the hypothesis $AX - XB = 0$. $\square$

Theorem 16. *Let the system* (9.13) *be observable and let*

$$
\mathrm{Obs}_k = \{(K, L, M, V) \in \mathbb{F}^{k \times (k+p+m+n)} \mid VA - KV = LC,\ M = VB\}
$$

be the set of all order k tracking observer parameters for system (9.13). *Obs_k is a smooth submanifold of $\mathbb{F}^{k\times(k+p+m+n)}$ of dimension* $\dim \mathrm{Obs}_k = k^2 + kp$. *Its tangent space at the point* $(K, L, M, V) \in \mathrm{Obs}_k$ *is*

$$T_{(K,L,M,V)}\mathrm{Obs}_k = \{(\dot{K}, \dot{L}, \dot{M}, \dot{V}) \mid -\dot{K}V - \dot{L}C + \dot{V}A - K\dot{V} = \dot{M} - \dot{V}B = 0\}.$$

Proof. Consider the map

$$f : \mathbb{F}^{k\times(k+p+m+n)} \longrightarrow \mathbb{F}^{k\times(n+m)},$$
$$(K, L, M, V) \mapsto (VA - KV - LC, M - VB).$$

It will be shown that $(0,0)$ is a regular value of f, hence $\mathrm{Obs}_k = f^{-1}(0,0)$ is a smooth submanifold of $\mathbb{F}^{k\times(k+p+m+n)}$. The derivative of f at a point (K, L, M, V) is given by

$$Df : (\dot{K}, \dot{L}, \dot{M}, \dot{V}) \mapsto (-\dot{K}V - \dot{L}C + \dot{V}A - K\dot{V}, \dot{M} - \dot{V}B),$$

where $(\dot{K}, \dot{L}, \dot{M}, \dot{V}) \in T_{(K,L,M,V)}(\mathbb{F}^{k\times(k+p+m+n)})$.

An element $(\xi, \eta) \in T_{f(K,L,M,V)}(\mathbb{F}^{k\times(n+m)})$ is orthogonal to the image of Df if and only if

$$\operatorname{tr}\xi^*(-\dot{K}V - \dot{L}C + \dot{V}A - K\dot{V}) + \operatorname{tr}\eta^*(\dot{M} - \dot{V}B) = 0$$

for all $(\dot{K}, \dot{L}, \dot{M}, \dot{V}) \in T_{(K,L,M,V)}(\mathbb{F}^{k\times(k+p+m+n)})$. This is equivalent to

$$V\xi^* = 0$$
$$C\xi^* = 0 \tag{9.17}$$
$$\eta^* = 0$$
$$A\xi^* - \xi^* K = 0. \tag{9.18}$$

From (9.18) it follows by Lemma 2 $A^i\xi^* - \xi^* K^i = 0$ for all $i \in \mathbb{N}$. Together with (9.17) this yields

$$\begin{pmatrix} C \\ CA \\ \vdots \\ CA^{n-1} \end{pmatrix} \xi^* = 0.$$

Since (C, A) is observable this implies $\xi^* = 0$. It follows that Df is surjective, f is a submersion and hence $(0,0)$ is a regular value of f.

The dimension of $\mathrm{Obs}_k = f^{-1}(0,0)$ is $k(k+p+m+n) - k(n+m) = k^2 + kp$. From the fibre theorem it follows $T_{(K,L,M,V)}\mathrm{Obs}_k = (Df)^{-1}(0,0)$. $\square$

Corollary 3. *Being an open subset of Obs_k the set*

$$\mathrm{Obs}_{k,k} = \{(K, L, M, V) \in \mathrm{Obs}_k \mid \operatorname{rk} V = k\}$$

is a smooth submanifold of $\mathbb{F}^{k\times(k+p+m+n)}$ of dimension $k^2 + kp$.

Now consider the *similarity action* on $\mathrm{Obs}_{k,k}$

$$\sigma : \mathrm{GL}_k(\mathbb{F}) \times \mathrm{Obs}_{k,k} \longrightarrow \mathrm{Obs}_{k,k},$$
$$(S, (K, L, M, V)) \mapsto (SKS^{-1}, SL, SM, SV)$$

and the induced *similarity classes*

$$[K, L, M, V]_\sigma = \{(SKS^{-1}, SL, SM, SV) \mid S \in \mathrm{GL}_k(\mathbb{F})\}.$$

Note that σ is well defined since $VA - KV = LC$ and $M = VB$ imply $SVA - SKS^{-1}SV = S(VA - KV) = SLC$ and $SM = SVB$.

Theorem 17. *The orbit space*

$$\mathrm{Obs}^\sigma_{k,k} = \{[K, L, M, V]_\sigma \mid (K, L, M, V) \in \mathrm{Obs}_{k,k}\}$$

of similarity classes of order k tracking observer parameters for system (9.13) *is a smooth manifold of dimension* $\dim \mathrm{Obs}^\sigma_{k,k} = kp$.

Proof. Since V has full row rank k for $(K, L, M, V) \in \mathrm{Obs}_{k,k}$, the similarity action is free and has a closed graph mapping (cf. Section 9.2): $SV = V$ implies $S = I$, furthermore $V_j \to V$ and $S_j V_j \to W$ imply $S_j \to S$ and $W = SV$. Hence the orbit space of σ is a smooth manifold of dimension $\dim \mathrm{Obs}^\sigma_{k,k} = \dim \mathrm{Obs}_{k,k} - \dim \mathrm{GL}_k(\mathbb{F}) = k^2 + kp - k^2 = kp$. $\square$

Finally, we are in the position to prove that InvJ_k is smooth.

Theorem 18. *Let the system* (9.13) *be observable. For each k the set*

$$\mathrm{InvJ}_k = \{(J, \mathsf{V}) \in \mathbb{F}^{n \times p} \times \mathrm{Grass}(k, \mathbb{F}^n) \mid (A - JC)\mathsf{V} \subset \mathsf{V}\}$$

is a smooth manifold of dimension $\dim \mathrm{InvJ}_k = np$. *The map*

$$\bar{f} : \mathrm{InvJ}_{n-k} \longrightarrow \mathrm{Obs}^\sigma_{k,k},$$
$$(J, \mathsf{V}) \mapsto [K, L, M, V]_\sigma,$$

defined by $\ker V = \mathsf{V}$, $M = VB$, $L = VJ$ *and* $KV = VA - LC = V(A - JC)$ *is a smooth vector bundle with fiber* $\mathbb{F}^{(n-k) \times p}$.

Proof. Consider the set

$$\mathcal{M}_{n-k} = \{(J, V) \in \mathbb{F}^{n \times p} \times \mathrm{St}(k, n) \mid (A - JC)\ker V \subset \ker V\},$$

where $\mathrm{St}(k, n)$ denotes the set of full row rank $k \times n$ matrices (*Stiefel manifold*). Apparently, if $(J, V) \in \mathcal{M}_{n-k}$ then $\ker V$ is a codimension k (C, A)-invariant subspace with friend J. Consider the map

$$f : \mathcal{M}_{n-k} \longrightarrow \mathrm{Obs}_{k,k},$$
$$(J, V) \mapsto (K, L, M, V),$$

where $L = VJ$, $M = VB$ and K is defined as the unique solution of the equation $KV = VA - LC = V(A - JC)$ (cf. Theorem 15, Part 1). By Theorem 15, Part 2, the map f is surjective. Since $K = V(A - JC)V^*(VV^*)^{-1}$, the map f is continuous. Moreover, it is the restriction of a smooth map defined on $\mathbb{F}^{n \times p} \times \mathrm{St}(k, n)$, which is an open subset of $\mathbb{F}^{n \times p} \times \mathbb{F}^{k \times n}$. According to Corollary 3 the set $\mathrm{Obs}_{k,k}$ is a smooth submanifold of $\mathbb{F}^{k \times (k+p+m+n)}$.

Given V and $L = VJ$, the solution set of $VX = VJ$ is the affine space $V^*(VV^*)^{-1}(VJ) + \prod_{i=1}^{p} \mathrm{Ker}\, V$. Furthermore, $\dim \mathrm{Ker}\, V = n - k$. Therefore, for every $(K, L, M, V) \in \mathrm{Obs}_{k,k}$ the fiber $f^{-1}(K, L, M, V)$ is an affine space of dimension $(k - n)p$.

Let $V_0 \in \mathrm{St}(k, n)$. Since V_0 has full row rank there exists a permutation matrix P_0 such that $V_0 P_0 = (X_0\ Y_0)$ with $X_0 \in \mathbb{F}^{k \times k}$ invertible. Then $W = \{(X\ Y)\, P_0^{-1} \mid X \text{ invertible}\}$ is an open neighborhood of V_0 in $\mathrm{St}(k, n)$ and $\mathrm{Ker}\, V = \{P_0([-X^{-1}Yy]^\top, y^\top)^\top \mid y \in \mathbb{F}^{n-k}\}$ for every $V = (X\ Y)\, P_0^{-1} \in W$. But then

$$\varphi_W : W \times \mathbb{F}^{n-k} \longrightarrow \mathbb{F}^n \times W,$$

$$(V, y) \mapsto \left(P_0\left[I_n - \begin{pmatrix} I_k \\ 0 \end{pmatrix} \left(V P_0 \begin{pmatrix} I_k \\ 0 \end{pmatrix}\right)^{-1} V P_0\right] \begin{pmatrix} 0 \\ I_{n-k} \end{pmatrix} y, V\right) = $$

$$\left(P_0([-X^{-1}Yy]^\top, y^\top)^\top, V\right)$$

is a smooth injective map mapping $(V, \mathbb{F}^{n-k})$ onto $(\mathrm{Ker}\, V, V)$ for every $V \in W$. Hence φ_W is a homeomorphism onto its image. The inverse map

$$\varphi_W^{-1} : \varphi_W(W \times \mathbb{F}^{n-k}) \longrightarrow W \times \mathbb{F}^{n-k},$$

$$(z, V) \mapsto (V, (0\ I_{n-k})\, P_0^{-1} z)$$

is the restriction of a smooth map defined on all of $\mathbb{F}^n \times \mathbb{F}^{k \times n}$. If $V = (X_1\ Y_1)\, P_1^{-1} = (X_2\ Y_2)\, P_2^{-1} \in W_1 \cap W_2$ then the change of coordinate function $\varphi_{W_2}^{-1} \circ \varphi_{W_1}$ induces the invertible linear transformation

$$\vartheta_{W_2, W_1, V} : y \mapsto (0\ I_{n-k})\, P_2^{-1} P_1 \begin{pmatrix} -X_1^{-1} Y_1 \\ I_{n-k} \end{pmatrix} y$$

on $\mathbb{F}^{n-k}$.

Using p-fold products of φ_W it is now easy to construct local trivializations of f. Given $(K_0, L_0, M_0, V_0) \in \mathrm{Obs}_{k,k}$ choose the neighborhood

$$U := (\mathbb{F}^{k \times k} \times \mathbb{F}^{k \times p} \times \mathbb{F}^{k \times m} \times W) \cap \mathrm{Obs}_{k,k},$$

which is open in $\mathrm{Obs}_{k,k}$. Let pr_1 denote the projection $(z, V) \mapsto z$, pr_2 the projection $(V, y) \mapsto y$ and consider the map

$$\phi_U : U \times \mathbb{F}^{(n-k) \times p} \longrightarrow f^{-1}(U),$$

$$(K, L, M, V, (y_1\ \ldots\ y_p)) \mapsto$$

$$\left(V^*(VV^*)^{-1} L + (\mathrm{pr}_1(\varphi_W(V, y_1))\ \cdots\ \mathrm{pr}_1(\varphi_W(V, y_p)))\, , V\right),$$

where y_i, $i = 1, \ldots, p$, denotes the i-th column of the matrix $Y \in \mathbb{F}^{(n-k) \times p}$. Apparently, $f(\phi_U(K, L, M, V, Y)) = (K, L, M, V)$ for all $(K, L, M, V) \in U$ and all $Y \in \mathbb{F}^{(n-k) \times p}$. Furthermore, ϕ_U is bijective by construction. Since φ_W is smooth, so is ϕ_U. Let e_i, $i = 1, \ldots, p$, denote the i-th standard basis vector of $\mathbb{F}^p$. The inverse map

$$\phi_U^{-1} : f^{-1}(U) \longrightarrow U \times \mathbb{F}^{(n-k) \times p},$$

$$(J, V) \mapsto (f(J, V), \big(g_1(J, V) \, \ldots \, g_p(J, V)\big)),$$

where

$$g_i(J, V) = \mathrm{pr}_2(\varphi_W^{-1}([J - V^*(VV^*)^{-1}(VJ)]e_i, V)), \quad i = 1, \ldots, p,$$

is the restriction of a smooth map defined on $\mathbb{F}^{n \times p} \times \mathrm{St}(k, n)$, which is an open subset of $\mathbb{F}^{n \times p} \times \mathbb{F}^{k \times n}$. It follows that ϕ_U is a homeomorphism. Finally, if $(K, L, M, V) \in U_1 \cap U_2 = (\mathbb{F}^{k \times k} \times \mathbb{F}^{k \times p} \times \mathbb{F}^{k \times m} \times (W_1 \cap W_2)) \cap \mathrm{Obs}_{k,k}$ then the change of coordinate function $\phi_{U_2}^{-1} \circ \phi_{U_1}$ induces the invertible linear transformation

$$\theta_{U_2, U_1, (K, L, M, V)} : \big(y_1 \, \ldots \, y_p\big) \mapsto \big(\vartheta_{W_2, W_1, V}(y_1) \, \ldots \, \vartheta_{W_2, W_1, V}(y_p)\big)$$

on $\mathbb{F}^{(n-k) \times p}$.

According to Theorem 2 the set $\mathcal{M}_{n-k}$ is a smooth submanifold of $\mathbb{F}^{n \times p} \times \mathbb{F}^{k \times n}$ of dimension $\dim \mathrm{Obs}_{k,k} + (n-k)p = k^2 + kp + (n-k)p = k^2 + np$. Furthermore, the map f is a smooth vector bundle with fiber $\mathbb{F}^{(n-k) \times p}$.

As has been shown in the proof of Theorem 17, the similarity action σ on $\mathrm{Obs}_{k,k}$ is free and proper (cf. Theorem 1). By the same arguments this is also true for the similarity action on $\mathcal{M}_{n-k}$:

$$\sigma : \mathrm{GL}_k(\mathbb{F}) \times \mathcal{M}_{n-k} \longrightarrow \mathcal{M}_{n-k},$$

$$(S, (J, V)) \mapsto (J, SV).$$

As is well known, the quotient space $\mathrm{St}(k, n)/\mathrm{GL}_k(\mathbb{F})$ is diffeomorphic to $\mathrm{Grass}(n - k, \mathbb{F}^n)$ via $[V]_\sigma \mapsto \mathrm{Ker}\, V$, hence the quotient $\mathcal{M}_{n-k}/\mathrm{GL}_k(\mathbb{F})$ is diffeomorphic to InvJ_{n-k} and the latter is a smooth manifold of dimension $\dim \mathcal{M}_{n-k} - \dim \mathrm{GL}_k(\mathbb{F}) = k^2 + np - k^2 = np$.

Apparently, $(K, L, M, V) \in U$ implies $(SKS^{-1}, SL, SM, SV) \in U$ since S being invertible and $V \in W$ implies $SV \in W$. Furthermore,

$$\phi_U(\sigma(S, (K, L, M, V)), Y) =$$

$$\phi_U((SKS^{-1}, SL, SM, SV), Y) =$$

$$((SV)^*(SV(SV)^*)^{-1}SL + \big(\mathrm{pr}_1(\varphi_W(SV, y_1)) \, \ldots \, \mathrm{pr}_1(\varphi_W(SV, y_p))\big), SV) =$$

$$(V^*(VV^*)^{-1}L + \big(\mathrm{pr}_1(\varphi_W(V, y_1)) \, \ldots \, \mathrm{pr}_1(\varphi_W(V, y_p))\big), SV) =$$

$$\sigma(S, (V^*(VV^*)^{-1}L + \big(\mathrm{pr}_1(\varphi_W(V, y_1)) \, \ldots \, \mathrm{pr}_1(\varphi_W(V, y_p))\big), V)) =$$

$$\sigma(S, \phi_U((K, L, M, V), Y)).$$

But then Theorem 3 implies that $\bar{f}$ is a smooth vector bundle with fiber $\mathbb{F}^{(n-k) \times p}$, which completes the proof. $\square$

9.6 A Resolution of Singularities for Nilpotent Matrices

In this section we will put our previous results together to construct a resolution of singularities for the variety $\mathcal{N}_n(\mathbb{F})$ of nilpotent $n \times n$ matrices. In order to do so, we first have to construct suitable transversal sections to nilpotent similarity orbits. We give two different ways to do so. The first one is standard and has been introduced by Arnol'd [2]. The second one is inspired by system theory and uses tangent spaces to output injection orbits.

Fix a nilpotent element $A \in \mathcal{N}_n(\mathbb{F})$. The tangent space to the similarity orbit $\mathcal{O} = \{SAS^{-1} \mid S \in \mathrm{GL}_n(\mathbb{F})\}$ at A then is

$$T_A\,\mathcal{O} = \{[X, A] \mid X \in \mathfrak{gl}_n(\mathbb{F})\}.$$

Definition 6. *Let $\mathcal{S} \subset \mathfrak{gl}_n(\mathbb{F})$ be a linear subspace with*

$$\mathfrak{gl}_n(\mathbb{F}) = \mathcal{S} + T_A\mathcal{O}.$$

Then $A + \mathcal{S}$ is called an affine transverse section *of $\mathcal{O}$ at A.*

Our goal now is to construct such affine transverse sections for similarity orbits of nilpotent matrices. This is closely related to work of Arnol'd [2] on versal deformations of matrices. We briefly review his construction. Consider the positive definite inner product on $\mathfrak{gl}_n(\mathbb{F})$ defined as

$$(X, Y) := \mathrm{Re}\,\mathrm{tr}(X^*Y),$$

where $X^* := \bar{X}^{\top}$ denotes Hermitian transpose. Let $ad_A : \mathfrak{gl}_n(\mathbb{F}) \longrightarrow \mathfrak{gl}_n(\mathbb{F})$ denote the adjoint transformation by A, i.e., $ad_A(X) := [A, X] = AX - XA$. For any $A \in \mathfrak{gl}_n(\mathbb{F})$, a straightforward computation shows that $\mathrm{Im}(ad_A)^{\perp} = \mathrm{Ker}(ad_{A^*})$. Thus we have shown

Proposition 2. *For any $A \in \mathfrak{gl}_n(\mathbb{F})$, the affine subspace*

$$\mathcal{S}^{\mathrm{A}} := A + \mathrm{Ker}(ad_{A^*})$$

is an affine transverse section to the similarity orbit $\mathcal{O}$ at A. Here

$$\mathrm{Ker}(ad_{A^*}) = \{X \in \mathfrak{gl}_n(\mathbb{F}) \mid [A^*, X] = 0\}.$$

We refer to $\mathcal{S}^{\mathrm{A}}$ as the Arnol'd slice *to $\mathcal{O}$.*

The Arnol'd slice has the advantage that it is a transversal slice of smallest possible dimension. Thus we have the direct sum decomposition

$$\mathfrak{gl}_n(\mathbb{F}) = \mathrm{Ker}(ad_{A^*}) \oplus T_A\,\mathcal{O}.$$

We illustrate this construction by an example. Let A denote the sub-regular nilpotent matrix in Jordan canonical form

$$A = \begin{pmatrix} 0 & 1 & 0 & 0 \\ 0 & 0 & 1 & 0 \\ 0 & 0 & 0 & 0 \\ 0 & 0 & 0 & 0 \end{pmatrix}.$$

The tangent vectors $X \in T_A \mathcal{O}$ are of the form

$$X = \begin{pmatrix} a & b & c & d \\ e & f & g & h \\ 0 & -e & i & 0 \\ 0 & j & k & 0 \end{pmatrix}$$

with $f = -a - i$. The kernel $\mathrm{Ker}(ad_{A^*})$ consists of all matrices X of the form

$$X = \begin{pmatrix} a & 0 & 0 & 0 \\ b & a & 0 & 0 \\ c & b & a & f \\ d & 0 & 0 & e \end{pmatrix}$$

and thus the Arnol'd slice consists of all matrices of the form

$$\begin{pmatrix} a & 1 & 0 & 0 \\ b & a & 1 & 0 \\ c & b & a & f \\ d & 0 & 0 & e \end{pmatrix}.$$

The nilpotent matrices in this transversal section are characterized by $ec - df - abe = 0$. Note that the block-Toeplitz matrices $X \in \mathrm{Ker}(ad_{A^*})$ can be interpreted as partial reachability matrices $X = (g_1, Fg_1, F^2 g_1, g_2)$, where $F := A^\top$ and $g_1 := (a, b, c, d)^\top$, $g_2 := (0, 0, f, e)^\top$. Indeed, this is no coincidence, as can be seen from [10].

Let us now develop a system theoretic approach to the construction of transversal sections. Choose an output matrix $C \in \mathbb{F}^{p \times n}$, $p \leq n$ suitable, such that (C, A) is in dual Brunovsky canonical form.

Definition 7. *The* Brunovsky slice *is the affine subspace* $\mathcal{S}^{\mathrm{B}} := A + \mathcal{L}_C$, *where* $\mathcal{L}_C$ *denotes the Lie subalgebra of* $\mathfrak{gl}_n(\mathbb{F})$

$$\mathcal{L}_C := \{ JC \mid J \in \mathbb{F}^{n \times p} \}.$$

The first observation we make is that the Brunovsky slice is always an affine transversal section to the nilpotent similarity orbit. In fact, the following lemma is an immediate consequence of the Hermann-Martin Lemma 1.

Lemma 3. *Let* $(C, A) \in \mathbb{F}^{p \times n} \times \mathbb{F}^{n \times n}$ *be an observable pair in dual Brunovsky canonical form. Then the Brunovsky slice* $\mathcal{S}^{\mathrm{B}}$ *defines an affine transversal section for the nilpotent similarity orbit* $\mathcal{O}$ *at* A.

The linear subspace $\mathcal{L}_C$ has obvious invariance properties.

Proposition 3. *Let* $(C, A) \in \mathbb{F}^{p \times n} \times \mathbb{F}^{n \times n}$ *be an observable pair in dual Brunovsky canonical form and* $H := [A, A^T]$. *Then*

$$ad_{A^T}(\mathcal{L}_C) \subset \mathcal{L}_C, \ ad_H(\mathcal{L}_C) \subset \mathcal{L}_C, \ ad_A(\mathcal{L}_C) \subset \mathcal{L}_C + \mathcal{L}_{CA}.$$

Proof. The first formula is obvious, in view of the fact that $CA^T = 0$ holds for any for Brunovsky pair. The last formula is also straightforward to see. For the middle one we observe that H is a diagonal matrix. Note, that C has exactly one nonzero entry $(= 1)$ in each row. But then $CH = DC$ for a diagonal matrix D. Thus $ad_H(JC) = HJC - JCH = (HJ - JD)C$ and the result follows. $\square$

While the Brunovsky slice intersects the similarity orbit transversally at A, it is not one of smallest dimension. It is therefore of interest to see, if one can reduce the number of parameters in the transversal subspace $\mathcal{L}_C$ to obtain a transversal section of minimal dimension. The idea here is to replace $\mathcal{L}_C$ by the intersection

$$\mathcal{L}_C^* := \mathcal{L}_C \cap (\mathcal{L}_C \cap T_A \mathcal{O})^\perp$$

of dimension $\dim \mathcal{L}_C^* = \dim \mathcal{L}_C - \dim(\mathcal{L}_C \cap T_A \mathcal{O}) = n^2 - \dim T_A \mathcal{O}$. We refer to

$$\mathcal{S}_{\min}^{\mathrm{B}} := A + \mathcal{L}_C^*$$

as the *minimal Brunovsky slice*. Before presenting an explicit description of $\mathcal{L}_C^*$ in the general case, let us return to the previous example. With C chosen as

$$C = \begin{pmatrix} 1 \ 0 \ 0 \ 0 \\ 0 \ 0 \ 0 \ 1 \end{pmatrix}$$

the Brunovsky slice $\mathcal{L}_C$ consists of matrices of the form

$$X = \begin{pmatrix} a \ 0 \ 0 \ e \\ b \ 0 \ 0 \ f \\ c \ 0 \ 0 \ g \\ d \ 0 \ 0 \ h \end{pmatrix}.$$

Comparing this with the above formula for the tangent space elements we conclude that $\mathcal{L}_C^*$ consists of matrices

$$X = \begin{pmatrix} a \ 0 \ 0 \ 0 \\ b \ 0 \ 0 \ 0 \\ c \ 0 \ 0 \ g \\ d \ 0 \ 0 \ h \end{pmatrix}.$$

Note that this has exactly the same number of parameters as in the Arnol'd slice, as it should be.

We will now show that our Conjecture 1 holds for this specific pair (C, A). In the terminology of Section 9.4 we have $L = \{J \in \mathbb{F}^{n \times p} \mid JC \in \mathcal{L}_C \cap T_A \mathcal{O}\}$

and $S^* = \{J \in \mathbb{F}^{n \times p} \mid JC \in \mathcal{L}_C^*\}$. We have to show that $\mathcal{L}_C^*$ is transversal to all the spaces $T_J := \{[X, A - JC] \mid X \in \mathfrak{gl}_n(\mathbb{F})\}$ for *all* $J \in S^*$, not just for $J = 0$ where $T_0 = \mathrm{T}_A \mathcal{O}$. Note that T_J is the tangent space to the similarity orbit of $A - JC$ at $A - JC$. For

$$X = \begin{pmatrix} a & e & i & m \\ b & f & j & n \\ c & g & k & p \\ d & h & l & q \end{pmatrix} \quad \text{compute} \quad [X, A] = \begin{pmatrix} -b & a-f & e-j & -n \\ -c & b-g & f-k & -p \\ 0 & c & g & 0 \\ 0 & d & h & 0 \end{pmatrix}$$

and note that the matrix elements in $[X, A]$ that correspond to the zeros in the elements of $\mathcal{L}_C^*$ can be arbitrarily assigned by choosing values for a, b, c, d (second column), e, f, g, h (third column) and n and p (fourth column) and hence $\mathcal{L}_C^*$ is transversal to T_0. We will focus on these elements and see how they are affected by introducing $J \in S^*$. We have

$$JC = \begin{pmatrix} \alpha & 0 & 0 & 0 \\ \beta & 0 & 0 & 0 \\ \gamma & 0 & 0 & \epsilon \\ \delta & 0 & 0 & \phi \end{pmatrix}$$

and hence looking at the second column of $[X, A]$ the $a - f$ entry is modified by adding $-\alpha e$, the $b - g$ entry by $-\beta e$, the c entry by $-\gamma e - \epsilon h$ and the d entry by $-\delta e - \epsilon h$. The decisive thing is that the modifications do not depend on a, b, c, d, and hence we can compensate for them by choosing those variables accordingly. The analogous argument applies to the third column. In the fourth column, $-n$ is modified by adding $i\epsilon + m\phi - \alpha m$ which is again independent of n, and the analogous argument applies to the remaining entry. Together we get that $\mathcal{L}_C^*$ is transversal to all T_J, $J \in S^*$, and Conjecture 1 follows for this pair (C, A).

A second example is given by a Brunovsky pair (C, A) with generic observability indices $(n, \cdots, n)$. Thus consider the nilpotent $pn \times pn$ block matrix

$$A = \begin{pmatrix} 0 & I_p & \cdots & 0 \\ \vdots & \ddots & \ddots & \vdots \\ \vdots & & \ddots & I_p \\ 0 & \cdots & \cdots & 0 \end{pmatrix}$$

and

$$C = \begin{pmatrix} I_p & 0 & \cdots & 0 \end{pmatrix}.$$

In this case the Brunovsky slice $\mathcal{S}^{\mathrm{B}}$ coincides with the minimal Brunovsky slice $\mathcal{S}_{\min}^{\mathrm{B}}$ and their elements are of the form

$$X = \begin{pmatrix} X_{11} & I_p & \cdots & 0 \\ \vdots & \vdots & \ddots & \vdots \\ X_{n-1,1} & 0 & \cdots & I_p \\ X_{n1} & 0 & \cdots & 0 \end{pmatrix}$$

for suitable $p \times p$ matrices X_{i1}, $i = 1, \ldots, n$.

Note, that in this case $\mathcal{L}_C^* = \mathbb{F}^{n\times p}C$ and hence $S^* = \mathbb{F}^{n\times p}$ in the terminology of Section 9.4. But in this case Conjecture 1 is a direct consequence of the Hermann-Martin Lemma 1, cf. Theorem 9, and is hence also true for the case of generic observability indices.

For a general pair $(C, A) \in \mathbb{F}^{p\times n} \times \mathbb{F}^{n\times n}$ in dual Brunovsky form with observability indices $\mu_1 \geq \cdots \geq \mu_p$, i.e.,

$$C = \left(\begin{array}{c|c|c} \begin{array}{c} 0 \ldots 0\, 1 \end{array} & & \\ \hline & \ddots & \\ \hline & & 0 \ldots 0\, 1 \end{array}\right) \underbrace{}_{\mu_1} \cdots \underbrace{}_{\mu_p} \quad \text{and} \quad A = \left(\begin{array}{c|c|c} \begin{array}{c} 0 \\ 1 \ddots \\ \ddots \ddots \\ 1 \ 0 \end{array} & & \\ \hline & \ddots & \\ \hline & & \begin{array}{c} 0 \\ 1 \ddots \\ \ddots \ddots \\ 1 \ 0 \end{array} \end{array}\right) \underbrace{}_{\mu_1} \cdots \underbrace{}_{\mu_p},$$

an explicit formula for $\mathcal{L}_C \cap T_A\mathcal{O}$ has been given in [30, Lemma 1]. A straightforward calculation then shows $\mathcal{L}_C^* = \{JC \mid J = (J_{ij})_{i,j=1}^p\}$, where

$$J_{ij} \in \mathbb{F}^{\mu_i \times 1}, \quad J_{ij} = \begin{cases} \left(y_{ij}^1 \; \cdots \; y_{ij}^{\mu_i}\right)^\top & \text{if } \mu_i \leq \mu_j, \\ \left(y_{ij}^1 \; \cdots \; y_{ij}^{\mu_j} \; 0 \; \ldots \; 0\right)^\top & \text{if } \mu_i > \mu_j. \end{cases}$$

Moreover, we get the dimension formula

$$\dim \mathcal{S}_{\min}^{\mathrm{B}} = \sum_{i,j=1}^p \min\{\mu_i, \mu_j\}.$$

In our first example ($\mu_1 = 3$, $\mu_2 = 1$) we would get

$$\mathcal{L}_C^* = \left\{ \begin{pmatrix} 0 & 0 & a & e \\ 0 & 0 & b & 0 \\ 0 & 0 & c & 0 \\ 0 & 0 & d & f \end{pmatrix} \;\middle|\; a, b, c, d, e, f \in \mathbb{F} \right\}$$

which coincides with our previous result after accounting for the permutation of variables that relates the two (C, A) pairs. Note that a change of variables does not affect transversality of $\mathcal{L}_C^*$ and T_J, since $[X, S(A - JC)S^{-1}] - YCS^{-1} = S([S^{-1}XS, A - JC] - S^{-1}YC)S^{-1}$ for all $S \in \mathrm{GL}_n(\mathbb{F})$.

Algebraic geometers have early found interest in the deformation analysis of similarity orbits of semisimple groups. In his address at the International Congress of Mathematicians in Nice 1970, Brieskorn outlined a program how the singularities of nilpotent matrices may contribute to a deeper understanding of classical geometric problems, such as isolated singularities for complex surfaces. We briefly recall the most important results in this direction, specializing to the simplest case of the general linear group $\mathrm{GL}_n(\mathbb{F})$. Consider the complete flag manifold $\mathrm{Flag}(\mathbb{F}^n)$, given as in Section 9.3. We have already shown in Theorem 6 and Corollary 2 that

$$\mathrm{T}^*\mathrm{Flag}(\mathbb{F}^n) \simeq N(\mathbb{F}^n) =$$
$$\{(A,(V_1,\ldots,V_n)) \in \mathfrak{gl}_n(\mathbb{F}) \times \mathrm{Flag}(\mathbb{F}^n) \mid AV_i \subset V_{i-1}, i=1,\ldots,n\} =$$
$$\{(A,(V_1,\ldots,V_n)) \in \mathcal{N}_n(\mathbb{F}) \times \mathrm{Flag}(\mathbb{F}^n) \mid AV_i \subset V_i, i=1,\ldots,n-1\}$$

is a smooth manifold. Its dimension is $n(n-1)$, i.e. twice the dimension of $\mathrm{Flag}(\mathbb{F}^n)$. The dualized moment map for the natural $\mathrm{GL}_n(\mathbb{F})$-action on the cotangent bundle $\mathrm{T}^*\mathrm{Flag}(\mathbb{F}^n)$ coincides with the projection on the first factor

$$\mathrm{pr}_1 : \mathrm{T}^*\mathrm{Flag}(\mathbb{F}^n) \longrightarrow \mathfrak{gl}_n(\mathbb{F}), \ (A,(V_1,\ldots,V_r)) \mapsto A.$$

Moreover, its image set is equal to the singular algebraic variety $\mathcal{N}_n(\mathbb{F})$ of nilpotent matrices. Note, that the dimension of $\mathcal{N}_n(\mathbb{F})$ is also equal to $n(n-1)$ and therefore the two sets have equal dimension. Now suppose that A is a nilpotent matrix with a single Jordan block. Thus we assume that A is cyclic. Then A has a unique A-invariant flag. This shows

Proposition 4 (Steinberg [27]). *Let*

$$\mathrm{T}^*\mathrm{Flag}^{\mathrm{reg}}(\mathbb{F}^n) := \{(A,(V_1,\ldots,V_n)) \in N(\mathbb{F}^n) \mid A \ cyclic\}$$

and let $\mathcal{N}_n^{\mathrm{reg}}(\mathbb{F})$ denote the set of cyclic nilpotent $n \times n$ matrices. Then $\mathrm{T}^*\mathrm{Flag}^{\mathrm{reg}}(\mathbb{F}^n)$ and $\mathcal{N}_n^{\mathrm{reg}}(\mathbb{F})$ are smooth manifolds that are open and dense in $\mathrm{T}^*\mathrm{Flag}(\mathbb{F}^n)$ and $\mathcal{N}_n(\mathbb{F})$, respectively. Moreover, the dualized moment map restricts to a diffeomeorphism

$$\mathrm{pr}_1|_{\mathrm{T}^*\mathrm{Flag}^{\mathrm{reg}}(\mathbb{F}^n)} : \mathrm{T}^*\mathrm{Flag}^{\mathrm{reg}}(\mathbb{F}^n) \longrightarrow \mathcal{N}_n^{\mathrm{reg}}(\mathbb{F}).$$

The result shows that, indeed, the dualized moment map defines a desingularization of the nilpotent variety $\mathcal{N}_n(\mathbb{F})$. Next, consider the subregular case of nilpotent matrices with a subgeneric Jordan structure, i.e., one nilpotent block of size $(n-1) \times (n-1)$ and a second (zero block) of size 1×1. These matrices thus constitute a single similarity orbit of nilpotent ones. In contrast to the regular case, the fibres $\mathrm{pr}_1^{-1}(A)$ of a subregular nilpotent matrix are not single points, but form a two-dimensional variety of A-invariant flags. We quote the following result that answers a conjecture of Grothendieck.

Theorem 19 (Brieskorn [4]). *Let A be a subregular nilpotent matrix and $\mathcal{S}$ an $n+2$-dimensional transversal section to the similarity orbit $\mathcal{O}$. Then*

1. *The intersection $\mathcal{S} \cap \mathcal{N}_n(\mathbb{C})$ is a two-dimensional complex surface with an isolated singularity at the point A. The singularity is Kleinian and in fact isomorpic to the surface singularity $\mathbb{C}^2/G$, where G is the cyclic subgroup of $\mathrm{SU}(2)$ of order n.*

2. If $\mathcal{S}$ is chosen sufficiently small then

$$\mathrm{pr}_1^{-1}(\mathcal{S} \cap \mathcal{N}_n(\mathbb{C}))$$

is a smooth two-dimensional manifold and the projection map pr_1 restricts to a resolution of singularities of $\mathcal{S} \cap \mathcal{N}_n(\mathbb{C})$.

As mentioned in the introduction, our goal is to develop a system theoretic approach to such results. This is motivated by the attempt to obtain a better understanding of the transversal slices for subregular nilpotents, constructed in [28]. Note that, although all minimal dimensional transversal sections at a point are conjugate, the right choice of a transversal section still becomes an issue. The situation here is similar to the search for good normal forms in linear algebra. Thus, instead of using the above desingularization via the dualized moment map of the flag manifold, we construct an alternative one where the manifold consists of pairs of conditioned invariant flags together with their friends. The motivation behind this is that it might lead to easier constructions of resolutions for nilpotent orbits of arbitrary co-dimension. So far we have however not achieved that purpose and therefore only explain two partial results.

The first result is an analogon of Proposition 4 for certain unions of nilpotent similarity orbits that arise out of Rosenbrock's theorem, see [20]. Recall from Theorem 11 that

$$\mathrm{NilTJ}(\mathbb{F}^n) =$$

$$\{(T, J, (V_1, \ldots, V_n)) \in P(\mathbb{F}^n) \,|\, T(A - JC)T^{-1}V_i \subset V_{i-1}, i = 1, \ldots, n\} =$$
$$\{(T, J, \mathsf{V}) \in P(\mathbb{F}^n) \,|\, T(A - JC)T^{-1}\mathsf{V} \subset \mathsf{V}, T(A - JC)T^{-1} \in \mathcal{N}_n(\mathbb{F})\}$$

is a smooth manifold of dimension $n^2 + (p-1)n$.

Theorem 20. *Let $(C, A) \in \mathbb{F}^{p \times n} \times \mathbb{F}^{n \times n}$ denote an observable pair in dual Brunovsky canonical form with observability indices $\mu_1 \geq \cdots \geq \mu_p$. Let $\mathcal{N}^{\mu_1, \ldots, \mu_p}(\mathbb{F}) \subset \mathcal{N}_n(\mathbb{F})$ denote the set of nilpotent matrices whose nilpotency indices $n_1 \geq \cdots \geq n_p$ fullfill the Rosenbrock conditions*

$$\sum_{i=1}^{j} n_i \geq \sum_{i=1}^{j} \mu_i \quad for\ j = 1, \ldots, p-1\ and$$

$$\sum_{i=1}^{p} n_i = \sum_{i=1}^{p} \mu_i. \tag{9.19}$$

Then $\mathcal{N}^{\mu_1, \ldots, \mu_p}(\mathbb{F})$ is a disjoint union of nilpotent similarity orbits which contains $\mathcal{N}_n^{\mathrm{reg}}$. Furthermore, we have the surjective smooth map

$$f : \mathrm{NilTJ}(\mathbb{F}^n) \longrightarrow \mathcal{N}^{\mu_1, \ldots, \mu_p}(\mathbb{F}), \quad (T, J, \mathsf{V}) \mapsto T(A - JC)T^{-1}.$$

Proof. The nilpotency indices form a complete set of invariants for similarity on $\mathcal{N}_n(\mathbb{F})$, hence $\mathcal{N}^{\mu_1,\ldots,\mu_p}(\mathbb{F})$ is a disjoint union of nilpotent similarity orbits. Since $n_1 = n$ and $n_2 = \cdots = n_p = 0$ fullfills (9.19), it contains $\mathcal{N}_n^{\mathrm{reg}}$. According to Rosenbrock's theorem, the elements of the Brunovsky orbit $\Gamma_{(C,A)} = \{T(A - JC)T^{-1} \in \mathfrak{gl}_n(\mathbb{F}) \,|\, T \in \mathrm{GL}_n(\mathbb{F}), J \in \mathbb{F}^{n \times p}\}$ are precisely characterized by (9.19), where the n_i have to be interpreted as the degrees of the invariant factors of $sI - A$. For nilpotent A, the latter are equal to the nilpotency indices of A, thus f maps indeed into $\mathcal{N}^{\mu_1,\ldots,\mu_p}(\mathbb{F})$. Since for every nilpotent map there exists an invariant flag, the map f is surjective. It is clearly smooth as a map into $\mathfrak{gl}_n(\mathbb{F})$. $\square$

The second result generalizes Theorem 19. Recall from Theorem 11 that

$$\mathrm{NilJ}(\mathbb{F}^n) =$$
$$\{(J, (V_1, \ldots, V_n)) \in \mathbb{F}^{n \times p} \times \mathrm{Flag}(\mathbb{F}^n) \,|\, (A - JC)V_i \subset V_{i-1}, i = 1, \ldots, n\} =$$
$$\{(J, \mathsf{V}) \in \mathbb{F}^{n \times p} \times \mathrm{Flag}(\mathbb{F}^n) \,|\, (A - JC)\mathsf{V} \subset \mathsf{V},\ A - JC \in \mathcal{N}_n(\mathbb{F})\}$$

is a smooth manifold of dimension $(p - 1)n$.

Theorem 21. *Let $(C, A) \in \mathbb{F}^{p \times n} \times \mathbb{F}^{n \times n}$ denote an observable pair in dual Brunovsky canonical form. Let $\mathcal{S}^{\mathrm{B}}$ be the Brunovsky slice for the nilpotent similarity orbit $\mathcal{O}$ through A. Then the surjective smooth map*

$$g : \ \mathrm{NilJ}(\mathbb{F}^n) \longrightarrow \mathcal{S}^{\mathrm{B}} \cap \mathcal{N}_n(\mathbb{F}), \ (J, \mathsf{V}) \mapsto A - JC$$

restricts to a surjective map

$$\mathrm{NilJ}^*(\mathbb{F}^n) = g^{-1}(\mathcal{S}_{\mathrm{min}}^{\mathrm{B}} \cap \mathcal{N}_n(\mathbb{F})) \longrightarrow \mathcal{S}_{\mathrm{min}}^{\mathrm{B}} \cap \mathcal{N}_n(\mathbb{F}).$$

Proof. By definition, g maps into $\mathcal{S}^{\mathrm{B}} \cap \mathcal{N}_n(\mathbb{F})$ and is clearly smooth as a map into $\mathfrak{gl}_n(\mathbb{F})$. Since for every nilpotent map there exists an invariant flag, the map g is surjective. From the definitions of $\mathrm{NilJ}^*(\mathbb{F}^n)$ and $\mathcal{S}_{\mathrm{min}}^{\mathrm{B}}$ it follows that $\mathrm{NilJ}^*(\mathbb{F}^n) = g^{-1}(\mathcal{S}_{\mathrm{min}}^{\mathrm{B}} \cap \mathcal{N}_n(\mathbb{F}))$. $\square$

A consequence of Conjecture 1 would be that $\mathrm{NilJ}^*(\mathbb{F}^n)$ is a smooth submanifold of $\mathrm{NilJ}(\mathbb{F}^n)$, cf. Theorem 12. Hence it is reasonable to formulate the following second conjecture.

Conjecture 2. $\mathrm{NilJ}^*(\mathbb{F}^n)$ *is a smooth submanifold of* $\mathrm{NilJ}(\mathbb{F}^n)$ *and*

$$g : \ \mathrm{NilJ}^*(\mathbb{F}^n) \longrightarrow \mathcal{S}_{\mathrm{min}}^{\mathrm{B}} \cap \mathcal{N}_n(\mathbb{F}), \ (J, \mathsf{V}) \mapsto A - JC$$

defines a resolution of singularities of $\mathcal{S}_{\mathrm{min}}^{\mathrm{B}} \cap \mathcal{N}_n(\mathbb{F})$.

In our first example ($\mu_1 = 3$, $\mu_2 = 1$) from above, which is a subregular case, we end up with the same transversal slice as Steinberg in [28]. Note, that in this case we have a proof of Conjecture 1. However, as has been pointed out before, the manifold appearing in the above desingularization differs from that used by Steinberg.

References

1. G. S. Ammar and C. F. Martin. The geometry of matrix eigenvalue methods. *Acta Applicandae Math.* 5: 239–278, 1986.
2. V. I. Arnol'd. On matrices depending on parameters. *Uspekhi Mat. Nauk.* 26: 101–114, 1971.
3. W. Borho and R. MacPherson. Partial resolutions of nilpotent varieties. *Analysis and Topology on Singular Spaces, II, III, Asterisque,* Vol. 101-102, Soc. Math. France, Paris, 1983, pp. 23–74.
4. E. Brieskorn. Singular elements of semi-simple algebraic groups. *Intern. Congress Math. Nice* 2: 279-284, 1970.
5. J. Dieudonné. *Treatise on analysis. Vol. III.* Academic Press, New York, 1972. Translated from the French by I. G. MacDonald, Pure and Applied Mathematics.
6. J. Ferer, F. Puerta and X. Puerta. Differentiable structure of the set of controllable $(A, B)^t$-invariant subspaces. *Linear Algebra and its Appl.* 275-276: 161–177, 1998.
7. P. A. Fuhrmann. *Linear Systems and Operators in Hilbert Space.* McGraw–Hill, New York, 1981.
8. P. A. Fuhrmann. *A Polynomial Approach to Linear Algebra.* Springer Verlag, New York, 1996.
9. P. A. Fuhrmann and U. Helmke. A homeomorphism between observable pairs and conditioned invariant subspaces. *Systems and Control Letters,* 30: 217–223, 1997.
10. P. A. Fuhrmann and U. Helmke. On the parametrization of conditioned invariant subspaces and observer theory. *Linear Algebra and its Applications,* 332–334: 265–353, 2001.
11. I. Gohberg, P. Lancaster and L. Rodman. *Invariant Subspaces of Matrices with Applications.* Wiley, New York, 1986.
12. M. Hazewinkel and C. F. Martin. Representations of the symmetric groups, the specialization order, Schubert cells and systems. *Enseignement Mat.* 29: 53–87, 1983.
13. U. Helmke. *The cohomology of moduli spaces of linear dynamical systems.* Regensburger Math. Schriften, Vol. 24, 1992.
14. U. Helmke. Linear dynamical systems and instantons in Yang-Mills theory. *IMA J. Math. Control Inf.* 3: 151–166, 1986.
15. R. Hermann and C. F. Martin. Applications of algebraic geometry to systems theory. I. *IEEE Trans. Automatic Control,* AC-22(1):19–25, 1977.
16. R. Hermann and C. F Martin. Lie and Morse theory for periodic orbits of vector fields and matrix Riccati equations. I. General Lie-theoretic methods. *Math. Systems Theory* 15: 277–284, 1982.
17. D. Hinrichsen, H. F. Münzner and D. Prätzel-Wolters. Parametrizations of (C, A)–invariant subspaces. *Systems and Control Letters* 1: 192–199, 1981.
18. C. F. Martin. Grassmannian manifolds, Riccati equations and feedback invariants of linear systems. *Geometrical methods for the study of linear systems,* NATO Adv. Study Inst. Ser. C: Math. Phys. Sci., Vol.62, Reidel, Dordrecht-Boston, Mass., pp. 195–211, 1980.
19. X. Puerta and U. Helmke. The topology of the set of conditioned invariant subspaces. *Systems and Control Letters* 40: 97–105, 2000.

20. H. H. Rosenbrock. *State-space and multivariable theory*. John Wiley & Sons, Inc. [Wiley Interscience Division], New York, 1970.

21. M. A. Shayman. On the variety of invariant subspaces of a finite-dimensional linear operator. *Trans. of the Amer. Math. Soc.*, 274: 721–747, 1982.

22. M. A. Shayman. Geometry of the algebraic Riccati equation. I, II. *SIAM J. Control Optim.* 21: 375–394, 395–409, 1983.

23. E. D. Sontag. A remark on bilinear systems and moduli spaces of instantons. *Systems and Control Letters* 9: 361-368, 1987.

24. N. Spaltenstein. On the fixed point set of a unipotent transformation on the flag manifold. *Nederl. Akad. Wetensch. Indag. Math.* 38: 452–458, 1976.

25. T. A. Springer. The unipotent variety of a semisimple group. *Proc. Bombay Colloq. on Alg. Geom.* Oxford Press, London, pp. 452–458, 1969.

26. T. A. Springer. A construction of representations of Weyl groups. *Invent. Math.* 44: 279–293, 1978.

27. R. Steinberg. Desingularization of the unipotent variety. *Invent. Math.* 36: 209–224, 1976.

28. R. Steinberg. Kleinian singularities and unipotent elements. *Proc. of Symposia in Pure Math.* 37: 265–270, 1980.

29. J. Trumpf. *On the geometry and parametrization of almost invariant subspaces and observer theory*. Ph.D. Thesis, Universität Würzburg, 2002.

30. J. Trumpf, U. Helmke, and P. A. Fuhrmann. Towards a compactification of the set of conditioned invariant subspaces. *Systems and Control Letters*, 48: 101–111, 2003.

31. W. M. Wonham. *Linear Multivariable Control: A Geometric Approach*. Springer, New York, 1979.

10

Local Robustness of Hyperbolic Limit Cycles

Ulf T. Jönsson[1] and Alexandre Megretski[2]

[1] Optimization and Systems Theory, Royal Institute of Technology, 10044
   Stockholm, Sweden
   ulfj@math.kth.se
[2] Laboratory for Information and Decision Systems, Massachusetts Institute of
   Technology, Cambridge, MA 02139, USA
   ameg@mit.edu

Summary. Local robustness of limit cycles are investigated for systems that can
be modeled as a feedback interconnection of an exponentially stable linear system
with a nonlinear function. Conditions are given under which the limit cycle and the
number of unstable modes persist for sufficiently small dynamic perturbations.

10.1 Introduction

Stability and robustness of limit cycle oscillations are properties of fundamental importance in many applications in electronics, mechanics, biology, and
physics. Limit cycle oscillation is crucial in control applications such as biological locomotion [6], rhythmic mechanical motion [7] and auto-tuning [1].
Tools for rigorous analysis of stability and robustness of limit cycle oscillations is important in the design and verification of such systems. The classical
literature provides several useful results but few, if any of them, extends directly to system descriptions that are subject to various forms of unmodelled
dynamics. One problem is that the classic results were derived in a state space
formalism which does not extend easily to systems with unknown possible infinite dimension. Another problem is that the introduction of uncertainty in
the system dynamics perturbs both the period time and the orbit of the limit
cycle which is in stark contrast to the traditional problems in robust control
where the equilibrium solution remains fixed when the system is perturbed.
This makes robust stability analysis of limit cycles a challenging problem.

In this paper we briefly review and extend some of the results in [3]. There
we proved that the well-known condition on the characteristic multipliers for
robustness of finite dimensional systems extends to a class of systems with
dynamic uncertainties. We also showed how bounds on a robustness margin
can be estimated. Here we extend the local result to hold also in the case when
the limit cycle is hyperbolic. In particular, we show that the limit cycle and the
number of unstable modes persist for sufficiently small dynamic perturbations.

U.T. Jönsson and A. Megretski: *Local Robustness of Hyperbolic Limit Cycles*, LNCIS **321**, 165–
180 (2005)
www.springerlink.com   © Springer-Verlag Berlin Heidelberg 2005

Notation

We will let $C(1)$ denote the set of continuous one periodic functions equipped with the norm $\|v\|_{C(1)} = \sup_{t \in [0,1]} |v(t)|$. The exponentially weighted $\mathbf{L}_2$ space $\mathbf{L}_{2\alpha}[0,\infty) = \{e(t) : \int_0^\infty e^{2\alpha t} |e(t)|^2 dt < \infty\}$ will be used to define and prove exponential stability. The norm on the usual $\mathbf{L}_2[0,\infty)$ space is denoted $\|\cdot\|$ while the norm on $\mathbf{L}_{2\alpha}[0,\infty)$ is denoted and defined as $\|v\|_\alpha = (\int_0^\infty e^{2\alpha t} |v(t)|^2 dt)^{1/2}$. The spatial norm will always be the Euclidean norm $|v| = (\sum_{i=1}^n v_i^2)^{1/2}$. At several places we consider the space $C(1) \times \mathbf{R}$ with the norm $\|(v,T)\|_{C(1)\times\mathbf{R}} = (\|v\|_{C(1)}^2 + |T|^2)^{1/2}$. If X denotes a normed vector space then its dual X^* is the Banach space of all bounded linear functionals on X. If $g \in X^*$ and $x \in X$, then we use the notation $g(x) = \langle x, g \rangle$ for the functional.

We use that the *characteristic multipliers* of a periodic matrix $A(t) = A(t + T_0)$ are the eigenvalues of the monodromy matrix $\Phi(T_0, 0)$, where

$$\frac{d}{dt}\Phi(t,0) = A(t)\Phi(t,0), \qquad \Phi(0,0) = I.$$

10.2 Model Assumptions

We consider systems consisting of a feedback interconnection of an exponentially stable linear time-invariant (LTI) plant and a memoryless nonlinearity

$$y(t) = \int_{-\infty}^t h(t - \tau, \theta)\varphi(y(\tau))\mathrm{d}\tau, \quad \forall t. \tag{10.1}$$

This system equation is suitable for representing stationary solutions such as equilibrium solutions or stationary periodic solutions. The parameter θ is a scaling of the size of the uncertainty in the system and we assume it belongs to an open interval I_θ, which contains 0.

We next summarize the assumptions on (10.1).

Assumption 1 *For the system in (10.1) we assume*

(i) The nonlinearity $\varphi(\cdot)$ is C^1 (continuously differentiable).

(ii) For some exponential decay rate $\alpha > 0$ and all $\theta \in I_\theta$ (an open interval containing $\theta = 0$) we have $e^{\alpha t} h(t, \theta) \in \mathbf{L}_1[0, \infty)$ and furthemore that $h(t, \theta)$ is C^1 with respect to θ and has time differential $\mathrm{d}h(t, \theta) = \dot{h}_c(t, \theta)\mathrm{d}t + \sum_{k=0}^\infty h_k(\theta)\delta(t - t_k)\mathrm{d}t$, where $\delta(\cdot)$ denotes the dirac implulse, $e^{\alpha t}\dot{h}_c \in \mathbf{L}_1[0, \infty)$, $\sum_{k=0}^\infty e^{\alpha t_k} |h_k| < \infty$, $t_0 = 0$ and $t_k > 0$. Under these assumptions the Laplace transforms $H(s, \theta)$ and $sH(s, \theta)$ are (i) analytic in $\mathrm{Re}\, s > -\alpha$, (ii) continuous on $-\alpha + i\mathbf{R}$, and (iii) bounded such that for $\mathrm{Re}\, s \geq -\alpha$ we have $\max(|sH(s, \theta)|, |H(s, \theta)|) \leq b$ for some number b.

(iii) The system is called nominal when $\theta = 0$ and our assumption is that the nominal system has a T_0-periodic solution y_0. The periodic solution is called a limit cycle *when it is isolated.*

We will derive conditions under which there remains a limit cycle when θ is perturbed from zero. Conditions for stability of the limit cycle are also derived.

A more concise operator notation for (10.1) is

$$y = H(s,\theta)\varphi(y). \qquad (10.2)$$

The uncertain dynamics is often represented as a linear fractional transformation (LFT) (see Figure (10.1))

$$H(s,\theta) = H_{11}(s) + \theta H_{12}\Delta(s)(I - \theta H_{22}(s)\Delta(s))^{-1}H_{21}(s). \qquad (10.3)$$

Here we normally assume that the nominal dynamics $H(s)$ is a finite dimensional transfer function with all poles in $\operatorname{Re} s < -\alpha$ and with H_{11} and either H_{12} or H_{21} strictly proper. If $\Delta(s)$ is a transfer function with impulse response function satisfying

$$\Delta(t) = \Delta_c(t) + \sum_{k=0}^{\infty} \Delta_k \delta(t - t_k) \qquad (10.4\text{a})$$

$$e^{\alpha t}\Delta_c(t) \in \mathbf{L}_1[0\infty), \quad t_0 = 1, \ t_k > 0, \quad \sum_{k=0}^{\infty} |e^{\alpha t_k}\Delta_k| < \infty \qquad (10.4\text{b})$$

then Assumption 1 holds for $I_\theta = (-\hat{\theta}, \hat{\theta})$ if the small gain condition $\hat{\theta}\|\Delta(s - \alpha)\|_{\mathbf{H}_\infty} \cdot \|H_{22}(s - \alpha)\|_{\mathbf{H}_\infty} < 1$ is satisfied.

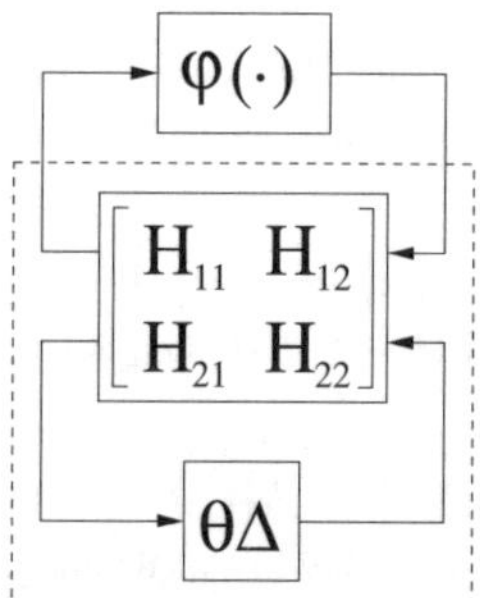

Fig. 10.1. Block diagram corresponding to the perturbed system in (10.2)-(10.3).

Example 1. Consider Van der Pol's equation with a dynamic uncertainty

$$\ddot{u}(t) + m(u(t)^2 - 1)\dot{u}(t) + u(t) = \theta(\Delta u)(t)$$

where $\Delta(s)$ is a transfer function with impulse response satisfying (10.4). To represent this system on the form (10.1) we introduce the new coordinates

$$x_1 = -\dot{u} - m(u^3/3 - u)$$
$$x_2 = u$$

Differentiation gives

$$\dot{x}(t) = Ax(t) + B\varphi(y(t)) + \theta B_\Delta(\Delta y)(t)$$
$$y(t) = Cx(t)$$

where

$$A = \begin{bmatrix} 0 & 1 \\ -1 & -2 \end{bmatrix}, \qquad B = \begin{bmatrix} 0 \\ 1 \end{bmatrix}, \qquad B_\Delta = \begin{bmatrix} -1 \\ 0 \end{bmatrix}, \qquad C = \begin{bmatrix} 0 & 1 \end{bmatrix}$$

and $\varphi(y) = -my^3/3 + (2+m)y$. Now the system can be represented on the LFT form in (10.2)-(10.3) with $\varphi(y) = -my^3/3 + (2+m)y$ and (where $C_\Delta = C$)

$$H(s) = \begin{bmatrix} H_{11}(s) & H_{12}(s) \\ H_{21}(s) & H_{22}(s) \end{bmatrix} = \left[\begin{array}{c|cc} A & B & B_\Delta \\ \hline C & 0 & 0 \\ C_\Delta & 0 & 0 \end{array} \right].$$

We will next discuss a condition from [3] for the existence of a periodic solution when θ is sufficiently close to 0. Later we derive conditions for local stability or more generally hyberbolicity of this limit cycle solution.

10.3 Existence of Solution

It is no restriction to assume that the period time $T_0 = 1$ since we can always re-scale the time axis by the transformation $t/T_0 \to t$, which gives the nominal dynamics

$$y_0(t) = \int_{-\infty}^{t} T_0 h(T_0(t-\tau), 0)\varphi(y_0(\tau))d\tau, \qquad \text{for } t \in [0, 1].$$

Hence, by redefining $T_0 h(T_0 t, 0) \to h(t, 0)$ we can assume $T_0 = 1$. A general periodic solution to (10.1) can thus be written

$$y(t) = \int_{-\infty}^{t} Th(T(t-\tau), \theta)\varphi(y(\tau))d\tau, \quad \text{for } t \in [0, 1]. \qquad (10.5)$$

The advantage of this reformulation is that a periodic solution can be represented as a pair $z = (y, T) \in C(1) \times \mathbf{R}$ of a 1-periodic trajectory and a period time. This will simplify our work considerably since the perturbation of the period time and the orbit are separated. In this section, the assumption is thus that (10.5) has a solution $z_0 = (y_0, 1)$ when $\theta = 0$. We often use the following concise notation for the system equation in (10.5)

$$y = H(s/T, \theta)\varphi(y) \qquad (10.6)$$

where the nominal transfer function $H(s, 0)$ in this paper often is assumed to have a finite dimensional state space realization.

With $Z = C(1) \times \mathbf{R}$ and $Y = C(1)$ we define the operator

$$F : Z \times I_\theta \to Y \quad \text{as} \quad (z, \theta) \mapsto F(z, \theta) = y - H(s/T, \theta)\varphi(y)$$

for any $z = (y, T) \in Z$. A solution to the equation $F(z_\theta, \theta) = 0$ corresponds to a periodic solution $z_\theta = (y_\theta, T_\theta) \in C(1) \times \mathbf{R}$ of (10.5) or similarly a T_θ-periodic solution $y_\theta(t/T_\theta)$ of (10.1).

We will use an implicit function theorem to derive conditions for the existence of a solution of the perturbed system. The Frechét derivative of F with respect to the trajectory at a periodic solution $z_\theta = (y_\theta, T_\theta)$ has the block structure

$$F_z'(z_\theta, \theta) = \begin{bmatrix} F_y'(z_\theta, \theta) & F_T'(z_\theta, \theta) \end{bmatrix} \tag{10.7a}$$
$$= \begin{bmatrix} I - L_{st}(z_\theta, \theta) & \frac{1}{T_\theta}(I - L_{st}(z_\theta, \theta))(t\dot{y}_\theta) \end{bmatrix} \tag{10.7b}$$

where

$$L_{st}(z_\theta, \theta) = H(s/T_\theta, \theta)\varphi'(y_\theta).$$

The notation L_{st} is used to indicate that this operator has to do with the stationary behavior of the system. The last component of the derivative follows after some calculation which is left to Appendix 1. Note that the argument $t\dot{y}_\theta$ of $I - L_{st}(z_\theta, \theta)$ does not belong to $C(1)$ while our claim is that the value does. The claim follows from the proof in Appendix 1.

It is interesting to note that the variational system corresponding to the nominal 1-periodic solution y_0 can be written $(I - L_{st}^0)v = 0$, where $L_{st}^0 = L_{st}(z_0, 0)$. The next proposition shows that $\dot{y}_0$ is in the kernel of $(I - L_{st}^0)$, i.e. 1 is an eigenvalue of L_{st}^0 with $\dot{y}_0$ as the corresponding eigenfunction. This follows since the periodic solution is unique only modulo arbitrary time translations.

Proposition 1. *We have $\dot{y}_0 \in \mathrm{Ker}\,(I - L_{st}^0)$, where $L_{st}^0 = L_{st}(z_0, 0)$.*

Proof. Let $h_0(t) = h(t, 0)$. By definition,

$$y_0(t) = \int_{-\infty}^{t} h_0(t - \tau)\varphi(y_0(\tau))\mathrm{d}\tau$$

for all $t \in \mathbf{R}$. Differentiation of this identity gives

$$\dot{y}_0(t) = h_0(0)\varphi(y_0(t)) + \int_{-\infty}^{t} \mathrm{d}h_0(t - \tau)\varphi(y_0(\tau))$$

$$= h_0(0)\varphi(y_0(t)) + \lim_{T \to -\infty} [-h_0(t - \tau)\varphi(y_0(\tau))]_T^t$$

$$+ \int_{-\infty}^{t} h_0(t - \tau)\varphi'(y_0(\tau))\dot{y}_0(\tau)\mathrm{d}\tau$$

$$= \int_{-\infty}^{t} h_0(t - \tau)\varphi'(y_0(\tau))\dot{y}_0(\tau)\mathrm{d}\tau$$

where we used that $h_0(t) \to 0$ as $t \to \infty$. This concludes the proof. $\square$

Theorem 1. *If the operator $F_z'(z_0, 0)$ in (10.7) has a bounded right inverse then for each sufficiently small $|\theta|$ there exists $y_\theta \in C(1)$ and $T_\theta > 0$ that satisfies (10.5). The perturbed solution $y(\theta) = y_\theta$, $T(\theta) = T_\theta$ are C^1 functions of θ such that $y(0) = y_0$ and $T(0) = 1$. The solution is unique modulo time translation of $y_\theta(t)$.*

Proof. We sketch a proof. Let $\hat{F}(x, \theta) = F(z_0 + G_0 x, \theta)$, where G_0 is a bounded right inverse of $F_z'(z_0, 0)$. We have $\hat{F}(0, 0) = 0$ and $\hat{F}_x'(0, 0) = I$. By the implicit function theorem there exists a unique solution C^1 function $x_\theta := x(\theta)$ such that $\hat{F}(x_\theta, \theta) = 0$ for all sufficiently small $|\theta|$. This implies that $z_\theta = z_0 + G_0 x_\theta$ satisfies $F(z_\theta, \theta) = 0$. The only nonuniqueness is due to the choice of G_0 and it can be shown that this corresponds to a time translation. $\square$

We will next show how to construct a right inverse using the following lemma.

Lemma 1. *Let X be a normed vector space and consider a bounded linear operator $F : X \times \mathbf{R} \to X$ with block decomposition*

$$F = \begin{bmatrix} F_1 & f_2 \end{bmatrix}$$

where $F_1 = I - L$, with $L : X \to X$ being a compact operator with a simple eigenvalue at one and $f_2 \notin \operatorname{Im} F_1$ is a nonzero vector. Then a right inverse can be constructed as

$$F^\dagger = \begin{bmatrix} I \\ g \end{bmatrix} (F_1 + f_2 g)^{-1}$$

where $g \in X^*$ is any vector such that $(F_1 + f_2 g) : X \to X$ has a bounded inverse. In fact, any $g \in X^*$ such that $|g(e)| = |\langle e, g \rangle| > 0$ for a unit length vector $e \in \operatorname{Ker} F_1$ can be used.

Proof. The first claim follows immediately since

$$\begin{bmatrix} F_1 & f_2 \end{bmatrix} \begin{bmatrix} I \\ g \end{bmatrix} (F_1 + f_2 g)^{-1} = (F_1 + f_2 g)(F_1 + f_2 g)^{-1} = I.$$

Let g be defined as suggested in the second claim and suppose there exists $x \in X$ such that $(F_1 + f_2 g)x = 0$. If $y \in (\operatorname{Im} F_1)^\perp$ is nonzero, then

$$\langle (F_1 + f_2 g)x, y \rangle = \langle f_2, y \rangle \langle x, g \rangle = 0.$$

Since $\langle f_2, y \rangle \neq 0$ it follows that $\langle x, g \rangle = 0$. Hence, $x \in \operatorname{Ker} F_1 \cap \operatorname{Ker} g = \{0\}$. This shows that $\operatorname{Ker}(F_1 + f_2 g) = \{0\}$. Since $L - f_2 g$ is a compact operator and $\operatorname{Ker}(I - L + f_2 g) = \operatorname{Ker}(F_1 + f_2 g) = \{0\}$ it follows that $\operatorname{Im}(F_1 + f_2 g) = X$ (see e.g. Theorem 8.4-5 in [5]). Hence, $F_1 + f_2 g : X \to X$ is a bijection and it follows from Banach's isomorphism theorem that $(F_1 + f_2 g)^{-1}$ is a bounded operator. $\square$

We next use this lemma to construct a right inverse for the nominal operator.

Theorem 2. *Consider the operator $F_z'(z_0, 0)$ defined in (10.7) in the finite dimensional case when $h(t, 0) = Ce^{At}B\nu(t)$, where $\nu(t)$ is the unit step function and $A \in \mathbf{R}^{n \times n}$. Let us define*

$$A_{cl}(t) = A + B\varphi'(y_0(t))C, \qquad B_{cl}(t) = B\varphi'(y_0(t))$$

and let $x_0(t)$ be the 1-periodic solution of the nominal state space representation of (10.5)

$$\dot{x}_0(t) = Ax_0(t) + B\varphi(Cx_0(t)).$$

If $n - 1$ of the characteristic multipliers of $A_{cl}(t)$ are different from 1 then $F_z'(z_0, 0)$ has a bounded right inverse. One possible right inverse

$$F_z'(z_0, 0)^\dagger = \begin{bmatrix} I + G_1 \\ G_2 \end{bmatrix} : C(1) \to C(1) \times \mathbf{R}$$

is defined as

$$(F_z'(z_0, 0)^\dagger(w))(t) = (w(t) + \int_0^1 g_1(t, \tau)w(\tau)\mathrm{d}\tau, \int_0^1 g_2(1, \tau)w(\tau)\mathrm{d}\tau)$$

where

$$g_1(t, \tau) = \begin{cases} (\Gamma(t)\Phi_{cl}(1, t) + C)\Phi_{cl}(t, \tau)B_{cl}(\tau), & t > \tau \\ \Gamma(t)\Phi_{cl}(1, \tau)B_{cl}(\tau), & t < \tau \end{cases}$$

$$g_2(t, \tau) = k(I - \Phi_{cl}(1, 0) - kk^T)^{-1}\Phi_{cl}(t, \tau)B_{cl}(\tau)$$

and

$$\Gamma(t) = C(\Phi_{cl}(t, 0) + \dot{x}_0(t)k)(I - \Phi_{cl}(1, 0) - kk^T)^{-1}.$$

Here $\Phi_{cl}(t, 0)$ is the transition matrix corresponding to A_{cl} and $k = \dot{x}_0(0)^T$.

Proof. See Appendix 2.   □

10.4 Stability

The system in (10.1) is generally of unknown or infinite dimension and the definition of stability needs extra care. We define local stability in terms of the variational system corresponding to the following *non-steady-state* version of (10.1)

$$y(t) = f(t) + \int_0^t h(t - \tau)\varphi(y(\tau))\mathrm{d}\tau, \quad t \geq 0 \qquad (10.8)$$

where the dependence on θ is suppressed for notational convenience. In (10.8), $f(\cdot)$ represents initial conditions and external disturbances. The choice

$$f_0(t) = \int_{-\infty}^{0} h(t - \tau)\varphi(y_0(\tau))\mathrm{d}\tau \tag{10.9}$$

gives the T_0-periodic solution $y_0(t)$, since (10.8) has a unique solution for any locally integrable function $f(\cdot)$. A linearization of (10.8) along the nominal periodic solution gives rise to the variational system

$$v = Lv + w \tag{10.10}$$

where we define

$$L : \mathbf{L}_2[0, \infty) \to \mathbf{L}_2[0, \infty) \quad \text{as} \quad v \mapsto Lv = H(s)\varphi'(y_0)v. \tag{10.11}$$

Note that we have defined the operator to act on $\mathbf{L}_2[0, \infty)$ and not $\mathbf{L}_\infty[0, \infty)$, which would be more natural for the linearization. However, this will allow an simple yet natural definition of stability in terms of the variational system (10.10) and the operator (10.11). Stability can also be defined in terms of the non-steady state system (10.8). This requires a more elaborate analysis but can be done, see [3].

The next proposition shows that the variational equation in (10.10) cannot be solved for arbitrary $w \in \mathbf{L}_2[0, \infty)$ unless $y_0 \equiv 0$. This follows because the input-output map $w \mapsto v$ defined by (10.10) is unbounded on $\mathbf{L}_2[0, \infty)$, since there is a finite energy input which maps to an infinite energy output. For finite dimensional systems this observation corresponds to the fact that the periodic linear system obtained as a result of linearization around a limit cycle always has a neutrally stable mode corresponding to a characteristic multiplier at unity.

Proposition 2. *If $y_0 \neq const$ is a T-periodic solution of (10.1) then*

$$w(t) = \int_{-\infty}^{0} h(t - \tau)\varphi'(y_0(\tau))\dot{y}_0(\tau)\mathrm{d}\tau$$

produces a periodic solution $v(t) = \dot{y}_0(t)$ of the variational system (10.10).

Proof. Let us differentiate $y_0(t)$. This gives

$$\dot{y}_0(t) = \frac{d}{dt}\int_{-\infty}^{t} h(t - \tau)\varphi(y_0(\tau))\mathrm{d}\tau = h(0)\varphi(y_0(t)) + \int_{-\infty}^{t} \mathrm{d}h(t - \tau)\varphi(y_0(\tau))$$

$$= h(0)\varphi(y_0(t)) + \lim_{T \to -\infty} [-h(t - \tau)\varphi(y_0(\tau))]_T^t$$

$$+ \int_{-\infty}^{t} h(t - \tau)\varphi'(y_0(\tau))\dot{y}_0(\tau)\mathrm{d}\tau$$

$$= \int_{0}^{t} h(t - \tau)\varphi'(y_0(\tau))\dot{y}_0(\tau)\mathrm{d}\tau + w(t)$$

where we used that $\lim_{T \to -\infty} h(t - T)\varphi(y_0(T)) = 0$ since h is exponentially stable and continuous. $\square$

In order to get around this problem we notice that the non-steady-state system, if stable, generally converges to $y_0(t + d)$, where $d \in \mathbf{R}$ is a nonzero *phase lag*. In fact, this is the reason for the neutrally stable mode of L, which implies that the image of the *return difference* $(I - L)$ has nonzero codimension. The lost term can be compensated for by considering the system

$$(I - L)v + ed = w \qquad (10.12)$$

where $e = (I - L)(\dot{y}_0)$. Under Assumption 1 (ii) it can be shown that $\mathbf{L}_{2\alpha}[0, \infty) \ni e \notin \operatorname{Im}(I - L)$. The next step is to consider (10.12) as a system on the space of exponentially converging signals $\mathbf{L}_{2\alpha}[0, \infty)$. The neutral mode is now moved to the unstable and if the equation (10.12) can be proven to have an exponentially bounded solution for all exponentially bounded inputs then the limit cycle y_0 is said to be exponentially stable. We also consider the case when in addition to the neutral mode derived in Proposition 2 there are a finite number of unstable solutions of (10.10). We state this as a definition.

Definition 1. *If the system (10.12) has a unique solution $(v, d) \in \mathbf{L}_{2\alpha}[0, \infty) \times \mathbf{R}$ for all $w \in \mathbf{L}_{2\alpha}[0, \infty)$ then the limit cycle y_0 is called locally exponentially stable and α corresponds to the rate of exponential decay. Otherwise, if the subspace $W \subset \mathbf{L}_{2\alpha}[0, \infty)$ of codimension n_u is the largest subspace such that (10.12) has a unique solution $(v, d) \in \mathbf{L}_{2\alpha}[0, \infty) \times \mathbf{R}$ for all $w \in W$, then the limit cycle y_0 is said to have n_u unstable modes.*

This stability definition can be verified by computing the *stability defect* of the open loop operator L in (10.11). The stability defect is introduced as an equivalent of the notion "number of unstable closed-loop poles", which can be applied to time-varying systems. In the following definition, an open loop plant is represented by a linear operator on some normed space of signals.

Definition 2. *Let L be a bounded linear operator on a Banach space X, which is denoted $L \in \mathcal{L}(X, X)$. The feedback system with open loop operator L is called non-singular if there exists $\varepsilon > 0$ such that*

$$\|(I - L)u\| \geq \varepsilon \|u\|, \quad \forall u \in X. \qquad (10.13)$$

The stability defect $\operatorname{def}(L)$ *of a non-singular system with the open loop operator L is defined as the codimension of the subspace*

$$\operatorname{Im}(I - L) = \{(I - L)u : u \in X\} \subset X.$$

The stability defect, if well-defined on $X = \mathbf{L}_{2\alpha}$, will be called the α-defect of L in (10.11), denoted $\operatorname{def}_\alpha(L)$.

The stability defect corresponds to the number of unstable modes of $(I - L)^{-1}$, i.e., the unstable closed loop poles, while condition (10.13) means that $I - L$ does not have zeros on the stability boundary. The motivation for working on $\mathbf{L}_{2\alpha}$ is that the neutrally stable mode in Proposition 2 is moved from the stability boundary to an unstable mode.

An important feature of the stability defect is the *zero exclusion principle*, which says that the stability defect $\text{def}(L)$ remains constant as L changes continuously and condition (10.13) is satisfied. This is a robustness condition, which is used in the proof of Theorem 5.

Proposition 3 (Zero Exclusion Principle). *Let $\mathfrak{L} = \{L \subset \mathcal{L}(X, X) : \exists \varepsilon > 0 \text{ s.t. } \|(I - L)u\| \geq \varepsilon\|u\|, \ \forall u \in X\}$. Then any connected component $\mathbf{L}$ of $\mathfrak{L}$ containing an element with $\text{def}(L) < \infty$ has constant stability defect, i.e., every $\widetilde{L} \in \mathbf{L}$ has $\text{def}(\widetilde{L}) = \text{def}(L)$.*

Proof. The non-singularity and the finite codimension of the image implies that L is a Fredholm operator with index $n = \text{def}(L)$. The proof follows since the set of Fredholm operators with constant finite index is open [4]. A proof is given in Appendix 3. $\square$

Theorem 3. *Suppose $\text{def}_\alpha(L) = n_u + 1$ where L is defined in (10.11). Then y_0 is a hyperbolic solution with n_u unstable modes and the subspace W in Definition 1 is $W = R \oplus P_{R^\perp}\text{span}\{e\}$, where $e = (I - L)\dot{y}_0$, $R = \text{Im}(I - L)$ and $P_{R^\perp}$ is the orthogonal projection onto $R^\perp$.*

Proof. See Appendix 4. $\square$

The next results shows that the stability defect is easy to compute in the finite dimensional case.

Theorem 4. *Consider the operator L defined in (10.11) in the finite dimensional case when $h(t) = Ce^{At}B\nu(t)$, where $\nu(\cdot)$ is the unit step function and $\text{Re}\lambda(A) < -\alpha$. If the characteristic multipliers corresponding to $A_{cl}(t) = A + B\varphi'(y_0(t))C$ can be sorted as*

$$|\lambda_1| \geq |\lambda_2| \geq \cdots \geq |\lambda_{n_u}| > \lambda_{n_u+1} = 1 > |\lambda_{n_u+2}| \geq \cdots \geq |\lambda_n|$$

then $\text{def}_\alpha(L) = n_u + 1$ for $\alpha \in (0, -\frac{\log|\lambda_{n_u+2}|}{T})$.

Proof. See Appendix 5. $\square$

10.5 Main Result

By using Theorem 1 – Theorem 4 we obtain the following result.

Theorem 5. *Suppose the system in (10.1) has a T_0-periodic solution y_0 when $\theta = 0$. Assume further that the nominal system is finite dimensional with $h(t) = Ce^{At}B\nu(t)$, where $\nu(\cdot)$ is the unit step function and $\text{Re}\lambda(A) < -\alpha$. If the characteristic multipliers corresponding to $A_{cl}(t) = A + B\varphi'(y_0(t))C$ can be sorted as*

$$|\lambda_1| \geq |\lambda_2| \geq \cdots \geq |\lambda_{n_u}| > \lambda_{n_u+1} = 1 > |\lambda_{n_u+2}| \geq \cdots \geq |\lambda_n|$$

where $\alpha T < -\log(\lambda_{n_u+2})$, then for all sufficiently small $|\theta|$ there exists a unique (modulo time translation) hyperbolic limit cycle solution with n_u unstable modes to equation (10.1).

Proof. First note that the characteristic multipliers do not change if we normalize the nominal period time to $T_0 = 1$. Existence of a solution in a neighborhood of $\theta = 0$ follows from Theorem 1 if $F'_z(z_0, 0)$ has a bounded right inverse. From Theorem 2, we see that this is the case since $n - 1$ of the characteristic multipliers are different from 1.

To prove the stability statement we consider the operator L in (10.11), which becomes

$$(L(\theta)v)(t) = \int_0^t T(\theta)h(T(\theta)(t - \tau), \theta)\varphi'(y_\theta(\tau))v(\tau)d\tau.$$

It follows from Theorem 4 that $L(0)$ has α-defect $n_u + 1$. From Proposition 3 we conclude that the α-defect remains constant for sufficiently small $|\theta|$ since $L(\theta)$ depends continuously on θ. Hence, $\mathrm{def}_\alpha(L(\theta)) = n_u + 1$ for sufficiently small $|\theta|$, which by Theorem 3 proves the statement on stability. $\square$

Example 2. Theorem 5 shows that the characteristic multipliers of

$$A_{cl}(t) = A + B\varphi'(y_0(t))C = \begin{bmatrix} 0 & 1 \\ -1 & m(1 - y_0(t)^2) \end{bmatrix}$$

must be sorted as $|\lambda_2| < \lambda_1 = 1$ in order for the limit cycle of the Van der Pol oscillator to be robustly stable. From Liouvilles formula we have

$$\lambda_2 = \det(\Phi_{cl}(1, 0)) = e^{\int_0^1 \mathrm{tr}(A_{cl}(\tau))d\tau} = e^{\int_0^1 m(1 - y_0(\tau)^2)d\tau}.$$

If, for example $m = 0.2$, then a numerical integration shows that $\lambda_2 = 0.34$ and the Van der Pol system thus has a robustly stable limit cycle for this value of m. This gives a new interpretation to the same condition in [2].

Appendix 1

We have

$$F'_T(z_\theta, \theta) = -\frac{s}{T_\theta^2}H'_s(s/T_\theta, \theta)\varphi(y_\theta)$$

which in time domain has the representation

$$(F'_T(z_\theta, \theta))(t) = \int_{-\infty}^t h(T_\theta(t - \tau), \theta))\varphi(y_\theta(\tau))d\tau$$

$$+ \int_{-\infty}^t (t - \tau)T_\theta dh(T_\theta(t - \tau), \theta))\varphi(y_\theta(\tau)).$$

This is a $C(1)$ function by our assumptions on the transfer function $H(s, \theta)$ and since $\varphi(y_\theta(\tau)) \in C(1)$.

After a partial integration of the second term we get

$$(F_T'(z_\theta, \theta))(t) = \int_{-\infty}^{t} (t - \tau)h(T_\theta(t - \tau), \theta)\varphi'(y_\theta(\tau))\dot{y}_\theta(\tau)\mathrm{d}\tau$$

$$= -\int_{-\infty}^{t} h(T_\theta(t - \tau))\varphi'(y_\theta(\tau))(\tau\dot{y}_\theta(\tau))\mathrm{d}\tau$$

$$+th(0, \theta)\varphi(y_0(t)) + t\int_{-\infty}^{t} T_\theta \mathrm{d}h(T_\theta(t - \tau), \theta)\varphi(y_\theta(\tau))$$

$$= \frac{1}{T_\theta}\left(t\dot{y}_\theta(t) - \int_{-\infty}^{t} h(T_\theta(t - \tau))\varphi'(y_\theta(\tau))(\tau\dot{y}_\theta(\tau))\mathrm{d}\tau\right) \quad (10.14)$$

where in the second equality we made a partial integration and the last equality follows because

$$\dot{y}_\theta(t) = T_\theta h(0, \theta)\varphi(y_\theta(t)) + \int_{-\infty}^{t} T_\theta^2 \mathrm{d}h(T_\theta(t - \tau), \theta)\varphi(y_\theta(\tau)).$$

A more concise formulation of (10.14) is $\frac{1}{T_\theta}(I - L_{st}(z_\theta, \theta))(t\dot{y}_\theta(t))$ which proves the statement.

Appendix 2

The operator $F_z'(z_0, 0) : v \mapsto w$ has the following state space realization[3]

$$\dot{x} = Ax + B\varphi'(y_0)v + \dot{x}_0\delta T$$
$$w = v - Cx$$

where $\dot{x}_0(t) = Ax_0(t) + B\varphi(y_0(t))$. In order to use Lemma 1 we identify $F_1 : v \mapsto w_1$ and $f_2 : gv \mapsto w_2$ as operators with the state space realizations

$$F_1 : \begin{cases} \dot{x}_1 = Ax_1 + B\varphi'(y_0)v \\ w_1 = v - Cx_1 \end{cases} \qquad f_2 : \begin{cases} \dot{x}_2 = Ax_2 + \dot{x}_0 gv \\ w_2 = -Cx_2. \end{cases}$$

Let $g : v \mapsto gv$ be defined by the state space realization

$$\dot{x}_3(t) = Ax_3(t) + B\varphi'(y_0(t))v(t) + \dot{x}_0(t)kx_3(0)$$
$$gv = kx_3(0).$$

If $x_3(0) = x_1(0) + x_2(0)$ then $F_1 + f_2g : v \mapsto w$ has the state space realization

$$\dot{x}_3(t) = Ax_3(t) + B\varphi'(y_0(t))v(t) + \dot{x}_0(t)kx_3(0)$$
$$w(t) = v(t) - Cx_3(t).$$

The inverse of $F_1 + f_2g$ can be derived by using $v = w + Cx_3$ in this equation. This gives the right inverse $F_z'(z_0, 0)^\dagger : w \mapsto (v, \delta T)$

---

[3] All state equations in this section has a periodicity constraint of the form $x(1) = x(0)$ on the state vector. This is not written out explicitly.

$$\dot{x}_3(t) = (A + B\varphi'(y_0(t))C)x_3(t) + B\varphi'(y_0(t))w(t) + \dot{x}_0(t)kx_3(0) \quad (10.15a)$$

$$(v(t), \delta T) = (w(t) + Cx_3(t), kx_3(0)). \quad (10.15b)$$

In order for (10.15) to be well defined and bounded on $C(1)$ it is necessary and sufficient that the following equation has a solution for all $w \in C(1)$

$$x_3(0) = (\Phi_{cl}(1,0) + \dot{x}_0(0)k)x_3(0) + \int_0^1 \Phi_{cl}(1,\tau)B\varphi'(y_0(\tau),0)w(\tau)\mathrm{d}\tau$$

where we used that $\int_0^1 \Phi_{cl}(1,\tau)\dot{x}_0(\tau)d\tau kx_3(0) = \dot{x}_0(0)kx_3(0)$. Since we have $\mathrm{span}\{\dot{x}_0(0)\} = \mathrm{Ker}\,(I - \Phi_{cl}(1,0))$ it follows that there exists a vector k such that $I - \Phi_{cl}(1,0) - \dot{x}_0(0)k$ is invertible. Indeed, one possible choice is $k = \dot{x}_0(0)^T$. System (10.15) has the equivalent convolution form given in the theorem statement.

Appendix 3

The set $\mathbf{L}$ is open and by assumption connected. We will prove that the set of operators with constant (finite) stability defect is open. This proves the claim of the proposition since connectedness of $\mathbf{L}$ otherwise would be contradicted.

Consider an operator L with $\mathrm{def}(L) < \infty$. Since L is non-singular, we know that there exists $\varepsilon > 0$ such that $\|(I - L)u\| \geq \varepsilon\|u\|$ for all $u \in X$. Hence, it follows that $H = I - L$ has $\mathrm{Ker}\,H = 0$ and $\mathrm{codim}\,\mathrm{Im}H = \mathrm{def}(L)$. This means that H is a Fredholm operator with index

$$\mathrm{Ind}H := \dim \mathrm{Ker}\,H - \mathrm{codim}\,\mathrm{Im}\,H = -\mathrm{def}(L).$$

Since the codimension of $X_L = \mathrm{Im}\,H$ is finite it follows that there is a direct sum decomposition

$$X = X_L \oplus X_C$$

where $\dim X_C = \mathrm{def}(L)$. Now let ΔL be any perturbation of L with $\|\Delta L\| < \varepsilon/2$ and consider the maps $\widehat{H} : X \to X/X_C$ and $\widehat{\Delta L} : X \to X/X_C$ induced by H and ΔL. Here X/X_C denotes the quotient space and $\widehat{H} = q \circ H$, where $q : X \to X/X_C$ is the quotient map. Then $\|\widehat{\Delta L}\| < \varepsilon/2$ and $\widehat{H}$ has a bounded inverse by Banach's isomorphism theorem with norm bound $\|\hat{H}^{-1}\| \leq 1/\varepsilon$. We have

$$\widehat{H} - \widehat{\Delta L} = \widehat{H}(I - \hat{H}^{-1}\widehat{\Delta L}),$$

from which it follows that $\hat{H} - \widehat{\Delta L}$ has a bounded inverse since $\|\hat{H}^{-1}\widehat{\Delta L}\| \leq \|\hat{H}^{-1}\| \cdot \|\widehat{\Delta L}\| < 1/2$. Hence, $\mathrm{Ind}(\hat{H} - \widehat{\Delta L}) = 0$, which gives the relation

$$\mathrm{Ind}(\widehat{H} - \widehat{\Delta L}) = \dim X_C + \mathrm{Ind}(H - \Delta L) = 0 \quad (10.16)$$

since the quotient map $q : X \to X/X_C$ has index $\dim X_C$ and the index of the composite map $\widehat{H} - \widehat{\Delta L} = q \circ (H - \Delta L)$ is additive. Furthermore, invertibility of $\widehat{H} - \widehat{\Delta L}$ implies that $L + \Delta L$ is nonsingular and thus $\text{Ker}\,(I - L - \Delta L) = 0$. Hence, from (10.16) we get

$$\text{codim}\,\text{Im}\,(I - L - \Delta L) = \dim X_C + \dim \text{Ker}\,(I - L - \Delta L) = \dim X_C = \text{def}(L)$$

which shows that $\text{def}(L + \Delta L) = \text{def}(L)$ for all $\|\Delta L\| < \varepsilon/2$.

Appendix 4

Let $R = \text{Im}\,(I - L)$. We will prove that

$$(I - L)v + de = w, \qquad e = (I - L)(\dot{y}_0)$$

has a unique solution $(v, d) \in \mathbf{L}_{2\alpha}[0, \infty)$ if and only if $w \in W = R \oplus P_{R^\perp}\text{span}\{e\}$. Here $P_{R^\perp} = I - P_R$, where P_R is the orthogonal projection onto $R = \text{Im}\,(I - L) \subset \mathbf{L}_{2\alpha}[0, \infty)$. The assumptions on H in Assumption 1 can be used to prove that $e \in \mathbf{L}_{2\alpha}[0, \infty)$, see [3]. This implies that $R \subset W$ is a strict inclusion since $e \notin R$, i.e. $P_{R^\perp}e \neq 0$. We have

$$(I - P_R)((I - L)v + de) = d(I - P_R)e = (I - P_R)w$$

which gives $d = (I - P_R)w/((I - P_R)e)$ and the norm bound $|d| \leq c_1\|w\|_\alpha$, where $c_1 = 1/\|(I - P_R)e\|_\alpha$. Using this d it follows that $(I - L)v = w - de$ has a unique solution in $\mathbf{L}_{2\alpha}[0, \infty)$ if and only if $w \in W$ because then $w - de \in R$. Since L is nonsingular (see definition of stability defect) there exists $\bar{c}_2$ such that

$$\|v\|_\alpha \leq \bar{c}_2\|w - de\|_\alpha \leq \bar{c}_2(1 + c_1\|e\|_\alpha)\|w\|_\alpha$$

and hence for each $w \in W$ we have found a unique solution satisfying the norm bound $\|v\|_\alpha^2 + |d|^2 \leq c\|w\|_\alpha^2$, where $c^2 = c_1^2 + c_2^2$ and $c_2 = \bar{c}_2(1 + c_1\|e\|_\alpha)$.

Appendix 5

For convenience we transform L to an equivalent operator L_α defined on $\mathbf{L}_2[0, \infty)$ as $L_\alpha = e_\alpha L e_\alpha^{-1}$ where e_α is defined by multiplication in the time domain with $e^{\alpha t}$. It can be shown that $\text{def}(L_\alpha) = \text{def}_\alpha(L)$, see [3]. We will show

(i) $\text{Ker}\,(I - L_\alpha) = 0$
(ii) $\text{codim}\,\text{Im}\,(I - L_\alpha) = n_{u+1}$

Condition (i) and (ii) shows that L_α is a Fredholm operator with index $n_u + 1$. From Banach's isomorphism theorem it follows that $I - L_\alpha$ is nonsingular. This proves the theorem.

To prove (i) we assume there exists nonzero $v \in \mathbf{L}_2$ such that $(I - L_\alpha)v = 0$. In state space domain this means that

$$\dot{x} = (A + \alpha I)x + B\varphi'(y_0)v, \ \ x(0) = 0$$
$$0 = v - Cx$$

which implies that $v = Cx$ and $\dot{x} = (A + \alpha I + B\varphi'(y_0)C)x$, $x(0) = 0$. This contradicts the assumption that v is nonzero. Hence, $\mathrm{Ker}\,(I - L_\alpha) = 0$.

To prove (ii) we use $(\mathrm{Im}\,(I - L_\alpha))^\perp = \mathrm{Ker}\,(I - L_\alpha)^*$. One possible state space representation of the adjoint system $v \mapsto w = (I - L_\alpha^*)v$ is

$$\dot{x} = -(A + \alpha I)^T x + C^T v, \ \ x(\infty) = 0$$
$$w = v + \varphi'(y_0)^T B^T x.$$

Any $v \in \mathrm{Ker}\,(I - L_\alpha^*)$ must satisfy $v = -\varphi'(y_0)^T B^T x$ where

$$\dot{x} = -(A + \alpha + B\varphi'(y_0)C)^T x, \quad x(\infty) = 0. \tag{10.17}$$

A result by Lyapunov shows that there exists a time-periodic coordinate transformation that turns system (10.17) into a linear system with constant coefficients [2]. It is no restriction to assume the new coordinates are chosen such that

$$\begin{bmatrix} \dot{z}_1 \\ \dot{z}_2 \end{bmatrix} = \begin{bmatrix} A_1 & 0 \\ 0 & A_2 \end{bmatrix} \begin{bmatrix} z_1 \\ z_2 \end{bmatrix}, \qquad \begin{bmatrix} z_1(\infty) \\ z_2(\infty) \end{bmatrix} = 0$$

where $A_1 \in \mathbb{C}^{(n_u+1)\times(n_u+1)}$ is stable and A_2 is unstable with $|\mathrm{eig}(e^{A_2 T})| \geq |e^{-\alpha T}/\lambda_{n_u+2}| > 1$. If the coordinates are related as

$$x(t) = \begin{bmatrix} P_1(t) & P_2(t) \end{bmatrix} \begin{bmatrix} z_1(t) \\ z_2(t) \end{bmatrix}$$

where $P(t) = \begin{bmatrix} P_1(t) & P_2(t) \end{bmatrix}$ is invertible and T periodic, then we see that

$$\mathrm{Ker}\,(I - L_\alpha)^* = \left\{ v(t) = -\varphi'(y_0(t))^T B^T P_1(t) e^{A_1 t} z_1(0) : z_1(0) \in \mathbf{R}^{n_u+1} \right\}.$$

This is an $n_u + 1$ dimensional space.

Acknowledgement

This extension of the local result in [3] to hyperbolic limit cycles was triggered by comments from R. Brockett and C. I. Byrnes during the conference New Directions in Control theory and Applications. The authors were supported by grants from Swedish Research Council, the NSF and the AFOSR.

References

1. K. J. Åström and T. Hägglund. *Automatic Tuning of PID Controllers*. Research Triangle Park, N.C.: Instrument Society of America, 1988.
2. M. Farkas. *Periodic Motions*. Springer-Verlag, New York, 1994.

3. U. Jönsson and A. Megretski. A small gain theory for limit cycles. Technical Report TRITA/MAT-03-OS07, Department of Mathematics, Royal Institute of Technology, August 2003. Accepted for publication in SIAM Journal of Control and Optimization.
4. T. Kato. *Perturbation Theory for Linear Operators.* Springer-Verlag, Berlin, 1976.
5. E. Kreyszig. *Introductory Functional Analysis.* Wiley Classics Library. John Wiley and Sons, New York, 1978.
6. G. Taga. A model of the neuro-musculo-skeletal system for human locomotion. *Biol. Cybern.*, 73(97-111):97–111, 1995.
7. M. M. Williamson. Neural control of rhythmic arm movements. *Neural Networks*, 1998.

11

Low Codimension Control Singularities
for Single Input Nonlinear Systems

Arthur J. Krener[1] *, Wei Kang[2], Boumediene Hamzi[1] and Issa Tall[1]

[1] Department of Mathematics, University of California, One Shields Avenue, Davis, CA 95616, USA.
[2] Department of Mathematics, Naval Postgraduate School, Monterey, CA 93943, USA.

11.1 Introduction

Nonlinear dynamical systems exhibit complicated performance around bifurcation points. As the parameter of a system is varied, changes may occur in the qualitative structure of its solution around a point of bifurcation. In order to study dynamical systems with bifurcations, the following methodology is adopted in the theory of dynamical systems. First, the codimension of the bifurcation is computed, i.e. the smallest dimension of a parameter space which contains the bifurcation in a persistent way. Then, the system is embedded into a parameterized family of systems, with the number of parameters being equal to the codimension of the bifurcation. This family of systems, called miniversal deformation, describes the dynamics in the neighborhood of the bifurcation point. Finally, the dynamics of these systems is studied [6].

In this paper, we extend this methodology to controlled dynamical systems. Consider the class $C^k(\mathcal{X} \times \mathcal{U}, \mathbb{R}^n)$ of control systems of the form

$$\dot{x} = f(x, u) \tag{11.1}$$

where $x \in \mathcal{X}$, an open subset of $\mathbb{R}^n$, $u \in \mathcal{U}$, an open subset of $\mathbb{R}$ and f is C^k where $0 \leq k \leq \infty$. The equilibrium set $\mathcal{Z}$ of the control system is the set of all $(x^0, u^0) \in \mathcal{X} \times \mathcal{U}$ such that $f(x^0, u^0) = 0$. We are interested in studying control bifurcations, i.e., equilibria that are more difficult to stabilize than some of their neighboring equilibria.

A linear system of the form

$$\dot{x} = Fx + Gu \tag{11.2}$$

*Corresponding author. Research supported in part by NSF DMS-0204390 and AFOSR F49620-01-1-0202.

A.J. Krener et al.: *Low Codimension Control Singularities for Single Input Nonlinear Systems*, LNCIS **321**, 181–192 (2005)
www.springerlink.com

is *controllable* if the smallest F invariant subspace containing the columns of G is $\mathbb{R}^n$. A controllable linear system can be steered from any state to any other state in any $t > 0$.

The *linear part* of the nonlinear system (11.1) around the equilibrium (x^0, u^0) is the system

$$\dot{x} = F(x - x^0) + G(u - u^0) \qquad (11.3)$$

where

$$F = \frac{\partial f}{\partial x}(x^0, u^0), \quad G = \frac{\partial f}{\partial u}(x^0, u^0).$$

If this linear system is controllable then the nonlinear system can be steered from any nearby state to any other nearby state in any $t > 0$.

The generic equilibrium of (11.1) has a controllable linear part so if an equilibrium is not linearly controllable then it is more difficult to control than some of its neighboring equilibria and hence a control singularity. The goal of this paper is to study and classify the low codimension control singularities of single input nonlinear control systems. Systems with multiple inputs will be treated in a longer version of this article.

The set of all equilibria of all control systems like (11.1) is infinite dimensional but the nature of control singularity frequently depends on the low degree terms of the Taylor series of f at the equilibrium. Therefore instead of studying the infinite dimensional object we study the Taylor series through degree k of all possible systems at all possible equilibria. The latter is called the system k-jet space which we introduce in the next section.

11.2 Equilibria in the k-Jet Space of Systems

The *system k-jet space* $\mathcal{S}^k(\mathcal{X} \times \mathcal{U}, \mathbb{R}^n)$ is the space of all tuples of the form

$$\left((x, u), f(x, u), f^{(1)}(x, u), \ldots, f^{(k)}(x, u) \right) \qquad (11.4)$$

where $f \in C^k(\mathcal{X} \times \mathcal{U}, \mathbb{R}^n)$ and

$$f^{(j)}(x, u) = \frac{\partial^j f}{\partial(x, u)^j}(x, u).$$

Frequently when there is no chance of ambiguity we shall use the shortened notation $\mathcal{S}^k$. The terminology is a bit misleading, this is not a collection of systems but is a vector bundle with base $\mathcal{X} \times \mathcal{U}$ and fiber a real linear space of dimension $N(n, k)$ where

$$N(n, k) = n \binom{n + k + 1}{k}.$$

Moreover the notation (11.4) is very convenient but can be misleading. Each $f^{(j)}(x, u) = (f_1^{(j)}, \ldots, f_n^{(j)})'$ and each $f_i^{(j)}$ is actually a symmetric tensor of degree j in $n + 1$ indices. There is a natural projection of $\mathcal{S}^k$ onto $\mathcal{S}^l$ when $k \geq l \geq 0$.

A system (11.1) *realizes* a point in $\mathcal{S}^k$ if it has those derivatives at $\mathcal{X} \times \mathcal{U}$. Any point in $\mathcal{S}^k$ can be realized by a polynomial system of degree k but there are many other realizations.

It is more convenient to work with the systems jet space $\mathcal{S}^k(\mathcal{X} \times \mathcal{U}, \mathbb{R}^n)$ which is finite dimensional than the space of systems $C^k(\mathcal{X} \times \mathcal{U}, \mathbb{R}^n)$ which is infinite dimensional, particularly when studying a system locally around a particular (x, u) such as an equilibrium. The equilibrium set $\mathcal{Z}(f)$ of a system (11.1) is the set of all pairs $(x, u) \in \mathcal{X} \times \mathcal{U}$ such that $f(x, u) = 0$.

The equilibrium set $\mathcal{E}^k(\mathcal{X} \times \mathcal{U}, \mathbb{R}^n) \subset \mathcal{S}^k(\mathcal{X} \times \mathcal{U}, \mathbb{R}^n)$ is the space of all tuples of the form

$$\left((x, u), 0, f^{(1)}(x, u), \ldots, f^{(k)}(x, u) \right). \tag{11.5}$$

Again when there is no chance of ambiguity we shall use the shortened notation $\mathcal{E}^k$. The equilibrium set $\mathcal{E}^k$ is also a vector bundle with base $\mathcal{X} \times \mathcal{U}$ and fiber a real linear space of dimension $N(n, k) - n$. Clearly $\mathcal{E}^k$ is a subbundle of $\mathcal{S}^k$ and it is carried onto $\mathcal{E}^l$, $0 \leq l \leq k$ by the natural projection. Notice that $\mathcal{Z}(f) \subset \mathcal{X} \times \mathcal{U}$ and depends on the system (11.1) but $\mathcal{E}^k \subset \mathcal{S}^k$ and we don't need a system to define it.

A k-jet in $\mathcal{E}^k$ is a *control singularity* it is more difficult to control than some of its neighboring tuples in $\mathcal{E}^k$. Our goal is to study the classes of control singularities that are of low codimension in $\mathcal{E}^k$. Due to space limitations, we will study only two control singularities, the fold and the transcontrollable singularities. Other control singularities will be studied in the longer version of this article.

There is another jet bundle that is of interest. The feedback k-jet bundle $\mathcal{K}^k$ is the set of all tuples of the form

$$\left(x, \kappa(x), \kappa^{(1)}(x), \ldots, \kappa^{(k)}(x) \right) \tag{11.6}$$

where $\kappa : x \mapsto u$ is C^k mapping from $\mathcal{X}$ to $\mathcal{U}$ and

$$\kappa^{(j)}(x) = \frac{\partial^j \kappa}{\partial x^j}(x).$$

The maps $\kappa(x)$ are feedbacks and $\mathcal{K}^k$ is a fiber bundle with base $\mathcal{X}$ and fiber $\mathcal{U} \times \mathbb{R}^{M(n,k)}$ where

$$M(n, k) = \binom{n + k}{k} - 1.$$

Given an equilibrium (x^0, u^0) of the system (11.1), a typical goal is to find a smooth feedback such that the closed loop system

$$\dot{x} = f(x, \kappa(x)) \tag{11.7}$$

$$u = \kappa(x) \tag{11.8}$$

is locally asymptotically stable to (x^0, u^0). Frequently the stability of the closed loop system can be decided by its k-jet at x^0 for small k. And the k-jet of the closed loop system can be computed from the k-jet of the system at (x^0, u^0) and the k-jet of the feedback at x^0 assuming that $\kappa(x^0) = u^0$.

Therefore we say an equilibrium k-jet with base point (x^0, u^0) is *stabilizable* if there exists a feedback k-jet with base point x^0 so that every realization of the former makes every realization of the latter locally asymptotically stable to (x^0, u^0).

11.3 Linear Feedback Group and Linear Normal Form

Let (x^0, u^0) be an equilibrium of the system (11.1), and let

$$F = \frac{\partial f}{\partial x}(x^0, u^0), \quad G = \frac{\partial f}{\partial u}(x^0, u^0).$$

Then the controllability matrix of this pair is

$$\left[G\ FG\ \dots\ F^{n-1}G \right].$$

This is an $n \times n$ matrix of rank r, with $0 \leq r \leq n$. The span of the columns of this matrix is an F invariant subspace of dimension r denoted by $\mathcal{V}$. If $r = n$ then the pair F, G is said to be *controllable* and the system (11.1) is said to be *linearly controllable* at (x^0, u^0).

If $r < n$, the system (11.1) is not linearly controllable at (x^0, u^0). Let $r_0 = n - r$, then r_0 is the number of state dimensions that can't be controlled by linear effects and r is the number of dimensions that can be controlled by the linear effects of u.

Now, consider a linear change of state coordinates and a linear feedback on (11.1)

$$\begin{aligned} x - x^0 &= Tz, \\ u - u^0 &= Kz + Lv, \end{aligned} \tag{11.9}$$

where T is an $n \times n$ invertible matrix, K is an $1 \times n$ matrix and L is a scalar. This change of coordinates and feedback is an element of the *linear feedback group* of the form

$$\begin{bmatrix} T & 0 \\ K & L \end{bmatrix} \tag{11.10}$$

This latter acts on the system (11.1) and so acts on its k-jets. Moreover, the transformation (11.9) takes the equilibrium (x^0, u^0) to the equilibrium at $(0, 0)$ of

$$\dot{z} = \bar{f}(z, v) = T^{-1} f(x^0 + Tz, u^0 + Kz + Lv). \tag{11.11}$$

This induces a mapping from $\mathcal{E}^k$ to $\mathcal{E}^k$. The k jet

$$\left((x^0, u^0), 0, \frac{\partial f}{\partial(x, u)}(x^0, u^0), \ldots, \frac{\partial^k f}{\partial(x, u)^k}(x^0, u^0) \right) \tag{11.12}$$

goes to

$$\left((0, 0), 0, \frac{\partial \bar{f}}{\partial(z, v)}(0, 0), \ldots, \frac{\partial^k \bar{f}}{\partial(z, v)^k}(0, 0) \right) \tag{11.13}$$

If the linear part of the k-jet is $[F \ \ G]$ then it is changed to

$$[\bar{F} \ \bar{G}] = T^{-1} [F \ G] \begin{bmatrix} T & 0 \\ K & L \end{bmatrix}.$$

Given any system (11.1) there is an element of the linear feedback group which takes the linear part of the system into linear normal form. A system

$$\dot{z} = Az + Bv + O(z, v)^2 \tag{11.14}$$

is in *linear normal form* at the equilibrium $(z, v) = (0, 0)$ if

$$A = \begin{bmatrix} A_0 & 0 \\ 0 & A_1 \end{bmatrix}, \quad B = \begin{bmatrix} 0 \\ B_1 \end{bmatrix} \tag{11.15}$$

where the $r_0 \times r_0$ matrix A_0 is in real Jordan form and the pair consisting of the $r \times r$ matrix A_1 and the $r \times 1$ matrix B_1 is in Brunovsky form.

The former means that A_0 is a block diagonal matrix with diagonal blocks of the form

$$\begin{bmatrix} \Lambda_1 & I & 0 & \ldots & 0 \\ 0 & \Lambda_2 & I & \ldots & 0 \\ & & \ddots & \ddots & \\ 0 & 0 & 0 & \ddots & I \\ 0 & 0 & 0 & \ldots & \Lambda_s \end{bmatrix} \tag{11.16}$$

where Λ_i is a scalar

$$\Lambda_i = a_i$$

or a 2×2 matrix of the form

$$\Lambda_i = \begin{bmatrix} a_i & -\omega_i \\ \omega_i & a_i \end{bmatrix}$$

with $\omega_i \neq 0$.

The latter means that

$$
A_1 = \text{Diag} \begin{bmatrix} 0\ 1\ 0\ \dots\ 0 \\ 0\ 0\ 1\ \dots\ 0 \\ \ddots \\ 0\ 0\ 0\ \dots\ 1 \\ 0\ 0\ 0\ \dots\ 0 \end{bmatrix}^{r \times r}
$$

$$
B_1 = \text{Diag} \begin{bmatrix} 0 \\ 0 \\ \vdots \\ 0 \\ 1 \end{bmatrix}^{r \times 1}
$$

An equilibrium 1-jet in normal form is

$$
\left((0,0), 0, \begin{bmatrix} A_0 & 0 & 0 \\ 0 & A_1 & B_1 \end{bmatrix} \right) \tag{11.17}
$$

where A_0, A_1, B_1 are as above.

There is also a nonlinear feedback group that acts on (11.1). It includes the linear feedback group. It consists of C^k changes of state coordinates and state feedback of the form

$$
\begin{aligned} x &= \theta(z) \\ u &= \kappa(z,v) \end{aligned} \tag{11.18}
$$

where $(z,v) \mapsto (x,u)$ is a local diffeomorphism. This induces a corresponding action on points of $\mathcal{E}^k$. The nonlinear feedback group is stratified. There are near identity transformations of degree d which are of the form

$$
\begin{aligned} x &= z + \theta^{[d]}(z) \\ u &= v + \kappa^{[d]}(z,v) \end{aligned} \tag{11.19}
$$

where the superscript $^{[d]}$ indicates a polynomial vector field homogeneous of degree d. These do not form a subgroup as the composition of two such transformations typically has terms of degree d through d^2. A near identity transformation of degree d does not change the $d-1$ jet but it does modify the d-jet and higher jets. This allows one to bring the higher degree terms to normal form [4]

11.4 The Codimension of Orbits of the Linear Feedback Group

Control singularities are invariant under the linear and nonlinear feedback groups. If the original system (11.1) has a control singularity at an equilibrium

(x^0, u^0) then the transformed system has the same type of control singularity at the transformed equilibrium.

A class of *linear control singularities* is most conveniently defined by conditions on the 1 jet of the system in normal form. For such singularities, we are only interested in the action of the linear feedback group. The fold $\mathcal{F}$, treated in section 11.5.2, is a linear control singularity.

To compute the codimension of a class of linear control singularities we proceed as follows. As we said before the singular class is most conveniently defined by certain conditions on the linear normal form at the equilibrium $(0,0)$. The normal form may depend on one or more parameters. All other elements of the singular class are obtained by a linear feedback transformation acting on a singularity in normal form.

Hence we must study the action of the linear feedback group

$$z = Tx \tag{11.20}$$

$$v = Kx + Lu \tag{11.21}$$

on systems in linear normal form (11.14). We partition (11.10) compatibly with (11.15)

$$\begin{bmatrix} T & 0 \\ K & L \end{bmatrix} = \begin{bmatrix} T_{00} & T_{01} & 0 \\ T_{10} & T_{11} & 0 \\ K_0 & K_1 & L \end{bmatrix} \tag{11.22}$$

The result is a new system

$$\dot{x} = Fx + Gu + O(x, u)^2$$

where

$$\begin{bmatrix} F & G \end{bmatrix} = T^{-1} \begin{bmatrix} A & B \end{bmatrix} \begin{bmatrix} T & 0 \\ K & L \end{bmatrix}$$

$$\begin{bmatrix} F_{00} & F_{01} & G_0 \\ F_{10} & F_{11} & G_1 \end{bmatrix} = \begin{bmatrix} S_{00} & S_{01} \\ S_{10} & S_{11} \end{bmatrix} \begin{bmatrix} A_1 & 0 & 0 \\ 0 & A_1 & B_1 \end{bmatrix} \begin{bmatrix} T_{00} & T_{01} & 0 \\ T_{10} & T_{11} & 0 \\ K_0 & K_1 & L \end{bmatrix}$$

with $S = T^{-1}$.

Tannenbaum [5] has descibed the action of the group of linear changes of state coordinates, i. e. $K_1 = 0, L = I$, acting on linearly controllable systems, i.e. $r_0 = 0$. But we are interested in the full feedback group acting on possibly linearly uncontrollable systems.

When computing the codimension of an orbit of this action, it is simpler to compute the codimension of the infinitesmal action which is a linear calculation. Consider a curve $T = T(\mu)$, $K = K(\mu)$, $L = L(\mu)$ in the linear feedback group parameterized by $\mu \in \mathbb{R}$ where $T(0) = I$, $K(0) = 0$, $L(0) = I$. Let $'$ denote differentiation with respect to μ at $\mu = 0$ then

$$\begin{bmatrix} F_{00} & F_{01} & G_0 \\ F_{10} & F_{11} & G_1 \end{bmatrix}' = \left(T^{-1} \begin{bmatrix} A & B \end{bmatrix} \begin{bmatrix} T & 0 \\ K & L \end{bmatrix} \right)'$$

$$= \begin{bmatrix} A_0 T'_{00} - T'_{00} A_0 & A_0 T'_{01} - T'_{01} A_1 & -T'_{01} B_1 \\ A_1 T'_{10} - T'_{10} A_0 + B_1 K'_0 & A_1 T'_{11} - T'_{11} A_1 + B_1 K'_1 & -T'_{11} B_1 + B_1 L' \end{bmatrix}$$

This action splits into four mappings

$$T'_{00} \mapsto F'_{11} = A_0 T'_{00} - T'_{00} A_0, \tag{11.23}$$

$$T'_{01} \mapsto \begin{bmatrix} F_{01} & G_0 \end{bmatrix}' = \begin{bmatrix} A_0 T'_{01} - T'_{01} A_1, & -T'_{01} B_1 \end{bmatrix}, \tag{11.24}$$

$$\begin{bmatrix} T'_{10} \\ K'_0 \end{bmatrix} \mapsto F'_{21} = A_1 T'_{10} - T'_{10} A_0 + B_1 K'_0, \tag{11.25}$$

$$\begin{bmatrix} T'_{11} & 0 \\ K'_1 & L' \end{bmatrix} \mapsto \begin{bmatrix} F_{11} & G_1 \end{bmatrix}' = \begin{bmatrix} A_1 T'_{11} - T'_{11} A_1 + B_1 K'_1, & -T'_{11} B_1 + B_1 L' \end{bmatrix}.$$
$$\tag{11.26}$$

Each mapping is from a real vector space to another. For each of them, we wish to compute the codimension of its range and find a maximal set of linearly independent vectors which are transverse to the range.

The first linear mapping (11.23) is the action of infinitesmal linear changes of the uncontrollable coordinates. It goes from $I\!\!R^{r_0^2}$ to $I\!\!R^{r_0^2}$. This is the same mapping that occurs when studying dynamical systems without controls. It is never an isomorphism and the codimension of its range depends on A_0. An analysis of this map can be found in Wiggins [6] on page 315. We state the results for the cases where A_0 is 1×1 or 2×2.

If A_0 is 1×1 then the map (11.23) is identically zero so the range is of codimension one. A 1×1 matrix transverse to the range is 1.

If A_0 is 2×2 then there are several possibilities. We enumerate those of codimension two.

If A_0 has distinct, nonzero real eigenvalues then the range of (11.23) is of codimension two. Two matrices transverse to the range are

$$\begin{bmatrix} 1 & 0 \\ 0 & 0 \end{bmatrix}, \quad \begin{bmatrix} 0 & 0 \\ 0 & 1 \end{bmatrix}.$$

If A_0 has two complex eigenvalues whose real and imaginary parts are both nonzero then the range of (11.23) is of codimension two. Two matrices transverse to the range are

$$\begin{bmatrix} 1 & 0 \\ 0 & 1 \end{bmatrix}, \quad \begin{bmatrix} 0 & -1 \\ 1 & 0 \end{bmatrix}.$$

If A_0 is a 2×2 Jordan block

$$A_0 = \begin{bmatrix} a & 1 \\ 0 & a \end{bmatrix}$$

where $a \neq 0$ then the codimesion is also two. Two matrices transverse to the range are

$$\begin{bmatrix} 1 & 0 \\ 0 & 1 \end{bmatrix}, \quad \begin{bmatrix} 0 & 0 \\ 1 & 0 \end{bmatrix}.$$

Next consider the linear mapping (11.24). This is a mapping from $I\!\!R^{r_0 r}$ to $I\!\!R^{r_0(r+1)}$ which is clearly not onto, but it is one to one. Let $X = T'_{01}$ and $X_{\cdot,j}$ denote the jth column of the matrix X. Suppose

$$A_0 X - X A_1 = 0$$
$$X B_1 = 0$$

then these equations become

$$A_0 X_{\cdot,j} - X_{\cdot,j-1} = 0$$
$$X_{\cdot,r} = 0$$

so $X = 0$. Therefore the range of the mapping (11.24) has codimension r_0. One choice of r_0 independent $r_0 \times (r+1)$ matrices transverse to the range is

$$\begin{bmatrix} F_{01} & G_0 \end{bmatrix} = \begin{bmatrix} \mathbf{e_i} & 0 & \dots & 0 \end{bmatrix}, \quad i = 1, \dots, r_0 \tag{11.27}$$

where $\mathbf{e_i}$ is the ith vector in $I\!\!R^{r_0}$.

Next consider the linear mapping (11.25). This is a mapping from $I\!\!R^{r r_0 + r_0}$ to $I\!\!R^{r r_0}$ which we now show is onto. Let $X = T'_{10}$ and $X_{i,\cdot}$ denote the ith row of the matrix X. For $1 \leq i \leq r-1$

$$(A_1 X - X A_0 + B_1 K'_0)_{i,\cdot} = X_{i+1,\cdot} - X_{i,\cdot} A_0$$
$$(A_1 X - X A_0 + B_1 K'_0)_{n,\cdot} = -X_{n,\cdot} A_0 + K'_1$$

Hence we choose $X_{1,\cdot}$ arbitrarily, then for for $2 \leq i \leq r$ we choose $X_{i,\cdot}$ to arbitrarily fix the $i-1^{th}$ row and K'_1 to arbitrarily fix the rth row.

Finally we look at the linear mapping (11.26). This is a mapping from $I\!\!R^{r^2+r+1}$ to $I\!\!R^{r^2+r}$. If the controllability index $(0,r)$ is generic in the space of systems with r states and one input then this map is onto. If the pair is not generic then we must do a case by case study.

11.5 Versal Deformations and Low Codimension Linear Control Singularities of Scalar Input Sytems

11.5.1 Versal Deformations

Given a class of control singularities $\mathcal{G} \in \mathcal{E}^k$, one would like to study the types of equilibrium k jets that can be obtained by small perturbations. A family of control singularities is always invariant under the action of the feedback group.

Let $\mathcal{G} \subset \mathcal{E}^k$ which is invariant under the action of the feedback group. A C^k *versal deformation* of $\mathcal{G}$ is a C^k parametrized subset of $\mathcal{E}^k$ of the form

$$\phi : \mathcal{P} \to \mathcal{E}^k$$
$$\phi : \mu \mapsto \phi(\mu) = ((x, u), 0, \phi^{(1)}(\mu), \ldots, \phi^{(k)}(\mu)) \tag{11.28}$$

defined for μ in some neighborhood $\mathcal{P}$ of $0 \in I\!R^p$, which intersects $\mathcal{G}$ at $\mu = 0$ and which is transversal to $\mathcal{G}$. The versal deformation is said to be *miniversal* if the dimension p is minimal among all versal deformations. The minimal p is the *codimension* of $\mathcal{G}$.

A C^k *versal feedback* for a versal deformation (11.28) is a mapping

$$\psi : \mathcal{P} \to \mathcal{K}^k$$
$$\psi : \mu \mapsto \psi(\mu) = ((x, u), 0, \psi^{(1)}(\mu), \ldots, \psi^{(k)}(\mu)). \tag{11.29}$$

A versal feedback is *stabilizing* if at each $\mu \in \mathcal{P}$, $\psi(\mu)$ stabilizes $\phi(\mu)$.

11.5.2 Fold Control Singularities

Consider the scalar input system

$$\dot{z}_0 = az_0 + O(z, v)^2 \tag{11.30}$$
$$\dot{z}_1 = A_1 z_1 + B_1 v + O(z, v)^2 \tag{11.31}$$

where the $z_0 \in I\!R$, $z_1 \in I\!R^{n-1}$ and

$$A_0 = a \neq 0$$
$$A_1 = \begin{bmatrix} 0 & 1 & 0 & \ldots & 0 \\ 0 & 0 & 1 & \ldots & 0 \\ & & & \ddots & \\ 0 & 0 & 0 & \ldots & 1 \\ 0 & 0 & 0 & \ldots & 0 \end{bmatrix}^{r \times r}$$

$$B_1 = \begin{bmatrix} 0 \\ 0 \\ \vdots \\ 0 \\ 1 \end{bmatrix}^{r \times 1} . \tag{11.32}$$

Its 1-jet at the origin is

$$\left((0, 0), 0, \begin{bmatrix} a & 0 & 0 \\ 0 & A_1 & B_1 \end{bmatrix} \right) \tag{11.33}$$

This equilibrium 1-jet is the simplest example of a control singularity and is called a fold. The subset $\mathcal{F} \subset \mathcal{E}^k$ of *fold singularities* is the set of all 1-jets of the form

$$\left((x^0, u^0), 0, \begin{bmatrix} F_{00} & F_{01} & G_0 \\ F_{10} & F_{11} & G_1 \end{bmatrix} \right) \tag{11.34}$$

whose normal form is (11.33) for some $a \neq 0$. We shall show that $\mathcal{F}$ is of codimension one in $\mathcal{E}^k$.

Let us study the infinitesmal action of the linear feedback group on A, B (11.32). The range of the linear mapping (11.24) is of codimension one so from (11.27) we obtain a nonzero $n \times (n+1)$ matrix transverse to the orbit of A, B under the linear feedback group. The linear mapping (11.23) is identically zero so there is another linearly independent $n \times (n+1)$ matrix transverse to the orbit of A, B under the linear feedback group.

But perturbations in one of these directions can be accomplished by varying a and so it is not transverse to $\mathcal{F}$. Therefore $\mathcal{F}$ is a codimension one set of control singularities and a miniversal deformation of it is

$$\mu \mapsto \begin{bmatrix} F(\mu) & G(\mu) \end{bmatrix}$$

$$F(\mu) = \begin{bmatrix} a & \mu & 0 & 0 & \dots & 0 \\ \hline 0 & 0 & 1 & 0 & \dots & 0 \\ 0 & 0 & 0 & 1 & \dots & 0 \\ & & & & \ddots & \\ 0 & 0 & 0 & 0 & \dots & 1 \\ 0 & 0 & 0 & 0 & \dots & 0 \end{bmatrix}, \quad G(\mu) = \begin{bmatrix} 0 \\ 0 \\ 0 \\ 0 \\ 0 \\ 1 \end{bmatrix}$$

Now we see why the fold terminology. The controllabilty matrix of the versal deformation in reverse order is

$$\begin{bmatrix} F(\mu)^{n-1}G(\mu) & \dots & F(\mu)G(\mu) & G(\mu) \end{bmatrix} = \begin{bmatrix} \mu & 0 \\ \hline 0 & I \end{bmatrix}.$$

Notice that all these 1-jets are controllable except for $\mu = 0$ and the controllability reverses orientation (folds over) at $\mu = 0$.

There are two subclasses of fold singularities, those where $a < 0$ and those with $a > 0$. There is an important distinction between these subclasses. The former are linearly stabilizable, i.e., there exists a linear feedback $u = Kx$ so that all the poles of the linear part of the closed loop dynamics $A + BK$ are in the left half plane. In fact the versal deformation of a fold singularity with $a < 0$ is versally stabilizable by a versal linear feedback of the form

$$((x, u), 0, [K_0, K_1])$$

where $K_0 = 0$, and K_1 is such that $A_1 + B_1 K_1$ is Hurwitz.

But if $a > 0$ then the closed loop dynamics will always have at least one unstable eigenvalue a. The stabilization problem of systems with fold control singularities when $a > 0$ was treated in [2].

11.5.3 Transcontrollable Singularities

A transcontrollable singularity is a degenerate fold where $a = 0$. This means that the stabilizability of a system realizing this 1-jet is decided by its higher order terms. The class of transcontrollable singularities denoted by $\mathcal{TC}$ is of codimension two and a versal deformation of it is

$$(\mu_1, \mu_2) \mapsto \left[F(\mu_1, \mu_2) \; G(\mu_1, \mu_2) \right]$$

$$
F(\mu_1, \mu_2) = \left[\begin{array}{c|ccccc} \mu_1 & \mu_2 & 0 & 0 & \ldots & 0 \\ \hline 0 & 0 & 1 & 0 & \ldots & 0 \\ 0 & 0 & 0 & 1 & \ldots & 0 \\ & & & & \ddots & \\ 0 & 0 & 0 & 0 & \ldots & 1 \\ 0 & 0 & 0 & 0 & \ldots & 0 \end{array} \right], \quad G(\mu_1, \mu_2) = \left[\begin{array}{c} 0 \\ 0 \\ 0 \\ \\ 0 \\ 1 \end{array} \right]
$$

Notice that a transcontrollable singularity can be perturbed into a fold by changing μ_1 from zero and can be perturbed into a linearly controllable 1-jet by changing μ_2 from zero.

The stabilization problem of systems with transcontrollable singularities was treated in [3].

References

1. V. I. Arnold and D. Anosov. Dynamical systems I : ordinary differential equations and smooth dynamical systems. *Springer-Verlag, Berlin,* 1987.
2. B. Hamzi and A. J. Krener (2003). *Practical Stabilization of Systems with a Fold Control Bifurcation* in New Trends in Nonlinear Dynamics and Control and their Applications, W. Kang, C. Borges and M. Xiao eds., Springer, Berlin.
3. W. Kang (1998). *Bifurcation and Normal Form of Nonlinear Control Systems- part I/II.* SIAM J. Control and Optimization, **36**:193-212/213-232.
4. A. J. Krener, W. Kang, and D. Chang. Control Bifurcations. To appear *IEEE Trans. on Automatic Control.*
5. A. Tannenbaum. Invariance and Systems Theory: Algebraic and Geometric Aspects. *Springer-Verlag, Berlin,* 1981.
6. S. Wiggins. Introduction to Applied Nonlinear Dynamical Systems and Chaos. *Springer-Verlag, New York,* 1990.

Ellipsoidal Techniques for Hybrid Dynamics: the Reachability Problem

A.B. Kurzhanski[1] and P. Varaiya[2]

[1] University of California, Berkeley, CA 94720
   kurzhans@eecs.berkeley.edu
[2] University of California, Berkeley, CA 94720
   varaiya@eecs.berkeley.edu

Summary. This paper deals with the dynamics of hybrid systems under piecewise open-loop controls restricted by hard bounds. The system equations may be reset when crossing some prespecified domains ("the guards") in the state space. Therefore the continuous dynamics which govern the motion between the guards are complemented by discrete transitions which govern the resets. A state space model for such systems is proposed and reachability sets for such models are described. The computational side of reachability is treated through ellipsoidal techniques that indicate routes for numerical algorithms.

Keywords: hybrid systems, reachability sets, resets, discrete transitions, guards, control, ellipsoidal calculus.

12.1 Introduction

The standard reachability problem is an essential topic in control theory ([4], [6], [5], [9]). Recent applications require treating reachability in the more complicated setting of systems with hybrid dynamics [1], [10], [12]. The notion of hybrid system has various definitions, including those introduced in [2], [3], [11], [10] and [13].

This paper deals with a controlled process governed by an array of linear subsystems, one of which is switched on at each time and determines the system's on-line continuous dynamics. This switching is logically controlled: when the continuous state crosses some preassigned zones ("the guards"), the current subsystem may be switched to another subsystem of the array. The guards are taken here as hyperplanes, each of which allows a uniquely specified reset (a transition from the current subsystem to another one). In addition, the reset may or may not entail a change in the phase coordinates of the process.

A.B. Kurzhanski and P. Varaiya: *Ellipsoidal Techniques for Hybrid Dynamics: the Reachability Problem*, LNCIS **321**, 193–205 (2005)
www.springerlink.com

Despite the fairly complicated dynamics, efficient computation through ellipsoidal-valued approximations of reach sets for such branching processes appears to be available [9].

12.2 The Hybrid System

The overall system is governed by an array of subsystems indexed $i = 1, ..., k$,

$$\dot{x} \in A^{(i)}(t)x + B^{(i)}(t)u^{(i)}(t) + C^{(i)}(t)f^{(i)}(t), \tag{12.1}$$

with continuous matrix coefficients $A^{(i)}(t), B^{(i)}(t), C^{(i)}(t)$, and $x \in \mathbb{R}^n$. The $u^{(i)}$ are piecewise *open-loop controls* restricted by inclusions $u^{(i)}(t) \in \mathcal{P}^{(i)}(t) \subset \mathbb{R}^p$, and $\mathcal{P}^{(i)}(t)$ are continuous set-valued functions. The functions $f^{(i)}$ are given.

The bounds on the control values are *ellipsoidal,*

$$\mathcal{P}^{(i)}(t) = \mathcal{E}(p^{(i)}(t), P^{(i)}(t)). \tag{12.2}$$

The continuous functions $p(t)$ and $P(t) = P'(t)$ are the center and the "shape" matrix of the ellipsoid $\mathcal{E}(p(t), P(t))$. Recall that the *support function* of $\mathcal{E}(p, P)$ is

$$\rho(l \mid \mathcal{E}(p, P)) = \max\{(l, x) \mid x \in \mathcal{E}(p, P)\} = (l, p) + (l, Pl)^{1/2}.$$

In the phase space $\mathbb{R}^n$ are given k hyperplanes,

$$H_i = \{x \mid (c^{(i)}, x) - \gamma_i = 0\}, \ c^{(i)} \in \mathbb{R}^n, \ \gamma_i \in \mathbb{R} \ i = 1, \ldots, k.$$

These are *the enabling zones* (the guards).

Here is how the system operates.

At time t_0 the motion initiates from a point in the starting set $\mathcal{X}^0 = \mathcal{E}(x^0, X^0)$ and follows subsystem i (take $i = 1$, to be specific), with one of the controls $u^{(1)}(t)$ until (at time τ'_{i_1}) it reaches H_{i_1}, the first of the hyperplanes H_i along its route ($i_1 \neq 1$).

Here a binary operation interferes: The motion either continues along the "old" subsystem $i = 1$ or switches to (*is reset to*) a "new" subsystem $i_1 \neq 1$. (If $i_1 = 1$, we presume there is no reset.)

Before crossing H_{i_1}, the state of the system is denoted as $\{t, x, [1^+]\}$; after the binary operation the state is denoted as either $\{t, x, [1^+, i_1^+]\}$, if there was a reset, or $\{t, x, [1^+, i_1^-]\}$, if there was no reset.

The motion then develops according to the subsystem $i = 1$ or $i = i_1$ until crossing the next hyperplane H_{i_2}, when a similar binary operation takes place: The motion either follows the previous subsystem or is reset to subsystem i_2. After the second crossing the state of the system is $\{t, x, [1^+, i_1^s, i_2^s]\}$, where each boolean index "s" is either $s = +$ or $s = -$.

Thus the state has a part $\{t, x\}$ (without memory), which is the current position of the continuous-time variable, and a part $[1^+, i_1^s, i_2^s]$ with memory, related to the discrete event variable i_j^s, which describes the sequence of switchings made earlier by the system. At each new crossing a new term is added to this sequence.

Thus the following general rules are observed:

1. Crossing each hyperplane H_i results either in a reset to subsystem i or there is no reset.
2. The crossing takes place in the direction of the support vector $c^{(i)}$, and at points of crossing we have

$$\min\{(c^{(i)}(t), z) \mid z \in F^{(j)}(t, x)\} \geq \varepsilon > 0, \ \forall i, j = 1, ..., k, \qquad (12.3)$$

in which

$$F^{(j)}(t, x) = A^{(j)}(t)x + B^{(j)}(t)\mathcal{E}(p^{(j)}(t), P^{(j)}(t)) + C^{(j)}(t)f^{(j)}(t).$$

3. The state after j crossings is $\{t, x, [i_1^s, \ldots, i_j^s]\}$, and each index s is $+$ or $-$.
4. Upon crossing hyperplane H_m the sequence $[i_1^s, \ldots, i_j^+]$, describing the "discrete event" part of the state, is augmented by a new term, which is either m^+, if there is a switching to subsystem m, or m^-, if there is no switching.

This notation allows one to trace back the array of subsystems used earlier from any current position $\{t, x\}$. Thus, if the state is $\{t, x, [1^+, i_1^-, ..., i_j^-]\}$ with $s = -$ for all $i_1, ..., i_j$, the trajectory did not switch at any of the j crossings, having followed the initial subsystem $i = 1$ throughout the whole process.

Note that at each state $\{t, x, [1^+, i_1^s, ..., i_j^s]\}$ the system follows the subsystem whose number coincides with that of the last term with index $s = +$.

The hybrid system under consideration differs from so-called *switching systems* in that the time instants for crossing are not fixed but are determined by the course of each trajectory. Note, however, that resets result in an instantaneus change of velocity $\dot{x}(t)$, but with no change in the current position $x(t)$ of the system in the phase space.

The paper is concerned with reachability under piecewise open-loop controls with possible resets of controlled systems at given guards, and in between these resets the control is open-loop. The starting set $\mathcal{X}^0$ is ellipsoidal, $\mathcal{X}^0 = \mathcal{E}(x^0, X^0)$.

12.3 Reachability under Resets

The reachability problem has two versions.

Problem I. Find the set of all $\{x\}$ reachable from starting position $\{t_0, \mathcal{X}^0\}$ at **given time** t through all possible controls. This is the **reach set** $\mathcal{X}(t; t_0, \mathcal{X}^0)$ **at time** t from $\{t_0, \mathcal{X}^0\}$.

Problem II. Find the set of all $\{x\}$ reachable from starting position $\{t_0, \mathcal{X}^0\}$ at **some time** t within interval $t \in [t', t''] = T$ through all possible controls. This is the **reach set** $\mathcal{X}(t', t''; t_0, \mathcal{X}^0)$ **within interval** T.

One may observe that the problem will consist in investigating branching trajectory tubes, in describing their cross-sections ("cuts"), and the unions of such cross-sections. The reach sets may therefore be *disconnected*. We shall next discuss reach sets at given time t.

Let us first describe the reach set for a given sequence $[i_1^{s_1}, \ldots, i_r^{s_r}]$ of crossings, from position $\{t_0, \mathcal{X}^0, [j]\}$, taking $j = 1$ to be specific. The index s_1 is either $-$ or $+$.

The reach set after one crossing

(a) Before reaching H_{i_1}, $i_1 = j$ we have

$$\mathcal{X}^{(1)}[t] = \mathcal{X}^{(1)}(t; t_0, \mathcal{X}^0, [1^+]) = G^{(1)}(t, t_0)\mathcal{X}^0 +$$
$$+ \int_{t_0}^{t} G^{(1)}(t, s)[B(s)\mathcal{P}^{(1)}(s) + C(s)f^{(1)}(s)]\,ds,$$

in which $G^{(i)}(t, s)$ is the transition function for subsystem i.

(b) To be precise, suppose that before reaching H_j we have

$$\max\{(c^{(j)}, x) \mid x \in \mathcal{X}^{(1)}[t]\} = \rho_j^+(t) < \gamma_j.$$

The first instant of time when $\mathcal{X}^{(1)}[t] \bigcap H_j \neq \emptyset$ is τ_j', the smallest positive root of the equation

$$\gamma_j - \rho_j^+(t) = 0.$$

Introducing the function

$$\rho_j^-(t) = \min\{(c^{(j)}, x) \mid x \in \mathcal{X}^{(1)}[t]\},$$

we observe that the condition $\mathcal{X}^{(1)}[t] \bigcap H_j \neq \emptyset$ will hold so long as

$$\rho_j^-(t) \leq \gamma_j \leq \rho_j^+(t),$$

and the point of departure from H_j is the largest positive root τ_j'' of

$$\gamma_j - \rho_j^-(t) = 0.$$

Condition (12.3) above ensures that τ_j', τ_j'' are *unique*. Note that τ_j'' is the time instant when the whole reach set $\mathcal{X}^{(1)}[t]$ leaves H_j.

Denote $\mathcal{X}^{(1)}[t] \bigcap H_j = \mathcal{Z}_j^{(1)}(t)$.

(c) After the crossing we have to envisage two branches:
  $(-)$ with no reset—then nothing changes and

$$\mathcal{X}(t; t_0, \mathcal{X}^0, [1^+, j^-]) = \mathcal{X}^{(1)}(t; t_0, \mathcal{X}^0, [1^+]);$$

  $(+)$ with reset—then we consider the union

$$\mathcal{X}(t; t_0, \mathcal{X}^0, [1^+, j^+]) = \bigcup \{\mathcal{X}^{(j)}(t; s, \mathcal{Z}_j^{(1)}(s)) \mid s \in [\tau_j', \tau_j'']\}, \quad t \geq \tau_j''.$$

Thus, in case $(-)$ the reach tube develops further along the "old" subsystem (1), while in case $(+)$ it develops along the "new" subsystem $(j = i_1)$.

(d) For each new crossing we may now repeat this procedure. In this way we obtain the reach set

$$\mathcal{X}(t; t_0, \mathcal{X}^0, [1^+, i_1^{(s_1)}, \ldots, i_r^{(s_r)}])$$

for the branch $[1^+, i_1^{(s_1)}, \ldots, i_r^{(s_r)}]$.

We impose the following condition.

Assumption 12.3.1 *The intervals $[\tau_i', \tau_i'']$, $i = 1, \ldots, r$, do not intersect.*

For any interval $[t_0, t]$, with $\tau_{i_m}'' \leq t \leq \tau_{i_{m+1}}'$ and $\tau_{i_j}'' \leq t^* \leq \tau_{i_{j+1}}'$, $i_j < i_m$, one may observe the following semigroup-like property.

Lemma 1. *Each branch $[1^+, i_1^{s_1}, \ldots, i_m^{s_m}]$ has the following property:*

$$\begin{aligned}
&\mathcal{X}(t; t_0, \mathcal{X}^0, [1^+, i_1^{s_1}, \ldots, i_m^{s_m}]) \\
=&\mathcal{X}(t; t^*, \mathcal{X}(t^*; t_0, \mathcal{X}^0, [1^+, i_1^{s_1}, \ldots, i_j^{s_j}]), [i_{j+1}^{s_{j+1}}, \ldots, i_m^{s_m}])
\end{aligned} \qquad (12.4)$$

The one-stage crossing transformation

Recall that the continuous-time transition between crossings along subsystem j, from position $\{\tau, \mathcal{X}\}$ with $\tau \geq \tau_j''$ to the position at time $t \leq \tau_{j+1}'$ is $\mathcal{X}^{(j)}(t; \tau, \mathcal{X})$.

On the other hand, we may define a "one-stage crossing" transformation from position (state) $\{\tau_j', \mathcal{X}, [1^+]\}$ at the first time (τ_j') of crossing H_j to the last time (τ_j'') of crossing H_j:

$$T_j^s\{\tau_j', \mathcal{X}, [1^+]\} = \begin{cases} \{\tau_j'', \mathcal{X}^{(1)}(\tau_j''; \tau_j', \mathcal{X}, [1^+, j^-])\}, & \text{if } s = - \\ T_j^{(s)}\{\tau_j', \mathcal{X}, [1^+]\} = \{\tau_j'', \mathcal{Z}^{(j)}[\tau_j''], [1^+, j^+]\}, & \text{if } s = + \end{cases}$$

Above,

$$\mathcal{Z}^{(j)}[\tau_j''] = \cup\{\mathcal{X}^{(j)}(\tau_j''; t, H_j \cap \mathcal{X}^{(1)}(t; \tau_j', \mathcal{X})) \mid t \in [\tau_j'', \tau_j']\}.$$

We can now represent a branch $[1^+, i_1^{s_1}, \ldots, i_m^{s_m}]$ through a sequence of alternating operations of type T_j^s and $\mathcal{X}^{(j)}$.

For example, the reach set for the branch $[1^+, i_1^+, i_2^-]$ from starting position $\{\tau, \mathcal{X}, [1^+]\}$, $\tau \leq \tau_{i_1}'$, at time $t \in [\tau_{i_2}'', \tau_{i_3}']$ requires the following sequence of mappings:

$$T_{i_1}^+\{\tau_{i_1}', \mathcal{X}^{(1)}(\tau_{i_1}'; \mathcal{X}, \tau), [1^+]\} = \{\tau_{i_1}'', \mathcal{Z}^{(i_1)}[\tau_{i_1}''], [1^+, i_1^+]\},$$

$$\mathcal{X}^{(i_1)}[\tau_{i_2}'] = \mathcal{X}^{(i_1)}(\tau_{i_2}'; \tau_{i_1}'', \mathcal{Z}^{(i_1)}[\tau_{i_1}'']),$$

$$T_{i_2}^-\{\tau_{i_2}', \mathcal{X}^{(i_1)}[\tau_{i_2}'], [1^+, i_1^+]\} = \{\tau_{i_2}'', \mathcal{X}^{(i_1)}(\tau_{i_2}''; \tau_{i_2}', \mathcal{X}^{(i_1)}[\tau_{i_2}']), [1^+, i_1^+, i_2^-]\};$$

198   A.B. Kurzhanski and P. Varaiya

and then, for $t \in [\tau''_{i_2}, \tau'_{i_3}]$, the desired set of positions is given as

$$\{t; \mathcal{X}^{(i_1)}(t, \tau'_{i_2}, \mathcal{X}^{(i_1)}[\tau'_{i_2}]), [1^+, i_1^+, i_2^-]\}.$$

This is one branch of the overall reach set, whose continuous variables are

$$\mathcal{X}(t; \mathcal{X}, \tau) = \mathcal{X}^{(i_1)}(t; \tau'_{i_2}, \mathcal{X}^{(i_1)}[\tau'_{i_2}]).$$

Lemma 2. *A branch of type $\mathcal{X}(t; t_0, \mathcal{X}^0, [1, i_1^{(s_1)}, \ldots, i_k^{(s_k)}])$ is given by the composition (superposition) of alternating one-stage crossing transformations T_j^s and continuous maps $\mathcal{X}^{(j)}$, $j = 1, \ldots, k$.*

An alternative scheme for calculating reach sets is through value functions of optimization problems. Its advantage is that it is not restricted to linear systems.

12.4 Reachability Through Value Functions

As shown in [8], the reach sets for ordinary (non-hybrid) systems may be calculated as level sets of solutions to HJB (Hamilton-Jacobi-Bellman) equations for certain optimization problems. We will follow this scheme for the hybrid system under consideration. Consider first a one-stage crossing.

(a) *Before crossing* $H_{i_1} = H_j$, we assume as in Section 2, that the system operates from position $\{t_0, \mathcal{X}^0, [1]\}$. Then, for $t < \tau''_1$, we have

$$\mathcal{X}^{(1)}[t] = \{x \mid V^{(1)}(t, x) \leq 0\},$$

wherein

$$V^{(1)}(t, x) = \min_{u^{(1)}}\{d^2(x^{(1)}(t_0), \mathcal{X}^0) \mid x^{(1)}(t) = x\},$$

and $x^{(i)}(t) = x(t)$ is the trajectory of the subsystem i. We also write $V^{(1)}(t, x) = V^{(1)}(t, x_1, x_2, \ldots, x_n)$, for $x = (x_1, \ldots, x_n)$.

(b) *At the crossing* we have $\mathcal{X}^{(1)}[t] \cap H_j = \mathcal{Z}^{(1)}_{(j)}(t)$, which can be calculated as follows. Without loss of generality we may take $c_1^{(j)} = 1$. Then

$$\mathcal{Z}^{(1)}_{(j)}(t) = \{x \mid V^{(1)}(t, \zeta(x), x_2, \ldots, x_n)) \leq 0\} \cap H_j, \quad \zeta(x) = \gamma_j - \sum_{i=2}^{n} c_i^{(j)} x_i.$$

In particular, if the hyperplane $H_j = \{x \mid x_1 = \gamma_1\}$,

$$\mathcal{Z}^{(1)}_{(j)}(t) = \{x \mid x_1 = \gamma_1, V^{(1)}(t, \gamma_1, x_2, \ldots, x_n) \leq 0\},$$

and the set $\mathcal{Z}^{(1)}_{(j)}(t) \neq \emptyset$ iff $\rho_j^-(t) \leq \gamma_j \leq \rho_j^+(t)$, wherein

$$\rho_j^+(t) = \max\{(c^{(j)}, x) \mid V^{(1)}(t, x) \le 0\}, \ \rho_j^-(t) = \min\{(c^{(j)}, x) \mid V^{(1)}(t, x) \le 0\}.$$

This happens within the time interval $[\tau_j', \tau_j'']$, $\rho_j^-(t) = \tau_j'$, $\rho_j^+(t) = \tau_j''$.

(c) *After crossing $H_{i_1} = H_j$,* we envisage two branches:
   $(-)$ with no reset: Then $\mathcal{X}(t; t_0, \mathcal{X}^0, [1^+, j^-]) = \mathcal{X}^{(1)}(t; t_0, \mathcal{X}^0, [1^+])$,
   $(+)$ with reset: Then we have to calculate the union

$$\bigcup\{\mathcal{X}^{(i_1)}(t; s, \mathcal{Z}_{(j)}^{(1)}(s)) \mid s \in [\tau_j', \tau_j'']\}$$
$$= \mathcal{X}(t; t_0, \mathcal{X}^0, [1^+, j^+]), \ \tau_j'' \le t \le \tau_{j+1}'.$$

For $t > \tau_j''$ this union may be calculated as the level set for the function

$$\mathcal{V}(t, x \mid [1^+, j^+])$$
$$= \min_s \min_{u^{(i)}}\{V^{(j)}(s, \zeta(x), x_2, \ldots, x_n) + ((c^{(j)}, x) - \gamma_j)^2) \mid s \in [\tau_j', \tau_j'']\},$$

so that

$$\mathcal{X}(t; t_0, \mathcal{X}^0, [1^+, j^+]) = \{x \mid \mathcal{V}(t, x \mid [1^+, j^+]) \le 0\}.$$

The last set may be nonconvex.

(d) Repeating the procedure for each new crossing, we obtain the reach set $\mathcal{X}(t; t_0, \mathcal{X}^0, [1^+, i_1^{(s_1)}, \ldots, i_k^{(s_k)}])$ for the branch $[1^+, i_1^{(s_1)}, \ldots, i_k^{(s_k)}]$.
   The general schemes for applying the indicated transformations are given above. We now indicate some ellipsoidal techniques for these calculations.

12.5 The Ellipsoidal Techniques

Let us calculate the reach set after a one-stage crossing (a one-stage crossing transformation).

(a) Starting with $\mathcal{X}^0 = \mathcal{E}(x^0, X^0), i = 1$, the reach set $\mathcal{X}(t; t_0, \mathcal{X}^0, [1^+])$ is governed by

$$\dot{x} = A^{(1)}(t)x + B^{(1)}(t)\mathcal{E}(p^{(1)}(t), P^{(1)}(t)) + C^{(1)}(t)f^{(1)}(t),$$

for which the following internal and external *ellipsoidal approximations* hold (see [7], [9]),

$$\mathcal{E}(x^{(1)}(t), X_-^{(1)}(t)) \subset \mathcal{X}(t; t_0, \mathcal{X}^0, [1]) \subset \mathcal{E}(x^{(1)}(t), X_+^{(1)}(t)).$$

Above,

$$\dot{X}_+^{(1)} = A^{(1)}(t)X_+^{(1)} + X_+^{(1)}A^{(1)'}(t) + \pi(t)X_+^{(1)}$$
$$+ (\pi(t))^{-1}B^{(1)}(t)P^{(1)}(t)B^{(1)'}(t), \tag{12.5}$$

$$\dot{X}_-^{(1)} = A^{(1)}(t)X_-^{(1)} + X_-^{(1)}A^{(1)'}(t) + X_{-*}^{(1)-1}S(t)B^{(1)}(t)(P^{(1)})^{1/2}(t) +$$
$$+ (P^{(1)})^{1/2}(t)B^{(1)'}(t)S'(t)X_{-*}^{(1)-1'},$$

$$\dot{x}^{(1)} = A^{(1)}(t)x^{(1)} + B^{(1)}(t)p^{(1)}(t) + C^{(1)}(t)v^{(1)}(t),$$

$$X_+^{(1)}(t_0) = X_-^{(1)}(t_0) = X^0, \ x^{(1)}(t_0) = x^0.$$

where

$$\dot{X}_{-*}^{(1)} = S(t)P^{1/2}(t), \ \ X_{-*}^{(1)}(t_0) = S_0 P_0^{1/2}, \ S_0' S_0 = I.$$

and $\pi(t) > 0, S_0 = S_0', S(t) = S'(t)$ are parametrizing parameters. These approximations will be *tight* along a given *direction* $l(t) = G^{(1)'}(t_0,t)l, \ l \in \mathbb{R}^n$, ($G^{(i)}(t_0,t)$ is the transition matrix of the homogeneous subsystem i), if [9]

$$\pi(t) = (l(t), B^{(1)}(t)P^{(1)}(t)B^{(1)'}(t)l(t))^{1/2}(l, X_+^{(1)}(t)l)^{-1/2},$$

$$S(t)B^{(1)}(t)P^{(1)^{1/2}}(t)l(t) = \lambda(t)S_0 X^{0^{1/2}}l,$$

$$\lambda(t) = (l(t), B^{(1)}(t)P^{(1)}(t)B^{(1)'}(t)l(t))^{1/2}(l, X^0 l)^{-1/2},$$

$$S'(t)S(t) = I, \ S_0' S_0 = I.$$

(b) Let us now discuss the crossings $\mathcal{E}(x^{(1)}(t), X_+^{(1)}(t)) \cap H_j$, $\mathcal{E}(x^{(1)}(t), X_-^{(1)}(t)) \cap H_j$. Introducing the linear map

$$\mathbf{T}c^{(j)} = e^{(1)}; \ \mathbf{T}e^{(i)} = e^{(i)}, \ i = 1,\ldots,k, \ i \neq j, \ |\mathbf{T}| \neq 0,$$

we may transform the hyperplane H_j into a hyperplane of type $x_1 = \alpha_1$. (The unit vectors $e^{(i)}$ form the standard basis.) Without loss of generality we may also take $\alpha_1 = 0$, and denote H_j as $H_1(j)$ defined by the equality $x_1 = 0$.

Then

$$E_j^{(1)}(t) = \mathcal{E}(z_{(j)}(t), Z_{(j+)}^{(1)}(t)) = \mathcal{E}(x^{(1)}(t), X_+^{(1)}(t)) \cap H_1(j) = \{x \mid V_+^{(1)}(t,x) \le 1\},$$

in which

$$V_+^{(1)}(t,x) = \{\bar{x} \mid (\hat{x} - x^{(10)}(t), (X_+^{(1)})^{-1}(t))(\hat{x} - x^{(10)}(t))) \le 1\}, \qquad (12.6)$$

and $\hat{x} = \{0, x_2, \ldots, x_n\}$, $x^{(10)'}(t) = \{0, x_2^{(1)}(t), \ldots, x_n^{(1)}(t)\}$.

The intersection $E_j^{(1)}(t)$ is an ellipsoid whose support function is

$$\rho(l \mid E_j^{(1)}(t)) = (l, z_{(j)}(t)) + (l, Z_{(j+)}^{(1)}(t)l)^{1/2},$$

in which

$$z_{(j)}(t) = x^{(10)}(t) + x_1^{(1)}(t)q(t), \ q(t) = \{0, \bar{q}(t)\}.$$

$$\bar{q}(t) = \bar{\Gamma}^{-1}(t)\bar{\gamma}_c^{(1)}(t), \ \Gamma(t) = \{\gamma_{ij}\} = (X_+^{(1)})^{-1}(t), \ \gamma^{(1)'}_c(t) = \{\gamma_{12}(t), \ldots, \gamma_{1n}(t)\},$$

$$\bar{\Gamma} = \{\gamma_{ij} : \gamma_{jk} = \gamma_{jk}, \ j,k = 2,\ldots,n.\}$$

and

$$Z^{(1)}_{(j+)}(t) = (1 - h^2(t)) \begin{pmatrix} 0 & 0 \\ 0 & \bar{\Gamma}^{-1}(t) \end{pmatrix},$$

$$h^2(t) = x_1^{(1)2}(t)(\gamma_{11}(t) - (\bar{q}(t), \bar{\Gamma}^{-1}\bar{q}(t)).$$

Now, in the n-dimensional space $\mathcal{H}_1(j) = \{x \mid x_1 = 0\}$, we may consider an array of ellipsoids $\mathcal{E}(z_j(t), Z_{(j+)}(t))$, in which $t \in [\tau'_j, \tau''_j] = \mathcal{T}$ and

$$\mathcal{T} = \{t \ : \ (\hat{x} - x_1^{(1)}(t), (X_+^{(1)})^{-1}(t)(\hat{x} - x_1^{(1)}(t))) \leq 1\}.$$

When propagated after the reset, along the new subsystem j each of these ellipsoids turns into

$$\mathcal{X}^{(j)}[\tau''_j, t] = \mathcal{X}^{(j)}(\tau''_j, t, \mathcal{E}(z_{(j)}(t), Z_{(j)}(t))) =$$

$$G^{(j)}(\tau''_j, t)\mathcal{E}(z_{(j)}(t), Z_{(j)}(t)) +$$

$$+ \int_t^{\tau''_j} G^{(j)}(t, s)(B(s)\mathcal{E}(p^{(j)}(s), P^{(j)}(s)) + C(s)f^{(1)}(s))ds.$$

We are further interested in the union[3]

$$\mathcal{X}^{(j)}[\tau''_j \mid l] = \cup\{\mathcal{X}^{(j)}[\tau''_j, t \mid l] \mid t \in \mathcal{T}\},$$

and, after propagating $\mathcal{X}^{(j)}[\tau''_j \mid l]$ to $\mathcal{X}(\vartheta, \tau''_j, \mathcal{X}^{(j)}[\tau''_j \mid l])$, along subsystem j, in the union

$$\mathcal{X}(\vartheta, \tau''_j, \mathcal{X}^{(j)}[\tau''_j \mid l]) = \cup\{\mathcal{X}(\vartheta, \tau''_j, \mathcal{X}^{(j)}[\tau''_j, t \mid l]) \mid t \in \mathcal{T}\}, \tag{12.7}$$

for $t > \tau''_j$.

An external estimate of this union may be found as the level set of function

$$\mathcal{V}_+(\vartheta, x \mid l) = \min_t \min_{u^{(j)}}\{V_+(\vartheta, x \mid l) + x_1^2(t) \mid t \in \mathcal{T}, \ x(t) = x\}.$$

so that

$$\mathcal{X}(\vartheta, \tau''_j, \mathcal{X}^{(j)}[\tau''_j \mid l]) \subset \{x \mid \mathcal{V}_+(\vartheta, x \mid l) \leq 0\}.$$

The last inclusion is true for any l of unit norm. However, the knowledge of a parametrized family of functions $\mathcal{V}_+(\vartheta, x \mid l)$ allows us to calculate the exact reach set $\mathcal{X}(t, \tau''_j, \mathcal{X}^{(j)}[\tau'_j])$.

---

[3]From here on it is important to emphasize the dependence of the ellipsoids, the reach sets and the value functions on l. We henceforth include l in the arguments of the respective items.

Theorem 1. *The exact reach set after one crossing is*

$$\mathcal{X}(\vartheta, \tau_j'', \mathcal{X}^{(j)}[\tau_j']) = \{x \mid \mathcal{V}_+(\vartheta, x) \le 0\},$$

where $\mathcal{V}_+(\vartheta, x) = \max_l\{\mathcal{V}_+(\vartheta, x \mid l) \mid (l, l) \le 1\}.$

This follows from the property

$$\mathcal{X}(\vartheta, \tau_j'', \mathcal{X}^{(j)}[\tau_j'']) = \cap\{\mathcal{X}(\vartheta, \tau_j'', \mathcal{X}^{(j)}[\tau_j'' \mid l]) \mid (l, l) \le 1\},$$

which, in its turn, follows from the properties of external ellipsoids [9].

The value functions above thus allow both external approximations as well as exact description of the reach set. A similar scheme holds for internal approximations. However, the calculation of *unions* of reach sets is a cumbersome procedure. The generally nonconvex union of sets may however be substituted by a conservative estimate with *only one* external ellipsoid.

Let us therefore try to find one external ellipsoid first for the union $\mathcal{X}^{(j)}[\tau_j'' \mid l]$. We shall do this by approximating each element $\mathcal{X}^{(j)}[\tau_j'', t \mid l]$ of this union by a tight external ellipsoid $\mathcal{E}_l(x^{(j)}(\tau'', t \mid l), X_+^{(j)}(\tau'', t \mid l))$ taken along some direction $l(t)$ generated by vector $l \in \mathbb{R}^n$, so that

$$\mathcal{E}_l(x^{(j)}(\tau', t \mid l), X_+^{(j)}(\tau'', t \mid l)) \supset \mathcal{X}^{(j)}[\tau_j'', t \mid l]$$

and the vector $l \in \mathbb{R}^n$ would be one and the same for all $t \in \mathcal{T}$. The formulas for these ellipsoids are similar to those of (12.5) above.

We shall therefore seek an external ellipsoid

$$\mathbf{E}^{(j)}[\tau'', l] \supset \mathbf{X_1^{(j)}}[\tau''] = conv \cup \{\mathcal{E}_l(x^{(j)}(\tau', t \mid l), X_+^{(j)}(\tau'', t \mid l)) \mid t \in \mathcal{T}\},$$

wherein *convQ* stands for the convex hull of set Q. However,

$$conv \cup \{\mathcal{E}_l(x^{(j)}(\tau', t \mid l), X_+^{(j)}(\tau'', t \mid l)) \mid t \in \mathcal{T}\} =$$

$$\bigcup\left\{\int_{\tau'}^{\tau''} \alpha(s)\mathcal{E}_l(x^{(j)}(\tau'', s \mid l), X_+^{(j)}(\tau'', s \mid l))ds \,\middle|\, \alpha(s) \in \mathcal{A}\right\},$$

in which the set $\mathcal{A}$ consists of functions α, such that

$$\int_{\tau'}^{\tau''} \alpha(s)ds = 1, \quad \alpha(s) \ge 0, \ a.e.$$

Using the results of ([7, section 2.7]), we observe that *for any* continuous function

$$\int_{\tau'}^{\tau''} p(s)ds = 1, \quad p(s) > 0,$$

the following inequality is true for all $q \in \mathbb{R}^n$:

$$f(q, \alpha(\cdot) \mid l) = \rho\left(q \left| \int_{\tau'}^{\tau''} \alpha(s)\mathcal{E}_l(x^{(j)}(\tau'', s \mid l), X_+^{(j)}(\tau'', s \mid l))ds \right.\right) =$$

$$\int_{\tau'}^{\tau''} \left((q, \alpha(s)x^{(j)}(\tau'', s \mid l)) + (q, \alpha^2(s)X_+^{(j)}(\tau'', s \mid l)q)^{1/2}\right) ds$$

$$\leq \int_{\tau'}^{\tau''} \left((q, \alpha(s)x^{(j)}(\tau'', s \mid l)) + (q, X_+^{(j)}(\tau'', s \mid l)\alpha^2(s)p^{-1}(s)q)^{1/2}\right) ds$$

$$= \Phi(q, \alpha(\cdot), p(\cdot) \mid l),$$

whatever be the function $\alpha(s)$ and vector l of unit norm. For any such $p(\cdot)$ we therefore have

$$\max_\alpha \max_q \{f(q, \alpha(\cdot)|l)/\Phi(q, \alpha(\cdot), p(\cdot)|l)\} \leq 1, \quad q \in \mathbb{R}^n, \ \alpha(\cdot) \in \mathcal{A},$$

with maximizer $\alpha_0 = \alpha_0(\cdot \mid p(\cdot), l)$ and the external ellipsoid

$$\mathbf{E}_1^{(j)}[\tau'' \mid \mathbf{p}(\cdot)] = \mathcal{E}_l(x^{(j)}[\tau'', \alpha_0(\cdot \mid p(\cdot), l)], X^{(j)}[\tau'', \alpha_0(\cdot \mid p(\cdot), l)]) \supset$$

$$\supset conv \cup \{\mathcal{E}_l(x^{(j)}(\tau', t \mid l), X_+^{(j)}(\tau'', t \mid l)) \mid t \in \mathcal{T}\},$$

with

$$x_l^{(j)}[\tau'', p(\cdot)] = \int_{\tau'}^{\tau''} \alpha_0(s \mid p(\cdot), l)x^{(j)}(\tau'', s)ds,$$

$$X_l^{(j)}[\tau'', p(\cdot)] = \int_{\tau'}^{\tau''} \alpha_0^2(s \mid p(\cdot), l)p^{-1}(s)X_+^{(j)}(\tau'', s)ds$$

We may choose function $p(s)$ to optimize the external ellipsoid $\mathbf{E}^{(j)}{}_l[\tau'' \mid p(\cdot)]$.[4] Thus, for a given vector l we may first take $p(\cdot) = p_l(\cdot)$ to make this external ellisoid *tight*, according to [9], then select $l = l^0(\tau'', j)$, so that

$$l^0(\tau'', j) = \arg\min_l \{\varphi(\tau'', l, j)\}, \tag{12.8}$$

with criteria

$$\varphi(\tau'', l, j) = \{vol\mathbf{E}^{(j)}{}_l\}$$

or

$$\varphi(\tau'', l, j) = \{trX_l^{(j)}\},$$

or some other criteria (see e.g. [7]). Here *vol* stands for the volume and *tr* for the trace.

Using the described procedure, we may thus first find an external ellipsoid $\mathbf{E}_{(j)}(\tau'') \supset \mathcal{X}_{(j)}[\tau'' \mid l]$, for the union $\mathcal{X}_{(j)}[\tau'' \mid l]$. This ellipsoid will be independent of l.

---

[4]The functions $\alpha(s), p(s)$ may as well be selected to be just Lebesgue-integrable and positive almost everywhere.

Then, taking $\mathbf{E}_{(j)}(\tau'')$ as a starting set at time τ'', we may propagate the reach tube to an instant $t \in [\tau_1'', \tau_2']$. The reach set $\mathcal{X}(t, \tau_1'', \mathbf{E}_{(j)}(\tau''))$ may now be approximated by external ellipsoids through relations similar to (12.5).

Introducing a single internal ellipsoid for a nonconvex union of sets $\mathcal{X}_{(j)}[\tau'' \mid l]$, is possible, but this would be a very conservative estimate.

Further calculations follow the formulas above.

A (possible) algorithmic scheme

1. Starting with position $\{t_0, \mathcal{X}_0, [1^+]\}$, with nearest crossing H_j, find approximations

$$\mathcal{E}_-[t, l] = \mathcal{E}(x^{(1)}(t), X_-^{(1)}(t)) \subseteq \mathcal{X}(\tau_j'; t_0, \mathcal{X}^0, [1]) \subseteq \mathcal{E}(x^{(1)}(t), X_+^{(1)}(t)) = \mathcal{E}_+[t, l],$$
$$(12.9)$$

which are *tight* along a direction $l(t)$ generated by selected $l \in \mathbb{R}^n$, $(l, l) = 1$.

2. Calculate the one-stage crossing transformation $T_j^s \mathcal{E}$ for $\mathcal{E} = \mathcal{E}_-[t, l]$, $\mathcal{E} = \mathcal{E}_+[t, l]$:
with $s = -$ there is no reset, and we continue with the formulas of (12.8);
with $s = +$ we calculate the optimal $\mathbf{E}^{(j)}{}_l[\tau'' \mid p^0(\cdot))$ (due to a selected criterion) and if necessary, a similar internal ellipsoid.

3. Proceed further on:
with $s = -$, from continous-time positions $\{\tau'', \mathcal{E}_+[\tau'', l], [1^+, j^-]\}$ (external) and, if necessary, also from $\{\tau'', \mathcal{E}_-[\tau'', l], [1^+, j^-]\}$ (internal),
with $s = +$, from continous-time position $\{\tau'', \mathbf{E}^{(j)}{}_l[\tau'' \mid p^0(\cdot)), [1^+, j^+]\}$ (external) and if necessary, from a similar internal position.

4. Repeat the procedures of the above.

Remark This scheme relies on the direction l selected for the ellipsoidal approximations at stage 1. The procedure may extend in two ways.

(i) Deal with *one* ellipsoid, optimizing our ellpsoids $\mathcal{E}_l$ over all $l, (l, l) = 1$. This would produce conservative solutions of the problem.

(ii) Proceed *in parallel* with an *array of ellipsoids* corresponding to an array of directions $l, (l, l) = 1$. With increasing number of directions this would approach the exact solution with any degree of accuracy.

12.6 Conclusion

This paper studies a reachability problem for a hybrid system whose dynamics at each instant of time is governed by one of the linear subsystems in a given array, with possible resets from one subsystem to another when crossing any of the given "guards" specified as hyperplanes.

The paper introduces the state space variable for such a hybrid system and describes its reach sets through a branching process.

The computational scheme for finding the reach sets is described through external and internal ellipsoidal approximations.

These approximations may be based on calculating either one approximating ellipsoid at each time, which gives a conservative solution, or by a parallel calculation of an array of ellipsoidal approximations in which case, increasing the number of elements in the array, one may approach exact solutions with any degree of accuracy.

Acknowledgement

This research was supported by ONR-MURI program under Grant N00014-02-1-0720 and by National Science Foundation under Grant ECS 9725148.

References

1. Botchkarev O., Tripakis S. Verification of Hybrid Systems with Linear Differential Inclusions Using Ellipsoidal Approximations. *Hybrid Systems: Computation and Control*. N.Lynch and B.Krogh eds., LNCS, Springer, v.1790, 2000.
2. Branicky M., Borkar V., Mitter S., A unified framework for hybrid control:Model and optimal control theory. *IEEE Trans. Aut. Control*, 43(1), pp. 31-45.
3. Brockett R., Hybrid models for motion control systems. In Trentelman H., Willems J., eds.*Essays on Control: Perspectives in Theory and its Applications*, Birkhauser, Boston, pp. 29-53.
4. Krasovskii N.N. *The Theory of Control of Motion*. Nauka, Moscow, 1968.
5. Lee E.B., Markus L., *Foundations of Optimal Control Theory*. Wiley, New York, 1961.
6. Leitmann G., *Optimality and reachability with feedback controls*. In *Dynamic Systems and Microphysics*, Blaquiere A., Leitmann G., eds., Acad Press N.Y., 1982.
7. Kurzhanski A.B., Valyi I., *Ellipsoidal Calculus for Estimation and Control*. Birkhäuser, Boston, 1997.
8. Kurzhanski A.B., Varaiya P., Dynamic Optimization for Reachability Problems, *JOTA*, v.108, N2, pp. 227-251.
9. Kurzhanski A.B., Varaiya P., On ellipsoidal techniques for reachability analysis. Parts I, II, *Optimization Methods and Software*, pp. 187-237.
10. Lygeros J.C., Tomlin C. and Sastri S., Controllers for reachability specifications for hybrid systems. *Automatica*, v.35, N3, 1999,pp. 349-370.
11. Puri A., Varaiya P., Decidability of hybrid systems with rectangular inclusions. In: *Proc. CAV' 94, LNCS 1066*. Dill, D., ed,Springer, 1996.
12. Varaiya P., Reach set computation using optimal control. Proc. of KIT Workshop on Verification of Hybrid Systems, Verimag, Grenoble, 1998.
13. Van der Schaft A., Schumacher H., *An Introduction to Hybrid Systems*, Lect. Notes in Control and Info. Sciences, v.251, Springer, 2000.

13

Controlling the Shape of a Muscular Hydrostat: A Tongue or Tentacle[*]

William S. Levine[1], Caroline Essex Torcaso[2], and Maureen Stone[3]

[1] Dept. of ECE and ISR, University of Maryland, College Park, MD 20742
   `wsl@eng.umd.edu`
[2] ISR, University of Maryland, College Park, MD 20742
   `essex@isrmail.isr.umd.edu`
[3] University of Maryland Dental School, Departments of Biomedical Sciences and Orthodontics, 666 W. Baltimore St, Baltimore, MD, 21201
   `mstone@umaryland.edu`

Summary. A mathematical model of the human tongue, developed as part of a larger project to understand its movement, use in speech, and control, is described. The model has the mathematical form of an incompressible solid with a nonlinear, partially controllable, stress/strain relationship. The rationale for such a model, an analytic solution for a simplified version of the model, and indications of future work are included.

13.1 Introduction

The human tongue is an important component of the system that produces intelligible speech. To understand speech production, it is necessary to understand the dynamics and control of the tongue. There are several difficulties. First, the tongue is enclosed in the mouth during speech, greatly complicating the problem of obtaining accurate experimental data about its movements. Second, the tongue is one of a small collection of biomechanical systems known as muscular hydrostats. Such systems—other examples are the elephant's trunk and the squid's tentacles—are composed almost exclusively of muscle and do not include any rigid structure for the muscles to act upon.

Muscles can only shorten as a result of their own activation. To lengthen, a muscle must be acted upon by some force external to itself. In the case of the tongue, or any other muscular hydrostat, this force is created by the

[*]Acknowledgment: This work was supported in part by Grant No. R01 DC01758 from NIDCD. The authors would like to thank their coworkers on this research, Jerry Prince, Vijay Partharasarathy, Chandra Kambhamettu, and Min Li, who have provided the experimental data, considerable technical advice and comment, and an unusually enjoyable collaboration

W.S. Levine et al.: *Controlling the Shape of a Muscular Hydrostat: A Tongue or Tentacle*, LNCIS **321**, 207–222 (2005)
`www.springerlink.com`

interaction of other muscles within the tongue with the incompressibility of the tongue. Thus, the control of the tongue depends on exploiting the dynamical interaction between the contraction of individual muscles and the incompressibility. To understand this quantitatively, it is necessary to describe the tongue mathematically as an incompressible elastic structure that is controlled by temporally varying the stress/strain relationship within subregions of the overall structure. This article describes such a mathematical model of the tongue and the squid tentacle.

Before describing our model, we present a brief survey of previous attempts to model the tongue mathematically. We follow this with a demonstration that a linear model of the tongue that enforces incompressibility by means of a Poisson's ratio close to 1/2 does not both replicate reasonable movements of the tongue and satisfy the incompressibility constraint. We then give a new mathematical model of the tongue. The model is designed to handle large deformations because experimental data shows that the tongue normally undergoes large deformations. A crucial component of the model is a mathematical model of mammalian striated muscle, which is described in the next section. We conclude with a brief indication of the goals of this research and of the next steps towards achieving these goals.

13.2 Background

The first attempt to create a mathematical model of the human tongue is generally acknowledged to be by Perkell in 1969 [11]. Perkell used interconnected masses and springs to create a two-dimensional mathematical model of the mid-sagittal plane of the tongue. Kiritani et al.[9] substantially improved upon Perkell's work by creating a three-dimensional finite element model. Their model also included simple models of the lips and vocal tract and of tongue contact. Improved models of the tongue were later developed by Hashimoto and Suga in [7] and Sanguinetti et al. in 1997-1998 [12, 13]. At about the same time, Payan and Perrier [10] developed a two-dimensional model of the midsagittal plane of the tongue that included some geometrical nonlinearities. Their model was based on the somewhat controversial equilibrium point model of muscle. Dang and Honda [3] have continued and extended the development of two-dimensional models by including 2cm of transverse thickness in their essentially two-dimensional model. Their model was based on measurements from volumetric MRI and the tongue, jaw, hyoid bone, and vocal tract wall. However, their model was based on a mass-spring network that did not fully account for continuum effects. More recently, they have improved their model so that it includes the effects of Poisson's ratio and produces a "semicontinuum model" tissue model [4].

A somewhat different approach to the mathematical modelling of the tongue has been the work of Wilhelms-Tricarico [15]. Wilhelms-Tricarico's approach to modelling has been to create very detailed models that attempt to include as much as is possible of the known elements of the tongue. The models begin with a finite element grid that is tailored to the shapes of individual muscles within the tongue. A considerable amount of nonlinearity is included. For example, the model includes the nonlinear force/length and force/velocity relations for mammalian muscle developed by Zajac [16].

The reported aspects of virtually all of this work do not address the mathematical issues associated with the models. In particular, the models are almost always a single finite element approximation to the physical problem of describing the biomechanics of the tongue. Questions such as whether a refined set of elements would improve the approximation cannot be addressed. Furthermore, in almost all of these models, the finite element mesh is manually created so as to conform to the placement of muscles in the model. The resulting mesh may not satisfy the regularity criteria needed for good finite element computations. A notable exception is the recently developed finite element model of Gerard et al. [6] in which the regularity issue is addressed. However, it is still fair to say that the mathematical issues associated with models of the tongue have mostly been ignored in the published literature.

13.3 Volume Changes

In many finite element models of the tongue, a value of Poisson's ratio close to one-half is used to model the incompressibility constraint. A common measurement of how well incompressibility is enforced in this type of model is to compute a relative change in the overall volume, e.g.,

$$\text{Relative change in global volume} = \frac{\text{Initial global volume} - \text{Deformed global volume}}{\text{Initial global volume}} * 100$$

If the relative change is small in absolute value then enforcement of incompressibility in the model is considered to be acceptable.

However, incompressibility is a *local* phenomenon, and global volume preservation does not imply local volume preservation. We propose a new method for measuring incompressibility and present two examples where incompressibility appears to be accurately modelled when using the common global measurement, but our new method indicates significant local changes in volume.

Our new method for measuring incompressibility is to compute the relative change in volume at the finite element (FE) level; that is, for each finite element, compute

$$\text{Relative change in each FE volume} = \frac{\text{Initial FE volume} - \text{Deformed FE volume}}{\text{Initial FE volume}} * 100$$

Since there may be many finite elements, one may wish to save these in an array. Using this measurement, we can determine the maximum and minimum percentage of relative changes in local volume. In particular, we can graph the locations where these changes are above or below a threshold amount, and our model is failing to enforce the incompressibility constraint. We also compute a sum of the volume differences; e.g.,

$$\text{Sum of FE volume difference} = \sum_{FEs} |\text{Initial FE volume} - \text{Deformed FE volume}|$$

as an additional measurement tool.

Using our three-dimensional rectangular prism tongue model, we activated the first section of superior longitudinalis (SL1) and genioglossus (GG1) as shown in Figure 13.1. The relative change in global volume is .0488%; that

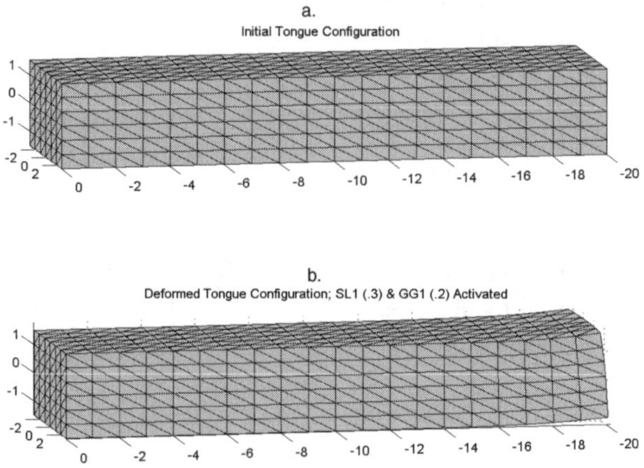

Fig. 13.1. a. Initial configuration with volume of 360 pixels^2. b. Deformed tongue after activating SL1 (.3) and GG1 (.2) with volume of 360.1759 pixels^2.

is, the overall volume increased slightly. Normally, this measurement would indicate that the enforcement of incompressibility is good. Using our new method, we find that the sum of the absolute values of volume differences in the finite elements is .3534, and the relative local volume changes are between 1.4645% and -1.0788% with the negative sign indicating a volume decrease. These numbers show there are areas where the incompressibility constraint is not being enforced.

In our second example, we activated the first section of genioglossus (GG1) in our two-dimensional tongue model. The initial and deformed configurations are almost identical as shown in Figure 13.2. The relative global volume change is .028287%, which indicates a reasonably good enforcement of incompressibility. However, the sum of the absolute values of finite element volume differences is 1.2350, and the local relative volume changes are between $-.1422\%$

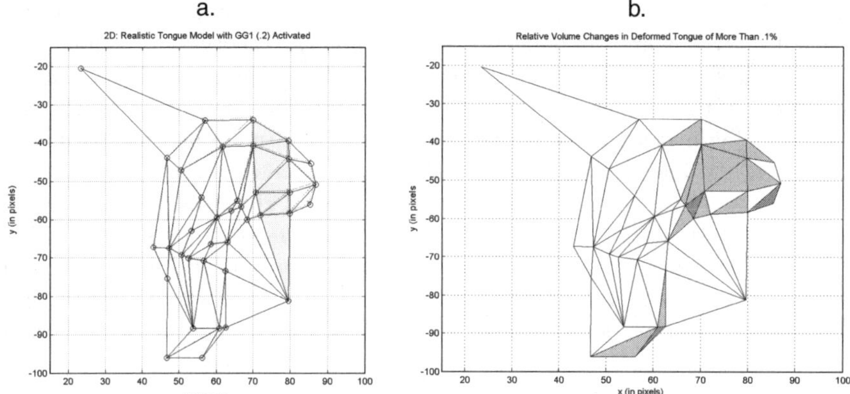

Fig. 13.2. a. The initial (dashed lines) and deformed (solid black lines) configurations for tongue with GG1 (.2) activated. GG1 lies in the shaded area. b. The shaded areas indicate local relative volume changes above .10% in absolute value. The darker areas indicate volume increases while the lighter areas indicate reductions in volume.

and .4795%. Figure 13.2 shows the locations where the incompressibility constraint is not satisfied.

The above examples were chosen because the activation parameters appear to create small deformations that satisfy the incompressibility constraint. However, on closer examination, there are regions within the deformed tongue where the incompressibility constraint has not been satisfied.

In the above examples, we were careful to keep our activation parameters small because the model used is a small deformation model. We also had to activate only one or two muscle sections at a time because activating more muscle sections even with very small parameters can result in large movements. This result supports our belief that a linear small deformation model cannot adequately model tongue movement. It is also interesting to note that both our new methods and the common technique for measuring incompressibility indicate the incompressibility constraint is not satisfied for these large deformations. This indicates the incompressibility constraint needs to be explicitly included in the finite element model. We propose using a mixed formulation [8] with both displacement and pressure finite elements to approximate a solution that satisfies the incompressibility constraint.

13.4 Tongue Model

The human tongue is incompressible and normally undergoes large deformations. A mathematical model that accurately describes an incompressible elastic structure that undergoes large deformations must be nonlinear. There are many modelling and mathematical issues that simply do not arise in the linear theory but are fundamental to the nonlinear theory. For example, the distinction between the material and spatial (Lagrangian and Eulerian) formulations of the dynamics is largely irrelevant for small deformations, the case described by linear models. In the case of large deformations, this distinction is very important. In fact, there are strong arguments for using the material formulation for the tongue model (Antman [1] p.ix). However, we do note that the spatial formulation can be useful for some inverse problems ([1] p. 451).

Similarly, there are many choices for the strain tensor. All of them give equivalent results in the linear case but this is not so in the nonlinear case. The main driver of our choice of strain tensor was the principle of frame indifference. This says that the stress/strain relationship of a material cannot change when the material undergoes rigid motions ([1] p. 420 et seq.). This principle implies that the stress/strain relationship should have a particular form that depends primarily on the right Cauchy-Green deformation tensor $\underline{C}(\underline{z}, t)$.

With the above-mentioned conditions in mind, our tongue model takes the following form. Let $\underline{z}$ denote a material point of the tongue belonging to a set of points, $\mathbf{B}$, in three dimensional space. Let $\underline{p}(\underline{z}, t)$ denote the position of material point $\underline{z} \subset \mathbf{B}$ at time t. Then,

$$\underline{F} = \underline{p}_{\underline{z}} \tag{13.1}$$

$$\underline{C} = \underline{F}^T \underline{F} \tag{13.2}$$

where $\underline{C}$ is the previously mentioned right Cauchy-Green deformation tensor, $\underline{F}$ is the deformation gradient, and the subscript indicates the partial derivative. Then, the fundamental dynamical equation has the form.

$$\nabla \cdot \underline{T}^T + \underline{f} = \underline{p}_{tt} \tag{13.3}$$

where $\underline{T}$ is the first Piola-Kirchoff stress tensor, $\underline{f}$ is the body force due to gravity, the double subscript denotes the second partial derivative with respect to t, and we have assumed the mass density of the tongue is a constant that we normalize to one. The actual modelling of the tongue consists primarily of specifying $\underline{T}$ in terms of $\underline{F}$ and $\underline{C}$.

Because the tongue is nearly incompressible we describe it by an incompressible model. This adds the condition

$$det\ \underline{C} = 1 \tag{13.4}$$

Incompressibility also imposes an important condition on the form of $\underline{T}$. Precisely,

$$\underline{T}(\underline{z}, t) = \underline{F}(\underline{z}, t)(-p\underline{C}^{-1}(\underline{z}, t) + \hat{\underline{S}}_A(\underline{C}(\underline{z}, t), p, \underline{z})) \tag{13.5}$$

Note that p without the underline denotes the pressure within the tongue, which is generally a function of $\underline{z}$ and t. The definition of $\hat{\underline{S}}_A$ is really part of the muscle model and will be given below. There is a further condition on $\hat{\underline{S}}_A$ that we give below. It is an additional component of the incompressibility constraint,

$$tr(\hat{\underline{S}}_A \underline{C}^{-1}) = 0 \tag{13.6}$$

Equations (13.3-13.6) constitute an abstract mathematical description of the tongue. At this level of abstraction, the equations characterize any incompressible elastic structure in either two or three dimensions. In addition, a model for a compressible elastic structure consists of Equations (13.3) and (13.5) with $p = 0$. It is the muscle model and the specific choice of $\mathbf{B}$ that makes these equations describe the tongue.

13.5 Muscle Model

The new muscle model was driven by three mathematical issues in addition to the obvious requirement that it fit the known physiological facts about mammalian muscle. The model had to be convenient to use, satisfy the frame indifference principle described earlier, and had to satisfy reasonable conditions for existence and uniqueness of solutions when incorporated into the overall tongue model. The resulting model is in two parts, the geometric description and the stress/strain characteristics.

a. Geometric description: Each muscle occupies a region of space, a subset of the material points of the tongue. This set changes as the muscle moves. We identify the muscle by the set of points, $\mathbf{M} \subset \mathbf{B}$, that it occupies in its reference configuration. A muscle consists of fibers. We denote these fibers by curves defined on $\mathbf{M}$ in its reference configuration. There is a unique fiber at every point $\underline{z} \in \mathbf{M}$. We define the fibers as a family of offset curves [2]. That is,

(i) A backbone curve, $\underline{c}(s)$, where s is the arc length along the backbone curve. For muscle fibers, s can always be chosen to be a coordinate direction, z_1 for instance. (ii) At each point of the backbone curve, which we can identify by s, there is a unit tangent vector, denoted by $\underline{\tau}(s)$ and, in three dimensions, a plane normal to the curve. Choose a pair of orthonormal basis vectors for this perpendicular plane, say $\underline{\eta}_1(s)$ and $\underline{\eta}_2(s)$. Then every other muscle fiber can be represented by an offset curve, $\underline{o}(s)$, defined by

$$\underline{o}(s) = z_{01}\underline{\eta}_1(s) + z_{02}\underline{\eta}_2(s) \tag{13.7}$$

for all $\underline{z}_{01}$ and $\underline{z}_{02}$ such that $\underline{o}(s) \in \mathbf{M}$. This form assumes the muscle fibers are all parallel in the reference configuration. This is not true for Genioglossus and Verticalis in the tongue. There is a simple generalization of Equation (13.7) that allows us to represent such fan-shaped muscles. Let $\underline{z}_{0i}$, $i = 1, 2$ depend on s.

b. Stress/strain description: Our basic assumption is that the stress produced by a muscle is aligned with the muscle fiber. Stress is described by a tensor so the previous assumption describes one of the principle axes of the stress tensor. We also assume that the stress/strain relationship in the plane normal to the muscle fiber is entirely passive elastic and independent of direction within the plane. Because this passive elasticity is entirely due to collagen fibers, which have no shear resistance, we assume there is no shear component. These assumptions imply that the stress tensor for a muscle has the form

$$\underline{S}(\underline{z},t) = diag(S_A(\underline{z},t), S_P(\underline{z},t), S_P(\underline{z},t)) \tag{13.8}$$

where S(z,t) is the second Piola-Kirchoff stress tensor. Note that the isolated muscle fibers would also have stress resulting from the incompressibility of the muscle. This is accounted for in the overall tongue model as a component of the new tongue model (see Eqn. (13.4). We emphasize the alignment of $\underline{S}(\underline{z},t)$ by noting that

$$\underline{S}(\underline{z},t)\underline{\tau}(\underline{z},t) = S_A(\underline{z},t)\underline{\tau}(\underline{z},t) \tag{13.9}$$

and

$$\underline{S}(\underline{z},t)\underline{\eta}_i(\underline{z},t) = S_P(\underline{z},t)\underline{\eta}_i(\underline{z},t) \tag{13.10}$$

for $i = 1, 2$. The stress actually depends on the deformation. The requirement that this dependence be frame indifferent implies that $S_A(\underline{z},t)$ and $S_P(\underline{z},t)$ can only depend on $\underline{C}(\underline{z},t)$, its derivatives, $\underline{z}$, and t. The next step is to give explicit expressions for $S_A(\underline{z},t)$ and $S_P(\underline{z},t)$. The passive elasticity is assumed to be given by

$$S_P(\underline{z},t) = -\hat{S}_P(\sqrt{\underline{\eta}^T(\underline{z},t)\underline{C}(\underline{z},t)\underline{\eta}(\underline{z},t)}) \tag{13.11}$$

for any unit vector, $\underline{\eta}(\underline{z},t)$ perpendicular to $\underline{\tau}(\underline{z},t)$, the tangent vector to the fiber. Note that we are assuming that any viscosity in the muscle can be neglected. This is an assumption that is very commonly made and it is certainly true for static problems and almost certainly true for simulations. It may not be true for questions related to stability. The active stress, that is, the stress produced by muscle activation, is assumed to be given by

$$S_A(\underline{z}, k(t), t) = -\hat{S}_A(\sqrt{\underline{\tau}^T(\underline{z},t)\underline{C}(\underline{z},t)\underline{\tau}(\underline{z},t)},$$
$$\sqrt{\underline{\tau}^T(\underline{z},t)(\partial\underline{C}(\underline{z},t)/\partial t)\underline{\tau}(\underline{z},t)}, \underline{z}, k(t)) \tag{13.12}$$

where $k(t)$ is the control, $0 \leq k(t) \leq 1$ i.e., the muscle activation.

Notice that $\hat{S}_P$ is a scalar function of a scalar argument. It has the functional form

$$\hat{S}_P(x) = \begin{cases} 0, & \text{if } x \leq 1; \\ \text{monotonic increasing to infinity} & \text{if } x > 1 \end{cases} \tag{13.13}$$

$\hat{S}_A$ is a scalar function of three scalar arguments. It has the functional form

$$\hat{S}_A(x,y,k) = kf_l(x)f_v(y) + \hat{S}_P(x)$$

Here, f_l corresponds to the force-length curve of the Hill model and f_v corresponds to the force velocity curve. Again, we impose the condition that $\hat{S}_A(x,y,k)$ is monotonic increasing in its first argument regardless of the other arguments. This is a common assumption in using Hill's model [16] because the system can be unstable otherwise.

13.6 Muscle Placement

The mathematical model of the tongue is not complete without a geometric description of the tongue. In [14], Takemoto developed a three-dimensional schemata of the musculature based on anatomical observations. In his work, he identified five distinct strata of muscles and provided a description of the muscle fiber locations within each stratum. Our geometric description is based primarily on his work.

Some key issues in developing a geometric description of the tongue are determining what muscles to include, identifying their location within the tongue, and defining their initial direction or backbone curve. The tongue has a complex musculature and is comprised of many muscles, both intrinsic and extrinsic. Intrinsic muscles are muscles that originate and terminate in the soft tissue within body of the tongue while extrinsic muscles originate on structures outside the body, such as the jaw, insert into the tongue, and terminate within the tongue body.

In his paper, Takemoto identified four intrinsic and four extrinsic muscles. Our geometrical description includes the four intrinsic muscles: superior longitudinalis (SL), inferior longitudinalis (IL), transverse (T), and verticalis (V). These muscles compress the tongue in three directions: front-to-back (SL, IL), top-to-bottom (V), and side-to-side (T). The longitudinal muscles also raise and lower the tip. We also include two extrinsic muscles: genioglossus (GG) and hyoglossus (HG) and are in the process of incorporating a third extrinsic muscle, styloglossus (SG), into the model, but this work is not completed. For simplicity, we have chosen not to include the fourth muscle, palatoglossus. The extrinsic muscles pull the tongue toward the muscle's origin or attachment location and deform the tongue according to their insertions. For example, genioglossus is a fan shaped muscle that originates at the jaw. Its fibers enter along the center underbody of the tongue and extend upward terminating along the entire upper tongue surface at midline and a few millimeters on either side. Genioglossus pulls local regions toward the jaw forming midline depressions and grooves.

A muscle is comprised of many individual muscle fibers, which have a cylindrical shape. Individual muscle fibers may extend from origin to insertion or may be much shorter. These shorter fibers are either aligned in parallel, i.e., running end-to-end and side-by-side, or may be in-series with short muscle fibers that overlap. The space in between the fibers contains an intricate collagen network or mesh. While extrinsic muscles have a clearly defined starting point, the termination of these fibers as well as the origination and termination of intrinsic muscle fibers are not as well defined. Takemoto's dissections[14] clearly indicate that the number of fibers from each muscle changes from slice to slice. In addition, Takemoto identified two alternating laminae of muscle fibers in the inner region or body of the tongue. These laminae were each approximately one fiber thick. One of the lamina is composed of genioglossus and verticalis muscle fibers while the other contains transverse fibers. For simplicity, we assume each muscle has a clearly defined boundary and that each muscle is homogeneous within that boundary, i.e. muscle fibers are evenly distributed. In particular, we assume all muscle fibers are aligned in parallel. We further assume that an increase in the activation of a muscular region results in an evenly distributed recruitment of fibers throughout the muscle location. We have currently made no attempt to model the alternating lamina but plan to study the effects of this muscle structure in future geometric descriptions after we have identified the limitations of our simplifying assumptions.

We present a brief description of the muscle locations within our tongue geometry. In order to define the tongue and individual muscle locations, we use a coarse finite element structure. In the future, these definitions will be independent of the finite element structure. SL is located within a thin region extending along the top of the tongue from the tip to the base. Ferrari et al.[5] hypothesized individual superior longitudinalis fibers, which have an in-series configuration, can have differing activation levels along their length allowing for independent movement regionally. In order to test this hypothesis, we divided SL into six sections along its length, each with its own activation parameter. We are, of course, assuming that SL has such clear boundaries along its length where independent muscle activation occurs. The shaded areas in Figure 13.3 show the region containing SL and our divisions. IL lies symmetrically on either side of the midsagittal plane and extends along the bottom of the tongue body from the tip to the base as shown in Figure 13.3.

T and V lie in the body of the tongue above IL. V lies symmetrically on either side of a region containing the midsagittal plane while T crosses this midsagittal region. We have divided T and V along their widths as shown in Figure 13.4 to allow for independent activation. Activation of the full muscle is achieved by choosing the same activation parameters for all of the shaded regions.

As stated earlier, GG is a fan shaped muscle that is located in the midsagittal plane and extends a few millimeters on either side. Its fibers originate at the jaw and extend up into the region containing SL. We have divided GG into five regions along its width to allow for independent activation as shown in Figure 13.5. The last muscle in our current model, HG, originates at the hyoid bone, extends upward on both sides of the tongue, and enters the tongue laterally. The anterior fibers run upward and then forward remaining below SL. The medial and posterior fibers run upward and then join with T

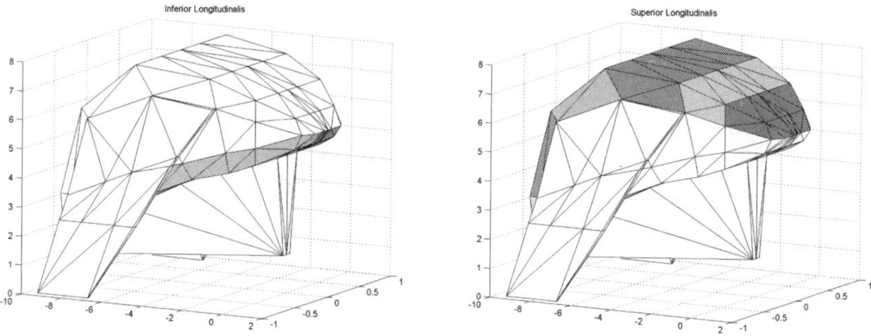

Fig. 13.3. The location of Inferior Longitudinalis is shown at the left and Superior Longitudinalis at the right. Shading indicates independently controllable sections of the muscle.

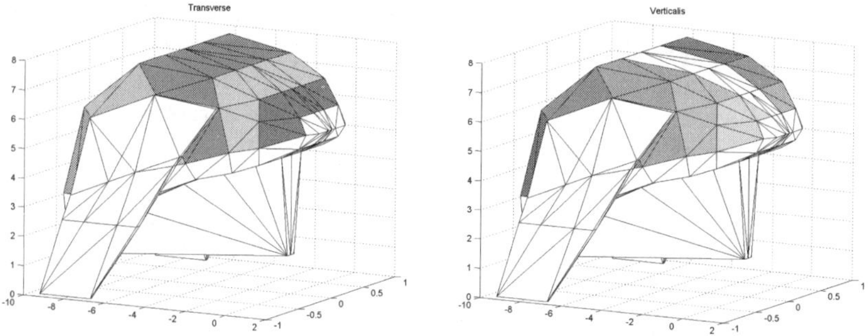

Fig. 13.4. The location of Transversalis is shown at the left and Verticalis at the right. Shading indicates independently controllable sections of the muscle.

fibers. We have separated this muscle into three sections, one consisting of the medial and posterior fibers, one with the upward portion of the anterior fibers and the last with the forward portion of the anterior fibers as shown in Figure 13.5. We are in the process of adding SG to our model. SG originates at the styloid process of the temporal bone, extends downward and forward partially intersecting with HG before entering the tongue. Once SG enters the tongue, posterior fibers run downward toward the base, middle fibers merge with T, and anterior fibers run forward along the sides of the tongue between SL and IL.

13.7 Example

To illustrate the model and its application, consider the problem of elongating a two-dimensional tentacle. We assume the tentacle has two muscle fibers

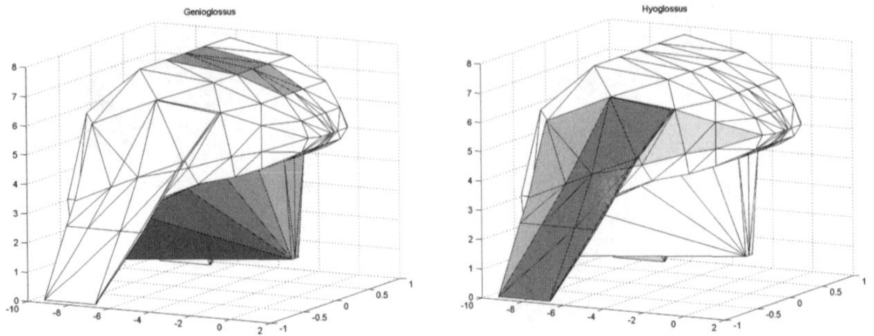

Fig. 13.5. The location of Genioglossus is shown at the left and Hyoglossus at the right. Shading indicates independently controllable sections of the muscle.

through each material point. One fiber is vertical and the other horizontal. We model incompressibility in two dimensions by requiring every infinitesimal area to be constant, regardless of muscle activation. The basic idea is illustrated in Figure13.6 where the initial position is shown dotted and the desired final position solid. We ignore gravity for simplicity.

Is it possible to control the muscle activation of this tentacle in such a way as to cause it to move from its initial position to its final position while maintaining its rectangular shape at all intermediate stages?

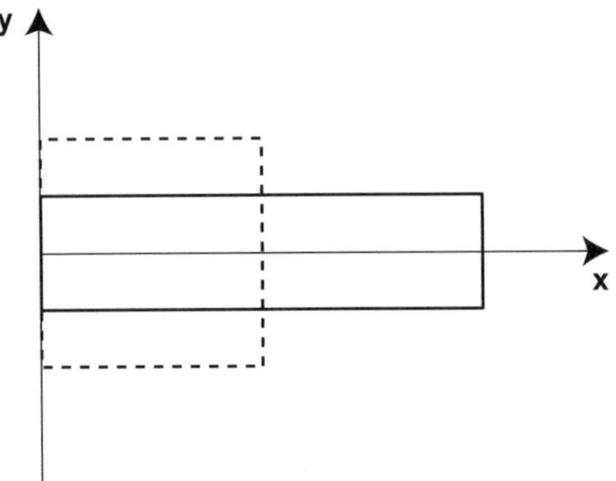

Fig. 13.6. A simple two-dimensional elongation of an abstract tentacle. Contracting the vertical muscles should cause the height to decrease and the length to increase.

To answer this question, denote material points of the tentacle by $\underline{z} = [x\ y]^T$. Let $\underline{p}(\underline{z}, t)$ denote the position of material point $\underline{z} \subset \{-1 \le x \le 1,\ 0 \le y \le 1\}$ at time t. Then,

$$\underline{F} = \underline{p}_{\underline{z}} \tag{13.14}$$

$$\underline{C} = \underline{F}^T \underline{F} \tag{13.15}$$

where $\underline{C}$ is the previously mentioned right Cauchy-Green deformation tensor and $\underline{F}$ is the deformation gradient and the subscript indicates the partial derivative. A precise characterization of the desired movement of the tentacle is

$$\underline{p}(\underline{z}, t) = [\alpha^{-1}(t)x \quad \alpha(t)y]^T \tag{13.16}$$

This makes

$$\underline{F}(\underline{z}, t) = \begin{bmatrix} \alpha^{-1}(t) & 0 \\ 0 & \alpha(t) \end{bmatrix}.$$

Then,

$$\underline{C}(\underline{z}, t) = \begin{bmatrix} \alpha^{-2}(t) & 0 \\ 0 & \alpha^2(t) \end{bmatrix}. \tag{13.17}$$

The incompressibility constraint is automatically satisfied by this $\underline{C}(\underline{z}, t)$. It is necessary to account for the effect of the constraint on the stresses. Physically, the constraint supplies whatever stress is required to insure that the constraint is satisfied. For a given $\underline{z}$, the incompressibility constraint (13.4) defines a two dimensional manifold in the three-dimensional inner product space of symmetric 2×2 matrices. The inner product of two such matrices, say $\underline{A}$ and $\underline{B}$ is just $tr(\underline{AB})$. An orthonormal basis for this space, aligned with the tangent plane to the manifold, consists of the three matrices, $\underline{B}_1 = \underline{C}^{-1}/\sqrt{\alpha^4 + \alpha^{-4}}$,

$$\underline{B}_2 = \begin{bmatrix} \alpha^{-2} & 0 \\ 0 & -\alpha^2 \end{bmatrix} / \sqrt{\alpha^4 + \alpha^{-4}} \tag{13.18}$$

and

$$\underline{B}_3 = \begin{bmatrix} 0 & 1/\sqrt{2} \\ 1/\sqrt{2} & 0 \end{bmatrix}$$

We will enforce the constraint (13.6) by projecting the stress tensor onto the tangent plane of the manifold.

The stress tensor is aligned with the muscle fibers. In this example, the vertical muscle fibers stay vertical and straight throughout the movement and the horizontal muscle fibers stay horizontal. A natural assumption is that only the vertical fibers are activated so as to decrease the height with minimum resistance from other forces. Activating the horizontal fibers would only increase the needed stress. With these assumptions, the stress tensor becomes

$$\underline{S}(\underline{z}, k(t), t) = \begin{bmatrix} -(k(t)f_l(\alpha^{-1}(t))f_v(d\alpha^{-1}(t)/dt) - \hat{S}_P(\alpha^{-1}(t))) & 0 \\ 0 & -\hat{S}_P(\alpha(t)) \end{bmatrix} \tag{13.19}$$

where $\hat{S}_P(\alpha)$ is defined in (13.13).

Taking the projection of $\underline{S}(\underline{z}, k(t))$ onto $\underline{B}_2$ gives

$$\hat{\underline{S}}_A(\underline{C}(\underline{z},t),p,\underline{z},k(t)) = tr(\underline{S}(\underline{z},k(t))\underline{B}_2)\underline{B}_2. \qquad (13.20)$$

Finally, Equation(13.5) becomes, explicitly,

$$\underline{T}(\underline{z},k(t),t) = (\frac{1}{\sqrt{\alpha^4 + \alpha^{-4}}})\left(\begin{bmatrix} -p\alpha & 0 \\ 0 & -p\alpha^{-1} \end{bmatrix}\right.$$

$$+ \frac{\alpha^{-2}(k(t)f_l(\alpha^{-1}(t))f_v(d\alpha^{-1}(t)/dt) + \hat{S}_P(\alpha^{-1}(t))) - \alpha^2 \hat{S}_P(\alpha(t))}{\sqrt{\alpha^4 + \alpha^{-4}}}\left.\begin{bmatrix} \alpha^{-3} & 0 \\ 0 & -\alpha^3 \end{bmatrix}\right) \qquad (13.21)$$

Note that p is given explicitly by the formula

$$p(\underline{z},t) = -tr(\underline{S}(\underline{z},k(t),t)\underline{B}_1(\underline{z},t))$$

Inspection of the formula for $\underline{T}(\underline{z},k(t),t)$ shows that there is no dependence on $\underline{z}$. Hence, $\bigtriangledown \cdot \underline{T}^T(\underline{z},k(t),t)$ must be 0. Thus, the hypothesized motion of the tentacle is impossible if there is only a single open-loop control of vertical muscle fibers. This would typically be true of muscles in the limbs. We believe it is unlikely to be true for tongue and tentacle muscles.

Suppose we allow the control of the vertical muscle fibers to depend on both the fiber and the position along its length. That is, the control is now $k(t,x,y)$. Choosing $k(t,x,y) = k(t)(x^2 + y^2)$ and substituting this into the equations above reduces the dynamics to a single nonlinear first-order differential equation in $\alpha(t)$ and $k(t)$. We are currently investigating whether this equation is solvable for some choice of $k(t)$.

We have so far said nothing about boundary conditions. The most natural boundary condition is that the component of stress perpendicular to the three edges of the tentacle other than the one on the y-axis be zero. We allow arbitrary vertical motion on the y-axis but zero horizontal motion. A more standard way to achieve the same result is to regard the tentacle as the right half of an object that is symmetric about the y-axis and has zero normal stress on all its boundaries. This leads to a pair of equations that are linear in p and k. Evaluation of the determinant shows the solution is unique for $\alpha > 1$ although the required k might not lie between 0 and 1. If we allow cocontraction, i.e., both sets of muscles to be active, then we have two linear equations in three unknowns and many solutions.

13.8 Further Research

The main purpose of this research is to determine how the activity of the muscles of the tongue controls its movements. Our collaborators have been using tagged cine-MRI and ultrasound to provide the motions of individual points within the tongue as well as lines on its surface. The data we already have is in two spatial dimensions. We, and they, expect to have three-dimensional data shortly. Given this data, we have an obvious inverse problem—given the movement of points within the tongue, compute the muscle activations that produced this movement. One important biological issue is that the extent to which individual muscle fibers can be independently controlled is unknown. In the muscles of the limbs, for example, there is normally no reason to control the shape; only the force needs to be controlled. Thus, fibers are recruited for a task in strict order of their size and force producing capabilities. Much more complicated recruitment strategies might be needed in the tongue, where shape must be controlled. Moreover, there is no reason to separately control different fibers that are connected in series. In the tongue, subtle control of the shape could require separate control of series connected fibers. Mathematical questions, such as whether a solution to the inverse problem exists and is unique can provide great insight into these, and other, issues in the control of the tongue.

References

1. S. S. Antman. *Nonlinear Problems of Elasticity*, Springer-Verlag, New York, NY, 1995.
2. G. S. Chirikjian. *Closed-form primitives for generating locally volume preserving deformations*, Journal of Mechanical Design, 117, pp.347–354, 1995.
3. J. Dang and K. Honda. *Estimation of vocal tract shapes from speech sounds with a physiological articulatory model* J. Phonetics 30, pp. 511–532, 2002.
4. J. Dang and K. Honda. *Construction and control of a physiological articulatory model* J. Acoustic. Soc. Am. 115(2), pp. 853–870, 2004.
5. O. Ferrari, T. J. Burkholder, and A. J. Sokoloff. In-series and in-parallel design in muscles of the human tongue, Program No. 857.11, 2002 Abstract Viewer/Itinerary Planner, Washington, DC: Society for Neuroscience, Online. 2002.
6. J. -M. Gerard, R. Wilhelms-Tricarico, P. Perrier, and Y. Payan. *A 3D dynamical biomechanical tongue model to study speech motor control*, Recent Res. Devel. Biomechanics, 1, pp. 49–64, 2003.
7. K.Hashimoto and S. Suga. *Estimation of the muscular tensions of the human tongue by using a three-dimensional model of the tongue*, Journal of the Acoustic society of Japan (E), 7 (1), pp. 39–46, 1986.
8. T. J. R. Hughes. *The Finite Element Method, Linear Static and Dynamic Finite Element Analysis*, Prentic-Hall, Englewood Cliffs, NJ, 1987.
9. S. Kiritani, K. Miyawaki, and O. Fujimura. *A computational model of the tongue*, Annual Report of the Research Institute of Logopedics and Phoniatrics, 10, pp. 243–252, Tokyo 1976.
10. Y. Payan and P. Perrier. *Synthesis of V-V sequences with a 2D biomechanical tongue model controlled by the equilibrium point hypothesis*, Speech Communication, 22, (2/3), pp. 185–205, 1997.
11. J. S. Perkell. *Physiology of speech production: results and implications of a quantitative cineradiographic study*, MIT Press, Cambridge, MA, 1969.
12. V. Sanguineti, R. Laboissiere, and Y. Payan. *A control model of human tongue movements in speech*, Biological Cybernetics, 77(11), pp. 11–22, 1997.

13. V. Sanguineti, R. Laboissiere, and D. J. Ostry. *A dynamic biochemical model for the neural control of speech production,* Journal of the Acoustical Society of America, 103, pp. 1615–1627, 1998.
14. H. Takemoto. *Morphological analyses of the human tongue musculature for three-dimensional modeling,* Journal of Speech, Language, and Hearing Research, 44: pp.95–107, 2001.
15. R. Wilhelms-Tricarico. *Physiological modeling of speech production: methods for modeling soft tissue articulators,* J. Acoust. Soc. Am. 97(5), pp. 3058–3098, 1995.
16. F. E. Zajac. *Muscle and tendon: properties, models, scaling, and application to biomechanics and motor control,* Critical Rev. Biomed. Eng., 17(4), pp. 359–411, 1985.

14

A Coordinate-Free Approach to Tracking for Simple Mechanical Systems On Lie Groups

D.H.S. Maithripala[1], J.M. Berg[2], and W.P. Dayawansa[3]

[1] Dept. of Mechanical Engineering, Texas Tech University, USA
`sanjeeva.maithripala@ttu.edu`
[2] Dept. of Mechanical Engineering, Texas Tech University, USA
`jordan.berg@ttu.edu`
[3] Dept. of Mathematics and Statistics, Texas Tech University, USA
`daya@math.ttu.edu`

Summary. A coordinate-free approach to the tracking problem for simple mechanical systems on Lie groups highlights the exploitation of geometrical and topological structure in controller design. First, the *algebraic structure* of the Lie group is used to construct an intrinsic and coordinate-free notion of configuration error. This construction leads through geometric ideas to a state feedback "inner" controller that gives the error dynamics the structure of a *simple mechanical system* on the original Lie group. This structure in turn allows the stability properties of the identity element of the error system to be characterized in terms of a potential energy-like *error function*, through implicit definition of a locally exponentially stabilizing state feedback "outer" controller. Stabilization of the identity element of the error system is equivalent to the original system tracking a smooth and bounded, but otherwise arbitrary, reference trajectory. Finally, suitable choices for the error function may be identified based on purely *topological* considerations. This means that a class of error functions may be defined for *any* simple mechanical system on the configuration space, greatly simplifying the design task. As an example of the power of this result, it is known that error functions corresponding to almost-global tracking may be found on any compact connected Lie group, and such functions have been explicitly constructed for many important special cases. The intrinsic, coordinate-free, structure-based approach is used to show a *separation principle* for this tracking problem, namely that the system velocities in the tracking controller may be replaced by their estimates, obtained from any locally exponentially convergent observer, without affecting the convergence properties of the closed-loop system.

14.1 Introduction

A progressively deeper understanding of the rich structure of mechanical systems has long driven advances in the geometric theory of dynamics and control. The texts of Abraham and Marsden [1], Arnold [5] and Marsden and

D.H.S. Maithripala et al.: *A Coordinate-Free Approach to Tracking for Simple Mechanical Systems On Lie Groups*, LNCIS **321**, 223–237 (2005)
`www.springerlink.com`   © Springer-Verlag Berlin Heidelberg 2005

Ratiu [18] provide an excellent overview of the mathematical foundations of geometric mechanics, while the collection *Geometry, Mechanics, and Dynamics* [20] highlights more recent developments, including a paper dedicated to control of mechanical systems. Results on the control of mechanical systems significantly lag those solely concerned with dynamics. Some sense of the evolution of geometric control theory, including its relation to mechanical systems, may be found in the collection [7]. Control of mechanical systems is specifically the subject of a recent monograph by Bloch [8] and text by Bullo and Lewis [10], which survey milestones in the development of this relatively young area, and present the current state of the art for nonholonomic and holonomic systems, respectively. The strategies pursued by these researchers are *intrinsic*. That is, they do not depend on the choice of coordinates. The intrinsic property is a necessity if one desires globally well-defined behavior. Furthermore, system structure that can be heavily exploited for purposes of controller design is only revealed by an intrinsic, geometric approach [8], [10]. As powerful as they are, the methodologies described in these references pertain only to stabilization and tracking with full-state feedback. They do not address the problems of output feedback stabilization or state estimation. This gap is addressed by Aghannan and Rouchon [2], who present an intrinsic, locally exponentially convergent observer for the velocities of a simple mechanical systems, given measurements of the configurations. This result paves the way for stabilization and tracking using dynamic output feedback of configuration measurements only. The present paper builds further upon these contributions in two ways. First, an almost-global stabilization scheme due to Koditschek [13] is extended to provide almost-global, locally-exponential tracking. Second, a separation principle for the tracking controller is presented, showing that the velocity terms in such a controller may be replaced by their estimates, obtained from configuration measurements by any locally exponentially convergent observer, without affecting the convergence properties of the closed-loop system. The derivation and the explicit expression of all the results presented here makes extensive use of the coordinate-free formulation of the problem.

Formally, a holonomic simple mechanical system consists of a smooth manifold, corresponding to the configuration space of the system; a smooth Lagrangian corresponding to kinetic energy minus potential energy; and a set of external forces or one-forms [1]. When some of these forces may be used for control, we refer to a simple mechanical control system [8], [10], [13]. We specifically restrict our attention to systems where the configuration space is a Lie group. This assumption provides substantial additional structure compared to, for example, simple mechanical systems on a general Riemannian manifold, but still includes many cases of interest. Among the numerous applications of engineering importance that take such a form are underwater vehicles, satellites, surface vessels, airships, hovercrafts, robots, and MEMS (see [8], [10], [14] and the references therein). Results developed in a general Riemannian framework will always apply in a more restrictive setting, but will typically not be as powerful as those that take full advantage of the features of the Lie group. This may be seen by comparing the tracking controller of Bullo and Murray [9], obtained by specializing an elegant Riemannian formulation, to the tracking controller presented below, which makes critical use of the algebraic and geometric structure of the Lie group. The result presented below is convergent for almost any initial configuration and velocity, while the performance of the controller of [9] can be guaranteed only locally, in some neighborhood of the reference trajectory.

The chief difficulty in the general tracking problem lies in defining the tracking error and associated tracking error dynamics. When the configuration space is a Lie group, we present a natural, globally well-defined, notion of error dynamics. Given two elements of the group, g, representing the desired configuration, and h, representing the actual configuration, the *configuration error* is defined to be gh^{-1}. The velocity error is naturally defined using left or right translation. Here the configuration error is itself an element of the Lie group. This has no analog in the general Riemannian approach.

The derivatives of the configuration and velocity errors define the tracking error dynamics on the tangent bundle of the Lie group. For fully-actuated simple mechanical systems on Lie groups there always exists a feedback control that transforms the tracking error dynamics to a simple mechanical control system, with kinetic energy, potential energy, and damping, arbitrarily assignable using additional state feedback. Solution of the tracking problem is thus reduced to the task of stabilizing the identity element of this transformed system. Now the results of Koditschek [13] on set point control can be used to assign a suitable potential energy and damping, and thereby stabilize the identity with "almost-global" asymptotic stability, that is, asymptotic stability for all initial conditions in an open and dense subset of the state space. The assigned potential energy should be a globally defined, smooth, proper Morse function, with a unique minimum at the identity. Koditschek [13] refers to such functions as navigation functions, but in this paper we adopt a more standard tracking nomenclature [9], [10] and refer to them as *error functions*. Thus the complex geometry and dynamics of the tracking problem are reduced to the purely topological task of finding suitable error functions. Observe that once such a function is found on a particular Lie group, it can be used to solve the tracking problem for all simple mechanical systems on that Lie group. Suitable functions are constructed for Lie groups of practical interest by Koditschek [13] and by Dynnikov and Vaselov [11]. A result of Morse [19] shows that such functions exist on any compact connected manifold. By a straightforward extension of these results, such functions also exist on any Lie group diffeomorphic to a Lie group of the form $G \times \mathcal{R}^n$, where G is any compact connected Lie group. If the Lie group is not of the form given above, the existence of a globally defined error function is, to our knowledge, an open question.

Unless the Lie group is homeomorphic to $\mathcal{R}^n$ the presence of anti-stable equilibrium points and saddle points and their stable manifolds will limit the achievable asymptotic stability of the identity to at best almost global [13]. This implies that the almost-global asymptotic stability conferred by the tracking controller of this paper is best convergence result possible.

Implementation of full state-feedback requires both configuration and velocity measurements. In application it is not unusual for only one of these to be available for measurement. In some cases it is the velocities [4],[21], while in others it is the configuration [3], [23], [22]. This paper is concerned with the latter case, where dynamic estimation of the velocities is necessary. We show that a "separation principle" applies for the dynamic configuration feedback tracking control obtained by composing the full state-feedback compensator with a velocity observer. Specifically, we have shown previously that if the configuration manifold is compact, then the dynamic configuration-feedback tracking controller almost-globally converges for any initial observer error for which the velocity converges exponentially [17] . A recent paper by Aghannan and Rouchon provides one such suitable observer [2].

In section 14.2 we briefly present notation and review mathematical background for simple mechanical control systems on Lie groups. Section 14.3.1 presents an intrinsic, globally valid, full state-feedback tracking control for any fully-actuated simple mechanical system on a wide class of Lie groups, guaranteeing almost-globally asymptotic stability with locally exponential convergence to an arbitrary twice differentiable configuration reference signal. No invariance properties are required of the kinetic energy, potential energy, or external forces. Section 14.3.2 derives our separation principle. In section 14.3.3 invariance properties of the kinetic energy and the external forces are exploited to considerably simplify the explicit expression of the controller. In section 14.4 this construction is demonstrated for any simple mechanical system with left-invariant kinetic energy on $SO(3)$, including simulations of the axisymmetric heavy top.

Lemma 1, provided in the Appendix, plays a crucial role in proving the separation principle.

14.2 Mathematical Background

This section provides necessary notation and background for the rest of the paper. For additional details the reader is referred [1], [12], [18]. Let G be a connected finite dimensional Lie group and let $\mathcal{G} \simeq T_e G$ be its Lie algebra. The left translation of $\zeta \in \mathcal{G}$ to $T_g G$ will be denoted $g \cdot \zeta = DL_g \zeta$. The adjoint representation $DL_g \cdot DR_{g^{-1}}$ will be denoted Ad_g. The Lie bracket on $\mathcal{G}$ for any two $\zeta, \eta \in \mathcal{G}$ will be denoted $[\zeta, \eta] = \mathrm{ad}_\zeta \eta$ and the dual of the ad operator will be denoted ad^*. Any smooth vector field $X(g)$ on G has the form $g \cdot \zeta(g)$ for some smooth $\zeta(g) \in \mathcal{G}$. Let $\{e_i\}$ be any basis for the Lie algebra $\mathcal{G}$ and let $\{E_i(g) = g \cdot e_i\}$ be the associated left invariant basis vector field on G. Now $[e_i, e_j] = C_{ij}^k e_k$, where C_{ij}^k are the structure constants of the Lie algebra $\mathcal{G}$ ($C_{ij}^k = -C_{ji}^k$), and $[E_i, E_j] = C_{ij}^k E_k$.

14.2.1 The Riemannian Structure

For each $g \in G$, $I(g) : \mathcal{G} \mapsto \mathcal{G}^*$ is an isomorphism such that the relation $\langle\langle \zeta, \eta \rangle\rangle_\mathcal{G} = \langle I(g)\zeta, \eta \rangle$ for $\zeta, \eta \in \mathcal{G}$ defines an inner product on $\mathcal{G}$. Here $\langle \cdot, \cdot \rangle$ denotes the usual pairing between a vector and a co-vector. Identifying $\mathcal{G}^*$ and $\mathcal{G}$ with $\mathcal{R}^n$, let $I_{ij}(g)$ and $I^{ij}(g)$ be the matrix representations of $I(g)$ and $I^{-1}(g)$ respectively. $I(g)$ is symmetric and positive definite. If $I(g)$ is globally smooth then such an $I(g)$ induces a unique metric on G by the relation $\langle\langle g \cdot \zeta, g \cdot \eta \rangle\rangle = \langle I(g)\zeta, \eta \rangle$. Further, it follows that every metric has such an associated family of isomorphisms. If the metric is left-invariant then I is a constant and any constant symmetric positive definite matrix induces a left-invariant metric on G.

In the remainder of the section we present expressions for the Levi-Civita connection and the Riemannian curvature corresponding to left-invariant metrics. These derivations are based on Cartan's structural equations [12]. We intensionally avoid the use of coordinate-frame fields to facilitate a coordinate-free formulation. As a result we obtain connection coefficients and the Riemannian curvature two-forms in a general left-invariant frame field in place of the more familiar Christoffel symbols and corresponding curvature coefficients in a coordinate frame field.

Associated with any metric there exists a unique torsion-free, metric, connection called the Levi-Civita connection. For a vector field $X = X^k E_k$ and a vector $v = v^k E_k$ the Levi-Civita connection is given by

$$\nabla_v X = (dX^k(v) + \omega_{ij}^k(g)v^i X^j)E_k, \tag{14.1}$$

where $\omega_{ij}^k(g)$ are the connection coefficients in the frame $\{E_k\}$. For left-invariant metrics the connection coefficients turn out to be constants, given by

$$\omega_{ij}^k = \frac{1}{2}\left(C_{ij}^k - I^{ks}(I_{ir}C_{js}^r + I_{jr}C_{is}^r)\right). \tag{14.2}$$

Since in general the E_k are not coordinate vector fields, ω_{ij}^k, are not the Christoffel symbols. The corresponding coefficients of the Riemannian curvature two-forms R_{jab}^k are also constant and can be shown to be [15],

$$R_{jab}^k = (-\omega_{rj}^k C_{ab}^r + 2\omega_{ar}^k \omega_{bj}^r). \tag{14.3}$$

The Riemannian curvature is then

$$R(\zeta, \eta)\xi = \{R_{jab}^k \xi^j(\zeta^a \eta^b - \zeta^b \eta^a) - \omega_{ij}^k C_{ab}^i \zeta^a \eta^b \xi^j\}e_k. \tag{14.4}$$

14.2.2 Simple Mechanical Control Systems on Lie Groups

A simple mechanical control system evolving on a Lie group G equipped with a metric $\langle\langle \cdot, \cdot \rangle\rangle$ is defined as a system with kinetic energy $E(\dot{g}) = \frac{1}{2}\langle\langle \dot{g}, \dot{g} \rangle\rangle$, conservative plus dissipative forces $f(g, \zeta) \in \mathcal{G}^*$ and a set of linearly independent forces $u_i f^i(g) \in \mathcal{G}^*$ for $i = 1, \cdots, m$, [1], [6]. The scalar $u_i \in \mathcal{R}$ are the controls. If $m = n = dim(G)$ the system is said to be fully actuated. In what follows we consider only fully-actuated simple mechanical systems.

Let $I(g) : \mathcal{G} \mapsto \mathcal{G}^*$ be the isomorphism associated with the kinetic energy metric; $\langle\langle g \cdot \zeta, g \cdot \eta \rangle\rangle = I(g)\zeta \cdot \eta$ for $\zeta, \eta \in \mathcal{G}$. Then the Euler-Lagrange equations of motion of the system give rise to a dynamical system on the left trivialization $G \times \mathcal{G}$ of TG as

$$\dot{g} = g \cdot \zeta, \tag{14.5}$$

$$\dot{\zeta} = \tilde{f}(g, \zeta) + I^{-1}(g)\left(f(g, \zeta) + \sum_i^m u_i f^i(g) \right), \tag{14.6}$$

where $\tilde{f}(g, \zeta) = -\omega_{ij}^k(g)\zeta^i\zeta^j e_k$ are the inertial forces arising from the curvature effects. Observe that $\nabla_{\dot{g}}\dot{g} = g \cdot (\zeta - \tilde{f})$. If the kinetic energy metric is left invariant then I is a constant and $\tilde{f}(g, \zeta) = I^{-1}\mathrm{ad}_\zeta^* I\zeta$.

14.3 Intrinsic Tracking for Simple Mechanical Systems

Let $(g_r(t), \zeta_r(t)) \in G \times \mathcal{G}$ be a desired once-differentiable reference trajectory to be tracked by the fully-actuated simple mechanical system (14.5)–(14.6). We introduce the *configuration error*,

$$e(t) = g_r(t)g^{-1}(t). \tag{14.7}$$

This object is intrinsic and globally defined. Most importantly, it is itself an element of the configuration space. Differentiating (14.7) and setting $\eta_e = \mathrm{Ad}_g(\zeta_r - \zeta)$, the error dynamics are found to be

$$\dot{e} = e \cdot \eta_e, \tag{14.8}$$

$$\dot{\eta}_e = \mathrm{Ad}_g\left(\dot{\zeta}_r - \dot{\zeta} + [\zeta, \zeta_r] \right), \tag{14.9}$$

where $\dot{\zeta}$ is given by (14.6). Observe that the error dynamics are defined on $TG \simeq G \times \mathcal{G}$ as well. As we now show, through a suitable choice of controls, the dynamics of the configuration error may be given the form of a fully-actuated simple mechanical system with arbitrarily assignable potential energy, kinetic energy, and damping.

14.3.1 Full State Feedback Tracking

Let $B = I^{-1}(g)[f^1(g) \ f^2(g) \ \cdots \ f^n(g)]$. Substituting

$$u = B^{-1}\left(\dot{\zeta}_r - \tilde{f}(g, \zeta) - I^{-1}f(g, \zeta) + [\zeta, \zeta_r] - \mathrm{Ad}_{g^{-1}}\nu \right) \tag{14.10}$$

in equation (14.9), we have the transformed error dynamics

$$\dot{e} = e \cdot \eta_e, \tag{14.11}$$

$$\dot{\eta}_e = \nu, \tag{14.12}$$

where $\nu \in \mathcal{G}$. These transformations reduce the problem of stably tracking the reference input to the problem of stabilizing $(id, 0)$ of (14.11)–(14.12). It is shown by Koditschek [13]. that any given point of a compact configuration space with or without boundary may be made an almost-globally stable equilibrium by the appropriate choice of an potential energy function. Thus the task is now to assign a suitable potential energy, $F : G \mapsto \mathcal{R}$, such that the equilibrium $(id, 0)$ is almost-globally stable. The value of the potential energy for the configuration error is therefore a measure of the size of that error, and we refer to it as the *error function*. The convergence properties are completely determined by the properties of the error function, and are independent of the specifics of the simple mechanical system. Thus for any given Lie group, solution of the tracking problem is reduced to the topological question of finding an appropriate error function for that space.

Let $\tilde{I}$ be any symmetric positive definite matrix. This induces an inner product $\langle\!\langle \cdot\, ,\, \cdot \rangle\!\rangle_\mathcal{G}$ on $\mathcal{G}$ and a left invariant metric on G. This metric need not be related to the kinetic energy metric of the mechanical system under consideration and is a design freedom. The choice of which can be made based on performance considerations. Let $\zeta_e = e^{-1} \cdot \operatorname{grad} F$, where $\langle dF, e \cdot \eta \rangle = \langle\!\langle e^{-1}\operatorname{grad} F, \eta \rangle\!\rangle_\mathcal{G}$ and

$$\nu = -\zeta_e - k\,\eta_e, \tag{14.13}$$

where k is a positive constant, yields the following error dynamics:

$$\dot{e} = e \cdot \eta_e, \tag{14.14}$$

$$\dot{\eta}_e = -\zeta_e - k\,\eta_e. \tag{14.15}$$

The complete state-feedback tracking control is

$$u = B^{-1}\left(\operatorname{Ad}_{g^{-1}}\zeta_e + k(\zeta_r - \zeta) + \dot{\zeta}_r - \tilde{f}(g,\zeta) - I^{-1}f(g,\zeta) + [\zeta, \zeta_r]\right), \tag{14.16}$$

Our approach implements a two-part composite control, in which the first component (14.10) is used to give the configuration error dynamics the structure of a simple mechanical control system and the second component (14.13) is used to assign a desired potential energy F and damping to the transformed system. The control (14.10) allows the full power of the results of [13] on set point control to be applied to tracking, via (14.13), in a very general setting.

We now investigate the properties required by the assigned potential energy function F for almost-global tracking. A function with only non-degenerate critical points is called *Morse*.

Definition 1. *An infinitely differentiable proper Morse function $F : G \mapsto \mathcal{R}$, bounded below by zero, and with a unique minimum at the identity is called an error function.*

The following theorem shows that such an error function used with the state-feedback control (14.16) provides the strongest possible convergence properties.

Theorem 1. *If $F : G \mapsto \mathcal{R}$ is an error function then the fully-actuated simple mechanical control system (14.5)–(14.6) with the control (14.16) almost-globally tracks any once-differentiable trajectory $(g_r(t), \zeta_r(t))$ with local exponential convergence.*

The above theorem is proved by the authors in [17]. The non-degeneracy is required to ensure the simplest possible limit behavior. As pointed out in [13] no smooth vector field can have a global attractor unless the configuration manifold is homeomorphic to $\mathcal{R}^n$. Thus, in general, the global stabilization of the identity of the error dynamics is impossible, so almost-global asymptotic tracking is the best possible outcome. Previous work that specialize results on the general Riemannian tracking problem to Lie groups, notably [9], does not take full advantage of the Lie group structure, and hence guarantees only local convergence.

It is shown by Morse [19] that error functions always exist on any smooth compact connected manifold. By an extension of these results it also follows that such functions exist on any manifold diffeomorphic to a manifold of the form $G \times \mathcal{R}^n$ where G is compact and connected. This class of Lie groups cover most of the situations of practical significance, including $SO(n)$, $SE(n)$, their subgroups and direct products. To apply theorem 1 on a general non-compact Lie group requires only a suitable error function. To the best of our knowledge it remains an open question as to whether such functions always exist in the more general setting.

Local exponential convergence requires only that the assigned potential energy has a non-degenerate local minimum. Such functions can be constructed on any Lie group, and so the control (14.16) may be applied in a very general setting.

A *perfect* Morse function has exactly as many critical points as the homology of the underlying manifold requires. To minimize the number of unstable equilibrium points, whenever possible we use a perfect Morse function as the error function. Examples of perfect Morse functions on certain symmetric spaces, including $SO(n)$, $SE(n)$, $U(n)$, and $Sp(n)$, may be found in the literature [11]. Koditschek [13] gives an example of an error function that is a perfect Morse function on $SO(3)$, which we use in Section 14.4.

14.3.2 Dynamic Output Feedback Tracking

The tracking control (14.16) involves both the configuration variable g and the velocity variable ζ. In this section we assume that only the configuration variables are available for measurement, and estimate the velocity. If ζ_o is the estimated value of ζ, the dynamic configuration-feedback tracking control obtained by composing a velocity observer with the state-feedback control (14.16) is

$$u = B^{-1}\left(\mathrm{Ad}_{g^{-1}}\zeta_e + k(\zeta_r - \zeta_o) + \dot{\zeta}_r - \tilde{f}(g,\zeta_o) - I^{-1}f(g,\zeta_o) + [\zeta_o,\zeta_r]\right),$$
(14.17)

where g is measured. The following theorem proved by the authors in [17] provides a separation principle for this dynamic configuration-feedback control.

Theorem 2. *Consider the fully actuated simple mechanical system (14.5)–(14.6) on a compact and connected Lie group G where the external forces are of the form $f(g,\zeta) = f^c(g) + f^d(g,\zeta)$ with $f^d(g,\zeta)$ linear in ζ. Then the dynamic configuration-feedback control (14.17) composed with any locally exponentially convergent velocity observer almost-globally tracks an arbitrary bounded once-differentiable reference trajectory $(g_r(t),\zeta_r(t))$ with local exponential convergence for sufficiently small initial observer errors.*

The compactness assumption in theorem 2 can be relaxed if Ad_g, $f^d(g,\zeta)$ and $\omega_{ij}^k(g)$ are bounded for all $g \in G$. This is automatically satisfied when G is compact.

One such velocity observer for simple mechanical systems is presented in [2], for the case $f(g,\zeta) = f^c(g)$, that is, in the absence of damping. It is shown by the authors in [16] that velocity dependent external forces do not in fact affect convergence. Because this locally exponentially convergent observer is also intrinsic [2], its use in conjunction with (14.17) yields a globally well defined configuration-feedback tracking control. As guaranteed by theorem 2 convergence is almost-globally asymptotic for sufficiently small initial observer errors.

To this point no invariance properties have been assumed for the simple mechanical system. In particular neither the statements nor the proofs of theorems 1 and 2 have any such restrictions.

14.3.3 Special case of Left-Invariant Kinetic Energy

For many cases of interest the above intrinsic and coordinate-free formulation of this dynamic configuration-feedback control can be considerably simplified and expressed explicitly. In particular if the kinetic energy of the system is left-invariant, the observer of [2] has a significantly simpler explicit coordinate-free form [15], [16],

$$\dot{g}_o = g_o \cdot (\zeta_o - 2\alpha\zeta_{oe}),$$
(14.18)

$$\dot{\zeta}_o = I^{-1}\left(\mathrm{ad}^*_{\zeta_o} I\zeta_o - \alpha(\mathrm{ad}^*_{\zeta_{oe}} I\zeta_o + \mathrm{ad}^*_{\zeta_o} I\zeta_o)\right)$$
$$+\alpha[\zeta_{oe},\zeta_o] + \Gamma(S(g,\zeta_0,u)) - R(\zeta_o,\zeta_{oe})\zeta_o - \beta\zeta_{oe},$$
(14.19)

where α,β are positive constants, and the configuration error $\zeta_{oe} \in \mathcal{G}$ is defined by $\exp(\zeta_{oe}) = g^{-1}g_o$ for g_o and g sufficiently close. Here $S(g,\zeta_o,u) = f(g,\zeta_o) + \sum_i^m u_i(g,\zeta_o)f^i(g)$ and,

$$\Gamma(S) = (S^k - \omega_{ij}^k S^i \zeta_{eo}^j)\, e^k. \tag{14.20}$$

The advantage of this formulation is that all the terms of the observer with the exception of the external forces S are independent of g. This leads to a compact and flexible representation that requires only changes to S and I to be adapted to different simple mechanical systems. Left-invariance of kinetic energy also allows the control (14.17) to be written as

$$u = B^{-1}\Big(\mathrm{Ad}_{g^{-1}}\zeta_e + k(\zeta_r - \zeta_o) + \dot{\zeta}_r - I^{-1}(\mathrm{ad}_{\zeta_o}^* I\zeta_o + f(g,\zeta_o)) + [\zeta_o,\zeta_r]\Big),$$
$$\tag{14.21}$$

where now the inertial forces $\tilde{f}$ may be written in terms of ζ only. This is itself a significant simplification, and if in addition all external forces are also left-invariant then only the error feedback term $\mathrm{Ad}_{g^{-1}}\zeta_e$ is dependent on g. This last assumption is fairly common, see for example [8], [10] and the references there in.

In the following section we explicitly compute the state- and dynamic-feedback tracking controller for any simple mechanical system on the Lie group $SO(3)$, with left-invariant kinetic energy. These expressions can be readily adapted to a particular application by specifying the inertia tensor I, and the external forces $f(R,\zeta)$. We make this specialization for the axi-symmetric top, and simulate the resulting performance.

14.4 Example: Tracking on $SO(3)$

The three-dimensional rotation group, $SO(3)$, is the Lie group of matrices $R \in GL(3,\mathcal{R})$ that satisfy $R R^T = R^T R = I$ and $det(R) = 1$. The Lie algebra $so(3)$ of $SO(3)$ is the set of traceless skew symmetric three-by-three matrices. Note that $so(3) \simeq \mathcal{R}^3$ where the isomorphism is defined by,

$$\xi \in \mathcal{R}^3 \mapsto \hat{\xi} = \begin{bmatrix} 0 & -\xi^3 & \xi^2 \\ \xi^3 & 0 & -\xi^1 \\ -\xi^2 & \xi^1 & 0 \end{bmatrix} \in so(3), \tag{14.22}$$

where $\xi = [\xi^1 \ \xi^2 \ \xi^3]^T$. We will use both ξ and $\hat{\xi}$ to mean the same element of $so(3)$, and identify $so(3)$ with $\mathcal{R}^3$ via the isomorphism (14.22). The adjoint representation $\mathrm{Ad}_R : so(3) \mapsto so(3)$ is explicitly given by $\mathrm{Ad}_R(\xi) = R\xi$, or $\mathrm{Ad}_R(\hat{\xi}) = R\hat{\xi}R^T$, respectively.

Define the isomorphism $I : so(3) \simeq \mathcal{R}^3 \mapsto so(3)^* \simeq \mathcal{R}^3$ by the positive definite matrix I. This induces a left invariant metric on $SO(3)$ by the relation, $\langle\langle R \cdot \xi, R \cdot \psi \rangle\rangle = \langle\langle \xi, \psi \rangle\rangle_{so(3)} = I\xi \cdot \psi$, for any two elements $R \cdot \xi, R \cdot \psi \in T_R SO(3)$. The Lie bracket on $so(3)$ is given by, $[\xi, \psi]_{so(3)} = \mathrm{ad}_\xi \psi = \xi \times \psi$. and the dual of the ad operator is given by, $\mathrm{ad}_\xi^* \Pi = \Pi \times \xi$, where $\Pi \in so(3)^* \simeq \mathcal{R}^3$.

From (14.5) – (14.6), a simple mechanical control system on $SO(3)$ with left-invariant kinetic energy takes the form,

$$\dot{R} = R\,\hat{\zeta}, \tag{14.23}$$

$$\dot{\zeta} = I^{-1}\left(I\zeta \times \zeta + f(R,\zeta) + \sum_i^m u_i f^i(R,\zeta)\right). \tag{14.24}$$

14.4.1 Construction of the Controller

Let $(R_r(t), \zeta_r(t))$ where $\dot{R}_r(t) = R(t)\hat{\zeta}_r(t)$ is a once differentiable reference trajectory to be tracked by (14.23)–(14.24). The intrinsic tracking error $e(t) \in G$ is given by $e(t) = R_r(t)R^T(t)$. Let $F(e)$ be an error function and let $\zeta_e = e^T grad\,F(e)$ with respect to the left-invariant kinetic energy metric induced by some three-by-three positive definite matrix $\tilde{I}$. The exact choice of $\tilde{I}$ is up to the designer.

Consider the function $F(e) = \frac{1}{2}trace\{K(I - e)\}$, where K is a symmetric, positive definite three-by-three matrix. It is shown in [13] that F is a Morse function with four critical points and a unique minima at the identity. It can also be shown that $\zeta_e = \tilde{I}^{-1}\Omega_e$, where $\hat{\Omega}_e = (Ke - e^T K^T)$. This implies that the tracking control (14.16)

$$u = B^{-1}\left(R^T\zeta_e + k(\zeta_r - \zeta) + \dot{\zeta}_r - I^{-1}(I\zeta \times \zeta + f(R,\zeta)) + \zeta \times \zeta_r\right), \tag{14.25}$$

achieves almost-global tracking with local exponential convergence. It is pointed out in [11], [13] that any Morse function on $SO(3)$ has at least four critical points. Thus this F is a perfect Morse function on $SO(3)$, and has the fewest possible unstable equilibria.

The intrinsic observer (14.18) – (14.19) takes the form,

$$\dot{R}_o = R_o \cdot (\zeta_o - 2\alpha\zeta_{eo}), \tag{14.26}$$

$$\dot{\zeta}_o = I^{-1}\left(I\zeta_o \times \zeta_o - \alpha(I\zeta_o \times \zeta_{eo} + I\zeta_{eo} \times \zeta_o)\right)$$
$$+ \alpha\zeta_{eo} \times \zeta_o + \Gamma(S) - R_c(\zeta_o, \zeta_{eo})\zeta_o - \beta\zeta_{eo}, \tag{14.27}$$

where ζ_{eo} satisfies $exp(\hat{\zeta}_{eo}) = R^T R_o$ and is given by,

$$\hat{\zeta}_{oe} = \frac{\psi_o}{2\sin\psi_o}(R^T R_o - R_o^T R), \tag{14.28}$$

where, $\cos\psi_o = (tr(R^T R_o) - 1)/2$, for $|\psi_o| < \pi$ [18]. The parallel transport term $\Gamma(S)$ is calculated from (14.20) where $S(R, \zeta_o) = f(R, \zeta_o) + \sum_i^m u_i f^i(R)$ and the curvature term $R_c(\tilde{\zeta}, \zeta_e)\tilde{\zeta}$ is calculated from (14.4).

With this observer the tracking feedback (14.25) can be implemented as,

$$u = B^{-1}\left(R^T\zeta_e + k(\zeta_r - \zeta_o) + \dot{\zeta}_r - I^{-1}(I\zeta_o \times \zeta_o + f(R,\zeta_o)) + \zeta_o \times \zeta_r\right), \tag{14.29}$$

and achieves almost-global tracking with only the measurement of the configuration R.

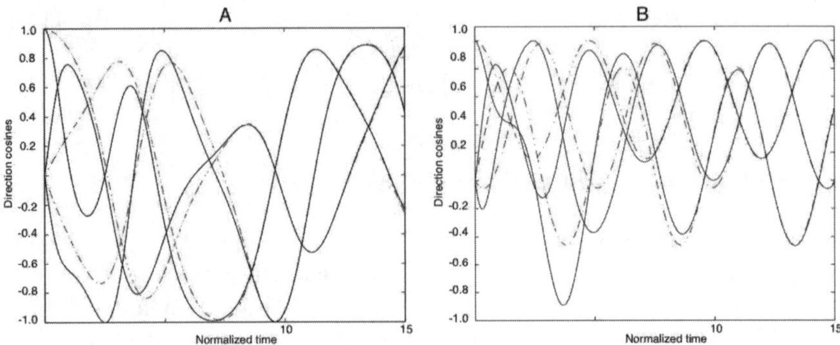

Fig. 14.1. Direction cosines of the unit vectors e_1 and e_3 with dynamic output feedback. Figure A corresponds to e_1 while figure B corresponds to e_3. The axisymmetric top values are the solid lines while the dotted lines are the reference values.

14.4.2 Simulation Results

In this section we apply the dynamic output feedback law (14.17) to a simulation of a simple mechanical control system in $SO(3)$. We consider the classical problem of a axisymmetric top in a gravitational field. Let $P = \{P_1, P_2, P_3\}$ be an inertial frame fixed at the fixed point of the top and let $e = \{e_1, e_2, e_3\}$ be a body-fixed orthonormal frame with the origin coinciding with that of P. At $t = 0$, the two frames coincide. Let the coordinates of a point p in the inertial frame P be given by x, and in the body frame e be given by X. The coordinates are related by $x(t) = R(t)X$ where $R(t) \in SO(3)$. Let $-P_3$ be the direction of gravity and let I be the inertia matrix of the axisymmetric top about the fixed point. The kinetic energy of the top is $K = I\zeta \cdot \zeta/2$, where ζ is the body angular velocity and the potential energy is $U(R) = mgl\, Re_3 \cdot P_3$. Here m is the mass of the top, g is the gravitational constant, l is the distance along the e_3 axis to the center of mass. For simplicity we assume the top to be symmetric about the e_3 axis, so $I = diag(I_1, I_1, I_3)$. The generalized potential forces $f(R)$ in the body frame are $< f(R), \zeta >= - < dU, R \cdot \zeta >= -mgl\, R\hat{\zeta}e_3 \cdot P_3$ for any $\zeta \in so(3)$, which yields $f(R) = mgl\, R^T P_3 \times e_3$. For convenience, let the desired reference trajectory $(R_r(t), \zeta_r(t))$ be generated by a simple mechanical system without external forces. It is not necessary for our results that the trajectory correspond to such a system.

In these simulations top parameters are $I_1 = I_2 = 1, I_3 = 2, mgl = 1$. The initial body angular velocity is $\zeta(0) = [1.3\,1.2\,1.1]$, and the initial configuration corresponds to a $\pi/2$ radian rotation about the $[1\ 1\ 1]^T$ axis. The reference trajectory $(R_r(t), \zeta_r(t))$ is generated by a model simple mechanical system without external forces. for the initial conditions $\zeta_r(0) = [-.8\ -.3\ -.5]$ and $R_r(0) = id$. The simulation results shown in Figure-14.1 correspond to the dynamic output feedback with $\alpha = \beta = 10$. The initial observer velocity is zero and the initial observer configuration corresponding to a $0.9\pi/2$ radian rotation about the $[1\ 1\ 1]^T$ axis.

14.5 Conclusions

We have presented a globally-defined state-feedback controller for a fully-actuated, simple mechanical control system on a general class of Lie groups. The state-feedback controller guarantees almost-global asymptotic stability, with locally exponential convergence. The requirement on the Lie groups is that a suitable error function must be found. Such functions exist on a wide class of Lie groups of interest, including compact Lie groups, such as $SO(n)$, Euclidian spaces, and their direct products, such as $SE(n)$. Extension of these results to infinite-dimensional and general non-compact Lie groups is a subject of future research. We have proved a separation principle, stating that the convergence properties of this controller are preserved when it is composed with any locally exponentially convergent velocity observer. In particular, the intrinsic velocity observer of Aghannan and Rouchon [2] provides configuration-feedback tracking for almost any initial configuration error. The estimator, however, guarantees only local convergence, and so the initial velocity and configuration estimates may be restricted to some neighborhood of the true values. No invariance properties are required for the above results, but if available they allow a considerably simplified explicit formulation.

Acknowledgments

The authors gratefully acknowledge support of National Science Foundation grants ECS-0218245 and ECS-0220314.

Appendix

Consider the system

$$\dot{e} = e \cdot \eta, \tag{14.30}$$

$$\dot{\eta} = \tilde{I}^{-1}\left(\mathrm{ad}^*_\eta \tilde{I}\eta + f(e,\eta)\right) + \psi(e,\eta,q), \tag{14.31}$$

with $(e,\eta) \in G \times \mathcal{G}$ and $q \in \mathcal{R}^n$. The matrix $\tilde{I}$ is a positive definite symmetric 3×3 matrix and consider the Left invariant metric $\langle\langle\cdot,\cdot\rangle\rangle$ induced by $\tilde{I}$. Consider $V = F(e) + \frac{1}{2}||\eta||^2$ where $F(e)$ is an error function and also consider the following assumptions.

Assumption 2 *The point $(id, 0)$ is an almost-globally stable equilibrium of (14.30)–(14.31) with $\psi \equiv 0$ and further more $\langle\langle e^{-1}\mathrm{grad}\,F, \eta\rangle\rangle + \langle\langle \tilde{I}^{-1}f, \eta\rangle\rangle \leq 0$.*

The condition $\langle\langle e^{-1}\mathrm{grad}\,F,\eta\rangle\rangle + \langle\langle\tilde{I}^{-1}f,\eta\rangle\rangle \leq 0$ is satisfied by any simple mechanical system with Rayleigh type dissipation and with $F(e)$ as the potential energy function.

Assumption 3 *The function $q(t) \in \mathcal{R}^n$ satisfies*

$$||q(t)|| \leq c||q(0)||e^{-\lambda t}, \tag{14.32}$$

for some $c > 0, \lambda > 0$ and all $t > 0$.

Assumption 4 *The interconnection term satisfies $\psi(e,\eta,0) \equiv 0$ and the following linear growth condition in η,*

$$||\psi|| \leq \gamma_1(||q||)||\eta|| + \gamma_2(||q||), \tag{14.33}$$

for two class $\mathcal{K}_\infty$ functions $\gamma_1(\cdot), \gamma_2(\cdot)$.

This assumption is automatically satisfied if the η dependence in $\psi(e,\eta,q)$ is linear and if either the Lie group is compact or $\psi(e,\eta,q)$ is independent of e.

Lemma 1. *If assumptions 2 – 4 are satisfied, then the equilibrium $(id, 0)$ of the system (14.30) – (14.31) is almost globally stable. If the inequality in assumption 2 is strict the convergence is asymptotic.*

References

1. R. Abraham and J. E. Marsden, *Foundations of Mechanics, Second Ed.* Westview, 1978.
2. N. Aghannan and P. Rouchon, "An Intrinsic Observer for a Class of Lagrangian Systems," *IEEE Transactions on Automatic Control*, (48) No. 6, pp 936–945, June 2003.
3. M. R. Akella, "Rigid Body Attitude Tracking without Angular Velocity Feedback," *Systems and Control Letters*, (42), pp 321–326, 2001.
4. M. R. Akella, J. T. Halbert and G. R. Kotamraju, "Rigid Body Attitude Control with Inclinometer and Low-cost Gyro Measurements," *Systems and Control Letters*, (49), pp 151–159, 2003.
5. V. I. Arnold, *Mathematical Methods of Classical Mechanics, Second Ed.*, Springer-Verlag, New York 1989.
6. J. Baillieul, "The Geometry of Controlled Mechanical Systems," *Mathematical Control Theory*, pp 322–354, Springer-Verlag, New York 1999.
7. J. Baillieul and J. C. Willems (Eds.), *Mathematical Control Theory*, Dedicated to R. W. Brockett on his 60th Birthday, Springer-Verlag, New York 1999.
8. A. M. Bloch, J. Baillieul, P. Crouch and J. E. Marsden, *Nonholonomic Mechanics and Control*, Springer-Verlag, New York 2003.
9. F. Bullo and R. M. Murray, "Tracking for Fully Actuated Mechanical Systems: A Geometric Framework," *Automatica*, (35), pp 17–34, 1999.
10. F. Bullo and A. D. Lewis, *Geometric Control of Mechanical Systems: Modeling, Analysis, and Design for Simple Mechanical Control Systems*, Springer-Verlag, New York 2004.

11. I. A. Dynnikov and A. P. Vaselov "Integrable Gradient Flows and Morse Theory," *eprint arXiv:dg-ga/9506004*, http://arxiv.org/abs/dg-ga/9506004, Sep. 1995.

12. T. Frankel, *The Geometry of Physics an Introduction,* Cambridge University Press, 1997.

13. D. E. Koditschek, "The Application of Total Energy as a Lyapunov Function for Mechanical Control Systems," *In J. E. Marsden, P. S. Krishnaprasad and J. C. Simo (Eds) Dynamics and Control of Multi Body Systems,* pp 131–157, Vol 97, Providence, RI:AMS, 1989.

14. D. H. S. Maithripala, R. O. Gale, M. W. Holtz, J. M. Berg and W. P. Dayawansa, "Nano-precision control of micromirrors using output feedback," *Proc. of the CDC,* Maui, HW, 2003.

15. D. H. S. Maithripala, J. M. Berg and W. P. Dayawansa, "An Intrinsic Observer for a Class of Simple Mechanical Systems on Lie Groups," *Proceedings of the 2004 American Control Conference,* Bostan, MA, 2004.

16. D. H. S. Maithripala, W. P. Dayawansa and J. M. Berg "Intrinsic Observer Based Control on Lie Groups," *SIAM Journal of Control and Optimization,* in review.

17. D. H. S. Maithripala, J. M. Berg and W. P. Dayawansa, "State and Configuration Feedback for Almost Global Tracking of Simple Mechanical Systems on a General Class of Lie Groups," *Submitted for Review: IEEE Transactions on Automatic Control,* 2004.

18. J. E. Marsden and T. S. Ratiu, *Introduction to Mechanics and Symmetry, Second Ed.* Springer-Verlag, New York 1999.

19. M. Morse, "The Existence of Polar Non-Degenerate Functions on Differentiable Manifolds," *Annals of Mathematics,* 71 (2), pp 352–383, 1960.

20. P. Newton, P. Holmes and A. Weinstein (Eds.), *Geometry, Mechanics and Dynamics: Volume in Honor of the 60th Birthday of J. E. Marsden* Springer-Verlag, New York 2002.

21. H. Rehbinder and X. Hu, "Nonlinear State Estimation for Rigid Body Motion with Low Pass Sensors," *Systems and Control Letters,* 40 (3), pp 183–191, 2000.

22. S. Salcudean, "A Globally Convergent Angular Velocity Observer for Rigid Body Motion," *IEEE Transactions on Automatic Control,* 36 (12), pp 1493–1497, December 1991.

23. G. V. Smirnov, "Attitude Determination and Stabilization of a Spherically Symmetric Rigid Body in a Magnetic Field," *International Journal of Control,* (74) No. 4, pp 341–347, 2001.

15

Efficient Autonomous Underwater Vehicle Communication Networks using Optimal Dynamic Flows

Joseph T. Napoli[1], Tzyh Jong Tarn[1], James R. Morrow, Jr.[1] and Edgar An[2]

[1] Department of Electrical and Systems Engineering, Washington University in St. Louis, One Brookings Drive, St. Louis, MO 63130, USA
`tarn@wuauto.wustl.edu,jmr3@cec.wustl.edu`

[2] Department of Ocean Engineering, Florida Atlantic University, Sea Tech Campus, Florida 33004, USA
`ean@oe.fau.edu`

Summary. Underwater vehicles are being increasingly employed in exploring the world's oceans. The communications mechanism used by these autonomous underwater vehicles (AUV's) plays a vital roll mission planning. This paper presents a procedure for constructing optimal controllers that limit and route information flow in a network of AUVs. Particular attention is paid to the objectives of maximizing throughput and bandwidth efficiency while minimizing power consumption and queue length. A receding horizon control approach is adopted and shown to yield piecewise continuous optimal controllers that are unique.

Introduction

Robotic networks for Autonomous Underwater Vehicles (AUV's) are an area where efficient use of the communications channel can increase both the mission time and the amount of information communicated between the AUVs, thus resulting in a more effective mission. The traffic in robotic networks must be managed in such a way that takes into account transmission delay, power efficiency, and traditional network measures such as queue length, throughput, and bandwidth efficiency. System optimization of all these criteria is possible because of the cooperative nature of robotic networks. Many missions, such as studying ocean processes, weather forecasting, and mapping the ocean floor, need a small, inexpensive craft to take measurements. Current missions require a ship to support one or two underwater vehicles, making widespread underwater exploration prohibitively expensive. Efforts are underway to network multiple autonomous underwater vehicles (AUVs) and make them independent of support vessels, thereby reducing cost and making underwater exploration more accessible. Researchers have acknowledged the

J.T. Napoli et al.: *Efficient Autonomous Underwater Vehicle Communication Networks using Optimal Dynamic Flows*, LNCIS **321**, 239–256 (2005)
`www.springerlink.com`

need for AUVs working in concert to provide more effective sampling of the ocean [16]. An Autonomous Ocean Sampling Network (AOSN) of underwater vehicles and support hardware has been proposed in [15]. Among the work to make such a network possible are advances in acoustic signal processing [10], [9]. Sophisticated signal processing is necessary to combat the Doppler and multipath effects, fading, absorption, and ambient noise that occur in underwater acoustic channels [5].

There are three distinguishing characteristics about the AUV communication networks considered in this paper. First, the ocean is a broadcast environment with very limited bandwidth, so some form of frequency reuse must be employed. This has important implications for the dynamics of the network. Second, the power consumption and storage capacity of each vehicle plays a pivotal role in mission runtime. An effective communication strategy must take this into account. The final characteristic, unlike most broadcast networks, is that vehicles act in a coordinated and cooperative manner.

This paper presents a procedure for constructing controllers that limit and route information flow in AUV networks. It uses a dynamic flow model from tailored to specifically address the unique challenges of the acoustic medium [2]. An optimal control problem is formulated on both a finite and infinite time horizon. Unique, piecewise continuous optimal controllers that limit and route traffic are con-structed for each case using a signal destination routing approach. A receding ho-rizon approach is proposed in the infinite time case along with a stability result [12], [13]. Finally, simulation results are presented.

Bandwidth Efficiency

The underwater environment is a severely bandwidth restricted environment. In order to network a large number of autonomous underwater vehicles together, a bandwidth efficient strategy is necessary. In the case of a network of vehicles moving in a fixed or slowly changing formation, the structure of the network can be exploited for bandwidth efficiency. Frequency reuse can be employed in these networks, much as it is in cellular networks.

Each vehicle in the network can be allocated a slice of the total bandwidth via frequency division multiplexing. Vehicles far apart from one another in the network can be allocated the same part of the frequency spectrum. In order to prevent interference, vehicles that use the same part of the spectrum can not both directly communicate with a common vehicle. Therefore the networks considered in this paper consist of underwater vehicles that communicate with a few neighbors, then store and forward packets until they reach their destination. Throughput can be measured by the percentage of new traffic accepted into the network. The network should accept as many packets as is reasonable.

Power Consumption and Storage Utilization

Power efficiency is a cost associated with the links of a network. Each time a packet is transmitted along a link, a cost in terms of power is incurred. Autono-mous vehicles operate by using some form of battery. Thus mission times are limited to the battery life of the system. Efficient use of the communication electronics onboard the AUV will increase the battery life, and thus the mission runtime.

Being an autonomous vehicle also limits the amount of onboard storage for communications. Therefore, storage capacity at each AUV is finite and the lengths of the buffers at each node should be minimized. Overflowing the storage will result in lost data and/or retransmissions which will use more power and waste more bandwidth of the network. Efficient use of the available onboard storage will optimize both power consumption and bandwidth.

Cooperation

Current research using flow models to direct network traffic considers users competing for network resources [6]. AUV networks differ in the respect that the vehicles need to cooperate to optimize system resources. These characteristics are best highlighted when contrasted with pure ALOHA networks where bandwidth efficiency is very low, transmission power does not factor into the model, and users act independently and competitively [17].

Notational Conventions

This section introduces notation and terminology for describing the networks in this paper. A network consists of a set of N nodes, denoted $\mathcal{N} = \{1, 2, ..., N\}$, and a set of L links, denoted $\mathcal{L} = \{(j, k)|j, k \in \mathcal{N}, j \neq k\}$. There need not be a link for every pair of nodes. The set of downstream nodes for node j is denoted $O(j) = \{k \in \mathcal{N}|(j, k) \in \mathcal{L}\}$. The set of upstream nodes for node j is denoted $I(j) = \{i \in \mathcal{N}|(i, j) \in \mathcal{L}\}$. Related concepts are the sets of outgoing and incoming links for node j, defined as $\{(j, k)|k \in O(j)\}$ and $\{(i, j)|(i \in I(j)\}$, respectively.

A cycle, C, refers to a set of nodes $\{j_c|c = 1, 2, ...C, (j_c, j_{c+1}) \in \mathcal{L}, j_1 = j_C\}$. The integer C is called the size of the cycle and is also denoted by $|C|$. The notation C is slightly abused by also denoting the set of links $\{(j_c, j_{c+1})|c = 1, 2, ..., C, (j_c, j_{c+1}) \in \mathcal{L}, j_1 = j_C\}$. The set to which C refers should be obvious from context by examining its elements.

When dealing with networks, the states and parameters in the system are denoted with subscripts whereas symbols without subscripts denote vectors. For example, α_{jk} is the proportion of the traffic arriving at node j that is routed to node k. When the subscript is dropped, it refers to a vector containing the subscripted elements. Therefore α_{jk} are the elements of the vector α.

Single Destination Routing

In this paper we consider single destination routing. In single destination routing, one and only one, AUV acts as a sink for all of the data entering the network. In a future paper, the more general case will be presented.

The AUVs can be modeled as the nodes of a communication network where packets queue while they wait to be transmitted. The model chosen for the queue at each node is based on the following scalar dynamic flow model from [2] $\dot{\xi}(\tau) = -\mu G(\xi(\tau)) + \lambda(\tau)$. The state $\xi(\tau)$ represents the average length of the queue at time τ and $\lambda(\tau)$ is the average rate or intensity at which traffic is entering the queue at time τ. The parameter μ is the capacity of the server and the function $G(\xi(\tau))$ is called the utilization function. The utilization function represents the average percentage of capacity being used at each instant. The function $\mu G(\xi(\tau))$ approximates the average rate of traffic leaving the queue at time τ.

The utilization $G(\cdot)$ is modeled as a function of the state. The utilization of a queue switches between 0 and μ depending on whether the queue is empty or not. Intuitively, if the queue contains many elements then its utilization will be equal to μ for a while and one would expect the average utilization to be near μ. If the queue contains few elements, one would expect the average utilization to be near zero. Analytical and empirical methods can be employed to find an appropriate form for $G(\xi)$ [2]. Typically, $G(\xi)$ is a monotonically increasing, concave function of satisfying $G(0) = 0$, $G'(0) = 1$, and $G(\xi) \to 1$ as $\xi \to \infty$.

The current problem of course is not managing a group of individual queues, but systemically controlling a network of queues. At the network level, the interaction dynamics of the queues in the network are primarily influenced by the fact that AUVs communicate in a broadcast environment. Specifically, the network of queues is modeled with the system of equations:

$$\dot{\xi}_j(\tau) = -\mu_j G(\xi_j(\tau)) + \phi_j(\tau) r_j(\tau) + \sum_{i \in I(j)} \alpha_{ij}(\tau) \mu_i G(\xi_i(\tau)). \qquad (15.1)$$

Where $\xi(t_0) = x_{j0}$ denotes initial state x_{j0} at the initial time t_0. The state ξ_j represents the length of the queue at node j. The incoming trafic to node j is the sum of two trafic flows; the flow $r_j(\tau)$ is new trafic to the network while $\sum \alpha_{ij}(\tau) \mu_i G(\xi_i(\tau))$ is the sum of trafic arriving from other nodes.

The cooperation between vehicles is used to implement a bandwidth efficient communication strategy. The capacity μ_j, of each vehicle is determined by the frequency allocated to it. It is assumed that the vehicles not only cooperate in sharing frequency, but also coordinate to maintain a formation. It is assumed that vehicle formation, if not maintained, changes slowly. Vehicles far away from one another can then use the same frequency channels since it is envisioned the vehicles will only communicate with neighbors. Communicating only with neighbors also increases power efficiency since power decreases quadratically with distance.

The control over the network is exercised by routing and limiting the traffic flow. The capacity at each node can be divided amongst outgoing links using time division multiplexing. In this manner, the flow from each node can be proportionally routed onto outgoing links. The routing control α_{jk} is the proportion of traffic flow emanating from node j that is directed to node k. Since all the traffic leaving node j must be directed somewhere, α must satisfy the conservation constraint

$$\sum_{k \in O(j)} \alpha_{jk} = 1, \quad \forall j \in \mathcal{N}. \tag{15.2}$$

The limiting control $\phi_j(\tau)$ is the proportion of $r_j(\tau)$, the new traffic entering at node j, that is accepted. Since both α and ϕ are proportions they must satisfy the following conditions:

$$0 \leq \phi_j \leq 1, \quad \forall j \in \mathcal{N} \tag{15.3}$$

$$0 \leq \alpha_{jk} \leq 1, \quad \forall (j,k) \in \mathcal{L}. \tag{15.4}$$

Finally, we have a set of cycle constraints to eliminate nodes from being disconnected from the rest of the network via cycles

$$\sum_{(j,k) \in C} \alpha_{jk} \leq |C| - 1 \quad \forall C. \tag{15.5}$$

This should hold for all possible cycles. In practice heuristics enforce these constraints since there may be too many constraints to enforce explicitly.

We seek a cost criterion that will maximize throughput while minimizing power consumption and queue length. Let d_{jk} represent the cost incurred per packet, in terms of power consumption, from using link (j, k). Let b_j represent the relative benefit of accepting traffic at node j and let a_j represent the cost incurred by long queue lengths. Then, an appropriate cost functional to maximize is

$$J(\alpha, \phi, T) := \int_t^{t+T} \sum_{i \in \mathcal{N}} b_j \phi_j(\tau) r_j(\tau) - \sum_{(j,k) \in \mathcal{L}} d_{jk} \alpha_{jk}(\tau) \mu_j G(\xi(\tau))$$
$$- \sum_{j \in \mathcal{N}} \frac{1}{2} a_j \xi_j^2(\tau) d\tau \tag{15.6}$$

over some time interval $[t, t+T]$. The first term rewards accepting traffic into the network while the second term penalizes the user for power consumption. The last term penalizes for a large queue length.

Finite Time Optimal Control

Maximizing the cost functional with respect to the dynamic and control constraints defines a finite time optimal control problem. For this problem there exist unique piecewise continuous optimal controllers. These controllers are unique in the sense that other optimal controllers differ only on a set of measure zero. In this section, the piecewise continuous optimal controllers are constructed.

Our finite time optimal control problem is then defined by

$$\max_{\alpha,\phi\in\Omega} J(\alpha,\omega,T) \tag{15.7}$$

subject to (15.1), where Ω is the set of ϕ and α which satisfy (15.2)-(15.5).

Traditional optimal control techniques can now be applied. Forming the Hamiltonian

$$
\begin{aligned}
H(\xi,\rho,\phi,\alpha) = &\sum_{j\in\mathcal{N}}(\rho_j - b_j)\phi_j r_j + \sum_{(j,k)\in\mathcal{C}}(d_{jk} + \rho_k)\alpha_{jk}\mu_j G(\xi_j) \\
&+ \sum_{j\in\mathcal{N}}\frac{1}{2}a_j\xi_j^2 - \sum_{j\in\mathcal{N}}\rho_j\mu_j G(\xi_j)
\end{aligned}
\tag{15.8}
$$

and differentiating with respect to ξ yields the costate equations along with its transversality condition

$$\dot\rho(\tau) = -a_j\xi_j(\tau) + \left(\mu_j\frac{dG}{d\xi_j}(\tau)\right)\left[\rho_j(\tau) - \sum_{k\in O(j)}(d_{jk} + \rho_k(\tau))\alpha_{jk}(\tau)\right] \tag{15.9}$$

$$\rho_j(t_0 + T) = 0 \tag{15.9a}$$

Since there are no cross terms between ϕ and α, the Hamiltonian can be split into two linear programming problems

$$\min_{\phi\in\Omega}\sum_{j\in\mathcal{N}}(\rho_j - b_j)\phi_j r_j := \sum_{j\in\mathcal{N}}\phi_j H_j \tag{15.10}$$

$$\min_{\alpha\in\Omega}\sum_{(j,k)\in\mathcal{C}}(d_{jk} + \rho_k)\alpha_{jk}\mu_j G(\xi_j) := \sum_{(j,k)\in\mathcal{C}}\alpha_{jk}H_{jk} \tag{15.11}$$

whose solutions are naturally binary [3]. This means that if the minimizer of the Hamiltonian is unique, then the value of the optimal controllers will be binary. In the case where the value of the controller is not uniquely determined by the Hamiltonian, the controller may not be binary and another constraint is needed to determine its value. For illustrative purposes, this paper considers the case where only one component of the control will be undetermined at any time. The more general case of multiple undetermined controls is more tedious but conceptually is not any more complex.

Consider the case where a component of ϕ is undetermined. It is not difficult to see that a necessary condition for ϕ^* to minimize (15.10) is

$$\phi^* = \begin{cases} 0 & \text{if } \rho_j - b_j > 0 \\ 1 & \text{if } \rho_j - b_j < 0 \end{cases} \tag{15.12}$$

In particular, (15.10) does not have a unique minimizer if $p_j = b_j$ for some j. The set

$$S = \{(\tau, \xi, \rho) \mid \gamma(\tau, \xi, \rho) = \rho_j - b_j = 0\} \tag{15.13}$$

can be interpreted as a smooth surface in (τ, ξ, ρ) space on which ϕ is not uniquely determined by the Hamiltonian. The necessary conditions for a trajectory $(\tau, \xi(\tau), \rho(\tau))$ to hit S at time τ^0, and to remain on this surface for some interval $I = [\tau^0, \tau^0 + \Delta]$, are

$$\frac{d^2}{d\tau^2}\gamma(\tau, \xi(\tau), \rho(\tau)) = \frac{d}{d\tau}\gamma(\tau, \xi(\tau), \rho(\tau)) = 0, \tau \in I \tag{15.14}$$

These additional constraints imply that ϕ_j satisfy the conditions

$$\dot{\rho}_j(\tau) = -a_j\xi_j(\tau) + \mu_j\frac{dG}{d\xi_j}(\tau)\times$$
$$[\rho_j(\tau) - \sum_{k\in O(j)} \alpha_{jk}(\tau)(d_{jk} + \rho_k(\tau))] = 0 \tag{15.15}$$

and

$$\ddot{\rho}_j(\tau) = \phi_j h_1(\xi, \rho, \alpha) + h_2(\xi, \rho, \alpha) = 0 \tag{15.16}$$

on I. Where h_1 and h_2 are defined as

$$h_1(\xi, \rho, \alpha) = -a_j + \mu_j\frac{d^2 G}{d\xi_j^2}(\tau)\left[\rho_j(\tau) - \sum_{k\in O(j)} \alpha_{jk}(\tau)(d_{jk} + \rho_k(\tau))\right] \tag{15.17}$$

$$h_2(\xi, \rho, \alpha) = h_1(\xi, \rho, \alpha)\left[-\mu_j G(\xi_j) + \sum_{i\in I(j)} \alpha_{ij}\mu_i G(\xi_i(\tau))\right] \tag{15.18}$$

Recall that it is assumed that only ϕ_j is unknown, therefore the functions h_1 and h_2 depend only on the value of known states and controls. Equation (15.16) is linear in ϕ_j and can be solved uniquely for $\phi_j(\tau)$ on I. In fact, $\phi_j(\tau)$ solved for in this way will be continuous on I.

If either (15.15) is violated or (15.16) can not be solved for $\phi_j \in [0, 1]$, then the trajectory will pass through S. According to (15.12), when the trajectory passes through S, there will be a switching in the control ϕ_j. Therefore, S is referred to as a switching surface. In this manner, one can construct a piecewise continuous optimal limiting controller, $\phi^*(t; t_0, x_0)$ for time $t \in [t_0, t_0 + T]$, that is uniquely defined given an initial condition (t_0, x_0). In order to maintain uniqueness at switching points (where any value of the control can be chosen without loosing optimality), continuity from the right can be enforced.

In a directly analogous manner, it is possible to construct the unique piecewise continuous optimal routing controller $\alpha^*(t; t_0, x_0)$ for the finite time problem [7]. In this case, the switching surface $\mathcal{T}$ for the routing controller can be constructed from knowledge about the nature of minimizers to linear programming problems. If α is not uniquely determined by the Hamiltonian, then there are at least two binary minimizers. Let α^1 and α^2 denote the two binary minimizers. Furthermore, assume that these two minimizers determine a conservation constraint. That is, assume that the line connecting them is a conservation constraint. (The line connecting them must be a constraint of the linear programming problem since it is an edge of the simplex formed by the constraints. The other possibility is that a cycle constraint connects them.) This being the case, α^1 and α^2 will be identical except for two components. The discrepancies in the control will be

$$\alpha^1_{jk_i} = 0 \quad \alpha^1_{jk_2} = 1 \tag{15.19}$$

and

$$\alpha^2_{jk_i} = 1 \quad \alpha^2_{jk_2} = 0 \tag{15.20}$$

for some node j. The conservation constraint that connects the binary minimizers is

$$\alpha^1_{jk_1} + \alpha^1_{jk_2} = 1 \tag{15.21}$$

Therefore, the switching surface $\mathcal{T}$ will be characterized by the fact that on an interval I (where $\tau \in I$),

$$\mathcal{T} = \{(\tau, \xi, \rho) \mid \psi(\tau, \xi, \rho) = H(\alpha^1) - H(\alpha^2) = 0\} \tag{15.22}$$

Receding Horizon Control

The running time and traffic load of an AUV sampling mission is generally not known in advance. This precludes an explicit offline application of a finite time approach to network control. Solving the infinite time problem, in the absence of steady state load conditions, also requires knowledge of the load for all time. An alternative solution is to take a receding horizon approach to construct a feedback controller [12], [13].

Receding horizon controllers are motivated by the following control procedure. At time t, estimate the traffic load over the interval $[t, t + T]$. Solve the finite time problem on this interval and apply the optimal controller until time $t + \delta$, where $\delta < T$. At time $t + \delta$, repeat the process by solving the finite time problem over $[t + \delta, t + \delta + T]$ and apply the optimal controller until time $t + 2\delta$. Using this iterative approach, one can control the network over an infinite interval without ever estimating the load more than T seconds into the future.

The controller obtained by letting $\delta \to 0$ is the receding horizon controller. In this section, it is shown that properly defined receding horizon controllers for our system are unique and piecewise continuous. Stabilizing properties of the receding horizon controller are also explored. The chief advantage of this approach is that only knowledge of the load in the near future is required. It is helpful to introduce vector notation for the finite time system

$$\dot{\xi} = f(t, \xi, \phi, \alpha), \quad \xi(t_0) = x_0 \tag{15.23}$$
$$\dot{\rho} = g(t, \xi, \rho, \phi, \alpha), \quad \rho(t_0 + T) = 0 \tag{15.24}$$

where (15.23) represents the state dynamics and (15.24) represents the costate dynamics. The receding horizon system is defined as

$$\dot{x} = \hat{f}(t, x) = f(t, x, \Phi(t, x), A(t, x)), \quad x(t_0) = x_0 \tag{15.25}$$

where $\Phi(t, x)$ and $A(t, x)$ denote the receding horizon controllers. Typically, the receding horizon controllers would be defined as

$$\Phi_j(t, x) = \phi_j^*(t; t_0, x_0) \tag{15.26}$$
$$A_{jk}(t, x) = \alpha_{jk}^*(t; t_0, x_0) \tag{15.27}$$

However, the receding horizon controller defined in this way does not guarantee existence or uniqueness of solutions. There are surfaces in the state space on which special care must be taken. A suitable definition for the receding horizon controller on these surfaces to guarantee existence and uniqueness is now presented. We will present the following lemmas and theorems, which will be useful in establishing existence and uniqueness results.

Lemma 1. *The function $p(t, x) = \rho^*(t; t_0, x_0)$ is continuous in (t, x).*

Proof. Consider the cost functional as a function of the initial conditions for the time, state, and costate (t_0, x_0, p_0). For a fixed t and x the trajectories of the system, and therefore the cost, will vary as the initial costate changes. One could then consider the finite time optimal control problem as that of finding the best initial costate [7]. Consequently,

$$p(t, x) = \rho^*(t; t_0, x_0) \in \arg\max_{p_0} J(t_0, x_0, p_0). \tag{15.28}$$

We can conclude that $J(t, x, p(t, x)) - J(t, x, p) > 0, \forall p \neq p(t, x)$.

Now assume to the contrary that $p(t, x)$ is not continuous. Consider a sequence $(t_k, x_k) \to (t, x)$ such that $p(t_k, x_k) \to p \neq p(t, x)$. Define the difference $d = J(t, x, p(t, x)) - J(t, x, p) > 0$ and construct the real sequence

$$J(t_k, x_k, p(t, x)) - J(t_k, x_k, p(t_k, x_k)) \to d. \tag{15.29}$$

This implies that there exists an integer K such that for all $k > K$

$$J(t_k, x_k, p(t, x)) - J(t_k, x_k, p(t_k, x_k)) > \frac{d}{2} > 0. \tag{15.30}$$

This is a contridiction and the result follows. $\square$

We now use this result to show a weak local existence and uniqueness result for the receding horizon system.

Lemma 2. *Given an initial condition (t_0, x_0) satisfying*

$$\hat{\gamma}(t_0, x_0) := \gamma(t_0, x_0, p(t_0.x_0)) \neq 0 \tag{15.31}$$

and

$$\hat{\Psi}(t_0, x_0) := \Psi(t_0, x_0, p(t_0.x_0)) \neq 0 \tag{15.32}$$

there exists a $\delta > 0$ such that a unique solution, $x(t)$, of (15.25) exists on the interval $[t_0, t_0 + \delta]$.

Proof. At time t_0, the receding horizon controller $\Phi(t_0, x_0)$ takes on the value of the finite time optimal controller $\phi^*(t; t_o, x_0)$ evaluated at $t = t_0$. Since (t_0, x_0) satisfies

$$\hat{\gamma}(t_0, x_0) = \gamma(t_0, \xi(t_0; t_0, x(t_0)), \rho(t_0; t_0, x(t_0))) \neq 0 \tag{15.33}$$

$\phi^*(t; t_o, x_0)$, and therefore $\Phi(t_0, x_0)$, will be binary. Furthermore, (15.33) will be satisfied in some small neighborhood of (t_0, x_0) since $\gamma(t, x)$ is continuous in (t, x). This implies that if $\Phi(t_0, x_0)$ is zero (or one), then $\Phi(t, x)$ will remain zero (or one) in some sufficiently small neighborhood of (t_0, x_0). That is, $\Phi(t, x)$ will be constant in some neighborhood of (t_0, x_0). The same argument for $A(t, x)$, with respect to (15.32), implies $A(t, x)$ is constant near (t_0, x_0). This implies local existence and uniqueness for trajectories of (15.28) for any initial condition satisfying (15.31) and (15.32). $\square$

In fact, the stronger statement that solutions can be continued up to the surfaces

$$\hat{\gamma}(t, x) = \gamma(t, x, p(t, x)) = 0 \tag{15.34}$$

$$\hat{\Psi}(t, x) = \Psi(t, x, p(t, x)) = 0 \tag{15.35}$$

is true. However, these are still fairly weak existence results since they do not hold for arbitrary initial conditions. The central question is whether a trajectory of the receding horizon system can be continued once it intersects one of these surfaces. The answer is affirmative and trajectories can be continued in much the same manner as they were in the finite time case.

Theorem 1. *Define the receding horizon controllers as*

$$\Phi(t,x) = \phi^*(t; t_0, x_0), \quad \forall (t,x) \notin \hat{S} \tag{15.36}$$

$$A(t,x) = \alpha^*(t; t_0, x_0), \quad \forall (t,x) \notin \hat{T} \tag{15.37}$$

where $\hat{S} = \{(t,x) \mid \hat{\gamma}(t,x) = 0\}$ *and* $\hat{T} = \{(t,x) \mid \hat{\Psi}(t,x) = 0\}$*. On the sets* $\hat{S}$ *and* $\hat{T}$ *define* $\Phi(t,x)$ *and* $A(t,x)$ *implicitly by*

$$\frac{d}{dt}\hat{\gamma}(t,x,\Phi,A) = 0, \forall (t,x) \in \hat{S} \Leftrightarrow$$
$$\left(\frac{\partial \gamma}{\partial t}\dot{r} + \frac{\partial \gamma}{\partial p}\frac{\partial p}{\partial t}\right) + \left(\frac{\partial \gamma}{\partial x} + \frac{\partial \gamma}{\partial p}\frac{\partial p}{\partial x}\right) f(t,x,\Phi,A) = 0 \tag{15.38}$$

and

$$\frac{d}{dt}\hat{\Psi}(t,x,\Phi,A) = 0, \forall (t,x) \in \hat{T} \Leftrightarrow$$
$$\left(\frac{\partial \Psi}{\partial t}\dot{r} + \frac{\partial \Psi}{\partial p}\frac{\partial p}{\partial t}\right) + \left(\frac{\partial \Psi}{\partial x} + \frac{\partial \Psi}{\partial p}\frac{\partial p}{\partial x}\right) f(t,x,\Phi,A) = 0 \tag{15.39}$$

respectively. If $p(t,x)$ *is* C^1 *continuous, then given an initial condition* (t_0, x_0)*, there exists a unique solution,* $x(t)$*, of (15.25) for all* $t \geq t_0$*.*

Remark 1. We have shown that $p(t,x)$ is continuous, but here it is assumed C^1 continuous so that (15.38) and (15.39) may be solved continuously for Φ and A. In the case of a linear system, where explicit knowledge of the solution yields a closed form solution of the two point boundary problem, $p(t,x)$ is C^∞. As an illustrative example of how one might solve (15.38), consider the case where we hit the surface

$$\hat{\gamma}(t,x) = \gamma(t,x,p(t,x)) = p_{\hat{j}}(t,x) - b_{\hat{j}} = 0. \tag{15.40}$$

In conjunction with the function

$$\hat{f}_j(t,x) = -\mu_j G(x_j) + \Phi_j r_j + \sum_{i \in I(j)} \alpha_{ij}\mu_i G(\xi_i(t)) \tag{15.41}$$

we see that these equations are linear in Φ and when solved yield

$$\Phi_{\hat{j}} = \left(-\frac{\partial p_{\hat{j}}}{\partial t} - \sum_{j \in \mathcal{N}, j \neq \hat{j}} \frac{\partial p_{\hat{j}}}{\partial x_j} \hat{f}_j(t,x) \right.$$
$$\left. -\frac{\partial p_{\hat{j}}}{\partial x_{\hat{j}}} \left(-\mu_{\hat{j}} G(x_{\hat{j}}) + \sum_{i \in I(\hat{j})} \alpha_{i\hat{j}}\mu_i G(\xi_i(t)) \right) \right) \Big/ \frac{\partial p_{\hat{j}}}{\partial x_{\hat{j}}} r_{\hat{j}}. \tag{15.42}$$

Although explicit expressions for the first order derivatives of p are not known, their values can be calculated numerically.

Remark 2. As in the finite time case, a solution of (15.38) for Φ (respectively, (15.39) for A) may not be in the range $[0, 1]$. If this is the case, then the trajectory will pass through $\hat{S}$ (respectively, $\hat{T}$). If the trajectory does remain on $\hat{S}$ for some interval, then the solution (15.41) is continuous on that interval. In general, (15.38) and (15.39) give linear equations in the unknown control and can be solved continuously for their values. This implies that, as in the finite time case, one can construct unique, piecewise con-tinuous receding horizon controllers. It is somewhat re-markable that the structure of the receding horizon controller is no more complicated than that of the finite case under the mild assumption that $p(t, x)$ is C^1 continuous. We now turn our attention to a stability result for the receding horizon system. We borrow the following concept of stability from [1].

Definition 1. *The solutions of $\dot{x} = \hat{f}(t, x)$ are said to be uniformly ultimately bounded if there exist positive constants b and c, and for every $\delta \in (0, c)$ there is a positive constant t_1 such that*

$$\|x(t_0)\| < \delta \Rightarrow \|x(t)\| \le b, \quad \forall t \ge t_0 + t_1 \tag{15.43}$$

where $x(t)$ is a solution of $\dot{x} = \hat{f}(t, x)$. The solutions are said to be globally uniformly ultimately bounded if this holds for arbitrarily large δ.

Theorem 2. *The solutions to the receding horizon system $\dot{x} = \hat{f}(t, x)$ are globally uniformly ultimately bounded.*

Proof. We will show that an arbitrary solution, $\bar{x}(t)$, with initial condition $\bar{x}(t_0) = \bar{x}_0$, is globally uniformly ultimately bounded. Let $x(t)$ be another solution of $\dot{x} = \hat{f}(t, x)$ with $x(t_0) = x_0$. We have for each node j that

$$\dot{x}_j(t) = -\mu_j G(x_j(t)) + \Phi_j(t, x(t)) r_j(t) + \sum_{i \in I(j)} A_{ij}(t, x(t)) \mu_i G(x_i(t)) \tag{15.44}$$
$$\le -\mu_j G(x_j(t)) + R + \mu$$

and

$$\dot{\bar{x}}(t) = -\mu_j G(\bar{x}_j(t)) + \Phi_j(t, \bar{x}(t)) r_j(t) + \sum_{i \in I(j)} A_{ij}(t, \bar{x}(t)) \mu_i G(\bar{x}_i(t)) \tag{15.45}$$
$$\ge -\mu_j G(\bar{x}(t))$$

where R is a bound on the flow $r(t)$ and μ is a bound on the capacities of the network. Bounding the difference using the Lipschitz property (with constant one) of G yields

$$|\dot{x}_j(t) - \dot{\bar{x}}(t)| = -\mu_j |G(x_j(t)) - G(\bar{x}_j(t))| + |R + \mu|$$
$$\le -\mu_j |x_j(t) - \bar{x}_j(t)| + |R + \mu| \tag{15.46}$$

By the comparison lemma,

$$|x_j(t) - \bar{x}_j(t)| \leq |y_j(t) - \bar{y}_j(t)| \tag{15.47}$$

where

$$\dot{y}_j(t) = -\mu_j y_j(t) + R + \mu \tag{15.48}$$

$$\dot{\bar{y}}_j(t) = -\mu_j \bar{y}_j(t). \tag{15.49}$$

We can solve for $\Delta y_j(t) := y_j(t) - \bar{y}_j(t)$ explicitly as

$$\Delta y_j(t) = [\Delta y_j(t_0) - \frac{R+\mu}{\mu_j}]e^{-\mu_j t} + \frac{R+\mu}{\mu_j}. \tag{15.50}$$

Letting

$$b = R + \mu + 1 \tag{15.51}$$

implies that there exists a t_1 such that

$$|x_j(t) - \bar{x}_j(t)| \leq |y_j(t) - \bar{y}_j(t)| \leq b, \forall t > t_0 + t_1. \tag{15.52}$$

Since this holds for all j, we have

$$\|x(t) - \bar{x}(t)\| \leq b \tag{15.53}$$

which completes the proof. $\square$

Remark 3. A closely related concept to that of ultimate boundedness is input-to-state stability. A corollary of the above theorem is that the receding horizon system is input-to-state stable when $r(t)$ is regarded as the input to the system.

Remark 4. Stability and asymptotic stability of receding horizon controller under certain conditions is reported in [13]. Reference [14] uses a condition that prohibits the trajectories of the receding horizon system to remain on the switching surfaces and also impose a terminal constraint on the finite time system, which is not practical here.

Simulation Results

In order to verify the above formulation in creating effective controllers, the example network of Fig. 1 was simulated for a routing controller. The routing controller was implemented in a receding horizon fashion. Every minute, the finite time optimal control problem is solved with a time interval of ten minutes. The optimal controller is used until the next finite time problem is solved.

The links of the network are labeled with their associated cost d_{jk}. This network might represent a mobile robotic network, such as a network of AUVs on a telemetry mission. As each vehicle gathers information, it must relay the

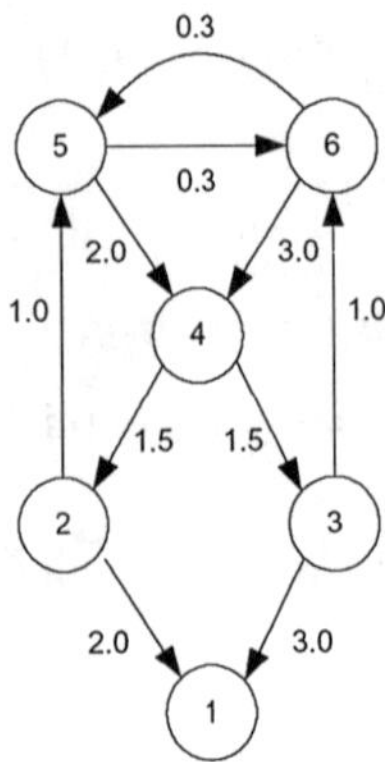

Fig. 15.1. Fig. 1. Six node simulation network.

data to a support ship or buoy for storage and analysis. A critical constraint
of AUV missions is the power limitations of the vehicles. Therefore the cost of
links would reflect the cost of transmission in terms of power. They might also
represent the cost in terms of the delay experienced in the acoustic channel.

This network was selected because it is small enough to serve as an example, yet large enough to exhibit properties that are of interest. One feature
of the network is that the costs of the links were designed to make the cycle
containing nodes 5 and 6 very attractive. Another is that node 4 will serve as
a bottleneck node when nodes 5 and 6 are busy. Lastly, we point out that the
outgoing links of node 4 are equal, so factors other than just the cost of the
links $(4, 2)$ and $(4, 3)$ must determine which link is used.

The first node of the network is the designated sink node so its queue
length will always be zero. One way of thinking of the sink node is that it has
a very large or infinite service capacity and does not send its packets to any
other node. The first costate is also identically zero.

Table 1. Parameters for the simulation network.

Node	1	2	3	4	5	6
μ_j	*	35	60	55	65	40

The parameters for the network are shown in Table 1. The load on the
network is modeled by a sum of two Gaussian functions,

$$r_j(t) = g_{j1} \exp\left(-\frac{t - g_{j2}}{g_{j3}^2}\right) + w_{j1} \exp\left(-\frac{t - w_{j2}}{w_{j3}^2}\right) \tag{15.54}$$

as presented in [2]. The parameters for the load are shown in Table 2 and
the loads are shown in Fig. 2 and Fig. 3. The parameters g_{j1} and w_{j1} will
determine the magnitude of two peaks that will occur at the times g_{j2} and w_{j2}

respectively. The parameters g_{j3} and w_{j3} control the duration of the peak. The receding horizon system was simulated for 24 hours in order to demonstrate network behavior under different scenarios.

Table 2. Load parameters for the receding horizon simulation network.

Node j	g_{j1}	g_{j2}	g_{j3}	w_{j1}	w_{j2}	w_{j3}
2	25	11	2	20	15	1.5
3	50	7	2	35	11	1.5
4	35	15	2	30	19	1.5
5	65	2	2	60	8	2.0
6	40	12	2	35	17	2.0

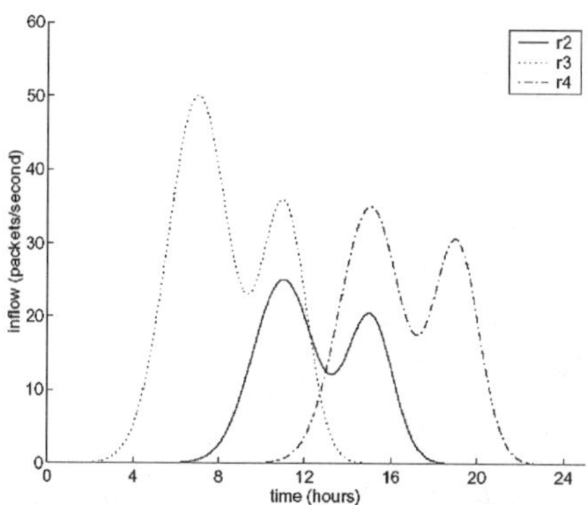

Fig. 15.2. The load for the network nodes 2, 3, and 4.

The resulting input and state trajectories are shown in Fig. 4 and Fig. 5. At 2 hours, node 5 is loaded to capacity. Not only do the queue lengths of nodes 5 and 4 increase as one would expect, but the queue at node 6 increases as well. The routing controller sends traffic along links (5,6) and (6,4) as well as along (5,4). This is because as the queue at node 4 increases, its costate increases and makes it less attractive. The net effect is that node 5 shares its congestion with neighboring nodes that are not busy.

At around 11 hours, nodes 2 and 3 are loaded and their queue lengths rapidly rise. Although they are loaded, they are not loaded at capacity. The load on nodes 2 and 3 peak at 25 and 35 respectively, while their capacities are 35 and 60. They cannot operate as close to capacity as the other nodes of the network because they constantly receive traffic from upstream.

From 15 to 19 hours, nodes 2, 4, and 6 are loaded. Again, the controller routes traffic from node 4 to node 2 to maintain its queue length. Any excess

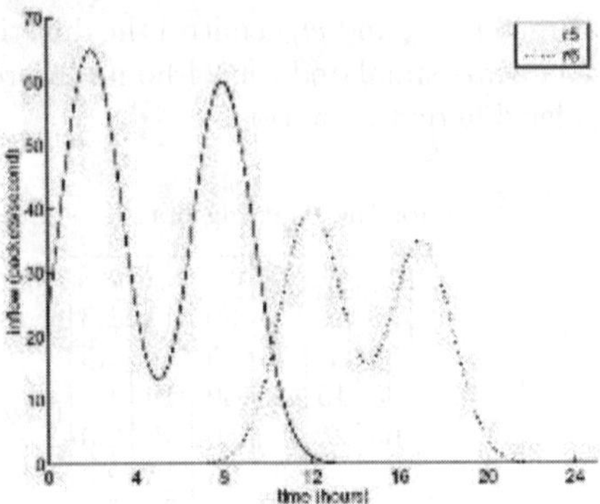

Fig. 15.3. The load for the network nodes 5 and 6.

flow is routed through node 3. The only time the queue length at node 2 rises significantly is when both node 2 and 3 are loaded at the same time.

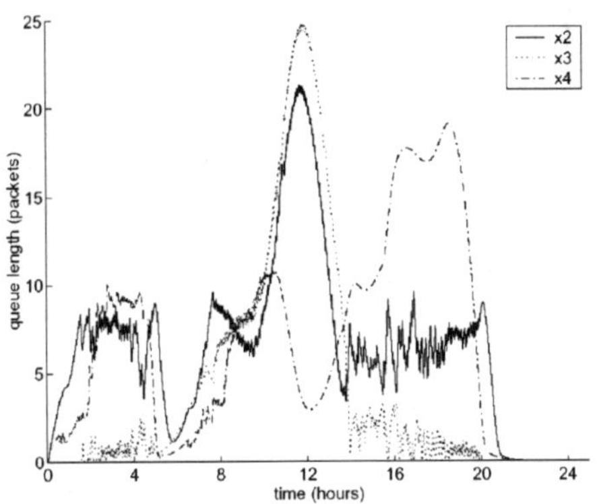

Fig. 15.4. The first three states of the receding horizon network.

This makes sense because the cheapest route, using only the costs of the links, from node 4 to the sink at node 1 is through node 2. However, once node 2 gets flooded, it becomes beneficial to route through node 3.

Conclusion

The general philosophy behind the modeling and control of the network is to leverage the cooperative structure of an AUV network against the challenges posed by bandwidth, power constraints, and limited storage. Towards that

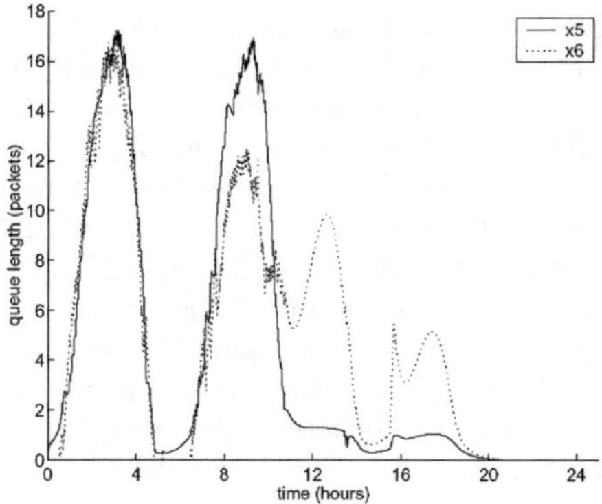

Fig. 15.5. The second two states of the receding horizon network.

end, this paper presented a model and unique piecewise continuous optimal controllers for directing information flow in AUV networks. Solutions were constructed for both the finite and infinite time horizon. A receding horizon approach was adopted on the infinite horizon and a stability result for it was presented. A chief advantage of receding horizon controllers is that the load need only be estimated into the near future.

Acknowledgement

This research has been supported in part by the National Science Foundation Grant Number IIS-0328378.

References

1. Hassan K. Khalil, *Nonlinear Systems*, Prentice Hall, Upper Saddle River, New Jersey, 1996.
2. Janusz Filipiak, *Modelling and Control of Dynamic Flows in Communication Networks*, Springer-Verlag, Berlin, 1988.
3. Alexander Schrijver, *Theory of Linear and Integer Programming*, John Wiley and Sons, Chichester, 1986.
4. A.F. Fillipov, *Differential Equations with Discontinuous Right Hand Side*, Kluwer, 1988.
5. H. Vincent Poor and Gregory W.Wornell, editors, *Wireless Com-munications: Signal Processing Perspectives*, chapter Underwater Acoustic Communications, pp. 330–379. Prentice Hall PTR, New Jersey, 1998.
6. Eitan Altman and Tamer Basar, "Multiuser Rate-Based Flow Control," *IEEE Trans-actions on Communications*, vol. 46, pp. 940–9, 1998.
7. Joseph T. Napoli, *Optimal Communication Strategies for Robotic Networks*, PhD thesis, Washington University in St. Louis, 2001.

8. Ravi Mazumdar, Lorne G. Mason, and Christos Douligeris, "Fairness in Network Optimal Flow Control: Optimality of Product Forms,D" *IEEE Transactions on Communications*, vol. 39, pp. 775–81, 1991.
9. Steven D. Gray, James C. Presig, and David Brady, "Multiuser Detection in a Horizontal Underwater Acoustic Channel Using Array Observations," *IEEE Transactions on Signal Processing*, vol. 45, pp. 148–60, 1997.
10. Milica Stojanovic, "Recent Advances in High-Speed Underwater Acoustic Communications," *IEEE Journal of Oceanic Engineering*, vol. 21, pp. 125–36, 1996.
11. James G. Bellingham, "New Oceanographic Uses of Autonomous Underwater Vehicles," *Marine Technology Society Journal*, vol. 31, pp. 34–47, 1997.
12. David Q. Mayne and Hannah Michalska, "Receding Horizon Control of Nonlinear Systems," *IEEE Transactions on Automatic Control*, vol. 35, pp. 814–24, 1990.
13. H. Michalska and D. Q. Mayne, "Robust Receding Horizon Control of Constrained Nonlinear Systems," *IEEE Transactions on Automatic Control*, vol. 11, pp. 1623–33, 1993.
14. H. Michalska and R. B. Vinter, "Nonlinear Stabilization Using Discontinuous Moving–horizon control", *IMA Journal of Mathematical Control and Information*, vol. 11, pp. 321–340, 1994.
15. Roy M. Turner, Elise H. Turner, and D. Richard Blidberg, "Organization and Reor-ganization of Autonomous Oceanographic Sampling Networks," *Proceedings of the Symposium on Autonomous Underwater Vehicle Technology, CA*, pp. 407–13, 1996.
16. James G. Bellingham and J. Scott Willcox, "Optimizing AUV Oceanographic Surveys," *Proceedings of the Symposium on Autonomous Underwater Vehicle Technology, CA*, pp. 391–8, 1996.
17. Leonard Kleinrock, *Queueing Systems* Vol II, Wiley, New York, 1975.

A Mathematical Model for the Progression of an Abnormality in the Hematopoietic System

Mandri N. Obeyesekere[1], Patrick P. Spicer[1], and Martin Korbling[2]

[1] Department of Biostatistics and Applied Mathematics, Unit 237
[2] Department of Blood and Marrow Transplantation and Laboratory Medicine,
The University of Texas M. D. Anderson Cancer Center, 1515 Holcombe Blvd.,
Houston, TX 77030.

Summary. Myelodysplastic Syndrome (MDS) is an abnormality of the hematopoietic stem cells. MDS affects the normal functions of the progenitor cells that lead to the development of the peripheral blood cells. The greatest impact that MDS has on a patient is usually anemia. This study tests the hypothesis that MDS is due to the birth of a mutated stem cell that has lost the ability to differentiate. We utilized a mathematical model to simulate the dynamics of a few of the hematopoietic components in order to predict the onset and progression of MDS. We modified this mathematical model as necessary in order to simulate the effect produced by a stem cell's loss of ability to differentiate. The output of this model reproduces, in good comparison, the length of time observed for the onset of neutropenia in an MDS patient. The model's simulations of a common treatment plan for neutropenia, an injection of a single dose of the granulocyte colony-stimulating factor (G-CSF) cytokine, predicted a short term relief of white blood cell count (WBC) deficiency and no effect on the length of time observed for the onset of neutropenia. This model, with the addition of more details of the hematopoietic dynamics, could prove to be extremely useful in future studies of MDS.

Introduction

A mathematical model is proposed to simulate the onset and progression of Myelo dysplastic Syndrome (MDS). MDS is an abnormality of the hematopoietic stem cells. This abnormality propagates down the progenitor lineages of these stem cells, specifically the myeloid lineages within the bone marrow. These changes affect the normal functions of the progenitor cells that lead to the development of the peripheral blood cells. A schematic representation of myeloid lineage is given by Figure 16.1. The last groups of cells are fully differentiated blood cells that cannot divide further. The first generations of the hematopoietic cells are predominantly located within the bone marrow.

In an MDS hematopoietic system, the cell counts of the whole marrow are not greatly affected. The largest impact is usually anemia (the decrease

M.N. Obeyesekere et al.: *A Mathematical Model for the Progression of an Abnormality in the Hematopoietic System*, LNCIS **321**, 257–267 (2005)
`www.springerlink.com`

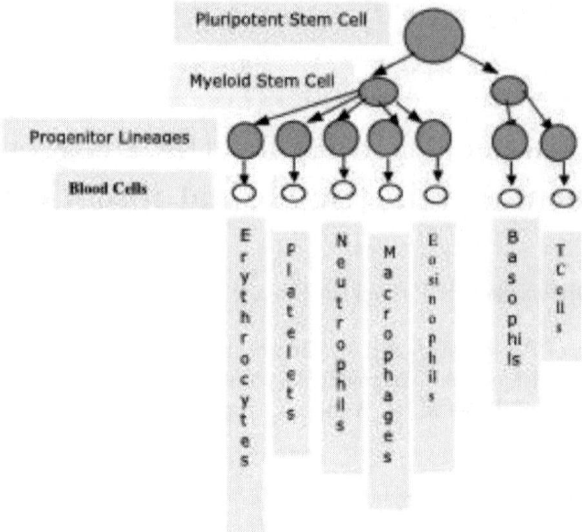

Fig. 16.1. Hematopoietic Cell lineage.

of the functional white blood count). MDS commonly transforms (progresses) into Acute Myeloid Leukemia (AML) with a trademark of increased immature white blood cells (blast cells). The transformation to AML is indicated by a large decrease in the number of normal functional blood stem cells. However, the overall count of the stem cells increase due to the increasing number of histologically abnormal stem cells. The timeline for this transformation varies but usually takes place in six to nine months.

In this manuscript, we hypothesize that MDS is due to the birth of a mutated stem cell that has lost the ability to differentiate. Hence, the abnormal cells will not leave the bone marrow compartment as the normal, fully differentiated, stem cells. Furthermore, this mutated stem cell will proliferate and give rise to a population of abnormal stem cells within the bone marrow. We propose a mathematical model to simulate the effect of this function of MDS and exhibit the possible outcomes under a single perturbation based on this model, which is clinically relevant.

Previous contributions of modeling the hematopoietic system have reproduced some other perturbations within the system, giving insight into the mechanisms of the hematopoietic system and how they can be modeled. Some of these include cyclic blood conditions, such as cyclic neutropenia (Wichmann and Loeffler [12]), (Haurie, Dale, Mack [4]), (Schmitz *et al.* [10]), (Santillan [8]), (Mackey [5]), and non-cyclic blood conditions such as certain leukemias (Fokas, Keller, Clarkson [1]), (Rubinow, Lebowitz 1976 [6, 7]). A recent study concentrated on modeling a specific treatment of neutropenia after HDC and PBPC transplantation (Scheding *et al.* [9]). In that study, a compartmental model was proposed, which is similar to the earlier work of Wichmann and

Loeffler (Wichmann, Loeffler 1985). A very detailed model, hence a complex model, was recently published by Shochat and colleagues (Shochat *et al.*, [11]). These models may consist of a number of different compartments within the bone marrow and blood peripheral. Recently, we have proposed a model that represents the dynamics of only a few components of this system with only two main compartments, the bone marrow and blood compartments. This simple mathematical model incorporates the dynamics of the stem cells of the marrow and the white blood cells (WBC) and the platelets of the peripheral of a normal hematopoietic system (Obeyesekere *et al.*, 2004). The proposed model in this manuscript is a modification of this previous mathematical model that was validated for the transient behavior of a normal stem-cell donor profile. These modifications allows the study of a possible change in the dynamics that leads to the malfunction of the hematopoietic system, specifically, MDS.

The following paragraphs in this manuscript will briefly describe this previous model, the necessary modifications that are implemented to simulate MDS, and the results obtained from solving the current mathematical model.

Model Development

The model presented by Obeyesekere *et al.*, reflected the general behavior of the hematopoietic system, especially the dynamics of the cluster of designation 34^+ ($CD34^+$) cells, a progenitor cell that is in the bone marrow as well as in the peripheral blood. This model utilized a minimal set of variables (five) of which most can be observed within the two main compartments, the bone marrow and the peripheral blood. The bone marrow compartment only consists of the bone marrow stem cells, $S(t)$ *cells/ml*. For the blood compartment, not all of the blood cells are modeled. The three components that are included are the peripheral $CD34^+$ (B), the WBC (W), and the platelets (P) *cells/ml*. Other than the specific cells, the effects of the cytokine, G-CSF (G-$\mu g/ml$), are modeled throughout the system. WBC and platelets are chosen because their counts are indicative of neutropenia and total hematopoietic recovery, respectively. It is assumed that both S, and B are hematopoietic stem cells, but their difference is in their location, which is necessary for this model. The interactions between the five components of this system are illustrated in Figure 16.2.

Through the flowchart in Figure 16.2, many of the basic mechanisms present within the hematopoietic system for maintaining homeostasis can be seen; e.g., the stem cell growth rate increases with a decrease in its own count (at rate f_1), white blood cell (f_2) or the platelet count (f_3), and with an increase in the G-CSF concentration (a_S) as depicted by the production loops along with the feedbacks. Stem cells are lost from the marrow due to differentiation (T_L), transfer of $CD34^+$ from the marrow to the blood (T_{SB}) and gained by the same cells transferring from the blood to the marrow (T_{BS}). In this model, it is assumed that the cells in the marrow are resistant to death.

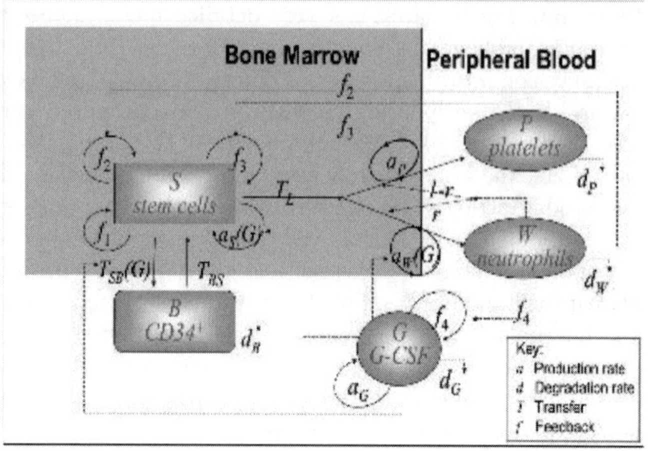

Fig. 16.2. Interactions between the bone marrow and the peripheral blood components.

The model from Obeyesekere *et al.*, proposed the 5 differential equations shown below:

$$\frac{dS}{dt} = \{a_S(G) + f_1 + f_2 + f_3\}S - T_L - T_{SB} + T_{BS} \tag{16.1}$$

$$\frac{dB}{dt} = T_{SB} - T_{BS} - d_B B \tag{16.2}$$

$$\frac{dW}{dt} = a_W(G)r(G)T_L - d_W W \tag{16.3}$$

$$\frac{dP}{dt} = a_P(1 - r(G))T_L - d_P P \tag{16.4}$$

$$\frac{dG}{dt} = a_G(1 + f_4) - d_G G \tag{16.5}$$

The parameter functions of the above system are given by

$$a_S(G) = a_S\left(1 + \frac{\alpha G}{c_1 + G}\right) \tag{16.6}$$

$$f_1 = \begin{cases} a_1\left(1 - \frac{S}{S_n}\right) & 0 \le S < S_n \\ 0 & S \ge S_n \end{cases} \tag{16.7}$$

$$f_2 = \begin{cases} a_2\left(1 - \frac{W}{W_n}\right) & 0 < W < W_n \\ 0 & W \ge W_n \end{cases} \tag{16.8}$$

$$f_3 = \begin{cases} a_3\left(1 - \frac{P}{P_n}\right) & 0 \le P < P_n \\ 0 & P \ge P_n \end{cases} \tag{16.9}$$

$$T_L = \begin{cases} 0 & 0 \le S < S_1 \\ \frac{a_T}{S_n}(S_2 - S_1)(S - S_1) & S_1 \le S < S_2 \\ \frac{a_T}{S_n}(S - S_1)^2 & S \ge S_2 \end{cases} \tag{16.10}$$

$$T_{SB} = \begin{cases} 0 & 0 \le S < S_2 \\ a_{TSB}(S - S_2)\frac{\kappa G}{c_3 + G} & S \ge S_2 \end{cases} \tag{16.11}$$

$$T_{BS} = \begin{cases} a_{TBS}(S_n - S)B + a_{TBS}S_n B & 0 \le S \le S_n \\ a_{TBS}S_n B & S \ge S_n \end{cases} \tag{16.12}$$

$$r(G) = \begin{cases} (a - g\frac{G}{G+G_n})(1 - \frac{W}{W_n}) + g\frac{G}{G+G_n} & 0 \le W \le W_n \\ g\frac{G}{G+G_n} & W \ge W_n \end{cases} \tag{16.13}$$

$$a_W(G) = a_W(1 + \frac{\beta G^{n1}}{c_2 + G^{n1}}) \tag{16.14}$$

$$f_4 = a_4(1 - \frac{W}{W_n}). \tag{16.15}$$

In the above equations, S_n, B_n, W_n, P_n, and G_n denote the normal steady state values for the functions S, B, W, P, and G in a "healthy" hematopoietic system, respectively. The parameter r is introduced to simulate the hierarchy of a white blood cell formation.

The model we present for the study addressed in this manuscript will contain Eq.s (16.1)-(16.15) with Eq. (16.10) modified as follows:

$$T_L = \begin{cases} 0 & S < S_1 \\ a_{TF}\frac{a_T}{S_n}(S_2 - S_1)(S - S_1) & S_1 \le S < S_2 \\ a_{TF}\frac{a_T}{S_n}(S - S_1)^2 & S \ge S_2 \end{cases} \tag{16.10a}$$

with

$$a_{TF} = e^{-.1t}.$$

As in the previous model, in the above equation, S is the number of stem cells in the bone marrow, S_1 and S_2 are threshold values governing the differentiation of the hematopoietic stem cells (simulating homeostasis within the stem cell compartment), and a_T is a transfer coefficient that mimics the process of cell transfer from the bone marrow to the peripheral due to differentiation. The modification of including a_{TF}, a time dependent variable, simulates a progressive change of the cell transfer rate, a_T, to simulate the dynamics of MDS. In this manuscript, we hypothesize that MDS is due to the birth of a mutated stem cell that has lost the ability to differentiate. Hence, the abnormal cells will not leave the bone marrow compartment as the normal, fully differentiated, stem cells. The effect of the growth of this abnormal population is simulated by including a_{TF} in equation (10a); i.e., this phenomenon is simulated by allowing the transfer rate of the total population to decrease exponentially.

262   M.N. Obeyesekere et al.

As we do not change the other terms in Equation (16.1), we have assumed that these abnormal cells will proliferate as the normal cells. Therefore, S in this model will reflect the sum of normal and abnormal stem cells in the marrow. However, W and P in this model represent only the normal functional units of their respective groups. Though it is known that in an AML patient there exists some abnormal WBC cells (blasts) in the peripheral, we do not model this specific event in this presentation. We have assumed that only the fully differentiated normal cells can enter the peripheral.

Results

The approximate solutions of the mathematical system was obtained by a numerical integration method (fourth order Runge Kutta routine from IMSL). The initial values of these variables at a healthy level were assumed to be the steady state values of the system. The parameter values for this model are given in Table 1.

The model output of the components of the hematopoietic system that simulates a normal human who begins MDS at day 0 and its progression till day 365 is presented by Figure 16.3. This figure depicts a system with MDS as it transforms into AML. The WBC profile of this figure reproduces the onset of neutropenia of a MDS patient; the WBC counts drops to neutropenic levels after day 204. Neutropenia is defined as a blood system with an absolute neutrophil count $\leq 0.1 \times 10^6/mL$, and as previously mentioned, onset of AML of a MDS patient happens between 6-9 months after the start of MDS. Notice the initially slight decrease in WBC and platelets then a sharp decrease by 6 months, signaling the onset of AML. These platelet levels also present qualitatively acceptable results.

Based on this initial model of an MDS development, we tested a commonly used treatment type for hematopoietic stabilization in clinical set-ups. G-CSF injection has been reported for regenerating and stabilizing a hematopoietic system after high dose chemotherapy treatment for cancer patients. This treatment is simulated with the use of our mathematical model for an MDS patient. A single dose of G-CSF injection, 10 μg/kg of body weight, on day 30 after MDS onset is simulated (no parameter changes) and the results from the model for the WBC profile and the G-CSF for MDS are shown by Figure 16.4a. We show the untreated WBC profile of the MDS case produced by the model in the same figure for comparison. In Figure 16.4b we present the simulation of a similar condition on a healthy person by injecting the same amount of G-CSF on day 1. It is interesting to observe that the WBC profile of the treated MDS case mimics that of the untreated after the 55th day. Furthermore, though the treatment of G-CSF elevated the WBC population for a short time it also went to levels lower than the untreated profile for a short. The profiles of all variables when treated with a single dose of G-CSF are given by Figure 16.4c. Together, these results depict a short term relief of

Table 1. Model parameter values

Parameters	Values	Source
a_S	$0.1\ d^{-1}$	
a_W	$40.0\ d^{-1}$	
a_P	2000.0	
a_G	$2.1\times10^{-5}\ \mu g\ ml^{-1}\ d^{-1}$	Eq. 20
a_T	$3.956\ d^{-1}$	Eq. 16
a_{TS0}	$800.0\ d^{-1}$	
a_{TS6}	$1.1\times10^{-7}\ ml\ Cells^{-1}\ d^{-1}$	
a_1	$0.1\ d^{-1}$	
a_2	$0.1\ d^{-1}$	
a_3	$0.1\ d^{-1}$	
a_4	0.1	
α	4.0	
β	5.0	
κ	0.003	
c_1	$0.04\ \mu g\ ml^{-1}$	
c_2	$2.3\times10^{-10}\ \mu g\ ml^{-1}$	
c_3	$1.3\times10^{-4}\ \mu g\ ml^{-1}$	
nI	2	
g	0.7	
d_S	$1.064\ d^{-1}$	
d_W	$0.7615\ d^{-1}$	
d_P	$0.4734\ d^{-1}$	
d_G	$1.4\ d^{-1}$	
S_1	$9.423\times10^5\ Cells\ ml^{-1}$	
S_2	$1.053\times10^6\ Cells\ ml^{-1}$	
S_a	$1.1086\times10^6\ Cells\ ml^{-1}$	
B_a	$11.6\times10^5\ Cells\ ml^{-1}$	
W_a	$6.3\times10^6\ Cells\ ml^{-1}$	
P_a	$271.0\times10^6\ Cells\ ml^{-1}$	
G_a	$1.94\times10^{-4}\ \mu g\ ml^{-1}$	

* d = day

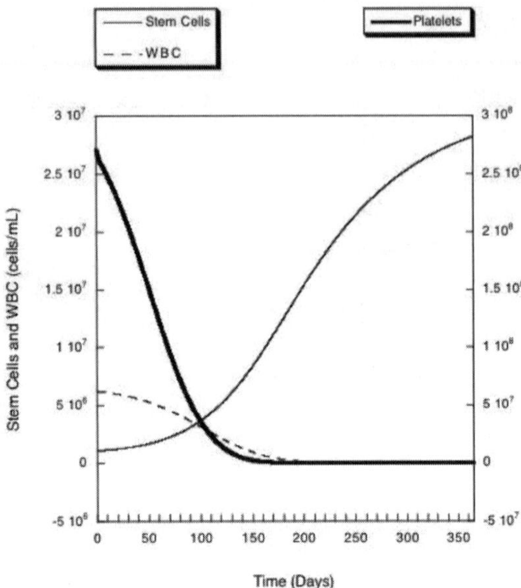

Fig. 16.3. Transformation of the hematopoietic system from normal to MDS to AML.

WBC deficiency by G-CSF treatment for an MDS patient where the benefit is questionable as it shows a subsequent short period of a severe WBC shortage. Another observation from this simulation is that the perturbation of the system did not result in postponing the onset of neutropenia; see WBC profiles in Figure 16.4a.

Discussion

We simulated the onset of MDS on day 0 of a normal (healthy) hematopoietic system. As seen in Figure 16.3, during the first two months the stem cell population has not increased much and also the WBC count is not very low. But, as time progresses, around the sixth month, the WBC count is extremely low, has reached the neutropenia levels. At this same time, the stem cells are crowded in the marrow as shown by an extreme increase by Figure 16.3. These results are a good reproduction of the changes in the hematopoietic system of a patient going through MDS to AML transition. We also presented results of our model to emphasize the differences of the profiles in a normal hematopoietic system compared to that of a MDS, subjected to external perturbations Figures 4a-c).

Based on this model we have shown that a usual treatment for neutropenia, injection of G-CSF, is of very limited benefit for MDS patients. After the treatment ends, the numbers return to normal and produce no lasting effects.

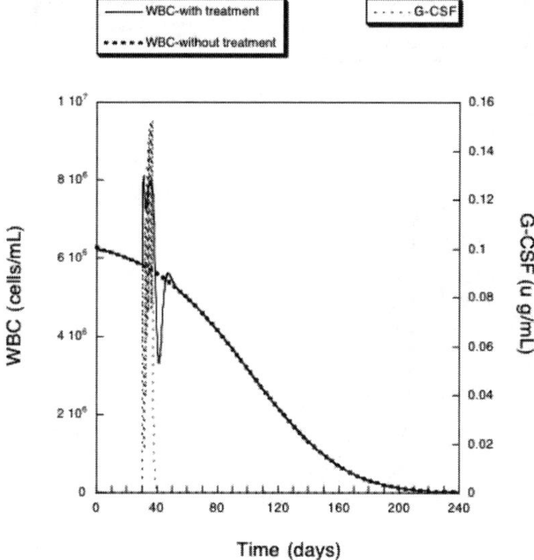

Fig. 16.4a. An injection of a single G-CSF dose on day 30 to an MDS case.

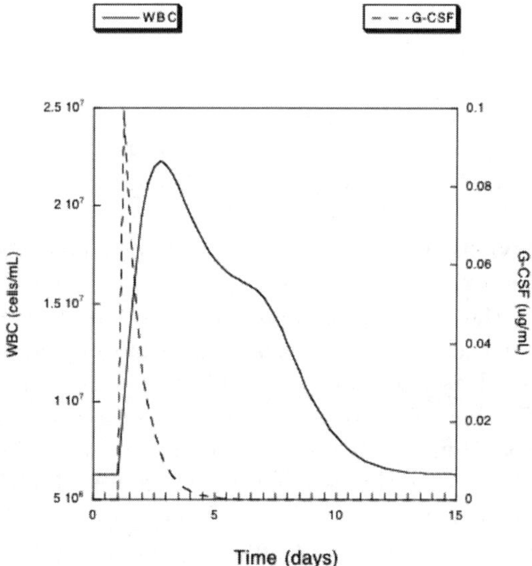

Fig. 16.4b. An injection of a single G-CSF dose on day1 to a normal hematopoietic system.

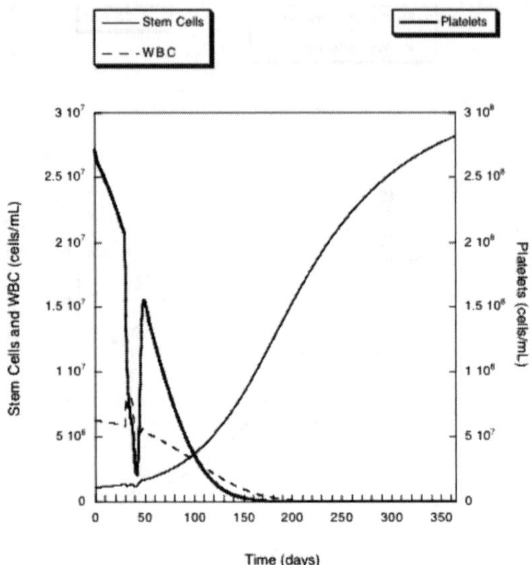

Fig. 16.4c. Profiles of a MDS case after a single injection of G-CSF on day 30.

Though results not shown, even repeated injections of G-CSF did not change the rate of neutropenia in the long term. This effect is due to the hypothesis that the abnormal stem cell has lost the capacity to differentiate and enter the peripheral blood as a normal blood cell. The effects of G-CSF as a treatment for neutropenia in MDS patients is still a puzzling question for clinicians (Freedman and Alter, [2]).

Using the previous model (Obeyesekere *et al.*, 2004), the application to MDS shows some accuracy but little can be known from the model due to the ambiguity of the stem cell compartment that is in a way a black box of undecipherable knowledge. For example, autologous stem cell transplantation is considered as a therapy for MDS to AML transforming patients (Gilliland and Gribben, [3]). To simulate the infusion of CD34+ progenitor cells, a comparable stem cell transplant treatment, using this current model will be impossible because the detailed information within the bone marrow compartment is not incorporated, though CD34+ is a variable in the model. This shows that further information may be obtained by altering the model to contain separate compartments for normal and abnormal stem cells. This will allow monitoring the abnormal cell dynamics by different variables. For example, as earlier stated, the number of overall WBC can increase during AML and in our model only the normal or non-cancerous cells are those that are included in the variable W, thus W should not increase during AML. With the proposed modification, it will allow to observe the undifferentiated WBCs that are in the peripheral. On the other hand, the stem cells S and B include the overall counts of their respective groups, meaning those counts essentially contain the normal and abnormal cells. This is true due to the way that the progression of MDS is programmed into the model.

In conclusion, our hypothesis of the underlying dynamics due to the onset of MDS does seem to be acceptable. However, the mathematical model needs to be expanded to obtain clinically relevant predictions, the ultimate goal of mathematical modeling for cancer research.

References

1. Fokas A S, Keller J B, Clarkson B D "Mathematical model of granulocytopoiesis and chronic myelogenous leukemia". *Cancer Res.* **51**, 2084.,1991
2. Freedman MH and Alter BP, "Risk of myelodysplastic syndrome and acute myloid leukemia in congenital neutropenias". *Semi. in Hematal.*, **39** (2), 128-133, 2002.
3. Gilliland DG and Gribben JG, "Evaluation of the risk of therapy-related MDS/AML after autologous stem cell transplantation". *Biol. of Blood and Marrow Transp.* **8** , 9-16, 2002.
4. Haurie C, Dale D C, Mackey M C, "Cyclical neutropenia and other periodic hematological diseases: a review of mechanisms and mathematical models". *Blood.* **92**, 2629, 1998.
5. Mackey M C, "Unified hypothesis for the origin of aplastic anemia and periodic haematopoiesis". *Blood.* **51**, 941, 1978.
6. Rubinow S I, Lebowitz J L , "A mathematical model of the acute myeloblastic leukemia state in man". *Biophys. J.* **16**, 897, 1976.
7. Rubinow S I, Lebowitz J L, "A mathematical model of the chemotherapeutic treatment of acute myeloblastic leukemia". *Biophys. J.* **16**, 1257,1976.
8. Santillan M, Mahaffy J M, Belair J, Mackey M C., "Regulation of platelet production: the normal response to perturbation and cyclical platelet disease". *J. Theor. Biol.* **206**, 585, 2000.
9. Scheding S, Franke H, Diehl V, Wichmann H E, Brugger W, Kanz L, Schmitz S, "How many myeloid post-progenitor cells have to be transplanted to completely abrogate neutropenia after peripheral blood progenitor cell transplantation? Results of a computer simulation". *Exp. Hematol.* **27**, 956, 1999.
10. Schmitz S, Franke H, Loeffler M, Wichmann H E, Diehl V., "Reduced variance of bone-marrow transit time of granulopoiesis: a possible pathomechanism of human cyclic neutropenia". *Cell Prolif.* **27**, 655, 1994.
11. Shochat E, Stemmer S M, Segel L, "Human haematopoiesis in steady state and following intense perturbations". *Bull. Math. Biol.* **64**, 861, 2002.
12. Wichmann H.E. and M. Loeffler, Structure of the Model". *Mathematical modeling of cell proliferation: Stem Cell Regulation" in Hemopoiesis*, **Vol. 1**, CRC Press, 1985.

Issues in the Identification of Physical Parameters in Non-Homogeneous Linear Elastic Solids

Stefano Perabò and Giorgio Picci[*]

Department of Information Engineering, University of Padova
`perabo,picci@dei.unipd.it`

Summary. Wave propagation in linear elastic solids of unknown space-varying physical characteristics is considered in this paper. We propose a two step procedure in which one first builds a linear parametric model of the system using suitable PDEs approximation and discretization tools, and then uses standard parametric identification techniques which compute the parameter estimates.

17.1 Introduction

We consider wave propagation in linear elastic solids of unknown space-varying physical characteristics. Recovering unknown parameters (say density, elasticity, etc) of a medium from various kinds of external measurements is a common problem in different disciplines such as structural mechanics and seismic exploration [1, 2].

This problem falls in the broad range of *inverse problems* for partial differential equations (PDEs) [3]. In particular, when second order hyperbolic PDEs are involved and wave propagation occurs, one speaks more appropriately of *inverse scattering problems*. These can be set mathematically as the problem of recovering unknown parameters of the PDE which are in general functions of the space variable, from the available sensor measurements and from knowledge of the input waveform of the source generating the waves.

Problems of this kind have been intensively studied for many decades. In practice they are traditionally tackled by means of asymptotic approximations or linearization of the original non-linear problem about some guessed initial values. The resulting algorithms can in general be viewed as regularized optimization schemes which attempt to fit the available noisy data to a model, by solving iteratively the differential or integral equations involving the unknown

---

[*]Paper dedicated to Clyde Martin in occasion of his 60-th birthday. This work has been supported by the MIUR project *Identification and Adaptive Control of industrial Systems* and in part by the RECSYS project of the European Community.

S. Perabò and G. Picci: *Issues in the Identification of Physical Parameters in Non-Homogeneous Linear Elastic Solids*, LNCIS **321**, 269–290 (2005)
`www.springerlink.com`

parameters. The estimates are refined iteratively using gradient-type methods [4, 5, 6].

The main drawback of these approaches is a high computational cost because the *forward problem* must be solved many times before reasonable estimates can be obtained.

In this paper we propose instead a two step procedure in which one first builds a linear parametric model of the system using suitable PDEs approximation and discretization tools, and then uses standard parametric identification techniques which compute the parameter estimates. In fact, in this paper we shall combine Kalman filtering and linear least-squares estimation into a procedure that turns out to be a variant of the well-known Expectation-Maximization (EM) algorithm [7, 8, 9]. Due to the reasonably fast convergence properties of this method we expect to be able to solve one dimensional problems essentially in real time.

17.2 Problem Statement

Wave propagation in solids is described by a second order partial differential equation (PDE) of the following kind [10, 11]:

$$\rho(x)\frac{\partial^2 q}{\partial t^2}(x,t) = \nabla \cdot \sigma(x,t) + f(x,t) \tag{17.1}$$

which relates the vector displacement $q(x,t)$ of a material point situated in position $x \in \mathbb{R}^3$, with the stress tensor $\sigma(x,t)$ and force field $f(x,t)$. Here ρ represents the density and, for simplicity, no dissipation is considered. Equation (17.1) is defined on some closed spatial domain $\mathcal{D} \subset \mathbb{R}^3$ and for a time interval $\mathcal{I} = [t_i, t_f] \subset \mathbb{R}$, together with initial conditions and appropriate boundary conditions on $\partial\mathcal{D}$, the choice of which we will discuss later.

When displacements and displacement gradients are sufficiently small, a linear approximation of the strain tensor $\varepsilon(x,t)$ is valid:

$$\varepsilon_{kl}(x,t) \approx \frac{1}{2}\left(\frac{\partial q_k}{\partial x_l}(x,t) + \frac{\partial q_l}{\partial x_k}(x,t)\right) \tag{17.2}$$

Moreover, assuming linear time-invariant, but possibly non-homogeneous, materials, the *generalized Hooke's Law* holds

$$\sigma_{ij}(x,t) = \sum_{k,l=1}^{3} c_{ijkl}(x)\varepsilon_{kl}(x,t) \tag{17.3}$$

where the fourth order tensor c_{ijkl} depends (by symmetry) on 21 parameters. By further restricting to isotropic materials, we get

$$c_{ijkl}(x) = \lambda(x)\delta_{ij}\delta_{kl} + \mu(x)(\delta_{ik}\delta_{jl} + \delta_{il}\delta_{jk}) \tag{17.4}$$

where λ and μ are known as the *Lamé moduli*. Introducing suitable 3×3 matrices $\mathbf{A}_{jl}$ and $\mathbf{A}_j$ containing the elastic parameters of the medium, equation (17.1) can be rewritten as

$$\rho(x)\frac{\partial^2 q}{\partial t^2}(x,t) - \sum_{j,l=1}^{3} \mathbf{A}_{jl}(x)\frac{\partial^2 q}{\partial x_j \partial x_l}(x,t) + \sum_{j=1}^{3} \mathbf{A}_j(x)\frac{\partial q}{\partial x_j}(x,t) = f(x,t) \quad (17.5)$$

which is a system of second order hyperbolic PDEs [12].

In the following we will restrict to a one dimensional medium; this is merely for ease of exposition and more general situations could be handled with the same procedures we are going to present. In this case we shall set $\lambda(x) = \lambda(x_3)$, $\mu(x) = \mu(x_3)$, and assume plane waves traveling along the x_3-direction, so that

$$\frac{\partial q}{\partial x_1}(x,t) = \frac{\partial q}{\partial x_2}(x,t) = 0 \quad \forall x \in \mathcal{D}, \ \forall t \in \mathcal{I} \qquad (17.6)$$

$$\frac{\partial f}{\partial x_1}(x,t) = \frac{\partial f}{\partial x_2}(x,t) = 0 \quad \forall x \in \mathcal{D}, \ \forall t \in \mathcal{I} \qquad (17.7)$$

From (17.5) the following decoupled equations are derived (we write z in place of x_3):

$$\left(\rho(z)\frac{\partial^2}{\partial t^2} - \frac{\partial}{\partial z}\mu(z)\frac{\partial}{\partial z}\right)\begin{bmatrix} q_1(z,t) \\ q_2(z,t) \end{bmatrix} = \begin{bmatrix} f_1(z,t) \\ f_2(z,t) \end{bmatrix} \qquad (17.8a)$$

$$\left(\rho(z)\frac{\partial^2}{\partial t^2} - \frac{\partial}{\partial z}[\lambda(z) + 2\mu(z)]\frac{\partial}{\partial z}\right) q_3(z,t) = f_3(z,t) \qquad (17.8b)$$

One can recognize in (17.8a) and (17.8b) the equations for *shear waves* and *compression waves* respectively, whose velocities of propagation are $\sqrt{\mu/\rho}$ and $\sqrt{(\lambda + 2\mu)/\rho}$.

Hereafter, due to the similarity of the two equations (17.8a) and (17.8b) we shall discuss only the to the second one. We define the function $\gamma(z) = \lambda(z) + 2\mu(z)$ and drop the subscript 3 everywhere.

The problem we want to solve is: *Given the system (17.8b), with the appropriate initial and boundary conditions and given some (to be specified later) measurements of the input $f(z,t)$ and of the state $q(z,t)$ in a given subset of the spatial domain $\mathcal{D} \times \mathcal{I}$, determine a "best" (in some sense) estimate of the function coefficients $\rho(z)$ and $\gamma(z)$.*

17.3 Mathematical Preliminaries

17.3.1 Non Reflecting Boundary Conditions

In many problems, the domain $\mathcal{D}$ is very large, often even unbounded, and the application of standard spatial discretization techniques would lead to models of a very large dimension. These models would be unpracticable from a computational point of view, especially in applications requiring real time or fast processing of data. One way to overcome this limitation is to impose artificial non-reflecting conditions at the boundary of a restricted size domain, by postulating that only outwards propagating waves are allowed to cross the boundary or, equivalently, that no sources or reflections occur outside.

Once a suitable such domain has been chosen, many techniques exist to define non-reflecting boundary conditions [13]. In the one-dimensional setting we are considering, outgoing waves can be simply selected by the following conditions:

$$\gamma(z)\frac{\partial q}{\partial z}(z,t) - \sqrt{\gamma(z)\rho(z)}\frac{\partial q}{\partial t}(z,t) = 0 \qquad z = z_{min}, \; \forall t \in \mathcal{I} \qquad (17.9a)$$

$$\gamma(z)\frac{\partial q}{\partial z}(z,t) + \sqrt{\gamma(z)\rho(z)}\frac{\partial q}{\partial t}(z,t) = 0 \qquad z = z_{min}, \; \forall t \in \mathcal{I} \qquad (17.9b)$$

17.3.2 Galerkin's Discretization Method

We shall introduce a finite-dimensional approximation of (17.8b) using Galerkin's method. Equations (17.8a) can be treated in a similar way and will not be considered explicitly in the following. A proper mathematical description of the approximation method should begin with a review of the concepts of weak solution and variational formulation of an initial boundary value problem for a PDE [14, 15], but here we shall limit to a quick explanation of the practical procedure. In practice we shall define a *least square approximation* in a suitable norm.

Let $\{\phi_j(z) : j = 1, \ldots, +\infty\}$ be a basis for a suitable space V_1 of real functions defined over $\mathcal{D}$, the space of *trial functions*. We consider the approximation

$$q(z,t) \approx q^N(z,t) = \sum_{j=1}^{N} q_j(t)\phi_j(z) \qquad (17.10)$$

Here $\{q_j(t)\}$ are a finite number of time-dependent scalar unknown parameters which describe the *state* of the system at time t. These quantities are calculated by determining the residual $r^N(z,t)$ which is produced when $q^N(z,t)$ is inserted in (17.8b) and then by imposing this residual to be *orthogonal* to another space V_2 of real functions defined over $\mathcal{D}$, the space of *test functions* (often $V_1 = V_2$). This condition is achieved approximately by chosing a basis of V_2, $\{\psi_i : i = 1, \ldots, +\infty\}$, and by imposing

$$\int_{\mathcal{D}} r^N(z,t)\psi_i(z)\, dz = 0 \qquad (17.11)$$

for all $i = 1, \ldots, N$. By computing the indicated $L^2(\mathcal{D})$ inner products, the residual thus belongs to the orthogonal complement of a finite dimensional approximation of V_2.

We assume that conditions guaranteeing the existence and uniqueness of the approximation are met and that $r^N(z,t) \to 0$ in some suitable norm when $N \to +\infty$, the rate of convergence depending on the particular choice of trial and test functions.

A theoretically sound and widely used Galerkin's method is the Finite Element Method (FEM) which is based on the partition of the domain $\mathcal{D}$ in a finite number of bounded disjoint sets, the *finite elements*. Over each of these elements the best polynomial approximation of a given order of the unknown solution is then calculated [16].

For the one-dimensional problem at hand we simply choose a partitioning of $\mathcal{D}$ into $N-1$ subintervals $K_i = [z_i, z_{i+1}]$, $i = 1, 2, \ldots, N-1$, such that $z_1 = z_{min}$, $z_N = z_{max}$ and $0 < z_{i+1} - z_i = h_i \le h$. The maximum element size h, and consequently the number of nodes N, controls the accuracy of the approximate solution and should be chosen according to well established rules [13]. Then a continuous approximation, which is linear over each element K_i, can be computed by solving a system of N second order ordinary differential equations in the unknown vector $\mathbf{q}(t) = [q^N(z_1,t)\ q^N(z_2,t)\ \ldots\ q^N(z_N,t)]^T$, which can be cast in the following matrix form:

$$\mathbf{M}(\rho)\ddot{\mathbf{q}}(t) + \mathbf{C}(\rho,\gamma)\dot{\mathbf{q}}(t) + \mathbf{K}(\gamma)\mathbf{q}(t) = \mathbf{f}(t) \tag{17.12}$$

The elements of the vector $\mathbf{q}$ are then the value of the approximating function at the set of *nodes* $\mathcal{Z} = \{z_i,\ i = 1, \ldots, N\}$. The elements of the $N \times N$ matrices $\mathbf{M}$, $\mathbf{C}$ and $\mathbf{K}$ (the mass, dissipation and stiffness matrices) depend on the unknown functions ρ and γ. If these functions were given, we could in principle compute these matrices through numerical quadrature techniques (in this case one then speaks more appropriately of generalized Galerkin methods). Note that the presence of matrix $\mathbf{C}$ (which derives from an integration by parts in equation (17.11) which describes the boundary conditions) introduces dissipation. This reflects the fact that energy is lost when waves cross the boundaries of the domain $\mathcal{D}$.

In our problem the density and elasticity parameters are unknown. For simplicity we shall approximate them by piecewise constant functions. This may be reasonable from physical considerations. Furthermore we shall assume that the functions are approximately constants exactly on the finite-element partition of $\mathcal{D}$ as:

$$\rho(z) = \sum_{i=1}^{N_e} \rho_i I(K_i) \qquad \gamma(z) = \sum_{i=1}^{N_e} \gamma_i I(K_i) \tag{17.13}$$

where $I(K_i)$ is the indicator function of the element K_i. Hence the values $\{\rho_i, \gamma_i\}$ become a set of $2N$ unknown parameters of the model (17.12). The FEM method outlined above leads to the following expressions:

$$\mathbf{M} = \begin{bmatrix} \rho_1\frac{h_1}{3} & \rho_1\frac{h_1}{6} & 0 & 0 & \cdots & 0 \\ \rho_1\frac{h_1}{6} & \rho_1\frac{h_1}{3}+\rho_2\frac{h_2}{3} & \rho_2\frac{h_2}{6} & 0 & \cdots & 0 \\ 0 & \rho_2\frac{h_2}{6} & \rho_2\frac{h_2}{3}+\rho_3\frac{h_3}{3} & \rho_3\frac{h_3}{6} & \cdots & 0 \\ 0 & 0 & \rho_3\frac{h_3}{6} & \rho_3\frac{h_3}{3}+\rho_4\frac{h_4}{3} & \cdots & 0 \\ \vdots & \vdots & \vdots & \vdots & \ddots & \vdots \\ 0 & 0 & 0 & 0 & \cdots & \rho_{N_e}\frac{h_{N_e}}{3} \end{bmatrix} \quad (17.14)$$

$$\mathbf{C} = \begin{bmatrix} \sqrt{\gamma_1\rho_1} & 0 & \cdots & 0 & 0 \\ 0 & 0 & \cdots & 0 & 0 \\ \vdots & \vdots & \ddots & \vdots & \vdots \\ 0 & 0 & \cdots & 0 & 0 \\ 0 & 0 & \cdots & 0 & \sqrt{\gamma_{N_e}\rho_{N_e}} \end{bmatrix} \quad (17.15)$$

$$\mathbf{K} = \begin{bmatrix} \gamma_1\frac{1}{h_1} & -\gamma_1\frac{1}{h_1} & 0 & 0 & \cdots & 0 \\ -\gamma_1\frac{1}{h_1} & \gamma_1\frac{1}{h_1}+\gamma_2\frac{1}{h_2} & -\gamma_2\frac{1}{h_2} & 0 & \cdots & 0 \\ 0 & -\gamma_2\frac{1}{h_2} & \gamma_2\frac{1}{h_2}+\gamma_3\frac{1}{h_3} & -\gamma_3\frac{1}{h_3} & \cdots & 0 \\ 0 & 0 & -\gamma_3\frac{1}{h_3} & \gamma_3\frac{1}{h_3}+\gamma_4\frac{1}{h_4} & \cdots & 0 \\ \vdots & \vdots & \vdots & \vdots & \ddots & \vdots \\ 0 & 0 & 0 & 0 & \cdots & \gamma_{N_e}\frac{1}{h_{N_e}} \end{bmatrix} \quad (17.16)$$

Note again that the matrix $\mathbf{C}$ introduces dissipation in the model even though we assumed a non-dissipative linear elastic medium.

Later on we shall discuss different parametrizations of the model other than via the parameters $\{\rho_i, \gamma_i : i = 1, \ldots, N_e\}$. The term $\mathbf{f}$ will be discussed in the next subsection.

17.3.3 Sources and Sensors

We assume that m input sources (forces) $u_l(t)$, $l = 1, 2, \ldots, m$, are concentrated at some known points in $\mathcal{D}$ belonging to the set $\mathcal{S} = \{s_1, s_2, \ldots, s_m\}$:

$$f(z,t) = \sum_{l=1}^{m} u_l(t)\delta(z - s_l) \quad (17.17)$$

Then from (17.11), $f_i(t) = \sum_{l=1}^{m} u_l(t)\psi_i(s_l)$ turns out to be the generic element of $\mathbf{f}(t)$. Since it is not restrictive to choose the set $\mathcal{Z}$ in such a way $\mathcal{S} \subset \mathcal{Z}$, i.e. the l-th force is applied at the nodal point $z_{i_l} = s_l$, the application of the FEM described above leads to the expression $\mathbf{f}(t) = \mathbf{G}u(t)$ where we have defined $\mathbf{u}(t) = [\, u_1(t)\ u_2(t)\ \ldots\ u_m(t)\,]^T$ and G is a $N \times m$ matrix whose elements are zero except those at column l and row i_l, $l = 1, \ldots, m$, which take value one.

In a similar fashion, assume p point-like sensors, $i = 1, 2, \ldots, p$, placed at certain known positions specified by the set $\mathcal{R} = \{r_1, r_2, \ldots, r_p\} \subset \mathcal{Z}$, measure either material point displacement, velocity or acceleration (for the moment continuously in time). Define the vector $\mathbf{y}(t) = [\, y_1(t)\ y_2(t)\ \ldots\ y_p(t)\,]^T$, where $y_i(t)$ is the output of the i-th sensor. By using the finite-dimensional approximation (17.10), $y_i(t) = q(r_i, t) \approx q^N(r_i, t)$ holds for displacement sensors and analogous relations hold for velocity and acceleration sensors. Then the FEM procedure explained above gives

$$\mathbf{y}(t) \approx \mathbf{H} \begin{bmatrix} \mathbf{q}(t) \\ \dot{\mathbf{q}}(t) \\ \ddot{\mathbf{q}}(t) \end{bmatrix} \tag{17.18}$$

where $\mathbf{H}$ is, analogously to $\mathbf{G}$, a $p \times 3N$ null matrix except for ones at predefined positions.

17.3.4 Time Discretization

Since in real world applications both input and output data consist of sampled version of continuous time signals $\mathbf{u}(t)$ and $\mathbf{y}(t)$, we must also consider a discretization of the time variable. Let T denote the chosen sampling period. A widely used discretization scheme is the Newmark method [17] which computes iteratively the discrete time approximations of $\mathbf{q}(kT)$, $\dot{\mathbf{q}}(kT)$ and $\ddot{\mathbf{q}}(kT)$, denoted by $\mathbf{d}(k)$, $\mathbf{v}(k)$ and $\mathbf{a}(k)$, from the following relations:.

$$\mathbf{d}(k + 1) = \mathbf{d}(k) + T\mathbf{v}(k) + T^2\left[(1/2 - \beta)\mathbf{a}(k) + \beta\mathbf{a}(k + 1)\right] \tag{17.19a}$$

$$\mathbf{v}(k + 1) = \mathbf{v}(k) + T\left[(1 - \gamma)\mathbf{a}(k) + \gamma\mathbf{a}(k + 1)\right] \tag{17.19b}$$

where the vector $\mathbf{a}(\cdot)$ is asked to solve the force equilibrium equation

$$\mathbf{Ma}(k) + \mathbf{Cv}(k) + \mathbf{Kd}(k) = \mathbf{Gu}(k) \tag{17.19c}$$

for each $k = 0, 1, \ldots, N_c - 1$, being N_c the number of time samples ($\mathbf{u}(k)$ is a shorthand for $\mathbf{u}(kT)$).

The parameters β and γ are chosen in order to guarantee the stability of the iteration and also control the accuracy of the approximation but will not be matter of discussion in the following.

17.4 Parametric Identification

It is clear that in order to build a finite dimensional discrete-time model from a physical description of the medium in term of partial differential equations, several approximations need to be introduced. When the objective is to simulate the behaviour of the physical system, i.e. to solve a forward problem, the effect of such approximations is usually estimated, as in Numerical Analysis, by deriving deterministic bounds on the errors. To this end specific functional analytic tools are usually employed and conditions guaranteeing asymptotic convergence in some suitable norm provided. This error analysis provides practical rules for choosing, for example, the maximum element size h and the sampling period T as a function of medium parameters and input signal properties.

Here, instead, we are dealing with an inverse (identification) problem and we shall instead have to take a statistical approach. This will be explained in the following subsections.

17.4.1 A Linear Stochastic Parametric Model

We shall first introduce a sequence $\{\mathbf{n}(k) : k = 0, \cdots, N_c - 1\}$ of random vectors describing the approximation errors incurred in using the discretized model (17.19c)

$$\mathbf{Ma}(k) + \mathbf{Cv}(k) + \mathbf{Kd}(k) = \mathbf{Gu}(k) + \mathbf{n}(k). \tag{17.20}$$

We shall model it as a sequence of uncorrelated zero mean vectors (white noise) with covariance $\mathbf{Q}$, i.e.:

$$\mathbb{E}[\mathbf{n}(k)] = 0 \qquad \mathbb{E}[\mathbf{n}(k)\mathbf{n}(j)^T] = \delta(k - j)\mathbf{Q} \tag{17.21}$$

The choice of a suitable covariance $\mathbf{Q}$ can be made on the basis of numerical estimates of the discretization error. We shall however not insist on this point.

Similarly, we shall introduce two white sequences of random zero mean vectors, $\{\mathbf{n}_0(k), \mathbf{n}_1(k) : k = 0, \ldots, N_c - 1\}$, describing the discretization errors (17.19a) and (17.19b):

$$\mathbf{d}(k+1) = \mathbf{d}(k) + T\mathbf{v}(k) + T^2\left[(1/2 - \beta)\mathbf{a}(k) + \beta\mathbf{a}(k+1)\right] + \mathbf{n}_0(k) \tag{17.22a}$$

$$\mathbf{v}(k+1) = \mathbf{v}(k) + T\left[(1 - \gamma)\mathbf{a}(k) + \gamma\mathbf{a}(k+1)\right] + \mathbf{n}_1(k) \tag{17.22b}$$

These two white sequences may in general be mutually correlated.

$$\mathbb{E}[\mathbf{n}_0(k)] = \mathbb{E}[\mathbf{n}_1(k)] = 0 \qquad \mathbb{E}\begin{bmatrix}\mathbf{n}_0(k)\\ \mathbf{n}_1(k)\end{bmatrix}\begin{bmatrix}\mathbf{n}_0(j)\\ \mathbf{n}_1(j)\end{bmatrix}^T = \delta(k - j)\mathbf{N} \tag{17.23}$$

where the covariance matrix $\mathbf{N}$ should be chosen according to a priori estimates of the discretization errors.

Introducing the state vector $\mathbf{x}(k)^T = [\mathbf{d}(k)^T \ \mathbf{v}(k)^T \ \mathbf{a}(k)^T]$, of dimension $n = 3N$, and the *model noise* $\boldsymbol{\xi}(k)^T = [\mathbf{n}_0(k)^T \ \mathbf{n}_1(k)^T \ \mathbf{n}(k)^T]$, the above equations can be recast in a (generalized) state space form:

$$\mathbf{Ex}(k+1) = \mathbf{Ax}(k) + \mathbf{Bu}(k) + \boldsymbol{\xi}(k) \tag{17.24}$$

where

$$\mathbf{E} = \begin{bmatrix} \mathbf{I} & \mathbf{O} & -\beta T^2\mathbf{I} \\ \mathbf{O} & \mathbf{I} & -\gamma T\mathbf{I} \\ \mathbf{K} & \mathbf{C} & \mathbf{M} \end{bmatrix} \qquad \mathbf{A} = \begin{bmatrix} \mathbf{I} & T\mathbf{I} & (1/2 - \beta)T^2\mathbf{I} \\ \mathbf{O} & \mathbf{I} & (1 - \gamma)T\mathbf{I} \\ \mathbf{O} & \mathbf{O} & \mathbf{O} \end{bmatrix} \qquad \mathbf{B} = \begin{bmatrix} \mathbf{O} \\ \mathbf{O} \\ \mathbf{G} \end{bmatrix} \tag{17.25}$$

The state noise covariance matrix is then block diagonal equal to $\mathrm{diag}\{\mathbf{N}, \mathbf{Q}\}$. The measurement equation (17.18) can be written as

$$\mathbf{y}(k) = \mathbf{H}\mathbf{x}(k) + \mathbf{w}(k) \tag{17.26}$$

where $\mathbf{w}(k)$ is the *measurement noise*, assumed a white sequence with known covariance $\mathbf{R}$, uncorrelated from $\boldsymbol{\xi}(k)$.

The model is parameterized by the parameter vector $\boldsymbol{\theta} := \{\rho_i,\ \gamma_i\} \in \mathbb{R}^{2N_e}$ describing the physical properties of each element of the FEM model. The dependence of system matrices on the parameters will be denoted, when needed, by $\mathbf{M}(\boldsymbol{\theta})$, $\mathbf{C}(\boldsymbol{\theta})$ and $\mathbf{K}(\boldsymbol{\theta})$. It is important to note that the unknown parameters enter only in the last block of N rows in model (17.24), compare (17.20), which is the true dynamical part of the system. We shall write it compactly as

$$\mathbf{L}(\boldsymbol{\theta})\mathbf{x}(k+1) = \mathbf{G}\mathbf{u}(k) + \mathbf{n}(k) \qquad \mathbf{L}(\boldsymbol{\theta}) = \begin{bmatrix}\mathbf{K}(\boldsymbol{\theta})\ \mathbf{C}(\boldsymbol{\theta})\ \mathbf{M}(\boldsymbol{\theta})\end{bmatrix} \tag{17.27}$$

Note also that $\mathbf{L}(\boldsymbol{\theta})$ depends linearly on the parameters.

17.4.2 Identifiability and Model Reduction

To deal with a well-posed identification problem we first need to check identifiability of the model derived in the previous section. Since we are actually interested in identifying only the parameters $\boldsymbol{\theta}$ in the deterministic input-output map, we shall not discuss identifiability of the unknown parameters of the stochastic noise model.

It is well-known that for a linear n-dimensional system with p outputs and m inputs a generically identifiable parametrization must be of dimension not greater than $\nu = n(p+m)$ [18]. Since in our case we have $2N_e$ free parameters describing our family of models and $n = 3N$, it is obvious that our model will in theory be identifiable even in the case of only one input and output ($m = p = 1$).

Unfortunately however, in order to guarantee a reasonable approximation, the number of finite elements must be taken relatively high, and could easily be of the order of a hundred. This automatically leads to a very high dimensional parametrization which can easily make the model practically useless and the parameter estimation extremely ill-conditioned. There is therefore an issue of *reducing the model order* and of suitably *reparametrizing the reduced model* in such a way as to preserve identifiability and to make the estimation problem well-posed.

Model reduction of linear continuous system can be accomplished by first bringing the system into a balanced state space representation as proposed in [19] and then retaining only the "most relevant" singular values of the system [20]. More refined techniques as described in [21] can be applied to do optimal Hankel norm approximation.

This procedure can in theory be applied also to the second order linear continuous system (17.12) by defining an augmented state $\mathbf{z}(t) = [\mathbf{q}(t)^T \ \dot{\mathbf{q}}(t)^T]^T$ and rewriting it as a $2N$ dimensional first order system. In this way however the resulting reduced system will no longer be a second order system and the reduced states will be mixtures of position and velocity components of the original state so that the physical interpretation will be lost. For this reasons specific model reduction techniques have been proposed for second order systems (see [22, 23]) which lead to a reduced second-order continuous time model with, say, r-dimensional state $\mathbf{q}_r$, of the form

$$\mathbf{M}_r(\boldsymbol{\theta})\ddot{\mathbf{q}}_r(t) + \mathbf{C}_r(\boldsymbol{\theta})\dot{\mathbf{q}}_r(t) + \mathbf{K}_r(\boldsymbol{\theta})\mathbf{q}_r(t) = \mathbf{G}_r(\boldsymbol{\theta})\mathbf{u}(t) \tag{17.28a}$$

$$\mathbf{y}(t) = \mathbf{H}_r(\boldsymbol{\theta}) \begin{bmatrix} \mathbf{q}_r(t) \\ \dot{\mathbf{q}}_r(t) \\ \ddot{\mathbf{q}}_r(t) \end{bmatrix} \tag{17.28b}$$

which approximates the original transfer function. Unfortunately the dependence of the reduced system matrices on the original parameters $\boldsymbol{\theta}$ is hard to sort out in an analytical form and an explicit reparameterization of the model (in terms of fewer new parameters) is in general not possible. For this reason standard model reduction techniques don't seem to be directly useful from the point of view of the design of parametric identification procedures. Classical order reduction methods based on modal decomposition for multi-degree-of freedom systems as described, for example, in [24] (see also [25] for more recent developments in this field) suffer from the same drawback. So far, the only practical way to go seems to be eigenmode truncation of the state equations without taking into account the input-output coupling. A generalized eigenvalue problem is first solved and the state vector $\mathbf{q}(t)$ is transformed into the new basis made of the generalized eigenvectors (modes), resulting in the decomposition of the original system into N autonomous second order subsystems. Reduction can then be accomplished by neglecting higher order modes, which are generally the state components which have little influence on the output (or on the overall transfer function), at least in the frequency band of interest. In this way the identification step can be implemented using a different parametrizations $\boldsymbol{\theta} \rightarrow \mathbf{p}$ with $\mathbf{p}$ of a smaller dimension.

17.4.3 Parameter Estimation

In the ideal case where we have available measurements of both the state and of input-output variables; say we measure realizations of the random processes $\mathbf{x}(k)$, $\mathbf{y}(k)$ and $\mathbf{u}(k)$, denoted

$$X = \{x(k) = [d(k)^T \ v(k)^T \ a(k)^T]^T, k = 0, \ldots, N_c\} \tag{17.29a}$$
$$Y = \{y(k), k = 0, \ldots, N_c - 1\} \tag{17.29b}$$
$$U = \{u(k), k = 0, \ldots, N_c - 1\} \tag{17.29c}$$

and the joint statistical description of the random vectors $\boldsymbol{\xi}(k)$ and $\mathbf{w}(k)$ is known, say assuming the joint probability density $p_{\boldsymbol{\xi},\mathbf{w}}$ is Gaussian, the log-likelihood function with respect to the parameters vector $\boldsymbol{\theta}$ can easily be written as:

$$l(\boldsymbol{\theta}|X,Y,U) = \sum_{k=0}^{N_c-1} \log p_{\boldsymbol{\xi},\mathbf{w}}\Big(\mathbf{E}(\boldsymbol{\theta})x(k+1)-\mathbf{A}x(k)-\mathbf{B}u(k+1); y(k)-\mathbf{H}x(k)\Big)$$

(17.30)

(compare with (17.24) and (17.26)). The maximum likelihood estimate of the parameter $\boldsymbol{\theta}$ would then be computed as

$$\hat{\boldsymbol{\theta}} = \max_{\boldsymbol{\theta}\in\Theta} \; l(\boldsymbol{\theta} \mid X,Y,U)$$

(17.31)

In the present case however the state of the system, is not available. The standard way to deal with incomplete data is to rewrite the system equations (17.24) and (17.26) in innovations form as

$$\hat{\mathbf{x}}(k+1) = \bar{\mathbf{A}}(\boldsymbol{\theta})\hat{\mathbf{x}}(k) + \bar{\mathbf{B}}(\boldsymbol{\theta})\mathbf{u}(k+1) + \bar{\mathbf{K}}(k,\boldsymbol{\theta})\mathbf{e}(k) \qquad (17.32\text{a})$$

$$\mathbf{e}(k) = \mathbf{y}(k) - \mathbf{H}\hat{\mathbf{x}}(k) \qquad (17.32\text{b})$$

In the above expression $\bar{\mathbf{A}}(\boldsymbol{\theta}) = \mathbf{E}(\boldsymbol{\theta})^{-1}\mathbf{A}$, $\bar{\mathbf{B}}(\boldsymbol{\theta}) = \mathbf{E}(\boldsymbol{\theta})^{-1}\mathbf{B}$, and the Kalman gain $\bar{\mathbf{K}}(k,\boldsymbol{\theta})$ and innovation covariance $\bar{\boldsymbol{\Lambda}}(k,\boldsymbol{\theta})$ sequences are obtained by solving a (transient) Riccati equation. It is well-known, see e.g. [26], that the exact likelihood function can then be written in terms of the probability density of the innovation process $\mathbf{e}$ as

$$l(\boldsymbol{\theta} \mid Y,U) = \sum_{k=0}^{N_c-1} \log p_{\mathbf{e}}\Big(y(k) - \mathbf{H}\hat{x}(k;\boldsymbol{\theta})\Big)$$

(17.33)

where the Kalman filter estimate of the state $\hat{x}(k;\boldsymbol{\theta})$, based on the data $\{u(t),y(t) \mid 0 \leq t \leq k\}$ is computed recursively from (17.32a). One may instead use a steady state Kalman filter estimate $\hat{x}_\infty(k;\boldsymbol{\theta})$ based on the infinite past data $\{u(t),y(t) \mid -\infty \leq t \leq k\}$ (which are actually not available) since the effect of the transient can usually be neglected if enough data points are available.

Even with this approximation however, maximizing the exact likelihood leads to a very complicated iterative algorithm for parameter estimation. We shall suggest a simpler route which is essentially an Expectation-Maximization (EM) procedure. The EM algorithm [7], widely used in statistical problems with incomplete data, is an iterative technique which refines the parameter estimate starting from an initial guess $\hat{\boldsymbol{\theta}}^{(0)}$. Under mild conditions the EM algorithm guarantees convergence to a local maximum of l. The original procedure consists in an iteration between two steps:

Expectation step: given a parameter estimate $\hat{\boldsymbol{\theta}}^{(i)}$, we can compute the conditional expectation of the unmeasurable state variable given all the available measurement data Y, U by, say, a Kalman smoother [27, 28]. Denote by $\hat{\mathbf{x}}(k, \hat{\boldsymbol{\theta}}^{(i)})$ these estimates and collect them in the vector

$$\hat{X}^{(i)} := \{\hat{\mathbf{x}}(k, \hat{\boldsymbol{\theta}}^{(i)}), k = 0, 1, \ldots, Nc - 1\} \tag{17.34}$$

Approximating $\mathbf{x}(k)$ in (17.30) by $\hat{\mathbf{x}}(k, \hat{\boldsymbol{\theta}}^{(i)})$ (i.e. the unknown X by $\hat{X}^{(i)}$) one obtains

$$q^{(i)}(\boldsymbol{\theta}) = l(\boldsymbol{\theta} \mid \hat{X}^{(i)}, Y, U) \tag{17.35}$$

which is computable with the available data. This may be interpreted as a rough estimate of the conditional expectation $\mathbb{E}_{\hat{\boldsymbol{\theta}}^{(i)}}\left[l(\boldsymbol{\theta} \mid \mathbf{X}, \mathbf{Y}, \mathbf{U}) \mid Y, U\right]$ assuming that $\hat{\boldsymbol{\theta}}^{(i)}$ is the "true" model describing the data.

Maximization step: maximize the "conditional likelihood" $q^{(i)}(\boldsymbol{\theta})$ with respect to $\boldsymbol{\theta}$, getting a new estimate

$$\hat{\boldsymbol{\theta}}^{(i+1)} = \mathrm{Argmax}_{\boldsymbol{\theta} \in \Theta}\, q^{(i)}(\boldsymbol{\theta}).$$

In practice, since the temporal dynamics of our system is very fast, we shall substitute the smoother with an ordinary Kalman filter. Hence at stage i, the estimation step in our algorithm consists of running the recursion,

$$\hat{\mathbf{x}}(k + 1, \hat{\boldsymbol{\theta}}^{(i)}) = \bar{\mathbf{A}}(\hat{\boldsymbol{\theta}}^{(i)})\hat{\mathbf{x}}(k, \hat{\boldsymbol{\theta}}^{(i)}) + \bar{\mathbf{B}}(\hat{\boldsymbol{\theta}}^{(i)})\mathbf{u}(k) + \bar{\mathbf{K}}(k, \hat{\boldsymbol{\theta}}^{(i)})\mathbf{e}(k) \tag{17.36a}$$

$$\mathbf{e}(k) = \mathbf{y}(k) - \mathbf{H}\hat{\mathbf{x}}(k, \hat{\boldsymbol{\theta}}^{(i)}) \tag{17.36b}$$

where $\bar{\mathbf{A}}(\hat{\boldsymbol{\theta}}^{(i)}) = \mathbf{E}(\hat{\boldsymbol{\theta}}^{(i)})^{-1}\mathbf{A}$, $\bar{\mathbf{B}}(\hat{\boldsymbol{\theta}}^{(i)}) = \mathbf{E}(\hat{\boldsymbol{\theta}}^{(i)})^{-1}\mathbf{B}$ are the model parameters estimates at stage i and

$$\bar{\mathbf{K}}(k, \hat{\boldsymbol{\theta}}^{(i)}) = \bar{\mathbf{A}}(\hat{\boldsymbol{\theta}}^{(i)})\bar{\mathbf{P}}(k, \hat{\boldsymbol{\theta}}^{(i)})\mathbf{H}^T \bar{\Lambda}(k, \hat{\boldsymbol{\theta}}^{(i)})^{-1} \tag{17.37a}$$

$$\bar{\Lambda}(k, \hat{\boldsymbol{\theta}}^{(i)}) = \mathbf{H}\bar{\mathbf{P}}(k, \hat{\boldsymbol{\theta}}^{(i)})\mathbf{H}^T + \mathbf{R} \tag{17.37b}$$

and $\bar{\mathbf{P}}(k, \hat{\boldsymbol{\theta}}^{(i)})$ solves the matrix Riccati equation

$$\bar{\mathbf{P}}(k + 1, \hat{\boldsymbol{\theta}}^{(i)}) = \bar{\mathbf{A}}(k, \hat{\boldsymbol{\theta}}^{(i)})\bar{\mathbf{P}}((k, \hat{\boldsymbol{\theta}}^{(i)}))\bar{\mathbf{A}}(k, \hat{\boldsymbol{\theta}}^{(i)})^T - \tag{17.38}$$

$$\bar{\mathbf{K}}(k, \hat{\boldsymbol{\theta}}^{(i)})\bar{\Lambda}(k, \hat{\boldsymbol{\theta}}^{(i)})\bar{\mathbf{K}}(k, \hat{\boldsymbol{\theta}}^{(i)})^T + \mathbf{E}(k, \hat{\boldsymbol{\theta}}^{(i)})^{-1}\mathbf{Q}\mathbf{E}(k, \hat{\boldsymbol{\theta}}^{(i)})^{-T} \tag{17.39}$$

We shall now turn to the optimization step. As noted earlier, the only part of the model which depends on the parameters is (17.27). Assuming multivariate Gaussian distributions of the noise sequences and recalling that the white noise $\mathbf{n}(k)$ is independent of all other noise terms in the model (17.24), (17.26), it is clear that the only parameter dependent part of the likelihood function which we need to consider is that induced by the distribution of $\mathbf{n}(k)$. Or objective will then be to minimize

$$\ell(\boldsymbol{\theta}) = N \log \det \mathbf{Q} + \frac{1}{2} \sum_{k=1}^{N_c} \left[\hat{\mathbf{e}}^i(k)\right]^\top \mathbf{Q}^{-1} \hat{\mathbf{e}}^i(k) \qquad (17.40)$$

where

$$\hat{\mathbf{e}}^i(k) := \mathbf{L}(\boldsymbol{\theta})\hat{\mathbf{x}}(k+1, \hat{\boldsymbol{\theta}}^{(i)}) - \mathbf{G}\mathbf{u}(k); \qquad k = 1, 2, \ldots, N_c \qquad (17.41)$$

are the prediction errors for the dynamic equation model (17.27), based on the parameter estimate $\hat{\boldsymbol{\theta}}^{(i)}$ at stage i. The cost function (17.40) depends on $\boldsymbol{\theta}$ and on $\mathbf{Q}$ which also should be regarded as an unknown parameter. The following proposition [29, 30], describes the minimizer.

Proposition 1 (Eaton,Mosca). *The cost function* (17.40) *admits a unique minimum for*

$$\hat{\mathbf{Q}} := \frac{1}{N_c} \sum_{k=1}^{N_c} \hat{\mathbf{e}}^i(k) \left[\hat{\mathbf{e}}^i(k)\right]^\top \qquad (17.42)$$

The minimizer with respect to $\boldsymbol{\theta}$ *is the parameter vector which minimizes the determinant* $\det\left[\hat{\mathbf{Q}}(\boldsymbol{\theta})\right]$, *i.e.*

$$\hat{\boldsymbol{\theta}}^{(i+1)} = \operatorname{Argmin}_{\boldsymbol{\theta}} \, \det\left[\hat{\mathbf{Q}}(\boldsymbol{\theta})\right] \qquad (17.43)$$

The minimization of the determinant of $\hat{\mathbf{Q}}(\boldsymbol{\theta})$ needs to be done numerically, say by a gradient iteration algorithm. Note that the gradient has a closed form expression involving the Jacobian of the matrix $\mathbf{E}(\boldsymbol{\theta})$ with respect to the parameters, which can be written by inspection. Nevertheless using the well-known inequality [31, p. 477] valid for positive semidefinite matrices

$$(\det A)^{1/n} \leq \frac{1}{n}\operatorname{trace}A$$

we may instead use the simple suboptimal alternative of minimizing the trace. In this way the problem is reduced to *ordinary least squares* i.e. to solving the overdetermined system

$$\mathbf{L}(\boldsymbol{\theta})\hat{X}_1^{(i)} \simeq \mathbf{G}U \qquad (17.44)$$

where $\hat{X}_1^{(i)} := \{\hat{\mathbf{x}}(k+1, \hat{\boldsymbol{\theta}}^{(i)}), k = 0, 1, \ldots, Nc - 1\}$, which can be done in closed form. Thus, it is suggested to adopt the suboptimal procedure of finding a $\hat{\boldsymbol{\theta}}^{(i+1)}$ such that

$$\mathbf{M}(\hat{\boldsymbol{\theta}}^{(i+1)})\mathbf{a}(k) + \mathbf{C}(\hat{\boldsymbol{\theta}}^{(i+1)})\mathbf{v}(k) + \mathbf{K}(\hat{\boldsymbol{\theta}}^{(i+1)})\mathbf{d}(k) - \mathbf{G}\mathbf{u}(k) \approx 0 \qquad (17.45)$$

for $k = 0, \ldots, N_c - 1$. This has been implemented in the algorithm. This task can be accomplished by solving a linear least squares problem if the matrices $\mathbf{M}$, $\mathbf{C}$ and $\mathbf{K}$ are linearly parametrized. The estimate $\hat{\mathbf{Q}}(\hat{\boldsymbol{\theta}}^{(i+1)})$ can then be used to compute the new Kalman filter estimates in the successive E -stage of the algorithm.

17.4.4 Generalizations of the EM Algorithm

In most cases maximizing the log-likelihood is equivalent to looking for stationary points solving the gradient equation

$$\nabla l(\hat{\boldsymbol{\theta}}|X,Y,U) = 0 \tag{17.46}$$

for the unknown $\hat{\boldsymbol{\theta}}$, where ∇l denotes the gradient (column) vector of l with respect to the parameters, also called the *score-function*. This is justified by the well-known result that for all $\boldsymbol{\theta} \in \Theta$

$$\mathbb{E}_{\boldsymbol{\theta}}\nabla l(\boldsymbol{\theta}|\mathbf{X},\mathbf{Y},\mathbf{U}) = 0 \tag{17.47}$$

where the actual random variables have been denoted by boldface symbols and the notation $\mathbb{E}_{\boldsymbol{\theta}}$ denotes expectation taken under the assumption that $\boldsymbol{\theta}$ is the true parameter value.

Further we have

$$\mathbb{E}_{\boldsymbol{\theta}}\left[\nabla l(\boldsymbol{\theta}|\cdot)\nabla l^{T}(\boldsymbol{\theta}|\cdot)\right] = -\mathbb{E}_{\boldsymbol{\theta}}\nabla\left(\nabla l(\boldsymbol{\theta}|\cdot)\right) = \mathcal{I}(\boldsymbol{\theta}) \tag{17.48}$$

$\nabla(\nabla l)$ being the Hessian matrix of l. It is a basic and well-known fact that the variance of any unbiased estimator has a lower bound (in the ordering of positive semidefinite matrices) given by $\mathcal{I}(\boldsymbol{\theta})^{-1}$, i.e. by the inverse of the so called *information matrix*. The inverse exists if and only if the parametrization is (locally) identifiable (at $\boldsymbol{\theta}$).

When no information about the underlying probability distributions are available, one needs to generalize the E-step of the algorithm in such a way that only hypotheses about the first and second moments of the random variables $\mathbf{n}(k)$ and $\mathbf{w}(k)$ are needed (see (17.21) and (17.23)).

To this end we substitute the exact score-function with a new function $\mathbf{g}_s^*(\boldsymbol{\theta}|\mathbf{X},\mathbf{Y},\mathbf{U})$ chosen optimally in a convenient space $\mathcal{G}$ of random vectors of dimension q with finite second moments. In strict analogy with (17.47) it is required that $\mathbb{E}_{\boldsymbol{\theta}}\mathbf{g}(\boldsymbol{\theta}|\mathbf{X},\mathbf{Y},\mathbf{U}) = 0$ for any $\mathbf{g} \in \mathcal{G}$ and $\boldsymbol{\theta} \in \Theta$. Moreover each $\mathbf{g} \in \mathcal{G}$ can be standardized in order to satisfy also a relation like (17.48). In fact $\mathbf{g}_s = -(\mathbb{E}_{\boldsymbol{\theta}}\nabla\mathbf{g})^{T}(\mathbb{E}_{\boldsymbol{\theta}}[\mathbf{g}\mathbf{g}^{T}])^{-1}\mathbf{g}$ satisfies $\mathbb{E}_{\boldsymbol{\theta}}[\mathbf{g}_s(\boldsymbol{\theta}|\cdot)\mathbf{g}_s(\boldsymbol{\theta}|\cdot)^{T}] = -\mathbb{E}_{\boldsymbol{\theta}}\nabla\mathbf{g}_s(\boldsymbol{\theta}|\cdot)$ for all $\boldsymbol{\theta} \in \Theta$ so that

$$\mathcal{E}_{\mathbf{g}_s}(\boldsymbol{\theta}) = \mathbb{E}_{\boldsymbol{\theta}}\left[\mathbf{g}_s(\boldsymbol{\theta}|\mathbf{X},\mathbf{Y},\mathbf{U})\mathbf{g}_s(\boldsymbol{\theta}|\mathbf{X},\mathbf{Y},\mathbf{U})^{T}\right] \tag{17.49}$$

is a natural generalization of the information matrix $\mathcal{I}(\boldsymbol{\theta})$.

The element $\mathbf{g}_s^*$ is then defined as the one which satisfies $\mathcal{E}_{\mathbf{g}_s^*}(\boldsymbol{\theta}) \geq \mathcal{E}_{\mathbf{g}_s}(\boldsymbol{\theta})$ for all $\mathbf{g}_s \in \mathcal{G}$; in this way $\mathbf{g}_s^*$ can also be interpreted as the *orthogonal projection* of ∇l onto the Hilbert space of random vectors with finite second moments generated by the elements of $\mathcal{G}$ (see [32] for details). The $\mathbf{g}_s^*$ is often called *quasi-score function*.

By making the choice that each $\mathbf{g} \in \mathcal{G}$ is of the form

$$\mathbf{g}(\boldsymbol{\theta}|\mathbf{X},\mathbf{Y},\mathbf{U}) = \sum_{k=0}^{N_c-1} \mathbf{P}(k,\boldsymbol{\theta}) \begin{bmatrix} \mathbf{n}(k) \\ \mathbf{w}(k) \end{bmatrix} \qquad (17.50)$$

the optimal $q \times (n+p)$ matrices $\mathbf{P}_s^*(k,\boldsymbol{\theta})$ can be determined using the conditions proved in [33]. Here computations are eased by exploiting the uncorrelation properties of model and measurement noises; however more general cases can be handled. The quasi-score function turns out to be:

$$\mathbf{g}_s^*(\boldsymbol{\theta}|\mathbf{X},\mathbf{Y},\mathbf{U}) = \sum_{k=0}^{N_c-1} \begin{bmatrix} \mathbb{E}_{\boldsymbol{\theta}}[\mathbf{M}_1(\boldsymbol{\theta})\mathbf{a}(k) + \mathbf{C}_1(\boldsymbol{\theta})\mathbf{v}(k) + \mathbf{K}_1(\boldsymbol{\theta})\mathbf{d}(k)]^T \\ \vdots \\ \mathbb{E}_{\boldsymbol{\theta}}[\mathbf{M}_q(\boldsymbol{\theta})\mathbf{a}(k) + \mathbf{C}_q(\boldsymbol{\theta})\mathbf{v}(k) + \mathbf{K}_q(\boldsymbol{\theta})\mathbf{d}(k)]^T \end{bmatrix} \cdot$$
$$\cdot \mathbf{Q}^{-1}\left[\mathbf{M}(\boldsymbol{\theta})\mathbf{a}(k) + \mathbf{C}(\boldsymbol{\theta})\mathbf{v}(k) + \mathbf{K}(\boldsymbol{\theta})\mathbf{d}(k) - \mathbf{G}\mathbf{u}(k)\right]. \qquad (17.51)$$

where, for each $j = 1,\ldots,q$,

$$\mathbf{M}_j(\boldsymbol{\theta}) = \frac{\partial \mathbf{M}}{\partial \theta_j}(\boldsymbol{\theta}) \qquad \mathbf{C}_j(\boldsymbol{\theta}) = \frac{\partial \mathbf{C}}{\partial \theta_j}(\boldsymbol{\theta}) \qquad \mathbf{K}_j(\boldsymbol{\theta}) = \frac{\partial \mathbf{K}}{\partial \theta_j}(\boldsymbol{\theta}) \qquad (17.52)$$

The E-step requires the computation of $h^{(i)}(\vartheta) = \mathbb{E}_{\hat{\boldsymbol{\theta}}^{(i)}}[\mathbf{g}_s^*(\vartheta|\mathbf{X},\mathbf{Y},\mathbf{U})|Y,U]$. Since $\mathbf{g}_s^*$ is a linear combination of random variables, this operation results in the substitution in (17.51) of $\mathbf{d}(\cdot)$, $\mathbf{v}(\cdot)$ and $\mathbf{a}(\cdot)$ (shortly $\mathbf{x}(\cdot)$) with $\mathbb{E}_{\hat{\boldsymbol{\theta}}^{(i)}}[\mathbf{d}(\cdot)|Y,U]$, $\mathbb{E}_{\hat{\boldsymbol{\theta}}^{(i)}}[\mathbf{v}(\cdot)|Y,U]$ and $\mathbb{E}_{\hat{\boldsymbol{\theta}}^{(i)}}[\mathbf{a}(\cdot)|Y,U]$ (shortly $\mathbb{E}_{\hat{\boldsymbol{\theta}}^{(i)}}[\mathbf{x}(\cdot)|Y,U]$) respectively, and of $\mathbf{y}(\cdot)$ and $\mathbf{u}(\cdot)$ with $y(\cdot)$ and $u(\cdot)$.

Moreover we shall consider in place of the random variables $\mathbf{x}(\cdot)|\mathbf{Y},\mathbf{U}$, the corresponding minimum variance linear estimates given $\mathbf{Y}$ and $\mathbf{U}$ or, in other words, the projection on the Hilbert space generated by $\mathbf{Y}$ and $\mathbf{U}$. For these variables we use the notation $\hat{\mathbf{x}}(\cdot)|\mathbf{Y},\mathbf{U}$. The advantage is that again only assumptions on the first and second moments of the random variables $\mathbf{n}(k)$ and $\mathbf{w}(k)$ are needed to perform this projection. The terms $\mathbb{E}_{\hat{\boldsymbol{\theta}}^{(i)}}[\hat{\mathbf{x}}(k)|Y,U]$ are then the *smoothed estimates* of the state and can be calculated by standard routines [27] from the state space representation given by equations (17.24) and (17.26). In order to limit the computational burden, however, these can be replaced by the *causal Kalman filter* estimates $\mathbb{E}_{\hat{\boldsymbol{\theta}}^{(i)}}[\hat{\mathbf{x}}(k)|Y_{k-1},U_{k-1}]$. Here $Y_{k-1} = \{y(i), i = 0,\ldots,k-1\}$, $U_{k-1} = \{u(i), i = 0,\ldots,k-1\}$ and the notation $\hat{x}^{(i)}(k|k-1)$ is also used for these estimates.

The M-step requires solving $h^{(i)}(\vartheta) = 0$. The vectors $\mathbb{E}_{\vartheta}[\mathbf{M}_j(\vartheta)\mathbf{a}(k) + \mathbf{C}_j(\vartheta)\mathbf{v}(k) + \mathbf{K}_j(\vartheta)\mathbf{d}(k)]$ in (17.51) can be computed iteratively by simply taking the expectation of equation (17.24) and recalling that $\mathbf{n}(\cdot)$ is a zero mean sequence. Unfortunately the dependence of $h^{(i)}(\vartheta)$ on the parameter would be non-linear even if the matrices $\mathbf{M}$, $\mathbf{C}$ and $\mathbf{K}$ were linearly parametrized.

17.4.5 Expression of $\mathcal{E}_{\mathbf{g}_s^*}(\boldsymbol{\theta})$

We now compute trace $\mathcal{E}_{\mathbf{g}_s^*}(\boldsymbol{\theta})$. Noting that $\mathbf{n}(k)$ is a white sequence, the expectation of the cross terms involving distinct sampling times vanishes, so that the double summation reduces to

$$\mathcal{E}_{\mathbf{g}_s^*}(\boldsymbol{\theta}) = \sum_{k=0}^{N_c-1} \mathbb{E}_{\boldsymbol{\theta}}\left[\ldots \ \mathbf{M}_j(\boldsymbol{\theta})\mathbf{a}(k) + \mathbf{C}_j(\boldsymbol{\theta})\mathbf{v}(k) + \mathbf{K}_j(\boldsymbol{\theta})\mathbf{d}(k) \ \ldots\right]^T \cdot$$
$$\cdot \mathbf{Q}^{-1}\mathbb{E}_{\boldsymbol{\theta}}\left[\ldots \ \mathbf{M}_j(\boldsymbol{\theta})\mathbf{a}(k) + \mathbf{C}_j(\boldsymbol{\theta})\mathbf{v}(k) + \mathbf{K}_j(\boldsymbol{\theta})\mathbf{d}(k) \ \ldots\right] \quad (17.53)$$

By using the simple relations $\text{trace}(\mathbf{A} + \mathbf{B}) = \text{trace}(\mathbf{A}) + \text{trace}(\mathbf{B})$ and $\mathbf{c}^T\mathbf{c} = \text{trace}(\mathbf{c}^T\mathbf{c}) = \text{trace}(\mathbf{c}\mathbf{c}^T)$ which are valid for arbitrary matrices $\mathbf{A}$, $\mathbf{B}$ and column vector $\mathbf{c}$, the following derivation can be easily proved:

$$\text{trace } \mathcal{E}_{\mathbf{g}_s^*}(\boldsymbol{\theta}) = \sum_{k=0}^{N_c-1} \sum_{j=1}^{q} \mathbb{E}_{\boldsymbol{\theta}}\left[\mathbf{M}_j(\boldsymbol{\theta})\mathbf{a}(k) + \mathbf{C}_j(\boldsymbol{\theta})\mathbf{v}(k) + \mathbf{K}_j(\boldsymbol{\theta})\mathbf{d}(k)\right]^T \cdot$$
$$\cdot \mathbf{Q}^{-1}\mathbb{E}_{\boldsymbol{\theta}}\left[\mathbf{M}_j(\boldsymbol{\theta})\mathbf{a}(k) + \mathbf{C}_j(\boldsymbol{\theta})\mathbf{v}(k) + \mathbf{K}_j(\boldsymbol{\theta})\mathbf{d}(k)\right] \quad (17.54a)$$
$$= \sum_{k=0}^{N_c-1} \sum_{j=1}^{q} \mathbb{E}_{\boldsymbol{\theta}}[\mathbf{x}(k)]^T \mathbf{L}_j(\boldsymbol{\theta})^T \mathbf{Q}^{-1}\mathbf{L}_j(\boldsymbol{\theta})\mathbb{E}_{\boldsymbol{\theta}}[\mathbf{x}(k)] \quad (17.54b)$$
$$= \sum_{k=0}^{N_c-1} \sum_{j=1}^{q} \text{trace}\left(\mathbf{L}_j(\boldsymbol{\theta})^T \mathbf{Q}^{-1}\mathbf{L}_j(\boldsymbol{\theta})\mathbb{E}_{\boldsymbol{\theta}}[\mathbf{x}(k)]\mathbb{E}_{\boldsymbol{\theta}}[\mathbf{x}(k)]^T\right)$$
$$(17.54c)$$
$$= \text{trace}\left(\sum_{k=0}^{N_c-1} \mathbb{E}_{\boldsymbol{\theta}}[\mathbf{x}(k)]\mathbb{E}_{\boldsymbol{\theta}}[\mathbf{x}(k)]^T \cdot \sum_{j=1}^{q} \mathbf{L}_j(\boldsymbol{\theta})^T \mathbf{Q}^{-1}\mathbf{L}_j(\boldsymbol{\theta})\right)$$
$$(17.54d)$$

where we have defined $\mathbf{L}_j(\boldsymbol{\theta}) = [\mathbf{K}_j(\boldsymbol{\theta}) \ \mathbf{C}_j(\boldsymbol{\theta}) \ \mathbf{M}_j(\boldsymbol{\theta})]$. This approximate expression of the information matrix could be used to compare the conditioning (or the asymptotic variance) of different parametrizations.

17.5 Conlusion and Numerical Results

The paper describes a number of issues in the identification of physical parameters in non-homogeneous linear elastic solids. Some of these issues are still to be addressed satisfactorily and will be object of future research. We believe nevertheless that the approach is promising. Several versions of the basic algorithm described in the paper have been tested in numerical simulations.

Here we shall report simulation results of an important benchmark problem which deals with the detection of a step discontinuity in the elastance function $\gamma(z)$ of the medium. In the first test a constant density (1000 Kg/m, 50 m wide) medium was considered where in the *left* subinterval ($z \in [0, 30)$ m) the elastic coefficient was of 4 GN/m^2 while in the *right* subinterval ($z \in [30, 50]$ m) of 9 GN/m^2. This choice corresponds to wave propagation velocities of 2000 m/s and 3000 m/s in the left and right media respectively. The source, a cosine pulse waveform with a frequency band of 200 Hz, was located at the distance $z = 14$ m from the origin. In this setting, a mesh of $N = 250$ equispaced nodes is adequate to model wave propagation with the finite element method.

The initial parameters estimate required by the algorithm was set equal to that of a homogeneous medium whose elasticity coincides with that of the left part of the true medium. Before running the Kalman filter in the E-step, the model has been transformed in modal coordinates and reduced by retaining only low natural frequency modes (precise figures will be given below). The estimates computed by the Kalman filter in the reduced modal coordinates have then been back-transformed in nodal coordinates and used in the M-step.

In Figure 17.1, the parameter estimates corresponding to the first four iterations of the algorithm are plotted. In both plots, the unknowns are the $N - 1 = 249$ values γ_i representing the elasticity in each finite element of the mesh (the ρ_i have been assumed known). These results have been obtained by placing sensors along the whole spatial domain. More precisely, Figure 17.1 (above) corresponds to the case of an equispaced array of 49 displacement sensors (i.e. one every 2 m) while Figure 17.1 (below) to a 1 m distance equispaced array. The choice of the spacing between sensors has consequences on the number of modes to be retained in the model reduction and also in the resolution capabilities of the algorithm. In general one would need fewer sensors to capture the modes with a low natural frequency (in the spatial coordinate), while the faster oscillating modes of high natural frequencies need closely spaced sensors. Accordingly, the Kalman filter performance was bad whenever the spacing between sensors was to coarse with respect to the wavelengths of the higher frequency modes which were retained in the reduced model. In the case of 2 m spacing only the modes (approximately 20) whose natural frequencies were below 400 Hz have been retained. This bound has been lifted to 800 Hz (approximately 45 modes) in the case of 1 m spacing, allowing the estimation of shorter spatial wavelengths and thus a closer approximation of the discontinuity in the M-step.

The presence of a spike at the source location point ($z = 14$) is an artifact which can be explained as follows: the true waveform is characterized by a cusp where the spatially concentrated input force is applied. However the waveform estimated by the Kalman filter is made by the superposition of a limited number of modes which can only approximate this cusp with a smooth curve. These smooth estimates are fed to the least squares optimization step (the M-step) which is trying to match data from the full order finite element model. Since the least squares algorithm can compensate the discrepancy between the data and the model-based estimates by only updating the parameters values,

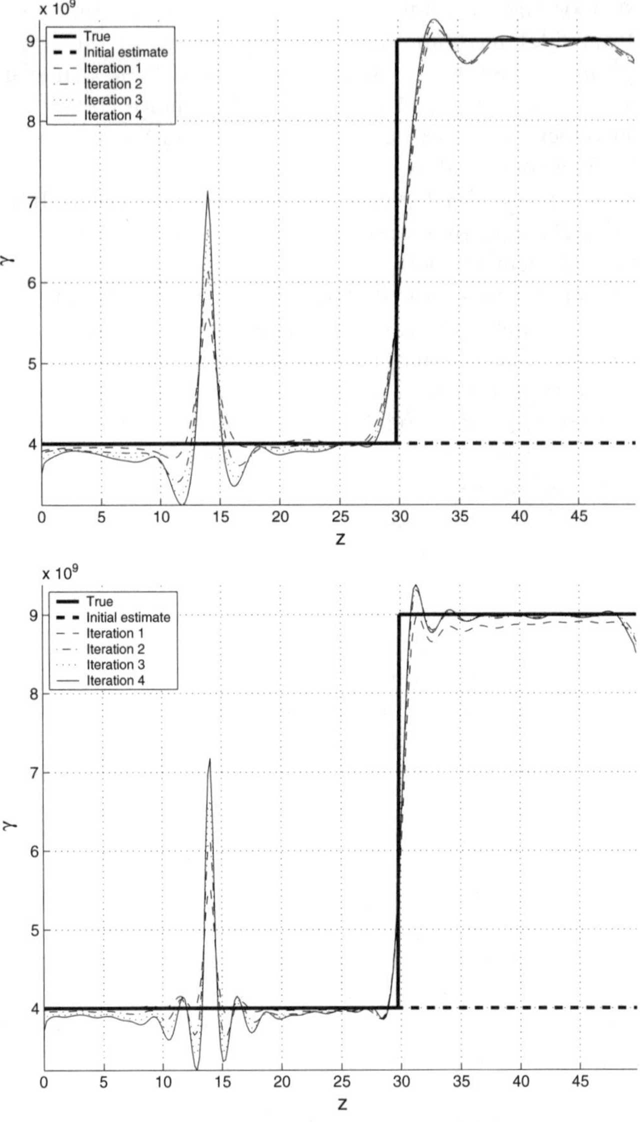

Fig. 17.1.

it is forced to introduce a sharp variation of the elasticity of the medium around the point where the cusp occurs.

Different parametrization were also considered for the function $\gamma(z)$. Figure 17.2 is the counterpart of Figure 17.1 being obtained in the same setting described above. The only modification was the reduction of the number of unknown parameters by grouping the finite elements in $249/3 = 83$ *super-*

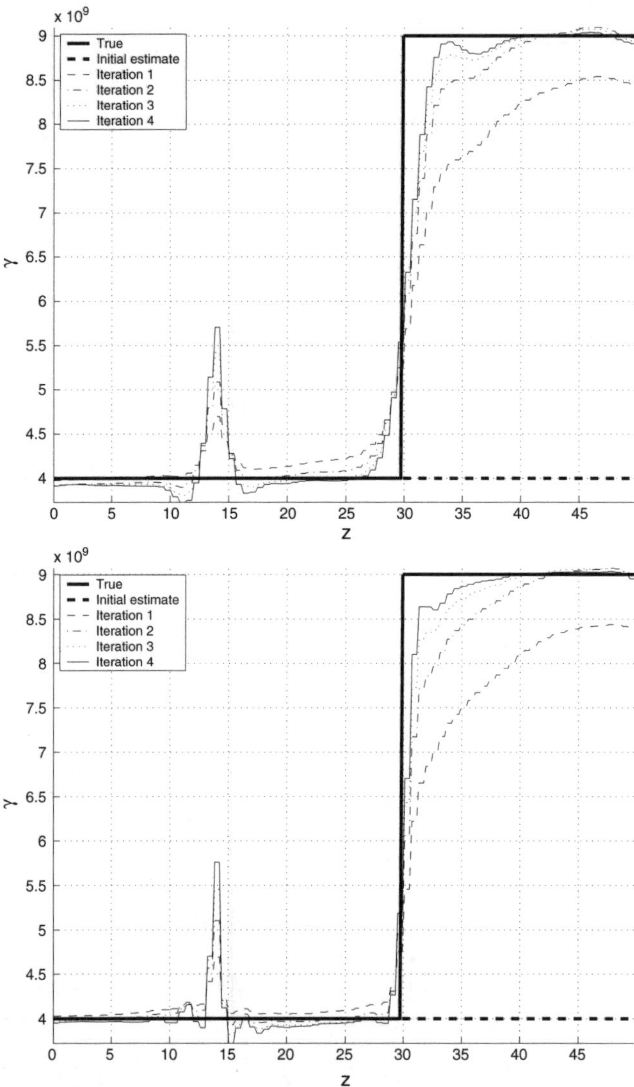

Fig. 17.2.

elements, each one consisting of three contiguous finite elements which have been constrained to assume the same value of the elasticity.

In Figure 17.3, instead, the function $\gamma(z)$ has been expanded using an orthogonal basis function formed by the modes selected in the order reduction step as explained above. In this last case the number of unknown parameter is then further reduced still keeping a good quality of the estimate.

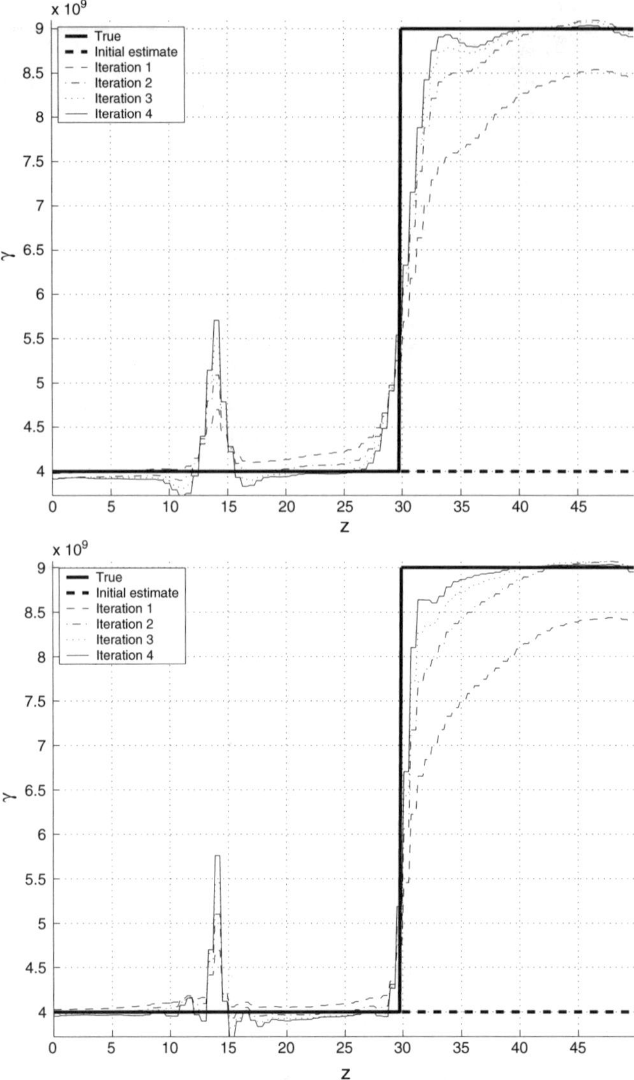

Fig. 17.3.

References

1. FRISWELL M.I. and MOTTERSHEAD J.E. *Finite element model updating in structural dynamics*, volume 38 of *Solid Mechanics and its Applications*. Kluver Academic Publisher, 1995.
2. SHERIFF R.E. and GELDART L.P. *Exploration Seismology*. Cambridge University Press, 1995.

3. ISAKOV V. *Inverse problems for partial differential equation*, volume 127 of *Applied Mathematical Sciences*. Springer, 1998.

4. BLEISTEIN N. *Mathematical methods for wave phenomena*. Academic Press, 1984.

5. ENGL H.W., HANKE M., and NEUBAUER A. *Regularization of Inverse Problems*. Kluwer, 1996.

6. COLTON D. and KRESS R. *Inverse acoustic and electromagnetic scattering theory*, volume 93 of *Applied Mathematical Sciences*. Springer, second edition, 1998.

7. DEMPSTER A.P., LAIRD N.M., and RUBIN D.B. Maximum likelihood from incomplete data via the EM algorithm. *J. Royal Stat. Soc.*, B-39(1):1–38, 1977.

8. HEYDE C.C. and MORTON R. Quasi-likelihood and generalizing the EM algorithm. *J. Royal Stat. Soc.*, B-58(2):317–327, 1996.

9. SOLO V. Identification of a noisy stochastic heat equation with the EM algorithm. In *Proceedings of the 41st IEEE Conference on Decision and Control*, December 2002.

10. ACHENBACH J.D. *Wave propagation in elastic solids*, volume 16 of *North-Holland series in Applied Mathematics and Mechanics*. North-Holland Publishing Company, 1973.

11. MIKLOWITZ J. *Elastic waves and waveguides*, volume 22 of *North-Holland series in Applied Mathematics and Mechanics*. North-Holland Publishing Company, 1978.

12. IKAWA M. *Hyperbolic partial differential equations and wave phenomena*, volume 189 of *Translations of Mathematical Monographs*. American Mathematical Society, 2000.

13. IHLENBURG F. *Finite element analysis of acoustic scattering*, volume 132 of *Applied Mathematical Sciences*. Springer, 1998.

14. HACKBUSCH W. *Elliptic differential equations: Theory and numerical treatment*, volume 18 of *Springer Series in Computational Mathematics*. Springer-Verlag, 1992.

15. QUARTERONI A. and VALLI A. *Numerical approximation of Partial Differential equations*, volume 23 of *Springer Series in Computational Mathematics*. Springer, second edition, 97.

16. ZIENKIEWICZ O.C. and TAYLOR R.L. *The finite element method - The basis*. Butterworth-Heinemann, fifth edition, 2000.

17. RAVIART P.A. and THOMAS J.M. *Introduction a l'analyse numerique des equations aux derivees partielles*. Collection Mathematiques Appliquees pour la Maitrise. Masson, 1983.

18. CLARK J.M.C. The consistent selection of parametrization in system identification. In *Proceedings of the Joint Automatic Control Conference*, 1976.

19. MOORE B.C. Principal component analysis in linear systems: controllability, observability and model reduction. *IEEE Trans. on Automatic Control*, 26(1):17–32, February 1981.

20. PERNEBO L. and SILVERMAN L.M. Model reduction via balanced state space representation. *IEEE Trans. on Automatic Control*, 27(2):382–387, April 1982.

21. GLOVER K. All optimal Henkel norm approximation of linear multivariable systems and their l_∞ error bounds. *International Journal of Control*, 39(6):1115–1193, 1984.

22. MEYER D.G. and SRINIVASAN S. Balancing and model reduction for second-order form linear systems. *IEEE Trans. on Automatic Control*, 41(11):1632–1644, November 1996.

23. CHAHLAOUI Y., LEMONNIER D., VANDENDORPE A., and VAN DOOREN P. Second-order balanced truncation. *Linear algebra and its applications*, 2005. article in press.

24. EWINS D.J. *Modal testing: theory and practice*. Research Studies Press, 1984.

25. GARVEY S.D., FRISWELL M.I., and PRELLS U. Co-ordinate transformations for second order systems. part i: general transformations. *Journal of sound and vibration*, 258(5):885–909, 2002.

26. LJUNG L. *System Identification. Theory for the user*. Information and System Sciences Series. Prentice Hall, second edition, 1999.

27. ANDERSON B.D.O. and MOORE J.B. *Optimal filtering*. Information and System Sciences Series. Prentice Hall, 1979.

28. JAZWINSKI A. *Stochastic processes and filtering theory*, volume 64 of *Mathematics in science and engineering*. Academic Press, 1970.

29. EATON J. Identification for control purposes. In *Proceedings of the 1967 IEEE Winter Meeting*, 1967.

30. MOSCA E. and ZAPPA G. Identificazione dei sistemi stocastici: l'approcio della massima verosimiglianza. In *Identificazione dei Sistemi Stocastici*, Collana di Automatica, pages 147–185. Consiglio Nazionale delle Ricerche, Roma, 1978.

31. HORN R.A. and JOHNSON C.R. *Matrix Analysis*. Cambridge University Press, 1985.

32. HEYDE C.C. *Quasi-likelihood and its applications*. Springer Series in Statistics. Springer, 1997.

33. HEYDE C.C. Fixed sample and asymptotic optimality for classes of estimating functions. *Contemp. Math.*, 89:241–247, 1988.

Genetic Regulatory Networks and Co-Regulation of Genes: A Dynamic Model Based Approach

Ashoka D. Polpitiya[1], J. Perren Cobb[1], and Bijoy K. Ghosh[2]

[1] Cellular Injury and Adaptation Laboratory, Department of Surgery, Washington University, St.Louis, MO 63110
`ashoka@cia.wustl.edu`
[2] Center for BioCybernetics and Intelligent Systems, Department of Electrical and Systems Engineering, Washington University, St.Louis, MO 63130

Summary. This paper describes a novel approach to obtain co-regulated genes and the underlying genetic regulatory network. Time evolution of gene expression levels is described using a discrete time classical state space dynamical system. The state transition matrix of the model can be used to obtain the adjacency matrix of the regulatory network. Gene expression values over time can also be written as a linear combination of the eigenmodes of the state transition matrix. Based on the relative participation of the eigenmodes, co-regulation is discussed.

Keywords: genes, microarrays, regulatory networks, dynamic models, reverse engineering, eigenvalues, co-regulation, Karhunen-Loève decomposition

18.1 Introduction

The recent advances in microarray techniques [8],[11] have enabled the measurement of expression values of many or all of an organism's genes. This has resulted in a large abundance of data. But the computational techniques available to handle this large-scale data are only a handful. Some challenges in analyzing this data are noisy measurements, stochastic nature of the data and the sheer number of genes involved [9]. There are some other inherent problems with the data itself too. Low time resolution and fewer replicates being the most significant accounting to the costs involved. Until one achieves the "lab-on-a-chip" technology, the techniques should be fine tuned for the available sparse and under-determined data in order to reveal some knowledge about the underlying regulatory interactions. The question of inferring the causal connectivity of the genes, i.e. genetic regulatory networks, using these expression profiles has been addressed by number of authors [6],[14],[4],[7]. But little attention is given to explore the characteristics of the models and the resulting networks.

A.D. Polpitiya et al.: *Genetic Regulatory Networks and Co-Regulation of Genes: A Dynamic Model Based Approach*, LNCIS **321**, 291–304 (2005)
`www.springerlink.com`

In this report, we first use Karhunen-Loève decomposition to explore the temporal behavior of the sparse time series data from microarrays. Then, we will discuss genetic regulatory networks emerging from the time series data. The idea behind the method is the use of a state-space model ([6], [10]). An algorithm to obtain the model parameters is outlined based on the minimum-norm least square method ([2]). Finally, a novel approach to look at genes that are "co-regulated" is being discussed based on the characteristics of the state-space model.

There are several papers where a simple interpolation scheme is used to obtain a continues representation of time series expression data (see [1], [4]). Here, a piecewise cubic interpolation was done on the data to obtain a finely interpolated time series. The underlying interpolating function $F(t)$, which is C^1 (i.e. continuous in the first derivative) was obtained for each interval of time $t_k < t < t_{k+1}$ and the slopes at t_k were chosen such that $F(t)$ preserves the shape of the data and respects monotonicity. This means that, on intervals where the data are monotonic, so is $F(t)$; at points where the data has a local extremum, so does $F(t)$ (see Figure 18.1).

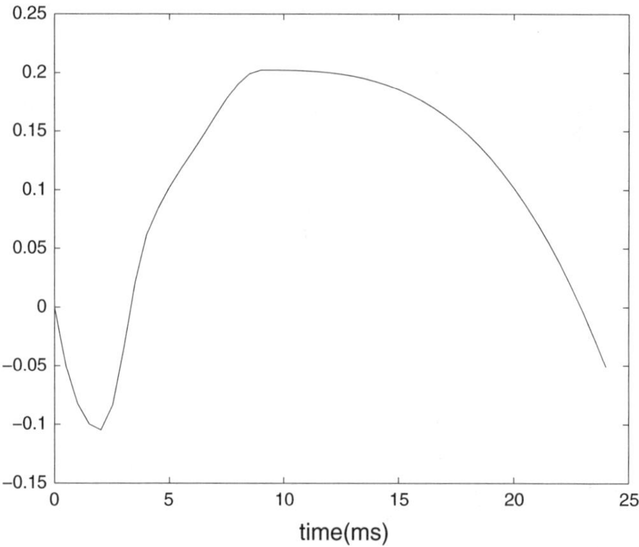

Fig. 18.1. Piecewise cubic interpolated data.

18.2 Temporal Patterns in Gene Expression Data

In order to visualize the temporal variation, the data was projected onto a smaller dimensional space using a series expansion method, the Karhunen-Loève decomposition (KLD) [13].

18.2.1 Series Representation of Gene Expression Data

Gene expression data over time is treated as sample functions of a random process. Let $\mathbf{x}^j(k)$ denote the $N \times 1$ dimensional expression vector at time point k, for the volunteer j. Therefore at kth time point, the expression vector can be viewed as

$$\mathbf{x}^j(k) = [x_1^j(k)\, x_2^j(k)\, \ldots\, x_N^j(k)]^T,$$

where $x_i^j(k)$ is the expression value of the ith gene at time k for volunteer j. The idea is to find a global basis for $\mathbb{R}^N$: ψ_1, ψ_2, $\ldots$, ψ_N and to expand $\mathbf{x}^j(k)$ as

$$\mathbf{x}^j(k) = \lim_{N \to \infty} \sum_{i=1}^{N} \alpha_i^j(k)\psi_i$$

where

$$\alpha_i^j(k) = \langle \mathbf{x}^j(k), \psi_i \rangle$$

and $\langle \cdot, \cdot \rangle$ stands for the standard inner product notation.

The orthonormal basis for KLD is selected as the eigenvectors of the average correlation matrix $C_1 \in \mathbb{R}^{N \times N}$, where

$$C_1 = \frac{1}{N_T M} \sum_{k=1}^{N_T} \sum_{j=1}^{M} (\mathbf{x}^j(k))(\mathbf{x}^j(k))^T, \tag{18.1}$$

and N_T is the total number of time samples. The matrix C_1 is symmetric and positive semidefinite, so its eigenvalues are all real and non-negative and the corresponding eigenvectors are orthonormal and forms the global basis for $\mathbb{R}^N$. The eigenvectors corresponding to the largest p eigenvalues of C_1 are called the principal eigenvectors (modes) and the pth order successive reconstruction $\hat{\mathbf{x}}^j(k)$ of the expression value $\mathbf{x}^j(k)$ is given by

$$\hat{\mathbf{x}}^j(k) = \sum_{i=1}^{p} \alpha_i^j(k)\psi_i.$$

It is observed that the first three principal modes capture 97% of the energy content in the expression values (see Figure 18.2). An interesting feature that one can clearly see from Figure 18.2 is that the volunteers treated with the toxic drug have a cyclic response over the 24hrs which correlates well with the fact that they all recover and become normal in 24hrs, and the ones who were given only Saline do not show much variation over time.

18.2.2 Classifying Time Profiles : 2nd KLD

The coefficients $\alpha_i^j(k)$ are uncorrelated with respect to i. The vector function

$$[\alpha_1^j(k)\, \alpha_2^j(k)\, \ldots\, \alpha_p^j(k)]$$

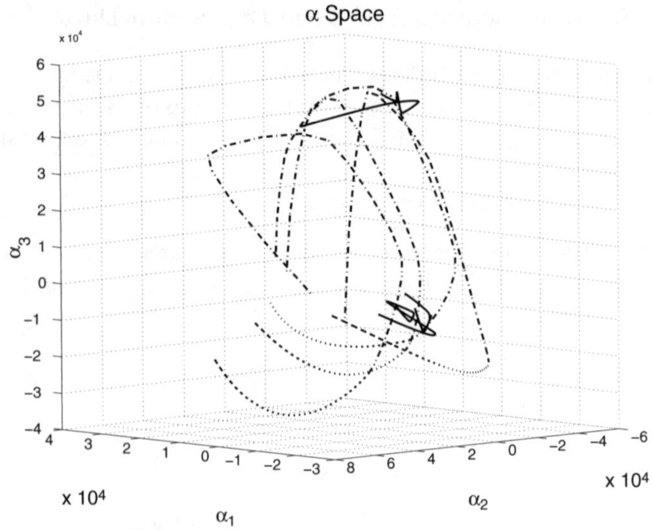

Fig. 18.2. "α-Space": α coefficients for three principal modes of seven human volunteer gene expression data for 5150 genes. Dotted lines correspond to four volunteers treated with a toxic drug (the dosage was administered such that they recover in 24 hrs) and the thick lines correspond to the three volunteers who only had Saline.

can be viewed as a sample function of a vector random process and a 2nd KLD can be performed. Let

$$\zeta_i^j = [\alpha_i^j(1)\ \alpha_i^j(2)\ \ldots\ \alpha_i^j(N_T)], \qquad \xi^j = [\zeta_1^j\ \zeta_2^j\ \ldots\ \zeta_p^j],$$

and it captures the pth order α-space representation of jth data set (i.e., j^{th} volunteer). The covariance matrix $C_2 \in \mathbb{R}^{(N_T \times p) \times (N_T \times p)}$ as in (18.1), will be

$$C_2 = \frac{1}{M} \sum_{j=1}^{M} (\xi^j)^T (\xi^j). \tag{18.2}$$

The qth order successive reconstruction $\hat{\xi}^j$ of the jth vector ξ^j is given by

$$\hat{\xi}^j = \sum_{i=1}^{q} \beta_i^j \varphi_i$$

where φ_i, $i = 1, 2, \ldots, q$ are the eigenvectors corresponding to the largest q eigenvalues of the matrix C_2 and the coefficients β_i^j are obtained by

$$\beta_i^j = \langle \xi^j, \varphi_i \rangle.$$

Figure 18.3 shows the plot of the "β-space" with the first three principal modes.

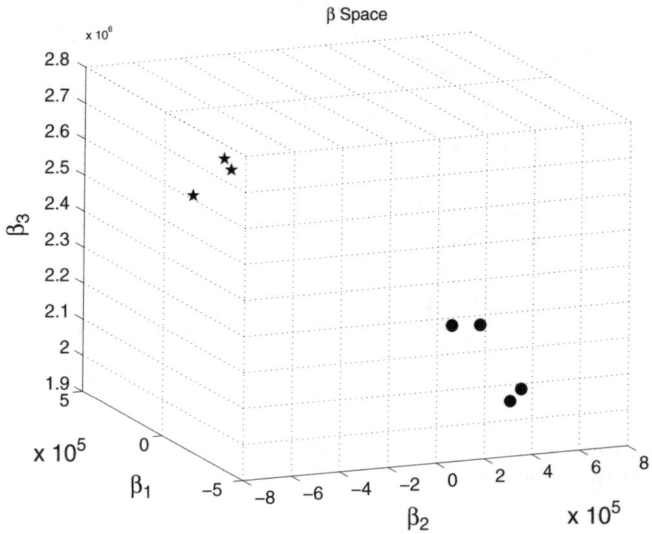

Fig. 18.3. "β-Space": β coefficients for three principal modes. Circles correspond to toxin treated volunteers and the stars correspond to Saline treated.

18.3 Dynamic Model of the Genetic Network

The idea of a dynamic model of the genetic network is that the expression value from one time point determines the expression value seen at the next time point. A complete realistic model should take into account various issues biologically relevant such as, for example, the inherent nonlinearities of bio-chemical reactions, internal and external noise, etc. But given the nature of the sparse time data available from microarrays, a simple linear time invari-ant (LTI) discrete time model treating the biological system as a simple state machine, would be a good starting point.

The general form of a discrete time LTI system is as follows:

$$\mathbf{x}(k+1) = A\mathbf{x}(k) + B\mathbf{u}(k)$$
$$\mathbf{y}(k) = C\mathbf{x}(k) + D\mathbf{u}(k)$$

where A, B, C and D are maps between suitable spaces (i.e. matrices of suit-able dimensions). u is the *input*, $\mathbf{x}$ is the state variable in an N–dimensional state space and finally, y is the *output* of the system. Here we consider the case with no input and $C = Id$. Hence the system becomes

$$\mathbf{x}(k+1) = A\mathbf{x}(k) \tag{18.3}$$
$$\mathbf{y}(k) = \mathbf{x}(k).$$

Figure 18.4 shows a typical network discussed in this paper. Here a_{ij}'s are the elements of the A matrix, i.e., $A = \{a_{ij}\}$. A positive element means that

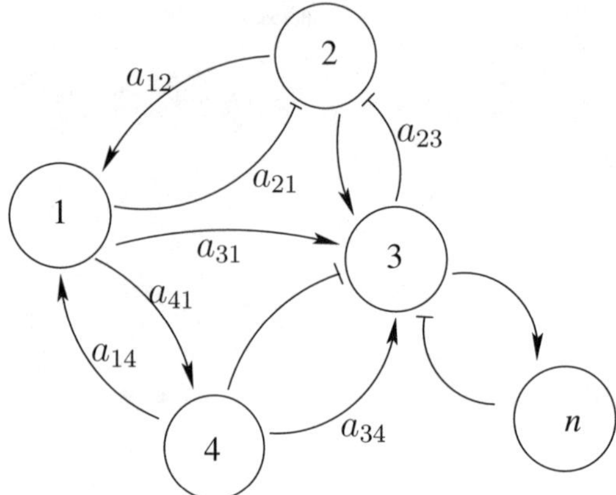

Fig. 18.4. Topology of the genetic regulatory network. Lines with arrow marks indicate positive reinforcing and lines with crossbars indicate negative reinforcing.

the ith gene being positively reinforced by the jth gene expression value at a previous time and vice-versa. In other words, one can write

$$x_i(k+1) = \sum_{j=1}^{N} a_{ij} x_j(k). \tag{18.4}$$

18.4 Reverse Engineering Algorithm for Genetic Networks

Problem 1 (Reverse engineering problem of genetic networks). For N genes, a noisy and sparse mRNA abundance data $\mathbf{y}(k)$ with "sufficiently" large k, is obtained for a known perturbation. Find the matrix A (in turn, find the adjacency matrix of the genetic regulatory network) described in Equation (18.3), such that it fits the data "best".

Fitting a model with $N \times N$ elements with $N \times M$, where $(N >> M)$, data points leads to a highly underdetermined system, i.e. many possible models that fit to the data almost perfectly, exist. However, one can find a optimum solution by imposing an additional constraint on smoothness. This will exclude models that may result erratic behaviors in between data points.

Problem can be formulated in the sense of a minimum norm least square solution[3]. Let

$$X_{k+1} = [\mathbf{x}(2)\,\mathbf{x}(3)\,\ldots\,\mathbf{x}(k+1)]$$
$$X_k = [\mathbf{x}(1)\,\mathbf{x}(2)\,\ldots\,\mathbf{x}(k)].$$

---

[3]Let's omit the superscript j which denotes each volunteer.

Then the problem becomes one of finding A where

$$X_{k+1} = AX_k$$

while minimizing the L_2 norm of A.

The solution to this problem is given by the well known *Moore-Penrose Pseudoinverse* (see [2]). Thus, A is obtained using

$$A = X_{k+1}X_k{}^\dagger \qquad (18.5)$$

where

$$X_k{}^\dagger = \left(X_k{}^T X_k\right)^{-1} X_k{}^T \qquad (18.6)$$

and $(\cdot)^\dagger$ denotes the Moore-Penrose Pseudoinverse. In other words, if

$$B = \sum_{i=1}^{r=rank(B)} \sigma_i u_i v_i^T \quad U = [u_i, \ \ldots, \ u_m], \ V = [v_1, \ \ldots, \ v_n] \qquad (18.7)$$

be the SVD of $B \in \mathbb{R}^{m \times n}$ $(m \geq n)$. Then the pseudoinverse of B is defined as

$$B^\dagger = V\Sigma^\dagger U^T,$$

where

$$\Sigma^\dagger = diag(\sigma_1^{-1}, \ldots, \sigma_r^{-1}, 0, \ldots, 0) \in \mathbb{R}^{n \times m}.$$

In order to obtain the regulatory interactions due to the toxic drug, we consider the average gene expression profiles over time of all the volunteers

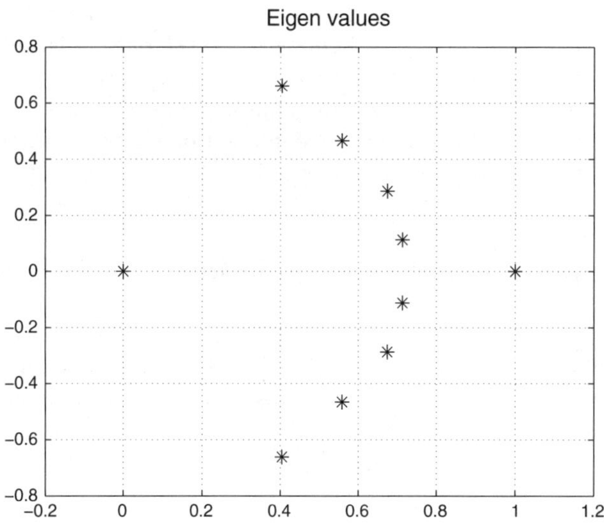

Fig. 18.5. Eigenvalues of the state transition matrix.

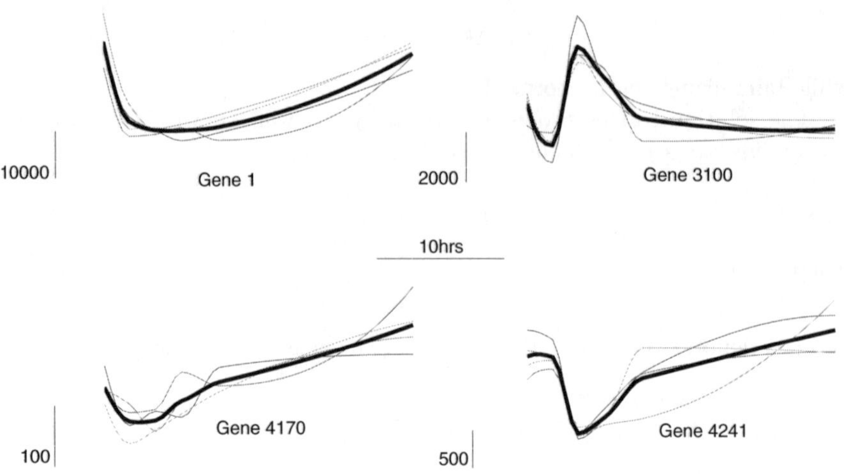

Fig. 18.6. Average gene expression profiles (centroids) for randomly selected four genes. Thick line is the centroid plot and the thin (color) lines are for 4 toxin treated volunteers. Axes are as usual expression value vs. time.

that were treated with the toxin. Figure 18.6 shows few randomly chosen gene profiles and their centroid plots. Figure 18.5 shows the eigenvalues of the resulting A matrix. This will enable one to find the adjacency matrix whose elements are '1', if the corresponding interaction is a positive reinforcement and '-1', for a negative reinforcement. One such network of 220 genes obtained for certain cutoff value and plotted using Cytoscape software[4] [12] is shown in Figure 18.7. The genes shown in light blue are the ones adjacent to NFKB1 (shown in darker green). Genes shown in light green are the ones which are common with the NFKB1 pathway found in Ingenuity©[5] shown in figure 18.8. This shows a significant overlap of gene interactions between what is reported in the literature and our proposed method.

18.5 Eigenmodes and Co-Regulation

Figure 18.5 shows that the state transition matrix of the model described in equation (18.3) has 10 distinct eigenvalues. In fact, the eigenvalue 1 repeats 4 times. Therefore the *Jordan canonical form* (JCF) [3] J of the matrix A has the following form:

$$J = \mathbf{diag}(J_0, J_1, \ldots, J_4, J_5) \tag{18.8}$$

---

[4]http://www.cytoscape.org
[5]http://www.ingenuity.com

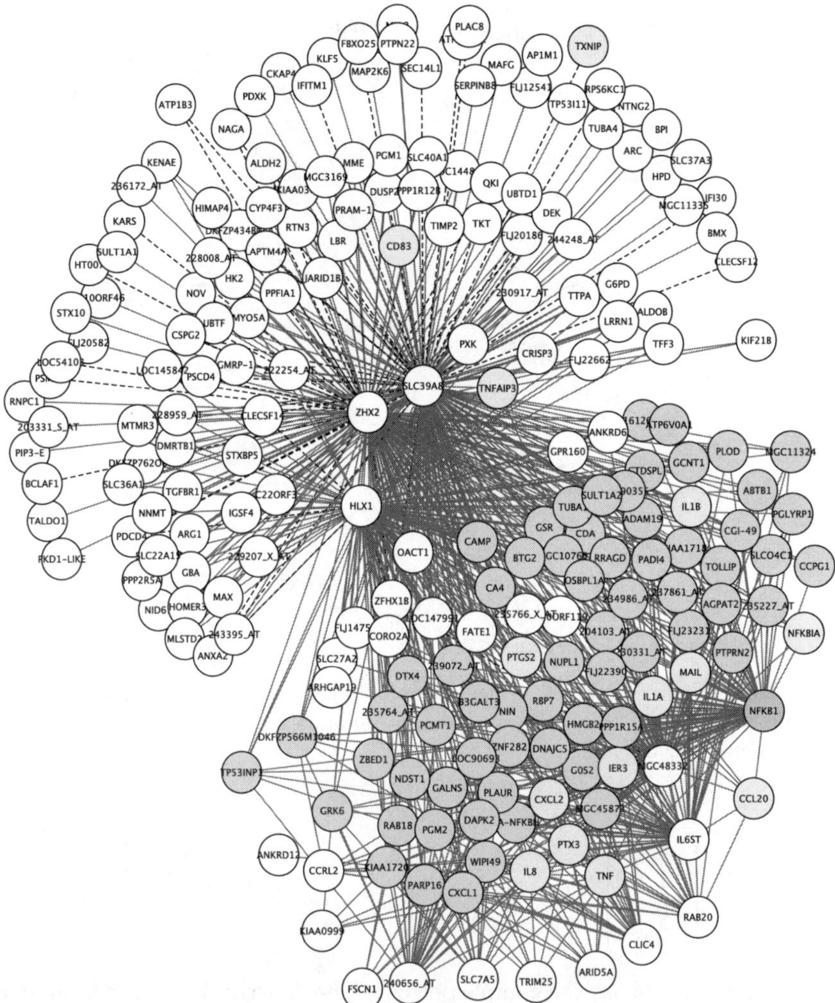

Fig. 18.7. Network of genes obtained from the state transition matrix. Threshold is chosen to be 60% of the maximum coefficient. Each label corresponds to a gene name. A thick edge corresponds to a positively reinforcing causal connectivity and a dotted edge corresponds to one of a negative effect. Shown in green color are the genes that are common with the ones found in the NFKB1 pathway found in Ingenuity© (see Figure 18.8). Plot created using Cytoscape.

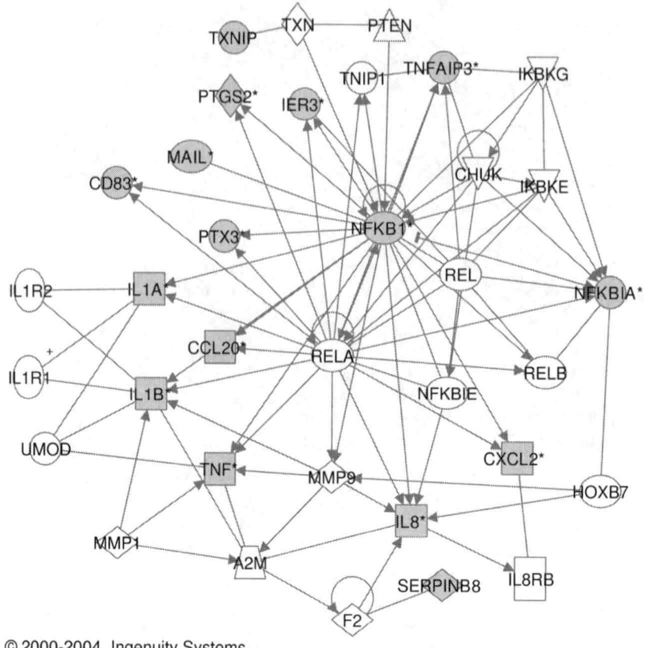

Fig. 18.8. Network obtained using Ingenuity$^{©}$ that shows the NFKB1 pathway. The top functions of this pathway includes immune response, tissue morphology and inflammatory disease.

$$\text{where } J_0 = \begin{bmatrix} 1 & 1 & 0 & 0 \\ 0 & 1 & 1 & 0 \\ 0 & 0 & 1 & 1 \\ 0 & 0 & 0 & 1 \end{bmatrix}, \; J_i = \begin{bmatrix} \sigma_1 & \omega_1 \\ -\omega_1 & \sigma_1 \end{bmatrix}, \; i = 1...,4, \; \begin{bmatrix} 0 & 1 & \dots & 0 \\ 0 & 0 & 1 & \vdots \\ \vdots & & \ddots & \vdots \\ 0 & \dots & \dots & 0 \end{bmatrix}, \text{ and } \lambda_i =$$

$\sigma_i \pm \omega_i$, $i = 1, \dots, 4$ are the complex eigenvalues. Matrix A can be put in JCF using the similarity transformation

$$J = T^{-1}AT.$$

Now, the solution of the system in equation (18.3) is

$$\mathbf{x}(k) = A^k \mathbf{x}(0)$$

and $A^k = TJ^kT^{-1}$. Due to the block structure, the k-th power of J is obtained by taking the k-th power of each individual Jordan cell. In particular, for a real eigenvalue λ_i,

$$J_i^k = (\lambda_i Id + \delta_{i+1,j})^k = \sum_{n=0}^{k} \binom{k}{n} \lambda_i^{k-n} \delta_{i+n}^j. \tag{18.9}$$

Let $T = [T_0\ T_1\ \ldots\ T_5]$ where T_i are the columns of T associated with ith Jordan block J_i, i.e. $AT_i = T_i J_i$. Now, with the change of coordinates $\mathbf{x} = T\tilde{\mathbf{x}}$, one can put the system in equation (18.3) into the form $\tilde{\mathbf{x}}(k+1) = J\tilde{\mathbf{x}}(k)$ and further decompose into independent '*Jordan block systems*' $\tilde{\mathbf{x}}_i(k+1) = J_i\tilde{\mathbf{x}}_i(k)$. Then

$$\mathbf{x}(k) = A^k\mathbf{x}(0) = TJ^k\tilde{\mathbf{x}}(0) = \sum_{i=1}^{n} T_i J_i^k (S_i^T \mathbf{x}(0)) \qquad (18.10)$$

where

$$T^{-1} = \begin{bmatrix} S_1^T \\ \vdots \\ S_n \end{bmatrix}.$$

Therefore we can identify the *generalized eigenmodes* of the system from J^k and according to equation (18.10), all solutions $\mathbf{x}(k)$ are linear combinations of these (generalized) modes.

From equations (18.8) and (18.9), we can list all the 12 modes corresponding to eigenvalues $1, \lambda_i = r_i e^{\pm j\theta_i}$, $(i = 1, \ldots, 4)$ as

$$\begin{aligned} &1, \\ &k \text{ terms}, \\ &k^2 \text{ terms}, \\ &k^3 \text{ terms}, \\ &r_i^k \cos(k\theta_i),\ i = 1, \ldots, 4 \\ &r_i^k \sin(k\theta_i),\ i = 1, \ldots, 4. \end{aligned} \qquad (18.11)$$

These modes can be treated as a set of basis functions (say, ϕ_j) and the gene expression profiles $x_i(k)$ can be written as,

$$x_i(k) = \sum_{j=1}^{12} \gamma_j \phi_j(k)$$

where γ_j are the corresponding coefficients. Figure 18.9 shows the coefficients of each of the 12 modes for six randomly chosen genes. Based on their relative participation one can cluster the genes. Such a clustering using the K-means algorithm on the coefficients is shown in Figure 18.10 and the corresponding gene expression profiles are shown in Figure 18.11.

18.6 Concluding Remarks

In this paper, we discussed in detail, various analysis approaches one could utilize in analyzing time-series microarray data. Methods are mostly geared toward a systems approach rather than one of a purely statistical in nature. A few interesting observations and ideas will be outlined here.

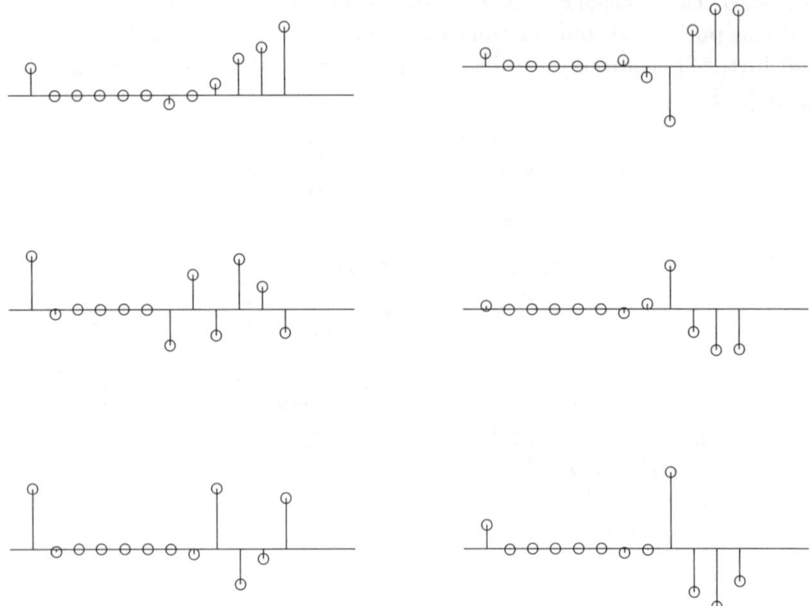

Fig. 18.9. Coefficients showing the relative participation of the 12 eigenmodes for six randomly selected genes. Horizontal axis shows the mode number and the vertical lines are proportional in height to the coefficient value.

The α-space trajectories shown in Figure 18.2 for toxin treated subjects, resemble to 'Homoclinic orbits'[5] found in chaos theory and nonlinear dynamics. Homoclinic orbits occur when the stable manifold and the unstable manifold of a dynamical system intersects. This analogy well explains the idea that the volunteers drift on to an unstable manifold when they get sick and transfers back on to the stable manifold when they become healthy.

The gene regulatory network shown in Figure 18.7 indeed reveals some interesting features. NFKB1 pathway overlaps with the one reported in Ingenuity$^{\circledR}$ well. The top functions of this pathway includes immune response, tissue morphology and inflammatory disease which are all related to the specific toxin response in human volunteers. There are few ESTs (Expressed Sequence Tags) that may worth giving attention.

References

1. J. Aach and G. Church. Aligning gene expression time series with time warping algorithms. *Bioinformatics*, 17(6):495–508, Jun 2001.

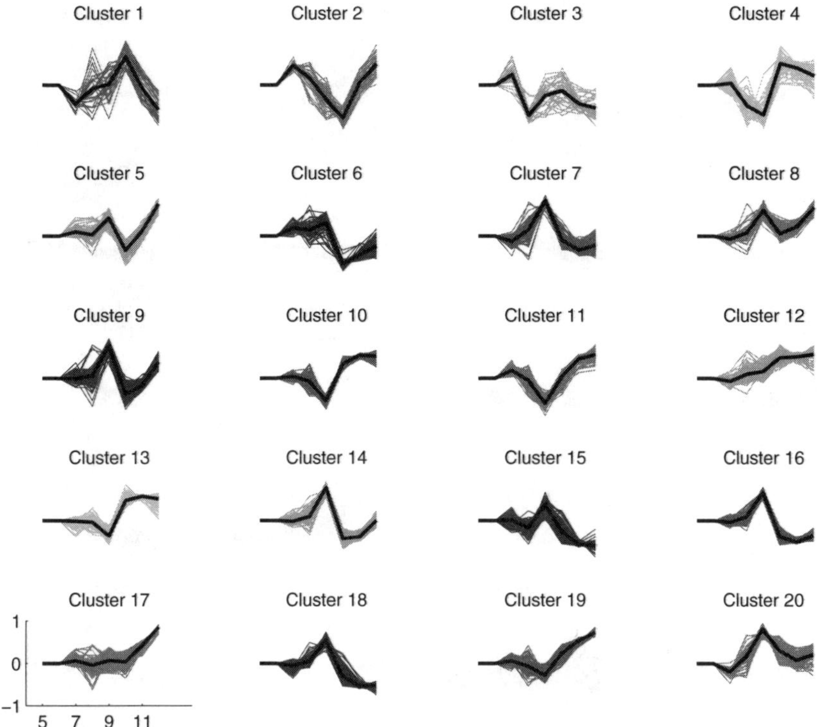

Fig. 18.10. Clusters of genes based on the relative eigenmode participation (only the modes corresponding to complex eigenvalues). Shown here are the coefficients corresponding to gene expression profiles. Horizontal axis shows the mode number (8 oscillatory modes correspond to indices 5-12) and the vertical axis shows the coefficient value.

2. A. Ben-Israel and T. N. E. Greville. *Generalized Inverses: Theory and Applications*. Wiley-Interscience [John Wiley & Sons], New York, 1974. (reprinted by Robert E. Krieger Publishing Co. Inc., Huntington, NY, 1980.).

3. C. T. Chen. *Linear Systems Theory and Design*. Oxford University Press, Oxford, 3 edition, 1999.

4. P. D'Haeseleer, X. Wen, S. Fuhrman, and R. Somogyi. Linear modeling of mrna expression levels during cns development and injury. *Pac Symp Biocomput*, pages 41–52, 1999.

5. J. Guckenheimer and P. Holmes. *Nonlinear Oscillations, Dynamical Systems, and Bifurcations of Vector Fields*. Number 42 in Applied Mathematical Sciences. Springer-Verlag, New York, NY, 1983.

6. N. S. Holter, A. Maritan, M. Cieplak, N. V. Fedoroff, and J. R. Banavar. Dynamic modeling of gene expression data. *Proc Natl Acad Sci U S A*, 98(4):1693–1698, Feb 2001.

7. S. Liang, S. Fuhrman, and R. Somogyi. Reveal, a general reverse engineering algorithm for inference of genetic network architectures. *Pac Symp Biocomput*, pages 18–29, 1998.

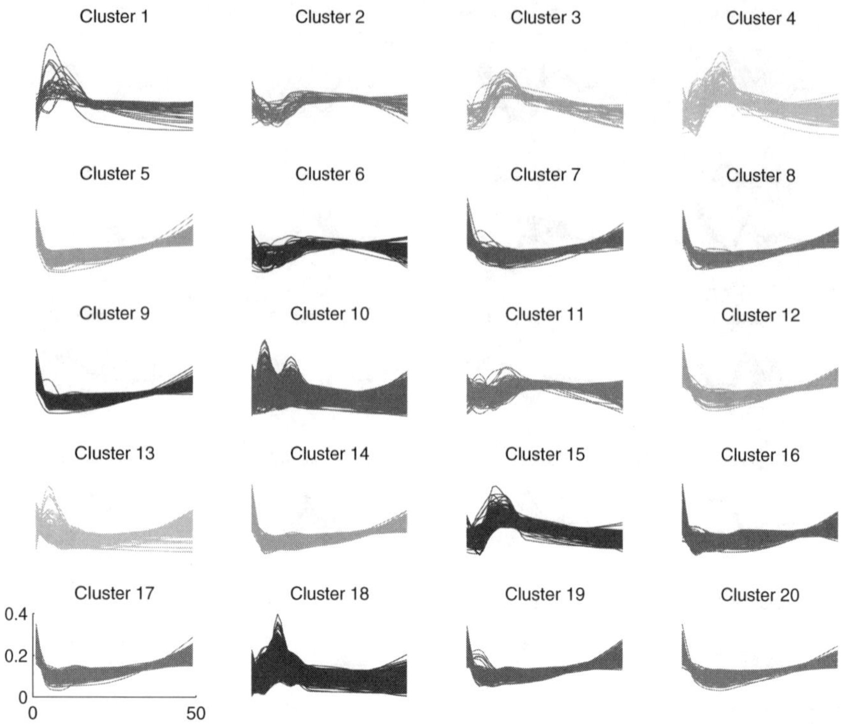

Fig. 18.11. Gene clusters corresponding to the eigenmode clusters shown in Figure 18.10. Axes are expression value vs. time.

8. A. C. Pease, D. Solas, E. J. Sullivan, M. T. Cronin, C. P. Holmes, and S. P. Fodor. Light-generated oligonucleotide arrays for rapid DNA sequence analysis. *Proc Natl Acad Sci U S A*, 91(11):5022–5026, May 1994.

9. T. J. Perkins, M. Hallett, and L. Glass. Inferring models of gene expression dynamics. *J Theor Biol*, 230(3):289–299, Oct 2004.

10. C. Rangel, J. Angus, Z. Ghahramani, M. Lioumi, E. Sotheran, A. Gaiba, D. L. Wild, and F. Falciani. Modeling T-cell activation using gene expression profiling and state-space models. *Bioinformatics*, 20(9):1361–1372, Jun 2004.

11. M. Schena, D. Shalon, R. W. Davis, and P. O. Brown. Quantitative monitoring of gene expression patterns with a complementary DNA microarray. *Science*, 270(5235):467–470, Oct 1995.

12. P. Shannon, A. Markiel, O. Ozier, N. S. Baliga, J. T. Wang, D. Ramage, N. Amin, B. Schwikowski, and T. Ideker. Cytoscape: a software environment for integrated models of biomolecular interaction networks. *Genome Res*, 13(11):2498–2504, Nov 2003.

13. H. Van-Trees. *Detection, Estimation, and Modulation Theory. Part I.* John Wiley & Sons, 1968.

14. D. C. Weaver, C. T. Workman, and G. D. Stormo. Modeling regulatory networks with weight matrices. *Pac Symp Biocomput*, pages 112–123, 1999.

New Results on Nonsmooth Output Feedback Stabilization of Nonlinear Systems

Chunjiang Qian[1]* and Wei Lin[2]†

[1] Department of Electrical Engineering, The University of Texas at San Antonio, San Antonio, TX 78249
cqian@utsa.edu

[2] Department of Electrical Engineering and Computer Science, Case Western Reserve University, Cleveland, OH 44106
linwei@nonlinear.cwru.edu

Dedicated to our friend Clyde F. Martin
on the occasion of his 60th birthday

Summary. This paper presents a *nonsmooth output* feedback framework for local and/or global stabilization of a class of inherently nonlinear systems, which may not be achieved by any smooth feedback controller. A systematic design algorithm is presented for the construction of stabilizing, dynamic output compensators that are nonsmooth but Hölder continuous. A new ingredient of the proposed output feedback control scheme is the introduction of a recursive observer design procedure, making it possible to construct a reduced-order observer step-by-step, in a naturally augmented fashion. Such nonsmooth design leads to a number of new results on output feedback stabilization of nonlinear systems with unstabilizable/undetectable linearization. One of them is the global stabilizability of a chain of odd power integrators by Hölder continuous output feedback. The other one is the local stabilization using nonsmooth output feedback for a class of nonlinear systems studied in [14], where global stabilizability by nonsmooth *state* feedback was proved to be possible.

19.1 Introduction

In this paper, we investigate the output feedback stabilization problem for a class of nonlinear systems described by equations of the form

$$\dot{x}_i = x_{i+1}^{p_i} + f_i(x_1, \cdots, x_i, x_{i+1}), \quad 1 \le i \le n - 1$$
$$\dot{x}_n = u + f_n(x)$$
$$y = x_1 \tag{19.1}$$

*C. Qian's work was supported in part by the U.S. NSF grant ECS-0239105.

†W. Lin's work was supported in part by the U.S. NSF grants ECS-9875273 and DMS-0203387.

C. Qian and W. Lin: *New Results on Nonsmooth Output Feedback Stabilization of Nonlinear Systems*, LNCIS **321**, 305–319 (2005)
www.springerlink.com

where $x = (x_1, \cdots, x_n)^T \in \mathbb{R}^n$, $y \in \mathbb{R}$ and $u \in \mathbb{R}$ are the system state, output and control input, respectively. For $i = 1, \cdots, n-1$, $p_i \geq 1$ is an *odd* positive integer, and $f_i : \mathbb{R}^{i+1} \to \mathbb{R}$, is a C^1 mapping with $f_i(0, \cdots, 0) = 0$. The function $f_n(x)$ is C^1 and $f_n(0) = 0$.

In the recent paper [3], it was proved that every smooth affine system, i.e.

$$\dot{\xi} = f(\xi) + g(\xi)v \qquad \text{and} \qquad y = h(\xi), \qquad (19.2)$$

is feedback equivalent to the nonlinear system (19.1) with $f_i(x_1, \cdots, x_i, x_{i+1})$ $= \sum_{k=0}^{p_i-1} x_{i+1}^k a_{i,k}(x_1, \cdots, x_i), 1 \leq i \leq n-1$, by a local diffeomorphism $x = T(\xi)$ and a smooth state feedback $v = \alpha(\xi) + \beta u$ if, and only if, a set of necessary and sufficient conditions, namely, the conditions (C1)—(C4) of Theorem 2.6 in [3] hold. Moreover, it was pointed out [3] that these conditions are nothing but a natural generalization of the exact feedback linearization conditions. Indeed, in the case when $(p_1, \cdots, p_{n-1}) = (1, \cdots, 1)$, (C1)—(C4) in [3] reduce to the well-known necessary and sufficient conditions for affine systems to be exactly feedback linearizable. Therefore, (19.1) with a suitable form of $f_i(\cdot)$ can be regarded as a *generalized normal form* of affine systems when exact feedback linearization is not possible.

For the nonlinear system (19.1) with $f_i(\cdot) = \sum_{k=0}^{p_i-1} x_{i+1}^k a_{i,k}(x_1, \cdots, x_i)$, the problem of global stabilization by state feedback has been addressed in [13, 14]. Due to the presence of uncontrollable unstable linearization, system (19.1) may not be stabilized by any smooth state feedback, even locally. It is, however, globally stabilizable by Hölder continuous state feedback [14]. Compared with significant advances in stabilization of nonlinear systems using state feedback, much less progress has been made in designing *output* feedback controllers for inherently nonlinear systems such as (19.1). As a matter of fact, output feedback stabilization of (19.1) is an extremely challenging problem, because the lack of observability of the linearized system of (19.1) makes the conventional output feedback design methods inapplicable. For highly nonlinear systems with uncontrollable/unobservable linearization, there are very few results in the literature addressing difficult issues such as output feedback stabilization. Even in some relatively simple cases, for instance, the *local* case, a fundamental question of whether the nonlinear system (19.1) is locally stabilizable by nonsmooth output feedback remains largely open and unanswered.

Over the past few years, attempt has been made to tackle this difficult problem and some preliminary results have been obtained towards the *output* feedback stabilization of lower-dimensional nonlinear systems (some elegant results on state feedback stabilization of two or three-dimensional systems can be found, for instance, in [1, 4, 5, 9]). The paper [15] considered a class of high-order planar systems in a lower-triangular form:

$$\dot{x}_1 = x_2^{p_1} + f_1(x_1), \quad \dot{x}_2 = u + f_2(x_1, x_2), \quad y = x_1. \qquad (19.3)$$

When $p_1 > 1$, the first approximation of system (19.3) is not controllable nor observable. As a result, the traditional "Luenberger-type" or "high-gain" observer proposed in [11, 6, 10, 12] cannot be applied to the high-order system (19.3). To solve the global stabilization problem by *smooth* output feedback, we proposed in [15] a one-dimensional nonlinear observer that is constructed using a feedback domination design combined with the tool of adding a power integrator. With the help of the reduced-order observer, a smooth output feedback stabilizer was designed for the planar system (19.3), under a high-order growth condition imposed on $f_i(\cdot)$ [15]. The growth condition was relaxed later in [16], by employing *nonsmooth* rather than smooth output feedback. Note that both papers considered only the output feedback stabilization for *planar* systems. A major limitation of the papers [15, 16] is that the proposed output feedback control schemes are basically an *ad hoc* design. They cannot be extended to higher-dimensional nonlinear systems with uncontrollable/unobservable linearization.

In the higher-dimensional case, there are fewer results available in the literature, which address systematically the question of how to stabilize inherently nonlinear systems via output feedback. In the recent paper [21], the problem of global stabilization by *smooth output* feedback was shown to be solvable for a chain of odd power integrators with same powers (i.e., $p_1 = \cdots = p_{n-1}$). This was done by developing a new output feedback control scheme that allows one to design both *homogeneous observers* and controllers explicitly. While the state feedback control law is constructed based on the tool of adding a power integrator, the observer design was carried out by using a newly developed machinery which can be viewed as a *dual* of the adding a power integrator technique. A novelty of the homogeneous observer design approach in [21] is that the observer gains can be assigned one-by-one, in an iterative manner. As a byproduct, it was also proved in [21] that the new output feedback design method, with a suitable twist, results in a solution to the global output feedback stabilization of a class of high-order non-homogeneous systems.

Despite the aforementioned progress, many important output feedback control problems remains open and unsolved. One of them, for instance, is whether a chain of odd power integrators with *different powers* can be stabilized by output feedback? The other is when the inherently nonlinear system (19.1), which is impossible to be handled by any smooth feedback, is locally or globally stabilizable by nonsmooth output feedback? These fundamental issues will be addressed in this paper, and some fairly satisfactory answers to these open questions will be given in the next two sections. Realizing that it is not possible to deal with genuinely nonlinear systems such as (19.1) by smooth feedback, even locally, we shall develop, throughout this paper, a *nonsmooth* output feedback control scheme to tackle the output feedback stabilization problem of the nonlinear system (19.1).

The objectives of the paper are twofold: to identify appropriate conditions under which a class of inherently nonlinear systems (19.1) with arbitrarily odd integers $p_i \geq 1$ is locally and/or globally stabilizable by *nonsmooth output* feedback, and to develop a nonsmooth feedback design approach for the explicit construction of a stabilizing, dynamic output compensator

$$\dot{z} = \eta(z, y), \qquad z \in \mathbb{R}^m, \quad m \le n$$
$$u = u(z, y), \tag{19.4}$$

where $\eta : \mathbb{R}^m \times \mathbb{R} \to \mathbb{R}^m$ is a C^0 but nonsmooth mapping.

Inspired by the observer design approach in [16], we shall develop in this paper a new methodology to construct systematically a reduced-order observer for a general class of nonlinear systems that go substantially beyond the systems studied in [15, 16, 21]. Combining the new reduced-order observer design with the tool of adding a power integrator [13, 14], we shall be able to solve the problem of global stabilization by *nonsmooth output* feedback, for a number of inherently nonlinear systems with unstabilizable/undetectable linearization. The nonsmooth output feedback stabilization theory developed in this work leads to several new and important conclusions. One of them, among the other things, is that every chain of odd power integrators is globally stabilizable by Hölder continuous output feedback. Another important result is that the local stabilization can be achieved by nonsmooth output feedback for the class of nonlinear systems considered in [13, 14], where only global stabilization via nonsmooth state feedback was shown to be possible.

19.2 The Case of a Chain of Power Integrators

In this section, we consider the problem of global output feedback stabilization for a relatively simple but rather important subclass of nonlinear systems (19.1). Specifically, we focus our attention on a nonlinear system with unstabilizable/undetectable linearization known as a chain of odd power integrators:

$$\dot{x}_i = x_{i+1}^{p_i}, \quad 1 \le i \le n - 1$$
$$\dot{x}_n = u$$
$$y = x_1. \tag{19.5}$$

When all the odd positive integers p_i are identical, global output feedback stabilization of system (19.5) has been shown to be solvable by *smooth* output feedback [21]. The solution in [21] was derived based on a newly developed output feedback control scheme that enables one to construct recursively a homogeneous observer and a smooth state feedback controller. The novelty of the work [21] lies in the design of a homogeneous observer, particularly, the iterative way of tuning the gains of the homogeneous observer. However, the output feedback design method proposed in [21] cannot be easily extended to the chain of power integrators (19.5) with different $p_i's$, due to the use of a smooth feedback design.

To achieve global output feedback stabilization for system (19.5), we now present a nonsmooth output feedback design method which provides a truly recursive way for the construction of dynamic output compensators. A new ingredient is the introduction of a recursive observer design algorithm, making it possible to construct a reduced-order observer step-by-step, in a naturally augmented manner. A combination of the new observer design with the tool of adding a power integrator [13, 14] (for the construction of nonsmooth state feedback controllers) leads to a globally stabilizing, nonsmooth output feedback controller. The main result of this section is the following theorem.

Theorem 1. There is a continuous output feedback controller of the form (19.4) globally stabilizing system (19.5).

Proof. We break up the proof into three parts. First, a *continuous* state feedback control law is designed via the adding a power integrator technique [13, 14]. Then, we construct a reduced-order observer step-by-step with a set of constant gains that will be determined in the last step. Finally, we show that a careful selection of the observer gains guarantees global strong stability of the closed-loop system. $\square$

I. Nonsmooth State Feedback Design

For a chain of power integrators (19.5), globally stabilizing nonsmooth state feedback controllers can be constructed using the method suggested in [13, 14]. The proposition below is a minor modification of the one in [14] and provides an explicit formula. The detailed proof is omitted for the reason of spaces.

Proposition 1. There is a Hölder continuous state feedback controller

$$u^* = -b_n \xi_n^{\frac{1}{p_1 \cdots p_{n-1}}}, \quad \xi_n = x_n^{p_1 \cdots p_{n-1}} + b_{n-1} x_{n-1}^{p_1 \cdots p_{n-2}} + \cdots + b_2 x_2^{p_1} + b_1 x_1,$$
$$(19.6)$$

with b_i, $1 \leq i \leq n$ being real constants, such that

$$\dot{W}(x)\Big|_{(19.5)-(19.6)} \leq -x_1^2 - x_2^{2p_1} - \cdots - x_n^{2p_1 \cdots p_{n-1}} + c_0 \xi_n^{2-1/(p_1 \cdots p_{n-1})}(u - u^*)$$
$$(19.7)$$

where $W(x)$ is a positive definite and proper Lyapunov function, whose form can be found in [14], and $c_0 > 0$ is a real constant.

II. Recursive Design of Nonlinear Observers

Although $y = x_1$ is available for feedback design, the state $(x_2, \cdots, x_n)$ of (19.5) is unmeasurable. As a result, the controller (19.6) cannot be implemented. To obtain an implementable controller, one must design an observer to estimate $(x_2, \cdots, x_n)$. Motivated by the observer design methods in [16, 21], we now develop a machinery that makes it possible to build a *nonsmooth* nonlinear observer step-by-step, in an augmented fashion. This is the basic philosophy to be pursued in the following design.

To see how a nonlinear observer can be recursively constructed, we first consider the case when $n = 2$ in the system (19.5). In this case, one can construct, similar to the design method in [16] with a suitable modification, the following one-dimensional nonlinear observer

$$\dot{\hat{z}}_2 = -\ell_1 \hat{x}_2^{p_1}, \qquad \hat{x}_2^{p_1} = \hat{z}_2 + \ell_1 x_1. \tag{19.8}$$

In other words, we use a reduced-order observer to estimate, instead of the state x_2, the unmeasurable variable $z_2 := x_2^{p_1} - \ell_1 x_1$, where ℓ_1 is a gain constant to be assigned later.

When $n = 3$, a two-dimensional observer need to be constructed for estimating the unmeasurable variables (z_2, z_3). Of course, a desirable way for the recursive observer design is to keep the one-dimensional observer already built for z_2 unchanged, and being a part of the two-dimensional observer. With this idea in mind, we simply augment a one-dimensional observer of the form

$$\dot{\hat{z}}_3 = -\ell_2 \hat{x}_3^{p_2}, \qquad \hat{x}_3^{p_2} = \hat{z}_3 + \ell_2 \hat{x}_2, \tag{19.9}$$

to the dynamic system (19.8), where $\hat{z}_3$ is an estimate of the unmeasurable variable $z_3 := x_3^{p_2} - \ell_2 x_2$, and ℓ_2 is a gain constant to be determined later.

In this way, we have obtained, in a naturally augmented manner, a two-dimensional observer consisting of (19.8)-(19.9) for the chain of integrators (19.5) with $n = 3$. There are three new ingredients in the construction of the observer (19.8)-(19.9): (i) no change of the structure is made to the observer for z_2 obtained in the previous step; (ii) the augmented observer estimates the unmeasurable variable z_3 that is related to $x_3^{p_2}$ and x_2, rather than a linear function of x_3 and x_1 as done in [21]; (iii) the observer (19.8)-(19.9) is *nonsmooth* while the homogeneous observer designed in [21] is *smooth*.

Such an augmented design method enables us to construct a nonlinear observer recursively, going from lower-dimensional systems to higher dimensional systems step-by-step. Indeed, for the n-dimensional chain of odd power integrators (19.5), applying the augmented design algorithm repeatedly, we arrive at the following $(n-1)$-dimensional observer

$$\dot{\hat{z}}_{i+1} = -\ell_i \hat{x}_{i+1}^{p_i}, \qquad \hat{x}_{i+1}^{p_i} = \hat{z}_{i+1} + \ell_i \hat{x}_i, \quad 1 \leq i \leq n-1, \tag{19.10}$$

where $\hat{x}_1 = x_1$ and $\hat{z}_{i+1}$ is the estimate of the unmeasurable variable $z_{i+1} := x_{i+1}^{p_i} - \ell_i x_i$.

Let $e_{i+1} = z_{i+1} - \hat{z}_{i+1}$, $i = 1, \cdots, n-1$, be the estimate errors. Then, a direct calculation yields

$$\dot{e}_2 = p_1 x_2^{p_1 - 1} x_3^{p_2} - \ell_1 \left(x_2^{p_1} - \hat{x}_2^{p_1} \right)$$

$$\vdots$$

$$\dot{e}_n = p_{n-1} x_n^{p_{n-1}-1} u - \ell_{n-1} \left(x_n^{p_{n-1}} - \hat{x}_n^{p_{n-1}} \right). \tag{19.11}$$

Note that

$$e_{i+1} = x_{i+1}^{p_i} - \ell_i x_i - \hat{x}_{i+1}^{p_i} + \ell_i \hat{x}_i. \tag{19.12}$$

Therefore,

$$x_{i+1}^{p_i} - \hat{x}_{i+1}^{p_i} = e_{i+1} + \ell_i(x_i - \hat{x}_i). \tag{19.13}$$

With this in mind, it is easy to see that

$$\begin{aligned}
\dot{e}_2 &= p_1 x_2^{p_1-1} x_3^{p_2} - \ell_1 e_2 \\
\dot{e}_3 &= p_2 x_3^{p_2-1} x_4^{p_3} - \ell_2\big(e_3 + \ell_2(x_2 - \hat{x}_2)\big) \\
&\ \ \vdots \\
\dot{e}_n &= p_{n-1} x_n^{p_{n-1}-1} u - \ell_{n-1}\big(e_n + \ell_{n-1}(x_{n-1} - \hat{x}_{n-1})\big).
\end{aligned} \tag{19.14}$$

Now, consider the Lyapunov function

$$V(e_2, \cdots, e_n) = \frac{e_2^2}{2} + \frac{e_3^{2p_1}}{2p_1} + \cdots + \frac{e_n^{2p_1 \cdots p_{n-2}}}{2p_1 \cdots p_{n-2}} \tag{19.15}$$

which is positive definite and proper. Clearly,

$$\begin{aligned}
\dot{V} &= \Sigma_{i=2}^n \big(e_i^{2p_1 \cdots p_{i-2}-1} p_{i-1} x_i^{p_{i-1}-1} x_{i+1}^{p_i} - \ell_{i-1} e_i^{2p_1 \cdots p_{i-2}}\big) \\
&\quad + \Sigma_{i=3}^n e_i^{2p_1 \cdots p_{i-2}-1} \ell_{i-1}^2 (x_{i-1} - \hat{x}_{i-1}).
\end{aligned} \tag{19.16}$$

In order to estimate the terms on the right-hand side of (19.16), we introduce two propositions whose proofs involve tedious calculations but nevertheless can be carried out using Lemmas 1.1–1.3 in the Appendix. The details are given in [17].

Proposition 2. For $i = 2, \cdots, n$, given any $N_i > 0$, there is a real constant $c_i > 0$ such that

$$\left| e_i^{2p_1 \cdots p_{i-2}-1} p_{i-1} x_i^{p_{i-1}-1} x_{i+1}^{p_i} \right| \le c_i e_i^{2p_1 \cdots p_{i-2}} + \frac{1}{N_i}\left(x_i^{2p_1 \cdots p_{i-1}} + x_{i+1}^{2p_1 \cdots p_i}\right) \tag{19.17}$$

where $p_n = 1$ and $x_{n+1} = u$.

Proposition 3. There exist constants $\beta_j(\ell_j, \cdots, \ell_{n-1}) > 0$ depending on the gain parameters $\ell_j, \cdots, \ell_{n-1}$, $j = 2, \cdots, n-1$, such that

$$\begin{aligned}
\sum_{i=3}^n e_i^{2p_1 \cdots p_{i-2}-1} \ell_{i-1}^2 (x_{i-1} - \hat{x}_{i-1}) &\le \sum_{i=3}^n \frac{1}{2} \ell_{i-1} e_i^{2p_1 \cdots p_{i-2}} \\
&\quad + e_2^2 \beta_2(\ell_2, \cdots, \ell_{n-1}) + \cdots + e_{n-1}^{2p_1 \cdots p_{n-3}} \beta_{n-1}(\ell_{n-1}).
\end{aligned} \tag{19.18}$$

With the help of Propositions 2 and 3 , the following inequality can be obtained by setting $N_i = 12$, $i = 2, \cdots, n-1$, and $N_n > 12$.

$$\begin{aligned}
\dot{V} &\le -\sum_{i=2}^n \frac{1}{2} \ell_{i-1} e_i^{2p_1 \cdots p_{i-2}} + \frac{1}{6}(x_2^{2p_1} + \cdots + x_n^{2p_1 \cdots p_{n-1}}) + \frac{u^{2p_1 \cdots p_{n-1}}}{N_n} \\
&\quad + e_2^2\big(c_2 + \beta_2(\ell_2, \cdots, \ell_{n-1})\big) + e_3^{2p_1}\big(c_3 + \beta_3(\ell_3, \cdots, \ell_{n-1})\big) \\
&\quad + \cdots + e_{n-1}^{2p_1 \cdots p_{n-3}}\big(c_{n-1} + \beta_{n-1}(\ell_{n-1})\big) + e_n^{2p_1 \cdots p_{n-2}} c_n.
\end{aligned} \tag{19.19}$$

III. Nonsmooth Output Feedback Design

In this part of design, we apply the certainty equivalence principle to obtain an implementable output feedback controller. Observe that the reduced-order observer (19.10) has provided an estimation for the unmeasurable states $(x_2, \cdots, x_n)$. Keeping this in mind, we simply replace $(x_2, \cdots, x_n)$ in the controller (19.6) by its estimate $(\hat{x}_2, \cdots, \hat{x}_n)$ generated from the observer (19.10). Thus,

$$u = u(\hat{x}) = -b_n \left[\hat{x}_n^{p_1 \cdots p_{n-1}} + \cdots + b_2 \hat{x}_2^{p_1} + b_1 x_1\right]^{\frac{1}{p_1 \cdots p_{n-1}}} \qquad (19.20)$$

Clearly, Substituting the implementable controller (19.20) into the inequality (19.7) results in a redundant term $c_0 \xi_n^{2-1/(p_1 \cdots p_{n-1})}(u(\hat{x}) - u^*)$, where $u^* = u(x)$ defined by (19.6). Using Lemmas 1 and 2, it is not difficult to prove that for any $M > 0$,

$$\left| c_0 \xi_n^{2-1/(p_1 \cdots p_{n-1})}\left(u(\hat{x}) - u(x)\right)\right| \leq \frac{1}{M}\xi_n^2(x) + c_{n+1}\left(\xi_n(\hat{x}) - \xi_n(x)\right)^2 \qquad (19.21)$$

where $\xi_n(x)$ is defined in (19.6) and $c_{n+1} > 0$ is a constant.

On the other hand, under the controller (19.20), the term $u = u(\hat{x})$ in (19.19) can be estimated as follows:

$$\frac{u^{2p_1 \cdots p_{n-1}}(\hat{x})}{N_n} \leq \frac{1}{M}\left((\xi_n(\hat{x}) - \xi_n(x))^2 + \xi_n^2(x)\right), \qquad (19.22)$$

where $N_n = 2b_n^{2p_1 \cdots p_{n-1}} M$.

Putting (19.19), (19.7) and (19.21)-(19.22) together, we have

$$\dot{W} + \dot{V} \leq -x_1^2 - x_2^{2p_1} - \cdots - x_n^{2p_1 \cdots p_{n-1}} + \frac{2}{M}\xi_n^2(x)$$

$$+(c_{n+1} + \frac{1}{M})(\xi_n(\hat{x}) - \xi_n(x))^2 - \sum_{i=2}^{n} \frac{1}{2}\ell_{i-1}e_i^{2p_1 \cdots p_{i-2}}$$

$$+\frac{x_2^{2p_1} + \cdots + x_n^{2p_1 \cdots p_{n-1}}}{6} + e_2^2(c_2 + \beta_2(\ell_2, \cdots, \ell_{n-1}))$$

$$+ \cdots + e_{n-1}^{2p_1 \cdots p_{n-3}}(c_{n-1} + \beta_{n-1}(\ell_{n-1})) + e_n^{2p_1 \cdots p_{n-2}} c_n. \qquad (19.23)$$

The next proposition gives an estimation for one of the terms in (19.23). Its proof can be found in [17].

Proposition 4. There exist constants $\gamma_i(\ell_i, \cdots, \ell_{n-1}) > 0$ depending on the gain parameters $\ell_i, \cdots, \ell_{n-1}$, $2 \leq i \leq n-1$ and a real constant $\gamma_n > 0$ independent of all the $\ell_i's$, such that

$$(c_{n+1} + \frac{1}{M})(\xi_n(\hat{x}) - \xi_n(x))^2 \leq \frac{1}{6}(x_1^2 + \cdots + x_n^{2p_1 \cdots p_{n-1}}) + e_2^2 \gamma_2(\ell_2, \cdots, \ell_{n-1})$$

$$+e_3^{2p_1} \gamma_3(\ell_3, \cdots, \ell_{n-1}) + \cdots + e_{n-1}^{2p_1 \cdots p_{n-3}} \gamma_{n-1}(\ell_{n-1}) + e_n^{2p_1 \cdots p_{n-2}} \gamma_n.$$

$$(19.24)$$

By definition (19.6), it is straightforward to show that for a sufficiently large constant $M > 0$,

$$\frac{2}{M}\xi_n^2(x) \leq \frac{1}{6}(x_1^2 + x_2^{2p_1} + \cdots + x_n^{2p_1 \cdots p_{n-1}}). \tag{19.25}$$

Substituting (19.24) and (19.25) into (19.23) yields

$$\dot{W}+\dot{V} \leq -\frac{1}{2}(x_1^2 + x_2^{2p_1} + \cdots + x_n^{2p_1 \cdots p_{n-1}}) - \sum_{i=2}^{n}\frac{1}{2}\ell_{i-1}e_i^{2p_1 \cdots p_{i-2}}$$
$$+e_2^2\big[c_2 + \beta_2(\ell_2, \cdots, \ell_{n-1}) + \gamma_2(\ell_2, \cdots, \ell_{n-1})\big] + \cdots$$
$$+e_{n-1}^{2p_1 \cdots p_{n-3}}\big[c_{n-1} + \beta_{n-1}(\ell_{n-1}) + \gamma_{n-1}(\ell_{n-1})\big] + e_n^{2p_1 \cdots p_{n-2}}(c_n + \gamma_n). \tag{19.26}$$

From (19.26), it is clear that choosing the gain parameters ℓ_i one-by-one, in the following manner:

$$\ell_{n-1} = 2(c_n + \gamma_n) + 1$$
$$\ell_{n-2} = 2(c_{n-1} + \beta_{n-1}(\ell_{n-1}) + \gamma_{n-1}(\ell_{n-1})) + 1$$
$$\vdots \tag{19.27}$$
$$\ell_2 = 2(c_3 + \beta_3(\ell_3, \cdots, \ell_{n-1}) + \gamma_3(\ell_3, \cdots, \ell_{n-1})) + 1$$
$$\ell_1 = 2(c_2 + \beta_2(\ell_2, \cdots, \ell_{n-1}) + \gamma_2(\ell_2, \cdots, \ell_{n-1})) + 1,$$

we immediately have

$$\dot{W} + \dot{V} \leq -\frac{1}{2}(x_1^2 + x_2^{2p_1} + \cdots + x_n^{2p_1 \cdots p_{n-1}}) - \sum_{i=2}^{n}\frac{1}{2}e_i^{2p_1 \cdots p_{i-2}}$$

which is negative definite. Therefore, the closed-loop system (19.5)-(19.10)-(19.20) is globally strongly stable in the sense of Kurzweil [14]. $\square$

19.3 The Case of a Class of Genuinely Nonlinear Systems

In this section, we use the nonsmooth output feedback design approach developed so far to investigate the output feedback stabilization problem for a significant class of nonlinear systems whose linearization is neither controllable nor observable, moreover, the uncontrollable modes are associated with eigenvalues on the open right-half plane. Although this kind of systems cannot be dealt with by any smooth feedback, even locally [2], we show that under appropriate conditions, they are stabilizable by nonsmooth output feedback.

We begin by introducing an important theorem that shows how the global output feedback stabilization result established for a chain of power integrators can be extended to the following class of homogeneous systems

$$\dot{x}_i = x_{i+1}^{p_i} + a_{i,i}\, x_i + a_{i,i-1}\, x_{i-1}^{1/p_{i-1}} + a_{i,i-2}\, x_{i-2}^{1/(p_{i-2}p_{i-1})}$$

$$+ \cdots + a_{i,1}\, x_1^{1/(p_1 \cdots p_{i-1})}, \qquad 1 \le i \le n,$$

$$y = x_1, \tag{19.28}$$

where $p_n = 1$, $x_{n+1} = u$ and $a_{i,j}, i = 1, \cdots, n, \ j = 1, \cdots, i$ are real constants with $a_{i,j} \equiv 0$ whenever $j \le 0$.

Theorem 2. The homogeneous system (19.28) is globally stabilizable by non-smooth output feedback.

Proof. The proof is similar to the one given in Theorem 1, with an appropriate modification. Due to the construction of a more complex nonlinear observer, the observer gains are needed to be designed in a more subtle manner.

First, it is straightforward to prove that the homogeneous system (19.28) can also be globally stabilized by a nonsmooth state feedback control law of the form (19.6), with different coefficients $b_i's$.

Next, we apply the recursive observer design method developed in the last section to construct a nonlinear observer. Since the homogeneous system (19.28) has lower-triangular structure, a more rigorous observer needs to be constructed. In the case when $n = 2$, observe that $\dot{x}_1 = x_2^{p_1} + a_{1,1}x_1$. Then, we design the following estimator for x_2.

$$\dot{\hat{z}}_2 = -\ell_1(\hat{x}_2^{p_1} + a_{1,1}\hat{x}_1), \quad \hat{x}_2^{p_1} = \hat{z}_2 + \ell_1\hat{x}_1, \tag{19.29}$$

where $z_2 := x_2^{p_1} - \ell_1 x_1$, and ℓ_1 is a gain constant to be determined.

To build estimators for $x_3, \cdots, x_n$, define

$$H_i = a_{i,i}x_i + a_{i,i-1}x_{i-1}^{1/p_{i-1}} + \cdots + a_{i,1}x_1^{1/(p_1 \cdots p_{i-1})}. \tag{19.30}$$

With the aid of this compact notation, a set of estimators can be constructed recursively as follows:

$$\dot{\hat{z}}_{i+1} = -\ell_i \left(\hat{x}_{i+1}^{p_i} + H_i(\hat{x}_1, \cdots, \hat{x}_i) \right), \tag{19.31}$$

where $\hat{x}_{i+1}^{p_i} = \hat{z}_{i+1} + \ell_i\hat{x}_i, \quad i = 1, \cdots, n-1, \ \hat{x}_1 = x_1$.

Similar to what we did in the pervious section, let $e_i = z_i - \hat{z}_i$. Then, the error dynamics can be represented as

$$\dot{e}_i = p_{i-1}x_i^{p_{i-1}-1} \left(x_{i+1}^{p_i} + H_i(x_1, \cdots, x_i) \right) - \ell_{i-1}[e_i + \ell_{i-1}(x_{i-1} - \hat{x}_{i-1})$$

$$+ H_{i-1}(x_1, \cdots, x_{i-1}) - H_{i-1}(\hat{x}_1, \cdots, \hat{x}_{i-1})], \qquad i = 2, \cdots, n. \tag{19.32}$$

Now, consider the Lyapunov function defined by (19.15). Following the same line of the argument in Theorem 2, one can show that the time derivative of the Lyapunov function (19.15) along the trajectories of (19.32) satisfies

$$\dot{V} \le -\sum_{i=2}^{n} \frac{1}{2}\ell_{i-1}e_i^{2p_1 \cdots p_{i-2}} + \frac{1}{6}(x_2^{2p_1} + \cdots + x_n^{2p_1 \cdots p_{n-1}}) + \frac{u^{2p_1 \cdots p_{n-1}}}{N_n}$$

$$+ e_2^2[c_2 + \hat{\beta}_2(\ell_2, \cdots, \ell_{n-1})] + \cdots + e_{n-1}^{2p_1 \cdots p_{n-3}}[c_{n-1} + \hat{\beta}_{n-1}(\ell_{n-1})] + e_n^{2p_1 \cdots p_{n-2}}c_n.$$

The remaining part of the proof is almost identical to the one in proving Theorem 2, and therefore left to the reader as an exercise. $\square$

As an immediate consequence, we have the following important conclusion.

Corollary 1. The analytic nonlinear system $(i = 1, \cdots, n)$

$$\dot{x}_i = x_{i+1}^{p_i} + a_{i,i} x_i + \cdots + a_{i,i-k} x_{i-k}, \qquad y = x_1, \qquad (19.33)$$

is globally stabilizable by nonsmooth output feedback, where $p_{i-1} = \cdots = p_{i-k} = 1$ and $p_{i-(k+1)} \neq 1$.

In the paper [20], a stabilizing nonsmooth *state* feedback control law was explicitly constructed for the homogeneous system (19.33). Corollary 1 has shown that there also exists a *Hölder continuous output* compensator, globally stabilizing the nonlinear system (19.33). Moreover, with the help of Theorem 2, one can construct a global *output feedback* stabilizer for (19.33). The following example illustrates the significance of Corollary 1.

Example 1. Consider a four-dimensional nonlinear system of the form

$$\begin{aligned}
\dot{x}_1 &= x_2 \\
\dot{x}_2 &= x_3^3 + x_2 + x_1 \\
\dot{x}_3 &= x_4 \\
\dot{x}_4 &= u + x_4 + x_3 \\
y &= x_1
\end{aligned} \qquad (19.34)$$

It is easy to verify that (19.34) is of the form (19.33). In fact, when $i = 2$ and 4, it is not difficult to see that $k = 1$. By Corollary 1, there is a nonsmooth output feedback controller globally stabilizing (19.34). Notably, system (19.34) is inherently nonlinear and cannot be dealt with by any existing output feedback scheme including the one suggested in [21], because the linearized system of (19.34) at the origin is uncontrollable and unobservable. Worse still, the uncontrollable mode has an eigenvalue whose real part is positive. As a such, no smooth state/output feedback controllers exist, even locally, for the nonlinear system (19.34) [2, 14].

In the remainder of this section, we show how Theorem 2 or Corollary 1 can lead to some important results on output feedback stabilization of a number of classes of highly nonlinear systems that are not smoothly stabilizable.

The first result is devoted to the output feedback stabilization of a chain of odd power integrators perturbed by a strict-triangular vector field (i.e. system (19.1) with $f_i(x_1, \cdots, x_i, x_{i+1}) \equiv f_i(x_1, \cdots, x_i)$), whose global stabilization has been proved to be possible by Hölder continuous state feedback [14].

Theorem 3. For the strict-triangular system (19.1), the problem of global stabilization is solvable by Hölder continuous output feedback under the following conditions:

(H1) For $i = 1, \cdots, n$, there are constants $a_{i,j} \geq 0$, $j = 1, \cdots i$ such that

$$|f_i(x_1, \cdots, x_i)| \leq a_{i,i}|x_i| + a_{i,i-1}|x_{i-1}|^{1/p_{i-1}} \tag{19.35}$$
$$+ a_{i,i-2}|x_{i-2}|^{1/(p_{i-2}p_{i-1})} + \cdots + a_{i,1}|x_1|^{1/(p_1 \cdots p_{i-1})}$$

(H2) For $i = 1, \cdots, n-1$, there are constants $b_{i,j}$, $j = 2, \cdots i$ such that

$$|f_i(x_1, x_2, \cdots, x_i) - f_i(x_1, \hat{x}_2, \cdots, \hat{x}_i)| \leq b_{i,i}|x_i - \hat{x}_i| \tag{19.36}$$
$$+ b_{i,i-1}|x_{i-1}^{\frac{1}{p_{i-1}}} - \hat{x}_{i-1}^{\frac{1}{p_{i-1}}}| + \cdots + b_{i,2}|x_2^{\frac{1}{p_2 \cdots p_{i-1}}} - \hat{x}_2^{\frac{1}{p_2 \cdots p_{i-1}}}|$$

The next result indicates that
without imposing any condition on $f_i(\cdot)$, *local* stabilization of the strict-triangular system (19.1), which may be not smoothly stabilizable due to the presence of unstabilizable/undetectable linearization, is always achievable by nonsmooth output feedback.

Theorem 4. There is a Hölder continuous output feedback controller of the form (19.4) locally stabilizing system (19.1) when $f_i(x_1, \cdots, x_i, x_{i+1}) \equiv f_i(x_1, \cdots, x_i)$, $i = 1, \cdots, n$.

Proof. Since $f_i(\cdot)$ is C^1 and vanishes at the origin, it follows from the Taylor expansion Theorem that

$$f_i(x_1, \cdots, x_i) = a_{i,1}x_1 + \cdots + a_{i,i}x_i + \mathcal{O}(\|(x_1, \cdots, x_i)\|^2). \tag{19.37}$$

Suppose $p_{i-1} = \cdots = p_{i-k} = 1$, and $p_{i-(k+1)} \neq 1$. Then,

$$H_i(x_{i-k}, \cdots, x_i) = a_{i,i}x_i + \cdots + a_{i,i-k}x_{i-k} \tag{19.38}$$

is a homogeneous approximation of (19.37) under the dilation

$$r_1 = 1, \quad r_2 = 1/p_1, \quad \cdots, \quad r_n = 1/(p_1 \cdots p_n). \tag{19.39}$$

Moreover, $R_i(x_1, \cdots, x_i) = f_i(x_1, \cdots, x_i) - H_i(x_{i-k}, \cdots, x_i)$ is a high-order term under this dilation. In fact,

$$\lim_{\varepsilon \to 0} \frac{R_i(\varepsilon^{r_1}x_1, \cdots, \varepsilon^{r_i}x_i)}{\varepsilon^{r_i}} = 0, \tag{19.40}$$

because $p_{i-(k+1)} > 1$ implies that $r_1 = 1 > r_i, \cdots, r_{i-(k+1)} = 1/(p_1 \cdots p_{i-(k+2)}) > r_i$, where $r_i = 1/(p_1 \cdots p_{i-(k+1)})$.

In summary, the relationships (19.38) and (19.40) imply that system (19.33) is a homogeneous approximation of (19.1). According to Corollary 1, the nonsmooth output feedback controller composed of (19.31) and (19.20) globally stabilizes the homogeneous system (19.33). Note that by construction, the resulted nonlinear observer

$$\dot{\hat{z}}_2 = -\ell_1 \left(\hat{x}_2^{p_1} + H_1(\hat{x}_1) \right) \qquad \hat{x}_2^{p_1} = \hat{z}_2 + \ell_1 x_1$$
$$\dot{\hat{z}}_3 = -\ell_2 \left(\hat{x}_3^{p_2} + H_2(\hat{x}_1, \hat{x}_2) \right) \qquad \hat{x}_3^{p_2} = \hat{z}_3 + \ell_2 \hat{x}_2$$
$$\vdots \qquad\qquad\qquad \vdots$$
$$\dot{\hat{z}}_n = -\ell_{n-1} \left[\hat{x}_n^{p_{n-1}} + H_{n-1}(\hat{x}_1 ... \hat{x}_{n-1}) \right] \quad \hat{x}_n^{p_{n-1}} = \hat{z}_n + \ell_{n-1} \hat{x}_{n-1}$$

is homogeneous of degree 0 with the dilation

$$r_{n+1} = 1, \quad r_{n+2} = 1/p_1, \quad \cdots, \quad r_{2n-1} = 1/(p_1 \cdots p_{n-2}). \tag{19.41}$$

In addition, $\hat{x}_i$ has a same degree of homogeneity as x_i. Therefore, the controller (19.20) is also homogeneous with the composite dilation (19.39) and (19.41). By Hermes' robust stability theorem [8, 18], it is concluded that the dynamic output compensator (19.31)-(19.20) renders the original of the triangular system (19.1) locally asymptotically stable. $\square$

The final result of this section shows how Theorem 4 can be further extended to the non-strictly triangular system (19.1) with

$$f_i(x_1, \cdots, x_{i+1}) = \Sigma_{k=0}^{p_i-1} x_{i+1}^k a_{i,k}(x_1, \cdots, x_i), \tag{19.42}$$

where $a_{i,j}(x_1, \cdots, x_i)$, $i = 1, \cdots, n-1$, $j = 0, \cdots, p_i-1$, with $a_{i,j}(0, \cdots, 0) = 0$, are C^1 functions.

Theorem 5. The nonlinear system (19.1) with (19.42) is locally stabilizable by Hölder continuous output feedback.

Proof. We give a sketch of the proof. Using the Taylor expansion formula and Young's inequality, one has

$$\left| \sum_{j=1}^{p_1-1} x_2^j a_{1,j}(x_1) \right| = \left| \sum_{j=1}^{p_1-1} x_2^j x_1 \hat{a}_{1,j}(x_1) \right| \leq \sum_{j=1}^{p_1-1} \left(|x_2|^{p_1+1} + |x_1 \hat{a}_{1,j}|^{1+\frac{j}{p_1+1-j}} \right),$$

which is obviously a higher-order term (in the sense of homogeneity) with respect to $x_2^{p_1}$ and x_1. As a result, the homogeneous approximation of $\sum_{j=0}^{p_1-1} x_2^j a_{1,j}(x_1)$ is simply equal to $\beta_{1,0} x_1$. Similarly, one can prove that a homogeneous approximation of $\sum_{j=0}^{p_k-1} x_{k+1}^j a_{k,j}(x_1, \cdots, x_k)$ is same as the homogeneous approximation of the function $a_{k,0}(x_1, \cdots, x_k)$, which is $\beta_{k1} x_1 + \cdots + \beta_{kk} x_k$. Keeping this in mind, it is immediate to conclude from Theorem 4 that there exists a nonsmooth output feedback controller locally stabilizing the nonlinear system (19.1) with (19.42). $\square$

We conclude this section by solving the open and challenging problem of global output feedback stabilization for the following benchmark example, which is a simplified version of the underactuated unstable two degree of freedom mechanical system studied in [19, 14].

Example 2. Consider the nonsmoothly stabilizable system

$$\dot{x}_1 = x_2$$
$$\dot{x}_2 = x_3^3 + \frac{1}{4}\sin x_1$$
$$\dot{x}_3 = u$$
$$y = x_1. \tag{19.43}$$

Obviously, $f_2(x_1) = \frac{1}{4}\sin x_1$ is not homogeneous but it satisfies the conditions (19.35)-(19.36). Therefore, we are able to design an output feedback controller globally stabilizing system (19.43). Using the tool of adding a power integrator [14], a globally stabilizing, nonsmooth state feedback controller can be derived. In fact, it is easy to show that

$$u(x) = -18\left(x_3^3 + \frac{3}{2}(x_2 + \frac{1}{2}x_1)\right)^{1/3}. \tag{19.44}$$

On the other hand, employing the recursive observer design method in Section 3, one can construct a reduce-order observer of the form

$$\dot{\hat{z}}_2 = -\ell_1\hat{x}_2, \quad \dot{\hat{z}}_3 = -\ell_2(\hat{x}_3^3 + \sin x_1/4) \tag{19.45}$$

where $\hat{x}_2 = \hat{z}_2 + \ell_1 x_1$ and $\hat{x}_3^3 = \hat{z}_3 + \ell_2\hat{x}_2$.

Then, it is not difficult to see that by choosing the gain parameters ℓ_1 and ℓ_2 suitably, the nonsmooth controller

$$u(x_1, \hat{x}_2, \hat{x}_3) = -18\left(\hat{x}_3^3 + \frac{3}{2}(\hat{x}_2 + \frac{1}{2}x_1)\right)^{1/3} \tag{19.46}$$

globally stabilizes the non-homogeneous system (19.43).

As a direct consequence, we conclude that the four-dimensional, underactuated unstable two degree of freedom mechanical system in the state-space form (after a change of coordinates and pre-feedback) [19, 14] is globally stabilizable by nonsmooth output feedback as well.

Appendix

This section collects three useful Lemmas from [14, 16].

Lemma 1. Suppose $p \geq 1$ is an odd integer. Then,

$$|a - b|^p \leq 2^{p-1}|a^p - b^p|, \quad \forall a,\, b \in \mathbb{R}. \tag{19.47}$$

Lemma 2. Suppose n and m are two positive real numbers, a, b, $\pi \geq 0$ are continuous scalar-value functions. Then, for any constant $c > 0$,

$$a^n b^m \pi \leq c \cdot a^{n+m} + \frac{m}{n+m}\left[\frac{n}{c(n+m)}\right]^{\frac{n}{m}} b^{n+m}\pi^{\frac{n+m}{m}} \tag{19.48}$$

Lemma 3. Suppose $p \geq 1$ is an odd integer. Then,

$$|(a + b)^p - a^p| \leq c(a^{p-1} + b^{p-1})|b|, \quad a,\, b \in \mathbb{R}. \tag{19.49}$$

References

1. Bacciotti, Local stabilizability of nonlinear systems, World Scientific, 1992.
2. R.W. Brockett, Asymptotic stability and feedback stabilization, in Differential Geometric Control Theory, R.W. Brockett, R.S. Millman and H.J. Sussmann, Eds., Birkäuser, Basel-Boston (1983), 181–191.
3. D. Cheng and W. Lin, On p-Normal Forms of Nonlinear Systems, *IEEE Trans. Aut. Contr.*, **48**, 1242-1248 (2003).
4. W. P. Dayawansa, Recent advances in the stabilization problem for low dimensional systems, *Proc. of 2nd IFAC NOLCOS,* Bordeaux (1992), 1-8.
5. W. P. Dayawansa, C. F. Martin and S. Samelson, Asymptoticstabilization of a class of smooth two dimensional systems, *SIAM. J. Optim. Contr.,* Vol. 28 (1990), 1321-1349.
6. J. P. Gauthier, H. Hammouri and S. Othman, A simple observer for nonlinear systems, applications to bioreactocrs, *IEEE TAC,* Vol. 37, 875-880 (1992).
7. W. Hahn, Stability of Motion, Springer-Verlag, (1967).
8. H. Hermes, Homogeneous coordinates and continuous asymptotically stabilizing feedback controls, in: S. Elaydi (Ed.), Differential Eqn. Stability and Contr., Lecture Notes in Applied Mathematics, Vol. 109, (1991), 249-260.
9. M. Kawski, Stabilization of nonlinear systems in the plane, *Syst. Contr. Lett.,* Vol. 12 (1989), 169-175.
10. H. K. Khalil, High-gain observers in nonlinear feedback control. *New Directions in Nonlinear Observer Design,* Ed. H. Nijmeijer and T.I. Fossen, Springer, 1999.
11. A. J. Krener and A. Isidori, Linearization by output injection and nonlinear observer, *Syst. Contr. Lett.,* Vol. 3, pp. 47-52, 1983.
12. A. J. Krener and W. Kang, Locally convergent nonlinear observers, *SIAM. J. Optimiz. and Contr.,* Vol. 42, No. 1, pp. 155-177, 2003.
13. C. Qian and W. Lin, Non-Lipschitz continuous stabilizers for nonlinear systems with uncontrollable unstable linearization, *Syst. Contr. Lett.,* Vol. 42, No. 3, 185-200 (2001).
14. C. Qian and W. Lin, A continuous feedback approach to global strong stabilization of nonlinear systems, *IEEE TAC.,* Vol. 46, 1061-1079 (2001).
15. C. Qian and W. Lin, Smooth Output Feedback Stabilization of Planar Systems without Controllable/Observable Linearization, *IEEE Trans. Automat. Contr.,* Vol. 47, No. 12 (2002), 2068-2073.
16. C. Qian and W. Lin, Nonsmooth Output Feedback Stabilization of a Class of Nonlinear Systems in the Plane, *IEEE TAC.,* Vol. 48(2003) 1824-1829.
17. C. Qian and W. Lin, Nonsmooth Output Feedback Design for Nonlinear Systems with Unstabilizable/Undetectable Linearization, *submitted for publication.*
18. L. Rosier, "Homogeneous Lyapunov function for homogeneous continuous vector field", *Syst. Contr. Lett.,* Vol. 19, (1992) 467-473.
19. C. Rui, M. Reyhanoglu, I. Kolmanovsky, S. Cho and N. H. McClamroch, Nonsmooth stabilization of an underactuated unstable two degrees of freedom mechanical system, *Proc. of the 36th IEEE CDC,* San Diego, (1997) 3998-4003.
20. M. Tzamtzi and J. Tsinias, Explicit formulas of feedback stabilizers for a class of triangular systems with uncontrollable linearization, *Syst. Contr. Lett.,* Vol. 38 (1999), 115-126.
21. B. Yang and W. Lin, Output feedback stabilization of a class of homogeneous and high-order nonlinear systems, *Proc. of the 42nd IEEE CDC,* Maui, Hawaii, (2003) 37-42.

Drift-free Attitude Estimation Using Quasi-linear Observers

Henrik Rehbinder[1] and Xiaoming Hu[2]

[1] RaySearch Laboratories, Sweden
`henrik.rehbinder@raysearchlabs.com`
[2] Optimization and Systems Theory, Royal Institute of Technology, Sweden
`hu@kth.se`

Summary. In this paper we study the attitude estimation problem for an accelerated rigid body using gyros and accelerometers. The application in mind is that of a walking robot and particular attention is paid to the large and abrupt changes in accelerations that can be expected in such an environment. We propose a state estimation algorithm that fuses data from rate gyros and accelerometers to give long-term drift free attitude estimates. The algorithm does not use any local parameterization of the rigid body kinematics and can thus be used for a rigid body performing any kind of rotations. The algorithm is a combination of two non-standard, but in a sense linear, Kalman filters between which a trigger based switching takes place. The kinematics representation used makes it possible to construct a linear algorithm that can be shown to give convergent estimates for this nonlinear problem. The state estimator is evaluated in simulations demonstrating how the estimates are long term stable even in the presence of gyrodrift.

20.1 Introduction

A prerequisite for mobile robot control is state estimation where the states typically are position, velocity and orientation. State estimation is especially important for walking robots in difficult terrain where a sense of balance is absolutely necessary as it is the basis of attitude control. With attitude we refer to the robot's orientation relative to the gravity vector, usually described by pitch and roll. Perhaps most important to walking robots, the problem still applies to any kind of robot moving in difficult terrain. Attitude estimation is usually performed by combining measurements from three kinds of sensors: rate gyros, inclinometers and accelerometers. It is possible to use a rate gyro to derive attitudes by integrating the rigid body kinematic equations. With high quality gyros and good initial values these estimates can be very accurate over long periods of time. However, if the aim is an autonomous vehicle then the attitude estimate should be reliable over an infinite time scale. It would also be desirable to use cheap gyros as high quality gyros are expensive. To

H. Rehbinder and X. Hu: *Drift-free Attitude Estimation Using Quasi-linear Observers*, LNCIS
321, 321–336 (2005)

provide an absolute reference of the attitude, inclinometers and accelerometers which relate the body to the gravity vector can be used. A problem is that both these sensors also are sensitive to translational accelerations. The sensor errors described are complementary in terms of frequency characteristics. Due to this complementarity, the estimation errors can be decreased by sensor fusion [1], [26]. If the rigid body motion is restricted to planar rotation, it would be possible to formulate the sensor fusion problem as a linear observer problem which can be solved by standard Kalman filtering techniques. Fundamental for walking robots is that the body motion is inherently three-dimensional, making most kinematics representations nonlinear. For nonlinear problems, there are no general state estimation algorithms that are guaranteed to work. Attitude estimation via different ensembles of the above mentioned sensors has been studied by many authors such as Baerveldt and Klang [2], Balaram [3], Barshan and Durrant-Whyte [4], Foxlin [8],[9], Greene [10] Lefferts *et. al* [13], Madni [14], Rehbinder and Hu [17], [18], Sakaguchi *et. al* [21], Smith *et. al* [22] and Vaganay *et.al* [25].

In this paper we provide a solution to fusing data from a 3-axis rate gyro and a 3-axis accelerometer that will provide stable estimates of the robot's attitude. By using a global description of rigid body rotation we are able to obtain a linear problem and can use a modified linear Kalman filter. Further, we show how the linear Kalman filter can be used to obtain a projected Kalman filter, where the states evolve on the unit sphere, a curved space. The global kinematics representation used eliminates all the nonlinear problems associated with Euler angles in a remarkably elegant and simple way. This is paid for by a slight increase in abstraction. As the kinematics representation used is global, there are no restrictions on the kind of motion the robot is allowed to perform, contrary to what would be the case if for example Euler angles were used. The projected Kalman filter can be given a geometric interpretation which is related to the observer presented by Rehbinder and Ghosh in [16], [15] where an observer for an implicit output system evolving on $SO(3)$ is presented. It would be possible to formulate large parts of this paper in a strict differential geometric framework but for the sake of readability, we choose not to do so. Further we discuss how to modify the observer gains in order to take accelerations that are not well modeled as white noise into account.

The main contributions of this article are: Two mathematically sound, quite simple algorithms that solves the important problem of fusing accelerometer and gyro data; The use of a global kinematics representation and a design idea for observers evolving on a curved space; An experimental evaluation of the observers, using a walking robot. The outline of the paper is as follows. In Section 20.2 the problem is formulated and the proposed algorithms are described and analyzed.

20.2 Problem Formulation and Solution

As is well known, rigid body kinematics have a somewhat delicate structure as rotations are most naturally described as elements of the rotation group $SO(3)$, a smooth manifold. It is common practice to consider some parameterization of $SO(3)$ such as various versions of Euler angles or quaternions. In this study we will work directly with the rotation matrix representation of $SO(3)$. We embed the manifold in a larger linear space and it is possible to obtain a linear problem formulation that is used for the observer design. Further, the kinematics representation is global, as opposed to the local Euler coordinates.

Apart from the above, the problem of distinguishing between inertial forces and the gravity vector is discussed. We will use an accelerometer as the attitude sensor. When the body is accelerated, which is the case for walking robots, the accelerometer noise can be very high. The straightforward Kalman filtering way of handling this problem is to assign a large covariance matrix to the output noise. We will investigate an approach where a variable output noise covariance is used. This variable noise covariance is designed to compensate for the variations in acceleration magnitude that originates from the feet impacts.

20.2.1 Mathematical Modeling

Consider a rigid body moving in inertial space. The body is undergoing both rotations and translations and our aim is to estimate attitude, the robots orientation relative to the gravity vector. For this, a 3-axis rate gyro and a 3-axis accelerometer is used. Introduce a coordinate system, N, fixed in inertial space and a coordinate system, B, fixed in the body. Let the coordinates of an arbitrary point ξ be denoted by ξ^N if expressed in the N-frame and by ξ^B if expressed in the B-frame. The relation between the two frames is

$$\xi^B = R(\xi^N - r^N) \tag{20.1}$$

where R is a rotation matrix, that is

$$R \in SO(3) = \left\{ R \in \mathbb{R}^{3\times3} : R'R = I, \det R = 1 \right\}. \tag{20.2}$$

The kinematics of a rigid body are

$$\begin{cases} \ddot{r}^N = a \\ \dot{R} = \Omega(t)R \end{cases} \tag{20.3}$$

where a is the acceleration expressed in the N-frame and where

$$\Omega = \begin{pmatrix} 0 & \omega_3 & -\omega_2 \\ -\omega_3 & 0 & \omega_1 \\ \omega_2 & -\omega_1 & 0 \end{pmatrix}. \tag{20.4}$$

ω_i are the components of the angular velocity expressed in the B-frame. The matrix Ω has the property that $\Omega x = -\omega \wedge x$ and is therefore called the wedge matrix. We consider the body to be equipped with a strap-down inertial measurement unit (IMU), that is, a body fixed rate-gyro/accelerometer package. The rate gyro measures the angular velocity in the body frame up to an unknown and slowly varying offset. In the analysis and observer design we will neglect this offset and therefore we take the angular velocity as a known entity in the problem.

The strap-down accelerometer measures the difference between the inertial forces and gravity, expressed in the body frame B. If the accelerometer output is denoted y, then the accelerometer model is

$$y = R(a - g^N). \tag{20.5}$$

From (20.5) it is obvious that the accelerometer may be used as an attitude sensor if the acceleration itself is considered as a disturbance and the gravity vector as the entity that we want to measure. As a matter of fact, attitude may be defined as the body's orientation relative to the gravity vector. It is perhaps more standard to consider attitude as being described by the two Euler angles pitch and roll. The Euler angle formulation of the problem is here reviewed for completeness and comparison. In the yaw(ψ)-pitch(θ)-roll(ϕ) parameterization of $SO(3)$

$$R = \begin{pmatrix} c\psi c\theta & s\psi c\theta & -s\theta \\ -s\psi c\theta + c\psi s\theta s\phi & c\psi c\phi + s\psi s\theta s\phi & c\theta s\phi \\ s\psi s\phi + c\psi s\theta c\phi & s\psi s\theta c\phi - c\psi s\phi & c\theta c\phi \end{pmatrix}, \tag{20.6}$$

where $c\psi = \cos\psi$ etc. As $g^N = -ge_3$ where $g = 9.81 m/s^2$, and where e_3 is the third unit vector, the accelerometer output may be written

$$y = \begin{pmatrix} -\sin\theta \\ \cos\theta \sin\phi \\ \cos\theta \cos\phi \end{pmatrix} g + Ra. \tag{20.7}$$

The pitch-roll kinematics can be derived by straightforward differentiation of (20.6) and (20.3) and are

$$\begin{pmatrix} \dot{\theta} \\ \dot{\phi} \end{pmatrix} = \begin{pmatrix} 0 & \cos\phi & -\sin\phi \\ 1 & \sin\phi\tan\theta & \cos\phi\tan\theta \end{pmatrix} \begin{pmatrix} \omega_1 \\ \omega_2 \\ \omega_3 \end{pmatrix}. \tag{20.8}$$

Therefore, the attitude kinematics and measurements may be written as the nonlinear system

$$\begin{cases} \begin{pmatrix} \dot{\theta} \\ \dot{\phi} \end{pmatrix} = \begin{pmatrix} 0 & \cos\phi & -\sin\phi \\ 1 & \sin\phi\tan\theta & \cos\phi\tan\theta \end{pmatrix} \begin{pmatrix} \omega_1 \\ \omega_2 \\ \omega_3 \end{pmatrix} \\[2em] y = \begin{pmatrix} -\sin\theta \\ \cos\theta \sin\phi \\ \cos\theta \cos\phi \end{pmatrix} g + Ra \end{cases} \tag{20.9}$$

This system is nonlinear, both in the dynamics and in the output equation. Further, the yaw-pitch-roll parameterization is not one-to-one globally and clearly θ and ϕ are undefined for $\theta = \pm\pi/2$. This makes the representation unsuitable for general robots. It would for example not be possible to use (20.9) for a climbing robot which operates around $\phi = \pm\pi/2$. Further, it is far from easy to design a convergent observer for (20.9) due to the nonlinearities. The observer design problem is further complicated by that $\dot{\phi} \to \infty$ when $|\theta| \to \pi/2$.

Consider instead the possibility of using the rotation matrix itself as state. This will in a remarkably clear way eliminate most of the problems associated with (20.9). The rotational kinematics $\dot{R} = \Omega(t)R$ may be written column-wise as

$$\dot{r}_i = \Omega(t)r_i, \quad i = 1, 2, 3 \tag{20.10}$$

where r_i is the ith column of R. In the same way, the accelerometer output can be written

$$y = r_3 g + Ra. \tag{20.11}$$

Denote now by

$$\begin{aligned} x &= r_3 \\ u &= a/g \end{aligned}, \tag{20.12}$$

and with a slight abuse of notation

$$y := y/g. \tag{20.13}$$

In this notation, the attitude kinematics and accelerometer measurements is described by the differential-algebraic equation

$$\begin{cases} \dot{x} = \Omega(t)x \\ ||x|| - 1 = 0 \\ y = x + w \end{cases} \tag{20.14}$$

where $w = Ru$. The algebraic constraint $||x||^2 - 1 = 0$, which captures that x is a rotation matrix column, is better viewed as the geometric constraint

$$x \in \mathcal{S}^2 = \left\{ x \in \mathbb{R}^3 : ||x||^2 - 1 = 0 \right\}, \tag{20.15}$$

where $\mathcal{S}^2$ is the unit sphere. As a matter of fact, $\mathcal{S}^2$ is a manifold of a special kind, a smooth hypersurface embedded in $\mathbb{R}^3$. It is well known that $\mathcal{S}^2$ is an invariant set under $\dot{x} = \Omega(t)x$, so the algebraic constraint can be replaced by an initial value constraint. The system (20.14) can therefore be written

$$\begin{cases} \dot{x} = \Omega(t)x \\ x(0) = x_0 \in \mathcal{S}^2 \\ y = x + w \end{cases}. \tag{20.16}$$

There is an appealing linear structure to this formulation, but, note that the vector field $\Omega(t)x$ viewed as a mapping $\kappa_t : \mathcal{S}^2 \to \mathbb{R}^3$ defined by

$$\kappa_t : x \mapsto \Omega(t)x \qquad (20.17)$$

is not linear. Linearity is only defined between linear spaces and $\mathcal{S}^2$ is not a linear space. However, if κ_t is extended to $\mathbb{R}^3$, $\kappa_t : \mathbb{R}^3 \to \mathbb{R}^3$, by the same definition (20.17) then it may be viewed as a linear mapping. This is perhaps not a very surprising observation but it still is the key insight needed to formulate the simple observers we will present.

The constraint $x \in \mathcal{S}^2$ adds a difficulty to the observer problem. It must be ensured that the state estimates $\hat{x} \in \mathcal{S}^2$. If this is not ensured, $\hat{x}$ can not be the column of a rotation matrix. However, the problem formulated as in (20.16) is still much easier to solve than formulated in the form (20.9). It will be shown that it is possible to base the observer design on the embedded linear system

$$\begin{cases} \dot{x} = \Omega(t)x \\ x(0) = x_0 \in \mathbb{R}^3 \\ y = x + w \end{cases} . \qquad (20.18)$$

and still obtain state estimates $\hat{x} \in \mathcal{S}^2$.

20.2.2 Two Linear Kalman Filter Based Observers

Consider the system model (20.18). It is linear and clearly it is observable as the entire state vector is measured. Therefore, the standard deterministic *linear* Kalman filter [24]

$$\dot{z} = \Omega(t)z + L(t)(y - z) \qquad (20.19)$$

where

$$\begin{aligned} L(t) &= P(t)Q_2^{-1} \\ \dot{P} &= \Omega(t)P + P\Omega(t)' + Q_1(t) - PQ_2(t)^{-1}P \end{aligned} \qquad (20.20)$$

is an exponentially convergent observer for properly chosen $Q_1(t)$, $Q_2(t)$. When using the deterministic framework, the matrices Q_1 and Q_2 have no interpretation as noise covariances. However, the usual heuristics that large measurement errors calls for a large Q_2 and that large system uncertainties calls for large Q_1 still applies. Now, (20.19) can actually be used to estimate $x \in \mathcal{S}^2$. In the noise free case, $||z(t) - x(t)|| \to 0$ as $t \to \infty$ so $z(t) \to \mathcal{S}^2$. For finite t, and in the presence of sensor noise, there are no guarantees that $z(t) \in \mathcal{S}^2$. But, this problem is solved easily by adding a projection $P_{\mathcal{S}^2}$ of z onto $\mathcal{S}^2$, outside the observer dynamics (20.19). The projection onto $\mathcal{S}^2$ is of course very straightforward

$$P_{\mathcal{S}^2}z = \frac{z}{||z||} \qquad (20.21)$$

if we assume that $z \neq 0$.

Remark 1. When implementing (20.23) in discrete time, the problem of handling $z = 0$ in (20.21) is easily solved by instead using

$$\hat{x}(t) = \begin{cases} z(t)/\|z(t)\|, & \text{if } \|z(t)\| \neq 0 \\ \hat{x}(t-1), & \text{if } \|z(t)\| = 0 \end{cases} . \tag{20.22}$$

The observer

$$\begin{cases} \dot{z} = \Omega(t)z + L(t)(y - z) \\ \hat{x} = P_{\mathcal{S}^2}z \end{cases} \tag{20.23}$$

will hereafter be referred to as the *pseudo-linear Kalman filter*. An illustration of the idea behind the pseudo-linear Kalman filter is given in Figure 20.1 where it is shown how the states x and the state estimates $\hat{x}$ evolve on the sphere while the observer states z evolve in $\mathbb{R}^3$. It must be pointed out that the reason

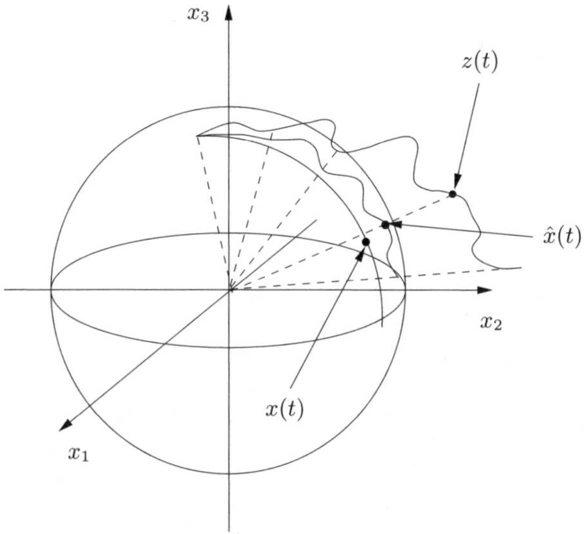

Fig. 20.1. Pseudo-linear Kalman filter states z and state estimates $\hat{x}$ along with true states x.

why this very simple observer works is that the system (20.16) is observable, *without* the information that $\hat{x}(0) \in \mathcal{S}^2$, that is, that (20.18) is observable. If, for example, the accelerometer was 2-axis, then the system (20.18) would not be observable and some other method would have to be used.

The idea presented here was to use an observer where the states did not evolve on $\mathcal{S}^2$ but anyway converged and then project the observer states onto $\mathcal{S}^2$. It would also be possible to design an observer where the states evolve on $\mathcal{S}^2$, just as for the underlying system. For an observer to evolve on $\mathcal{S}^2$, it must be so that the state estimate derivatives are tangential to the sphere , that is, that

$$\dot{\hat{x}} \in T_{\hat{x}}\mathcal{S}^2 \tag{20.24}$$

where

$$T_{\hat{x}}\mathcal{S}^2 = \{x \in \mathbb{R}^3 : \hat{x}'x = 0\} \tag{20.25}$$

is the tangent space of the sphere at the point $\hat{x}$. The observer (20.23) can be used to design such an observer. If the observer vector field is projected orthogonally onto $T_{\hat{x}}\mathcal{S}^2$, then an observer evolving on $\mathcal{S}^2$ is obtained. We note that the orthogonal projection onto $T_{\hat{x}}\mathcal{S}^2$ is given by

$$P_{T_{\hat{x}}\mathcal{S}^2} = (I - \hat{x}\hat{x}') \tag{20.26}$$

Proof: It is well known that $T_{\hat{x}}\mathcal{S}^2 = \ker \hat{x}'$, the orthogonal complement of $\{\hat{x}\}$. As $\ker \hat{x}' = (\mathrm{Im}\hat{x})^{\perp}$ and as $||\hat{x}|| = 1$, it is clear that $P_{T_{\hat{x}}\mathcal{S}^2} = (I - \hat{x}\hat{x}')$.$\square$
The projection of the vector field is

$$P_{T_{\hat{x}}\mathcal{S}^2}\left(\Omega(t)\hat{x} + L(t)(y - \hat{x})\right) = \Omega(t)\hat{x} + (I - \hat{x}\hat{x}')L(t)(y - \hat{x}) \tag{20.27}$$

due to that $\Omega(t)\hat{x} \in T_{\hat{x}}\mathcal{S}^2$. We can formulate the *projected Kalman filter*, defined by

$$\boxed{\begin{cases} \dot{\hat{x}} = \Omega(t)\hat{x} + (I - \hat{x}\hat{x}')L(t)(y - \hat{x}) \\ \hat{x}(0) = x_0 \in \mathcal{S}^2 \end{cases}} \tag{20.28}$$

with the property that $\hat{x}(t) \in \mathcal{S}^2 \,\forall t$. The projected Kalman filter can actually be shown to be convergent for a special choice of filter parameters. The proof of this is particularly simple if a natural error measure for this geometric setting is used. This error measure is based on the scalar product and is a simplified version of the error measure introduced in [12] for states on $SO(3)$. We note that $\xi = 1 - x'\hat{x}$ is an error measure for (20.28) in the sense that $\xi = 0 \Leftrightarrow x = \hat{x}$. Further $0 \leq \xi \leq 2$.

Proof: As x and $\hat{x}$ are of unit length, $x'\hat{x} = \cos\alpha$ for some $\alpha \in [0, 2\pi)$ and the proof readily follows.

Theorem 1. *Let $L(t) = l(t)I$ and let $l(t)$ be continuous, non-negative and bounded from above. Further, let there be $T < \infty$ and $\delta > 0$ such that for every t*

$$\int_t^{t+T} l(\tau)d\tau \geq \delta. \tag{20.29}$$

Then the error dynamics of the projected Kalman filter (20.28) is exponentially stable and $0 \leq \xi < 2$ is contained in the domain of attraction. The only feasible ξ for which the error dynamics are unstable is $\xi = 2$ which corresponds to $x(0) = -\hat{x}(0)$.

Proof: Consider the nominal dynamics of the error $\xi = 1 - x'\hat{x}$ where the noise is set to be zero.

$$\begin{aligned} \dot{\xi} &= -\dot{x}'\hat{x} - x'\dot{\hat{x}} = -x'(\Omega(t) + \Omega(t)')\hat{x} - l(t)x'(I - \hat{x}\hat{x}')(x - \hat{x}) \\ &= -l(t)x'(I - \hat{x}\hat{x}')x = -l(t)(1 - x'\hat{x}\hat{x}'x) \\ &= -l(t)\xi(2 - \xi). \end{aligned} \tag{20.30}$$

Since (20.30) is a Bernoulli equation, the solution can be written

$$\xi(t) = \frac{\xi_0 e^{-2\int_{t_0}^t l(s)ds}}{(1 - \frac{\xi_0}{2}) + \frac{\xi_0}{2} e^{-2\int_{t_0}^t l(s)ds}} \tag{20.31}$$

Consequently,

$$|\xi(t)| \le \frac{|\xi_0|}{(1 - \frac{|\xi_0|}{2})} e^{-2\int_{t_0}^t l(s)ds} \tag{20.32}$$

and by noting that from (20.29)

$$\int_{t_0}^t l(s)ds \ge \frac{\delta}{2T}(t - t_0), \quad t \ge t_0 + T \tag{20.33}$$

it is proven that (20.28) is exponentially stable and that $[0, 2)$ is contained in the domain of attraction. $\qquad\square$

Remark 2. The gain matrix $L(t) = l(t)I$ does not have to be a Kalman gain for Theorem 1 to hold. Further, the integral constraint (20.29) does not exclude the possibility that $l(t) = 0$ on intervals. This makes it possible to turn off the accelerometer during phases when the acceleration is extremely high and still obtain convergent estimates.

For the following simple set of Kalman filter parameters, $L(t) = l(t)I$ and the projected Kalman filter converges.

Corollary 1. *Let $P_0 = p_0 I$, $Q_1(t) = q_1(t)I$, $Q_2(t) = q_2(t)I$ and let there be $\underline{q}_1 > 0, \bar{q}_1 < \infty$, $\underline{q}_2 > 0, \bar{q}_2 < \infty$ such that $\underline{q}_1 \le q_1(t) \le \bar{q}_1$ and $\underline{q}_2 \le q_2(t) \le \bar{q}_2$. Let further $q_1(t)$ and $q_2(t)$ be continuous functions. Then the conditions in Theorem 1 are fulfilled.*

Proof: Take $P(t) = p(t)I$ and show that $P(\tau) = p(\tau)I$ for $\tau \ge t$.

$$\dot{P}(t) = p(t)\left(\Omega(t) + \Omega(t)'\right) + q_1(t)I - \frac{p(t)^2}{q_2(t)}I = \left(q_1(t) - \frac{p(t)^2}{q_2(t)}\right)I \tag{20.34}$$

as $\Omega(t) = -\Omega(t)'$. As $P(0) = p_0 I$, $P(t) = p(t)I \ \forall \ t$. From stochastic observability and reachability and the standard Riccati equation theory [6] it follows that $\exists l_l > 0$ and $l_u < \infty$ such that $l_l \le l(t) < l_u$. Therefore (20.29) is fulfilled. $\qquad\square$

In our experiment the two observers are found to have similar performances. The question if there are any conditions under which the two observers actually are equivalent may therefore be asked. The answer is that there is such a condition and that it is very special. However, it is in an approximate sense fulfilled in the experiments and therefore we discuss it here.

Remark 3. If $y = \gamma(t)\hat{x}$ then the two observers (20.23) and (20.28) are equivalent in the sense that

$$\{(z, \hat{x}) : z/||z|| - \hat{x} = 0\} \tag{20.35}$$

is an invariant set under the dynamics of the two observers

$$\begin{cases} \dot{z} = \Omega(t)z + l(t)(y - z) \\ \dot{\hat{x}} = \Omega(t)\hat{x} + l(t)(I - \hat{x}\hat{x}')(y - \hat{x}) \end{cases}. \tag{20.36}$$

Proof: $z/||z|| - \hat{x} = 0$ can equivalently be written $h(z, \hat{x}) = z - ||z||\hat{x} = 0$. By straightforward differentiation,

$$\frac{d}{dt}h(z, \hat{x}) = l(t)(1 + ||z||)[y - \hat{x}(\hat{x}'y)] \tag{20.37}$$

so for $h(z, \hat{x}) = 0$ to be invariant it must hold that

$$y = \gamma(t)\hat{x} \tag{20.38}$$

for some $\gamma(t)$. $\square$

20.2.3 A Geometric Interpretation

The projected Kalman filter was derived by taking the standard Kalman filter and projecting the observer vector field on $T_{\hat{x}}\mathcal{S}^2$. It turns out that the projected Kalman filter, for $L(t) = l(t)I$, can be given a quite precise geometric interpretation which clearly illustrates the observer dynamics. For any system $\dot{x} = f(x, t)$ evolving on $\mathcal{S}^2$, the vector field must be such that $x'f(x, t) \forall x \in \mathcal{S}^2, \forall t$ or in other words that $f(x, t) \in T_{\hat{x}}\mathcal{S}^2$. A spanning set for $T_{\hat{x}}\mathcal{S}^2$ is $\{-e_i \wedge x\}_{i=1...3}$ where e_i is the ith unit vector. Thus, the vector field can always be written

$$f(x, t) = -\lambda(x, t) \wedge x \tag{20.39}$$

for some vector function λ. Consequently, it must be possible to rewrite the observer (20.28) as

$$\dot{\hat{x}} = \Omega(t)\hat{x} - v(t, \hat{x}, y) \wedge \hat{x}. \tag{20.40}$$

Indeed, this is the case as the following simple calculation shows

$$\begin{aligned} (I - \hat{x}\hat{x}')(y - \hat{x}) &= (I - \hat{x}\hat{x}')y = y - \hat{x}(\hat{x}'y) = y\hat{x}'\hat{x} - (\hat{x}y')\hat{x} \\ &= (y\hat{x}' - \hat{x}y')\hat{x} = -(y \wedge \hat{x}) \wedge \hat{x} \end{aligned} \tag{20.41}$$

Apparently, the projected Kalman filter can be written as

$$\dot{\hat{x}} = \Omega(t)\hat{x} - l(t)(y \wedge \hat{x}) \wedge \hat{x} \tag{20.42}$$

which will be referred to as the *geometric attitude observer*. The geometric interpretation is depicted in Figure 20.2. Ideally, $y = x = \hat{x}$, and then $y \wedge x = 0$ so the estimate evolves according to a copy of the underlying systems dynamics $\dot{x} = \Omega(t)x$. If not, then a corrective angular velocity is added, driving $\hat{x}$ towards y. This angular velocity is directed along the vector $y \wedge \hat{x}$.

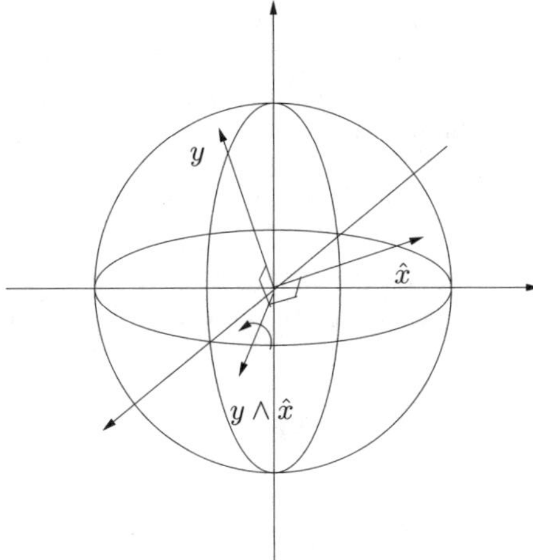

Fig. 20.2. Geometric interpretation of the projected Kalman filter.

20.2.4 Acceleration Handling

If the acceleration induced noise is modeled well by white noise then it is straightforward to apply either of the two filters presented with $q_2(t) = q_2$, the covariance of the acceleration noise. For a mobile robot such as a walking one, the acceleration pattern will typically consist not only of white noise but also of impulse like signals when the feet hit the ground. Other applications where non-white accelerations may occur is airplanes, cars or trains. When these vehicles turn, the centripetal acceleration will typically be non-white. Whatever characteristics the acceleration have, it might be argued that one may choose q_2 so high that the errors in the worst case scenario are small enough. The problem with this approach is that it is conservative and the filter performance will be unnecessarily bad. In an attempt to handle the problem of non-white accelerations we propose to use a variable $q_2 = q_2(y, \hat{x})$ that is a function of some acceleration estimate. We propose two different approaches, one switching approach and one continuous approach. In both cases the following structure

$$q_2(t) = (1 + \phi(y, \hat{x}))\hat{q}_2 \tag{20.43}$$

where $0 \leq \phi(y, \hat{x}) < \infty$ is used. In the following we only consider the scalar version of the filters but the same line of thinking can be applied to the general case. The different approaches presented below are equally applicable to the two filters. For both methods, a reliable acceleration estimate is of course necessary.

The most natural acceleration estimate is

$$\hat{w} = y - \hat{x} \tag{20.44}$$

as $y = x + w$ and in steady state, ideally $x = \hat{x}$. The question is now whether this estimate can safely be used or if there are false indications of low accelerations? Are there unfortunate combinations of x, $\hat{x}$ and w such that $\hat{w}$ is small even if w is not? The answer is yes. Consider the equation $||\hat{w}|| = 0$. As

$$||\hat{w}|| = ||x + w - \hat{x}|| \tag{20.45}$$

we have a false indication of zero acceleration if $w = \hat{x} - x$. This situation is depicted for a planar case in Figure 20.3. If $x = \hat{x}$ then there are no false

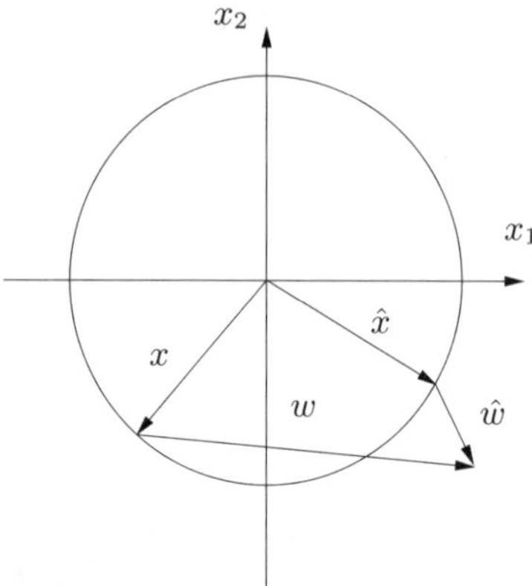

Fig. 20.3. False indications of low acceleration.

indications and for small differences $\hat{x} - x$ the acceleration must be small. Therefore, once the observer have converged, false indications will not be so fatal to performance. What though may happen is that combinations of state, state estimate and acceleration may prevent the observer from converging. However unlikely, it still must be considered. It should though be noted that in a typical robotics application, the robot would start from a stand still position with no acceleration. The observer can then be allowed to converge before the actual movement starts. To reduce the risk of false indications, the following time-window based acceleration norm estimate may be used

$$\widehat{||w||}(t) = \max_{\tau \in [t-T_0,t]} ||y(\tau) - \hat{x}(\tau)||. \tag{20.46}$$

Here T_0 is a parameter that needs to be tuned. The higher it is chosen the more certain it is that no false indications take place, but, the filter design is more conservative. It should be pointed out that a related idea has been proposed earlier by Foxlin ([8],[9]). In Figure 20.4, an illustrative example of $\widehat{||w||}$ and $||y - \hat{x}||$ shows how the expression (20.46) excludes the spurious occasions where $y - \hat{x} = 0$ but captures the longer phases of $y - \hat{x} = 0$.

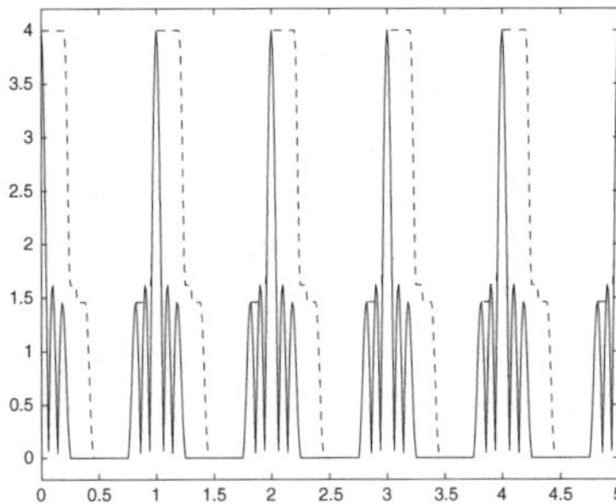

Fig. 20.4. Time-window approach for avoiding false detections.

If the accelerations are well modeled as consisting of phases of low acceleration and of phases of high acceleration, then a reasonable idea is to use a switching structure where

$$\phi(y, \hat{x}) = \begin{cases} 0, & \widehat{||w||} = 0 \\ \bar{\phi}, & \widehat{||w||} \neq 0 \end{cases} \tag{20.47}$$

Care need though to be taken when using this switching structure. The parameters in the Riccati equation (20.20) are now changing discontinuously. So is the observer vector field. How to prove stability for switching and hybrid systems is an open issue [5] and indeed, in the paper [6] where bounds on the Riccati equation is shown, the involved matrices are assumed to be continuous. Fortunately, these problems can be avoided by starting with a sampled-data model of (20.3) and working in discrete time. This approach is described by Rehbinder and Hu in [19].

In an environment where the acceleration is not well modeled as composed by high/low accelerations, a continuously tuned acceleration noise parameter may be used. Here we propose

$$\phi(y, \hat{x}) = \alpha^2 \widehat{||w||}^2 \tag{20.48}$$

where α is a parameter that needs to be tuned. To avoid the technical difficulties with the switching structure (20.47) we study the continuous approach (20.48). Further, our main application is walking robots and we argue that a switching structure is less suitable than the continuous structure (20.48). In the following experimental evaluation, the acceleration estimate (20.46) has not been used.

20.3 Summary and Discussion

In this paper we have presented an approach to solving the attitude estimation problem for a rigid body. The algorithm does in a theoretically consistent and easily implementable way solve this problem. The application in mind is a walking robot but the algorithms are in no way dependent on that application and could just as well be applied to other robots such as flying or climbing ones. For climbing robots, the use of the global representation instead of the more standard yaw/pitch/roll parameterization could prove very useful as such robots usually have an attitude out of the bounds for which that parameterization is valid. Further, the kinematics representation is very important as the linear Kalman filter can be used which makes it possible to easily obtain theoretical convergence results. We are also able to formulate a nonlinear observer that evolves on the sphere. It is worth emphasizing that the hard nonlinear problems often associated with the problem studied in this paper originate from the choice of coordinates and that it was possible to avoid them by a suitable representation. The observers presented have also been evaluated in experiments. The most challenging and interesting research topic that should follow this study is to study observer design for systems on Lie groups in a more general setting. Questions to ask are: Is it possible to find conditions guaranteeing that projected Kalman filters in some more general setting are convergent? Does the geometric attitude filter have some generality? Are there any fundamental connections to the observer presented in [16], which has a similar structure? From the applications perspective, more careful experimental testing is necessary. A discrete-time counterpart of the pseudo-linear Kalman filter, presented in [19], has been used in feedback loops for walking robot balancing. The next experimental step is to close the loop with the observers presented in this paper.

References

1. M. A. Abidi and R. C. Gonzales, editors. *Data Fusion in Robotics and Machine Intelligence*. Academic Press, 1992.
2. A. J. Baerveldt and R. Klang. A low-cost and low-weight attitude estimation system for an autonomous helicopter. In *IEEE International Conference on Intelligent Engineering Systems, Proceedings*, pages 391–395, 1997.
3. J. Balaram. Kinematic observers for articulated rovers. In *Proceedings of the 2000 IEEE International Conference on Robotics and Automation*, pages 2597–2604, San Francisco, April 2000.
4. B. Barshan and H. F. Durrant-Whyte. Inertial navigation systems for mobile robots. *IEEE Transactions on Robotics and Automation*, 11(3):328–342, June 1995.
5. M. S. Branicky. Multiple Lyapunov functions and other analysis tools for switched and hybrid systems. *IEEE Transactions on Automatic Control*, 43(4):475–482, April 1998.
6. R. S. Bucy. The Ricatti equation and its bounds. *Journal of computer and system sciences*, 6:343–353, 1972.
7. F. Bullo and R. M. Murray. Tracking for fully actuated mechanical systems: A geometric framework. *Automatica*, 35(1):17–34, 1999.

8. E. Foxlin. Inertial head-tracker sensor fusion by a complementary separete-bias Kalman filter. In *Proceedings of the IEEE 1996 Virtual Reality Annual International Symposium*, pages 185–94, 1996.

9. E. Foxlin, M. Harrington, and Y. Altshuler. Miniature 6-DOF inertial system for tracking HMD. In *Proceedings of the SPIE*, volume 3362, pages 214–228, 1998.

10. M. Greene. A solid state attitude heading reference system for general aviation. In *Proceedings.1996 IEEE Conference on Emerging Technologies and Factory Automation*, volume 2, pages 413–17, 1996.

11. J.Ingvast, C. Ridderström, F. Hardarson, and J. Wikander. Improving a trotting robot's gait by adapting foot reference offsets. In *Int. Conf. on Climbing and Walking Robots*, Karlsruhe,Germany, September 2001.

12. D. E. Koditschek. The application of total energy as a Lyapunov function for mechanical control systems. In J. E. Marsden, P. S. Krishnaprasad, and J. C. Simo, editors, *Dynamics and Control of Multibody Systems*, volume 97, pages 131–157. AMS, 1989.

13. E. Lefferts, F. Markley, and M. Shuster. Kalman filtering for spacecraft attitude estimation. *Journal of Guidance, Control and Dynamics*, 5(5):417–429, Sept-Oct 1982.

14. A. M. Madni, D. Bapna, P. Levin, and E. Krotkov. Solid-state six degree of freedom, motion sensor for field robotic applications. In *Proceedings 1998 IEEE/RSJ International Conference on Intelligent Robots and System*, volume 3, pages 1389–9, 1998.

15. H. Rehbinder and B. K. Ghosh. Pose estimation using line based dynamic vision and inertial sensors. Submitted to the IEEE Transactions on Automatic Control.

16. H. Rehbinder and B. K. Ghosh. Rigid body state estimation using dynamic vision and inertial sensors. In *Proceedings of the 40th IEEE Conference on Decision and Control*, 2001. (To appear).

17. H. Rehbinder and X. Hu. Nonlinear pitch and roll estimation for walking robots. In *Proceedings of the 2000 IEEE International Conference on Robotics and Automation.*, volume 3, pages 2617–2622, San Francisco, CA, USA, 2000.

18. H. Rehbinder and X. Hu. Nonlinear state estimation for rigid body motion with low-pass sensors. *Systems & Control Letters*, 40(3):183–190, 2000.

19. H. Rehbinder and X. Hu. Drift-free attitude estimation for accelerated rigid bodies. In *Proceedings of the 2001 IEEE International Conference on Robotics and Automation*, volume 4, pages 4244–4249, Seoul, Korea, 2001.

20. C. Ridderström, J. Ingvast, F. Hardarson, M. Gudmundsson, M. Hellgren, J. Wikander, T. Wadden, and H. Rehbinder. The basic design of the quadruped robot Warp1. In *International Conference on Climbing and Walking Robots*, Madrid, Spain, October 2000.

21. T. Sakaguchi, K. Tsutomu, H. Katayose, K. Sato, and S. Inokuchi. Human motion capture by integrating gyroscopes and accelerometers. In *Proceedings of the 1996 IEEE/SICE/RSJ International Conference on Multisensor Fusion and Integration for Intelligent Systems*, pages 470–475, 1996.

22. R. Smith, A. Frost, and P. Probert. Gyroscopic data fusion via a quartenion-based complementary filter. In *Proccedings of the SPIE - The International Society for Optical Engineering. vol.3067*, pages 148–59, 1997.

23. S. Soatto, R. Frezza, and P. Perona. Motion estimation via dynamic vision. *IEEE Transactions on Automatic Control*, 41(3):393–413, March 1996.

24. E. D. Sontag. *Mathematical Control Theory : Deterministic Finite Dimensional Systems.* Springer, New York, 1998.
25. J. Vaganay, M. Aldon, and A. Fournier. Mobile robot attitude estimation by fusion of inertial data. In *Proceedings of the 1993 IEEE International Conference on Robotics and Automation,* volume 1, pages 277–82, 1993.
26. P. K. Varshney, editor. *Proceedings of the IEEE, Special Issue on Data Fusion,* volume 85. IEEE, January 1997.

On Existence and Nonexistence of Limit Cycles for FitzHugh-Nagumo Class Models*

Mattias Ringkvist[1] and Yishao Zhou[2]

[1] Department of Mathematics, Stockholm University, SE-10691, Stockholm, Sweden
`mattias@math.su.se`

[2] Department of Mathematics, Stockholm University, SE-10691, Stockholm, Sweden
`yishao@math.su.se`

Dedicated to Clyde Martin on the occasion of his 60th birthday.

Summary. In this paper we discuss the existence and non-existence of limit cycles of FitzHugh-Nagumo class models. The purpose is to clarify some unclear facts in the literature. We show also that this class of model exhibits double cycle bifurcation in addition to Andronov-Hopf bifurcation.

21.1 Introduction

In this paper we consider FitzHugh-Nagumo class models common in literature:

$$\begin{cases} \dfrac{du}{dt} = -Cw + Au(B - u)(u - \lambda) + I \\ \dfrac{dw}{dt} = \varepsilon(u - \delta w) \end{cases} \tag{21.1}$$

where the variable u is the negative of the membrane potential, w is the quantity of refractoriness, and I is the magnitude of stimulating current, the parameter A is to scale the amplitude of the curve, the parameter C affects the coupling strength and ε is added to more easily control the speed of one variable relative to the other. The choice of these parameters together with other three would produce oscillations which are of primary interests in biological contexts. The oscillators produced by system (21.1) are called *FitzHugh-Nagumo type oscillators* and are common to many biological mechanism at the cellular level. In addition to producing oscillations in the barnacle muscle, the same dynamical structures will appear in mechanistic models of insulin secretion and Ca^{2+} oscillations, see details in [3].

---

*The work was supported in part by the Swedish Research Council (VR).

M. Ringkvist and Y. Zhou: *On Existence and Nonexistence of Limit Cycles for FitzHugh-Nagumo Class Models*, LNCIS **321**, 337–351 (2005)
`www.springerlink.com`

The system defined by (21.1) is originated from the simplified Hodgkin-Huxley nerve systems, the famous model of the squid giant axon, see e.g. [2, 3, 8]:

$$\begin{cases} \dfrac{du}{dt} = w - \dfrac{u^3}{3} + u + I \\[2ex] \dfrac{dw}{dt} = \rho(a - u - bw) \end{cases} \tag{21.2}$$

where the parameters ρ, a, b are assumed to satisfy the conditions: $b \in (0, 1)$, $a \in \mathbb{R}$, $\rho > 0$.

The existence and nonexistence of limit cycles for the nerve system (21.2) were investigated e.g. in [4, 10, 12] in case $I = 0$. It was shown in [9] that the dynamical system (21.1) with $\varepsilon = C = B = 1$ exhibits a rich structure of bifurcation when $I \neq 0$.

Although the systems (21.2) and (21.1) are polynomial systems of same degree, the dynamical behavior can be different, see Section 21.2. Moreover, it is not easy to show under which circumstances there are limit cycles and the question of how many limit cycles the systems have is also left open. Many attempts have been made in simulations in case a current input I is introduced. The purpose of this paper is to give a new criterion under which the system (21.1) has no limit cycles with three fixed parameters $B = C = \varepsilon = 1$ and with the quantity $I = 0$. Moreover, we shall give a rather complete phase portrait and bifurcation diagram for the system (21.1) with $A = B = C = 1$, and we demonstrate that this system exhibits the double-cycle bifurcation in addition to the Andronov-Hopf bifurcation.

The paper is organized as follows. In Section 21.2, we point out the differences between the two polynomial systems mentioned above. Then we reformulate (21.1) to two different types of Liénard equations. Then we analyze the nonexistence limit cycles in Section 21.3. In Section 21.4 we give a proof on existence of limit cycles for the system with a unique unstable fixed point, and we will show that the system undergoes Andronov-Hopf bifurcation (which is well-known), Bautin (or generalized Andronov-Hopf) bifurcation and double cycle bifurcation. And finally we conclude the paper by some further remarks.

21.2 Preliminaries

In this section we review the theory that is relevant in our study. First consider the system (21.1) with $B = C = \varepsilon = 1$

$$\begin{cases} \dfrac{du}{dt} = - w + Au(u - \lambda)(1 - u) \\[2ex] \dfrac{dw}{dt} = u - \delta w \end{cases} \tag{21.3}$$

where $0 < \lambda < 1$ and $A, \delta > 0$. The choice of these three parameters is motivated by a desire to get a finer criterion of non-existence of limit cycles than that stated in [9].

A straightforward calculation shows that the system has origin as its only fixed point if and only if

$$(1 - \lambda)^2 - \frac{4}{\delta A} < 0 \tag{21.4}$$

Remark 1. Note that (21.3) can have either one fixed point as stated above, or two fixed points, origin and $(-(\lambda+1)/2, -(\lambda+1)/(2\delta))$, if $(1-\lambda)^2 - \frac{4}{\delta A} = 0$, or three fixed points, origin and $(u_\pm, w_\pm)$, where

$$u_\pm = -\frac{\lambda}{2} \pm \sqrt{(\lambda - 1)^2 - \frac{4}{A\delta}}, \quad w_\pm = \frac{u_\pm}{\delta}$$

if $(1-\lambda)^2 - \frac{4}{\delta A} > 0$. However, (21.2) has only one fixed point (u_I, w_I) for each $I \in \mathbb{R}$:

$$u_I = \sqrt[3]{(3(I + \frac{a}{b}) + \sqrt{9(I + \frac{a}{b})^2 + 4(\frac{1}{b} - 1)^3})/2} +$$

$$+ \sqrt[3]{(3(I + \frac{a}{b}) - \sqrt{9(I + \frac{a}{b})^2 + 4(\frac{1}{b} - 1)^3})/2}$$

and

$$w_I = \frac{a - u_I}{b}$$

under the assumptions.

Remark 2. Let (u_e, w_e) be a fixed point, and $h(u) = u(u - \lambda)(1 - u)$. It is not hard to show that

(i) if $A\delta h'(u_e) < 1$, then (u_e, w_e) is locally asymptotically stable, for $Ah'(u_e) < \delta$ and a rebeller for $Ah'(u_e) > \delta$;
(ii) if $A\delta h'(u_e) > 1$, then (u_e, w_e) is a saddle point;
(iii) if $A\delta h'(u_e) = 1$, then (u_e, w_e) is unstable for $Ah'(u_e) > \delta$.

Now we make a variable change, to transform the system to a special type of Liénard system, as follows

$$\begin{cases} x = -u, \\ y = w + \delta x. \end{cases}$$

This yields the Liénard system

$$\begin{cases} \dfrac{dx}{dt} = y - F(x) \\ \dfrac{dw}{dt} = -g(x) \end{cases} \tag{21.5}$$

where

$$\begin{cases} F(x) = x(\delta + A(x + \lambda)(x + 1)) \\ g(x) = x(1 + \delta A(x + \lambda)(x + 1)) \end{cases} \tag{21.6}$$

Since the Liénard systems are well-studied planar polynomial systems, our idea is to apply the known results for this system in our particular case. First we note that (21.4) coincides with the condition

$$xg(x) > 0, \qquad \forall x \neq 0. \tag{21.7}$$

It is also obvious that

$$F(0) = 0. \tag{21.8}$$

Clearly, the functions F and g are continuous functions on $\mathbb{R}$ satisfying the Lipschitz condition. Let

$$G(x) = \int_0^x |g(\xi)| d\xi,$$

and

$$M = \min \left\{ \int_0^\infty g(x)dx, \int_0^{-\infty} g(x)dx \right\}.$$

It is easy to show that $M = \infty$ and $G(x)$ is strictly increasing. Therefore the inverse of G exists and we denote it by G^{-1}. In the sequel we shall make use of the following theorem for the Liénard systems.

Theorem 1 ([11]). *Suppose that the parameters are chosen such that the origin is the unique fixed point, $xg(x) > 0$, $\forall x \neq 0$ and*

$$F(G^{-1}(-w)) \neq F(G^{-1}(w)), \qquad \forall w > 0. \tag{21.9}$$

Then (21.5) has no limit cycles.

When nonlinear waves in the nerve or muscle fibers or in the heart collide with each other, they mutually annihilate. There are, however, cases where the experiment and theory have shown that the inelasticity of the collisions is not that drastic. [1] considers the collision properties of nonlinear waves in an excitable medium of FitzHugh-Nagumo type, which is paradigmatic to account for quite a variety of biological, biochemical and neurobiological phenomena. It shows that the system

$$\begin{cases} \dfrac{du}{dt} = -w - u(u - 1)(u - \lambda) \\ \dfrac{dw}{dt} = \varepsilon(u - \delta w) \end{cases} \tag{21.10}$$

exhibits bistability for certain parameters, that is one asymptotically stable fixed point and one stable limit cycle coexist. Note that we allow the parameter

λ to be negative. In order to investigate the nonexistence of limit cycles we make a variable change as before to get the type of Liénard system shown (21.5). Let $x = -u$ and $y = w + \varepsilon\delta x$: Then (21.10) is transformed into the Liénard system (21.5) with

$$F(x) = x^3 + (\lambda + 1)x^2 + (\lambda + \varepsilon\delta)x$$

$$g(x) = \varepsilon\delta(x^3 + (\lambda + 1)x^2 + (\lambda + \frac{1}{\delta})x$$

In order to prove the existence of limit cycles, we need another type of Liénard system to be able to apply a theorem by Lefschetz [7]. Now we change the variables u and w to

$$\begin{cases} x = w \\ y = \varepsilon(u - \delta w) \end{cases}$$

the system transforms to

$$\begin{cases} \dfrac{dx}{dt} = y \\ \dfrac{dy}{dt} = -f(x,y)y - \tilde{g}(x) \end{cases} \tag{21.11}$$

where

$$f(x,y) = \varepsilon(\frac{y^2}{\varepsilon^3} + \frac{3\delta x - (1 + \lambda)}{\varepsilon^2}y + \frac{3\delta^2}{\varepsilon}x^2 - \frac{2\delta(1 + \lambda)}{\varepsilon}x + \delta + \frac{\lambda}{\varepsilon}) \tag{21.12}$$

$$\tilde{g}(x) = \varepsilon(\delta^3 x^3 - (1 + \lambda)\delta^2 x^2 + (\lambda\delta + 1)x) \tag{21.13}$$

Before stating the theorem we introduce the function $\tilde{G}(x)$ defined by

$$\tilde{G}(x) = \int_0^x \tilde{g}(s)ds = \varepsilon(\frac{\delta^3}{4}x^4 - \frac{(1 + \lambda)\delta^2}{3}x^3 + \frac{1 + \lambda\delta}{2}x^2)$$

Theorem 2 ([7]). *If the following conditions are fulfilled the system* (21.11) *has at least one stable limit cycle:*

1. *The origin is the only critical point and it is unstable.*
2. *$f(x,y)$ and $\tilde{g}(x)$ are continuous and satisfies a Lipschitz condition $\forall x, y \in \mathbb{R}$.*
3. *$|\tilde{g}(x)| \to \infty$ as $|x| \to \infty$ and $\tilde{g}(x) > 0$ for $x > N$ for some $N > 0$.*
4. *$\frac{\tilde{g}(x)}{\tilde{G}(x)} = O(\frac{1}{|x|})$.*
5. *$\exists M, m, n > 0$ such that $f(x,y) \geq M$ for $|x| \geq n$ and $f(x,y) \geq -m$ for $|x| \leq n$.*

21.3 Analysis on Nonexistence of Limit Cycles

Now we turn to the analysis of limit cycles of (21.3).

Proposition 1. *Suppose the origin is the only fixed point and*

$$\left(\lambda - \frac{1}{2}\right)^2 + \frac{3}{4} - \frac{3\delta}{A} < 0. \tag{21.14}$$

Then the system defined by (21.3) has no limit cycles.

Proof. By Bendixson's criterion we know that the system has no limit cycles
if

$$\frac{\partial}{\partial u}(-w - Au(u-1)(u-\lambda)) + \frac{\partial}{\partial w}(u - \delta w)$$
$$= -3Au^2 + 2A(\lambda+1)u - \lambda A - \delta$$

is not equal to zero and does not change sign. This is equivalent to the discriminant of the above polynomial of u is negative, which is in turn (21.14). $\square$

From now on we assume that the parameters violate (21.14). To give a
finer criterion we turn our interest to the properties of the function F. A
straightforward calculation shows that $F(x)$ has three real zeros if and only if

$$A(1-\lambda)^2 - 4\delta > 0 \tag{21.15}$$

Denote the three zeros of F as

$$\begin{cases} \alpha = 0 \\[2mm] \beta = -\dfrac{1+\lambda}{2} + \dfrac{\sqrt{A(1-\lambda)^2 - 4\delta}}{2A} \\[4mm] \gamma = -\dfrac{1+\lambda}{2} - \dfrac{\sqrt{A(1-\lambda)^2 - 4\delta}}{2A} \end{cases}$$

Clearly $\gamma < 0$. A short calculation shows that $\beta < 0$:

$$-\frac{1+\lambda}{2} + \frac{\sqrt{A(1-\lambda)^2 - 4\delta}}{2A} < -\frac{1+\lambda}{2} + \frac{\sqrt{A^2(1-\lambda)^2}}{2A} = -\lambda < 0$$

To be able to use the Theorem 1, we will show that (21.9) is satisfied. First
we show that the following proposition holds.

Proposition 2. *The condition* (21.9) *is equivalent to*

$$\big(G(\alpha(\eta)) + G(\beta(\eta)) \neq 0\big) \wedge \big(G(\alpha(\eta)) + G(\gamma(\eta)) \neq 0\big) \quad \forall \eta \in (0, \eta^*], \tag{21.16}$$

where $\eta^* = F(\overline{x})$, *and* $\overline{x}$ *and* $\underline{x}$ *are the local maximum and local minimum of
the function* F, *which are*

$$\underline{x} = -\frac{1}{3}\left((\lambda+1) - \sqrt{(\lambda+1)^2 - 3(\lambda + \frac{\delta}{A})}\right)$$

$$\overline{x} = -\frac{1}{3}\left((\lambda+1) - \sqrt{(\lambda+1)^2 - 3(\lambda + \frac{\delta}{A})}\right)$$

and $\gamma \leq \overline{x} \leq \beta \leq \underline{x} < \alpha$.

Proof. This is an immediate consequence of the fact that $\gamma \leq \beta < \alpha = 0$ and the function G is strictly increasing. $\square$

Next we prove

Proposition 3. *If*

$$6 - 3\delta^2 - \delta A(\lambda + (1-\lambda)^2) > 0 \tag{21.17}$$

then condition (21.16) *is equivalent to*

$$(6 - 3\delta^2 - \delta A(\lambda + (1-\lambda)^2))(\alpha(\eta) + \beta(\eta))$$
$$< -3\delta\eta + \delta(1+\lambda)(\delta + A\lambda), \quad \forall \eta \in (0, \eta^*]. \tag{21.18}$$

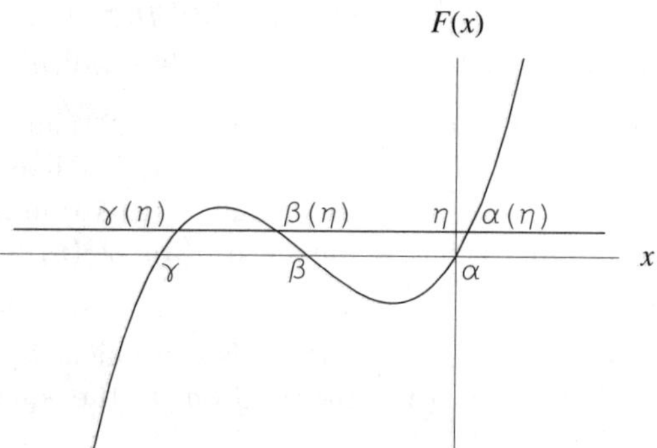

Fig. 21.1. The zeros of the function F.

Proof. From (21.6)

$$G(x) = \begin{cases} \dfrac{x^2}{12}(3\delta Ax^2 + 4\delta A(1+\lambda)x + 6(1+\delta A\lambda)), & x \geq 0, \\[3mm] -\dfrac{x^2}{12}(3\delta Ax^2 + 4\delta A(1+\lambda)x + 6(1+\delta A\lambda)), & x < 0. \end{cases} \tag{21.19}$$

Using

$$0 = F(\alpha(\eta)) - \eta = A\alpha^3(\eta) + A(1+\lambda)\alpha^2(\eta) + (\delta + \lambda A)\alpha(\eta) - \eta$$

and the similar computation for $\beta(\eta)$ and $\gamma(\eta)$ we get

$$G(\alpha(\eta)) = \frac{1}{12}((6 - 3\delta^2 - \delta A(\lambda + (1 - \lambda)^2))\alpha^2(\eta) +$$
$$+ (3\delta\eta - \delta(1 + \lambda)(\delta + A\lambda))\alpha(\eta) + \delta(1 + \lambda)\eta), \qquad (21.20)$$

$$G(\beta(\eta)) = \frac{1}{12}((6 - 3\delta^2 - \delta A(\lambda + (1 - \lambda)^2))\beta^2(\eta) +$$
$$+ (3\delta\eta - \delta(1 + \lambda)(\delta + A\lambda))\beta(\eta) + \delta(1 + \lambda)\eta), \qquad (21.21)$$

$$G(\gamma(\eta)) = \frac{1}{12}((6 - 3\delta^2 - \delta A(\lambda + (1 - \lambda)^2))\gamma^2(\eta) +$$
$$+ (3\delta\eta - \delta(1 + \lambda)(\delta + A\lambda))\gamma(\eta) + \delta(1 + \lambda)\eta). \qquad (21.22)$$

Using these equations (21.16) becomes

$$(6 - 3\delta^2 - \delta A(\lambda + (1 - \lambda)^2))(\alpha(\eta) + \beta(\eta))$$
$$\neq -3\delta\eta + \delta(1 + \lambda)(\delta + A\lambda), \qquad \forall \eta \in (0, \eta^*], \qquad (21.23)$$
$$(6 - 3\delta^2 - \delta A(\lambda + (1 - \lambda)^2))(\alpha(\eta) + \gamma(\eta))$$
$$\neq -3\delta\eta + \delta(1 + \lambda)(\delta + A\lambda), \qquad \forall \eta \in (0, \eta^*], \qquad (21.24)$$

where $\eta^* > 0$. When $\eta \to 0$ the right hand side of (21.23) and (21.24) becomes $\delta(1 + \lambda)(\delta + A\lambda) > 0$ while the expressions of the left hand side of (21.23) and (21.24) is less than zero due to (21.17). Thus (21.16) implies that (21.18) holds. The converse implication is true since $\beta(\eta) > \gamma(\eta)$ for all $\eta \in (0, \eta^*]$. $\square$

In order to analyze inequality (21.18) it is convenient to introduce a new parameter ξ instead of η. We study the solutions to the equation

$$F(x) - F(\xi) = 0, \quad \text{for } \xi \in [\gamma, \overline{x}). \qquad (21.25)$$

It is clear that ξ is a root of the equation (21.25). Denoting the other two roots by $\alpha(\xi), \beta(\xi)$ gives that

$$\begin{cases} \alpha(\xi) + \beta(\xi) = -(1 + \lambda + \xi), \\ \alpha(\xi)\beta(\xi) = (1 + \xi)(\xi + \lambda) + \dfrac{\delta}{A}. \end{cases}$$

Now (21.18) becomes

$$H(\xi) := -3\delta A\xi^3 - 3\delta A(1 + \lambda)\xi^2 - (6(\delta^2 - 1) + \delta A(1 + \lambda)^2)\xi$$
$$+ (1 + \lambda)(6 - 2\delta^2 - \delta A(1 - \lambda)^2) > 0, \quad \forall \xi \in (\gamma, \overline{x}]. \qquad (21.26)$$

Since the discriminant of $H'(\xi)$ is greater than zero, H has a local maximum and a local minimum, denoted by $\overline{\xi}$ respectively, $\underline{\xi}$:

$$\begin{cases} \underline{\xi} = -\dfrac{1 + \lambda}{3} - \dfrac{\sqrt{6\delta A(1 - \delta^2)}}{3\delta A} \\[3mm] \overline{\xi} = -\dfrac{1 + \lambda}{3} + \dfrac{\sqrt{6\delta A(1 - \delta^2)}}{3\delta A} \end{cases}$$

Now we are in the position to prove the following theorem.

Theorem 3. *The system (21.5), and hence the system (21.3) does not have limit cycles if (21.14) holds or if the following sets of inequalities are satisfied:*

$$(1 - \lambda)^2 - \frac{4}{\delta A} < 0 \tag{21.27}$$

$$3\delta - A\left(\left(\lambda - \frac{1}{2}\right)^2 + \frac{3}{4}\right) < 0 \tag{21.28}$$

$$4\delta - A(1 - \lambda)^2 < 0 \tag{21.29}$$

$$-6 + 3\delta^2 + \delta A\left(\left(\lambda - \frac{1}{2}\right)^2 + \frac{3}{4}\right) < 0 \tag{21.30}$$

$$-H(\gamma) < 0 \tag{21.31}$$

$$-H'(\gamma) < 0 \tag{21.32}$$

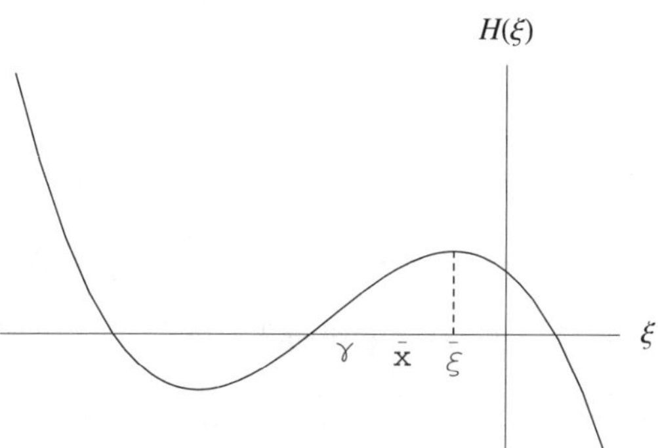

Fig. 21.2. Illustration of Theorem 3.

Note that $\underline{\xi} < \bar{x}$ due to (21.30).

Proof. Due to (21.27) theorem (1) can be applied. Condition (21.28) assures that Proposition 2 is valid and by (21.30) we can make use of Proposition 3. Thus, it remains to prove that (21.26) holds. But this is what (21.31) and (21.32) assure. □

Remark 3. It is worth pointing out that the parameter set formed by inequalities (21.27)-(21.32) is not empty. For example, take $\delta = \frac{1}{9}$ and $\lambda = \frac{2}{3}$ then $4 < A < 66$ satisfies the inequalities.

As a consequence of Theorem 3, Remark 2 and the boundedness of the trajectory ([9]), we have the following stronger statement than Proposition 3 in [9]:

Corollary 1. *Under the condition the conditions (21.14) and $4\delta > A$ or the conditions (21.27)-(21.32) in Theorem 3, the origin is a globally asymptotically stable fixed point.*

As for the nonexistence of limit cycles of the system defined by (21.10), we first find the fixed points, that are given by the equations

$$w = u(1 - u)(u - \lambda)$$
$$w = \frac{v}{\delta}$$

which gives that $u = w = 0$ is the only fixed point if and only if

$$u^2 - (1 + \lambda)u + \lambda + \frac{1}{\delta}$$

has no real roots, i.e

$$(1 + \lambda)^2 - 4(\lambda + \frac{1}{\delta}) < 0,$$

or equivalently,

$$(\frac{1 - \lambda}{2})^2 < \frac{1}{\delta}.$$

Then the characteristic equation of the Jacobian matrix is

$$s^2 + (\varepsilon\delta + \lambda)s + \varepsilon(1 + \lambda\delta) = 0. \tag{21.33}$$

Thus, the origin is locally asymptotically stable if

$$\mathrm{Re}(-(\varepsilon\delta + \lambda) \pm \sqrt{(\varepsilon\delta + \lambda)^2 - 4\varepsilon(\lambda\delta + 1)} < 0,$$

which is equivalent to, according to Routh test,

$$\varepsilon\delta + \lambda > 0 \quad \text{and} \quad 1 + \lambda\delta > 0, \tag{21.34}$$

and is unstable if

$$\mathrm{Re}(-(\varepsilon\delta + \lambda) \pm \sqrt{(\varepsilon\delta + \lambda)^2 - 4\varepsilon(\lambda\delta + 1)} > 0$$

Now we state the following theorem without detailed proof, since the analysis is similar.

Theorem 4. *Assume that (21.10) has only one fixed point, that is $(\frac{\lambda-1}{2})^2 < \frac{1}{\delta}$. Then system (21.10) has no limit cycles if either*

$$(\lambda - \frac{1}{2})^2 < 3(\delta\varepsilon - \frac{1}{4})$$

holds or one of the following sets of inequalities holds

$$\begin{cases} -(1+\lambda) < 0 \\ -(\lambda + \varepsilon\delta) < 0 \\ 3(\delta\varepsilon - \frac{1}{4}) - (\lambda - \frac{1}{2})^2 < 0 \\ 4\varepsilon\delta - (1-\lambda)^2 < 0 \\ (\lambda+1)^2 + 3(\lambda + \varepsilon\delta) - 6(\lambda + \frac{1}{\delta}) < 0 \\ -\tilde{H}(\tilde{\gamma}) < 0 \\ -\tilde{H}'(\tilde{\gamma}) < 0 \end{cases} \tag{21.35}$$

or

$$\begin{cases} 1 + \lambda < 0 \\ -(\lambda + \varepsilon\delta) < 0 \\ 3(\delta\varepsilon - \frac{1}{4}) - (\lambda - \frac{1}{2})^2 < 0 \\ 4\varepsilon\delta - (1-\lambda)^2 < 0 \\ (\lambda+1)^2 + 3(\lambda + \varepsilon\delta) - 6(\lambda + \frac{1}{\delta}) < 0 \\ \tilde{H}(\underline{\tilde{x}}) < 0 \\ \tilde{H}(\tilde{\gamma}) < 0 \\ -\tilde{H}'(\tilde{\gamma}) < 0 \end{cases} \tag{21.36}$$

where

$$\tilde{H}(\xi) = -3\xi^3 - 3(\lambda + 1)\xi^2 - (6(\frac{1}{\delta} - \varepsilon\delta) - (1+\lambda)^2)\xi$$

$$+ (\lambda + 1)(6(\lambda + \frac{1}{\delta}) - 2(\lambda + \varepsilon\delta) - (1+\lambda)^2)$$

$$\underline{\tilde{x}} = -\frac{\lambda+1}{3} + \frac{1}{3}\sqrt{(\lambda - \frac{1}{2})^2 + \frac{3}{4}(1 - 4\varepsilon\delta)}$$

$$\Xi = -\frac{\lambda+1}{3} + \frac{1}{3}\sqrt{6(\frac{1}{\delta} - \varepsilon\delta)}$$

$$\tilde{\gamma} = -\frac{\lambda+1}{2} - \frac{1}{2}\sqrt{(1 - \lambda)^2 - 4\varepsilon\delta}$$

$$\tilde{\mu} = -\frac{\lambda+1}{2} + \frac{1}{2}\sqrt{(1 - \lambda)^2 - 4\varepsilon\delta}$$

Remark 4. Note that the fifth inequality in the above sets of inequalities imply that $\Xi < \tilde{\underline{\xi}}$.

Remark 5. It follows that if the above sets of finequalities are satisfied then $\overline{x}, \tilde{\overline{x}} > \underline{\xi}$.

Remark 6. Note that neither of the sets of inequalities above is empty, e.g. (21.35) is satisfied for $\lambda = 2$, $\delta = \frac{1}{2}$ and $0 < \varepsilon < \frac{1}{2}$ while (21.35) is satisfied for $\lambda = -4$, $\delta = \frac{1}{8}$ and $0 < \varepsilon < 50$.

Corollary 2. *Let the conditions in Theorem 4 hold. Then the origin is globally asymptotically stable.*

We only need to show the following proposition.

Proposition 4. *The trajectories of system (21.10) are bounded.*

Proof. Define

$$V(x,y) = \frac{y^2}{2} + \tilde{G}(x) = \frac{y^2}{2} + \frac{\varepsilon}{12}x^2 h(x),$$

where $h(x) = 3\delta^3 x^2 - 4(1+\lambda)\delta^2 x + 6(1+\lambda\delta)$. This gives that

$$\dot{V}(x,y) = -y^2 f(x,y)$$

The idea is to show that V is a Lyapunov function outside a bounded set, which implies that the trajectories are bounded.

$$V(x,y) > 0 \quad \forall x, y \neq 0 \iff T = 9(1+\lambda\delta) - 2\delta(1+\lambda)^2 > 0.$$

If $T \leq 0$, then $h(x)$ has two zeros x_1, x_2 and a local minimum point x^* such that $|x^*| < \infty$. Define $\hat{y} = \inf\{y > 0 | y^2 + \frac{\varepsilon \hat{x}^2}{12}h(x^*) > 0\}$. Then $\hat{y} < \infty$. From this it follows that $\mathcal{A} = \{(x,y) \in \mathbb{R}^2 | V(x,y) < 0\}$ is a subset of $\mathcal{A}' = \{(x,y) \in \mathbb{R}^2 | x_1 < x < x_2, |y| < \hat{y}\}$ which is bounded. Turning our attention to $\dot{V}$ we see that

$$\dot{V} < 0 \quad \forall (x,y) \in \mathbb{R}^2 \setminus \{(x,y) | y = 0\}$$

This implies

$$S = f(\tilde{x}, \tilde{y}) = \varepsilon\delta - \frac{1}{3}(\lambda - \frac{1}{2})^2 - \frac{1}{4} > 0$$

Let $S < 0$ then f has a local minimum less than zero. But this combined with f being a paraboloid gives the existence of $\breve{x}, \breve{y} < \infty$ such that

$$f(x,y) > 0 \qquad (x,y) \in \mathcal{B}'$$

where $\mathcal{B}' = \{(x,y) \in \mathbb{R}^2 | |x| < \breve{x}, |y| < \breve{y}\}$. Thus $\mathcal{B} = \{(x,y) \in \mathbb{R}^2 | \dot{V} > 0\}$ is bounded by $\mathcal{B}'$. Further we have that $\dot{V} = 0$ if $(x,y) \in \mathcal{C}$ where $\mathcal{C} = \{(x,y) \in \mathbb{R}^2 | y = 0 \text{ or } f(x,y) = 0\}$. On this set the solutions to (21.10) lies on curves defined by $V(x,y) = c$ for constants c. The conclusion is that V is a Lyapunov function on whole of $\mathbb{R}^2$ except for a bounded set and thus the solutions to (21.10) must be bounded. $\square$

21.4 Proof of Existence of Stable Limit Cycles

In Theorem 4 we assumed $\lambda + \varepsilon\delta > 0$. Now we shall deal with the case $\lambda + \varepsilon\delta < 0$. It turns out that in this case the system has limit cycles. We follow the theorem by Lefschetz Theorem 2 to show the following theorem:

Theorem 5. *System* (21.10) *has at least one stable limit cycle if* $\varepsilon\delta + \lambda < 0$.

We verify the conditions in Theorem 2. Since $\varepsilon\delta + \lambda < 0$, $(0,0)$ is unstable by (21.34), and hence the first condition in Theorem 2 is satisfied.

Now both $f(x,y)$ and $\tilde{g}(x)$ are polynomials. So they are continuous and satisfy the Lipschitz condition $\forall x, y \in \mathbb{R}$. Thus condition 2 holds.

It is obvious that $|g(x)| \to \infty$ as $|x| \to \infty$ and that there exists an N such that $g(x) > 0$ for $x > N$ since $\varepsilon, \delta > 0$. From the fact that $\tilde{G}(x)$ is a polynomial of one degree higher than $\tilde{g}(x)$ it follows that $\frac{g(x)}{\tilde{G}(x)} = O(\frac{1}{|x|})$. Therefore, conditions 3 and 4 are also satisfied. In choosing $M, m, n > 0$ we have a lot of freedom. Taking first derivatives equal to zero yields

$$\tilde{x} = \frac{1+\lambda}{3\delta},$$
$$\tilde{y} = 0.$$

The quadratic form of $f(\tilde{x}, \tilde{y})$ is

$$Q(h,k) = 3\delta^2 h^2 + \frac{1}{\varepsilon^2} k^2 + \frac{3\delta}{\varepsilon} hk$$
$$= 3\delta^2 (h + \frac{1}{2\varepsilon\delta} k)^2 + \frac{1}{4\varepsilon^2} k^2 > 0 \qquad h, k \neq 0.$$

Thus $f(\tilde{x}, \tilde{y}) = \varepsilon\delta - \frac{1}{3}(\lambda - \frac{1}{2})^2 - \frac{1}{4}$ is a global minimal point of the paraboloid f. Thus, if $f(\tilde{x}, \tilde{y}) > 0$ we can choose $M = f(\tilde{x}, \tilde{y})$ and m, n arbitrary. On the other hand, if the converse is true there exist $M, n > 0$ such that $f(x,y) > M$ for $|x| > n$ and $f(x,y) > -m = f(\tilde{x}, \tilde{y})$ for $|x| < n$ since the graph of f is a paraboloid. Now the proof of Theorem 5 is complete.

21.5 Double Cycle Bifurcation

Let us assume that the origin is the only stable fixed point of the system defined by (21.10), which is the main topic of this section. Now we vary the parameter λ we see that an Andronov-Hopf bifurcation occurs when $\lambda + \varepsilon\delta = 0$. Then the fixed point becomes unstable and bifurcates to at least one stable limit cycle as shown in Theorem 5. However, the analysis in Section 21.3 showed that there is a gap in parameter space where the origin is the only globally asymptotically stable fixed point (nonexistence of limit cycles, see Theorem 4) and where limit cycles could possibly exist.

In the following we give a numerical example that shows that limit cycles occur in pair when we allow $\lambda < 0$ we see that there is a stable fixed point and a stable limit cycle, e.g. take $\varepsilon = 0.015$, $\delta = 3.5$ and $\lambda = -0.045$, and furthermore, there are at least two limit cycles. These parameters λ, δ and ε lie outside the parameter set formed by (21.35).

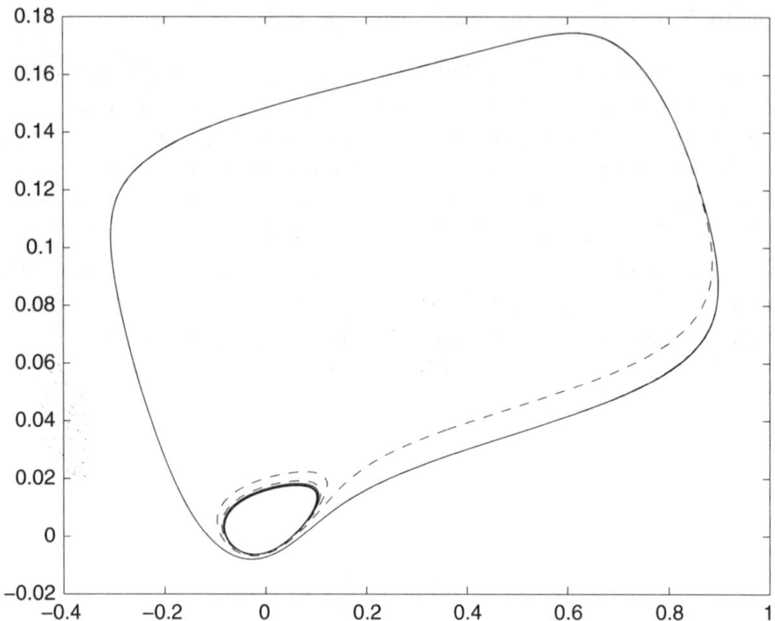

Fig. 21.3. Phase portrait of (21.10). There is bistability between the limit cycle and the asymptotically stable fixed point and two limit cycles occur. The inner cycle is unstable and the outer cycle is stable.

The bifurcation diagram of such a system is very interesting. It is possible to find a combination of the parameters where the first Lyapunov coefficient (see e.g. [6]) becomes zero, thus a Bautin (or generalized Andronov-Hopf bifurcation) occurs, in fact on the boundary of the parameter set formed by (21.35) and (21.35).. Furthermore, double-cycle bifurcation occurs when bistability exists. A completely theoretical analysis is given in a forthcoming paper, where both the first and second Lyapunov coefficients are investigated.

21.6 Conclusion

In this paper we gave some finer criteria for nonexistence of limit cycles for FitzHugh-Nagumo type of models, and hence a affirmative answer to the origin to be a globally asymptotically stable fixed point. Further, we proved the existence of stable limit cycles for the system (21.10). We also gave a numerical example that showed occurrence of double-cycle bifurcation in case bistability exists.

In our opinion it is a very hard problem to give the exact number of limit cycles of the FitzHugh-Nagumo system, since the analysis on how many limit cycles the FitzHugh-Nagumo class model has is in fact a problem that belongs to the second part of Hilbert's 16th problem.

References

1. M. Argentina, P. Coullet and V. Krinsky, Head-on collisions of wave in an excitable FitzHugh-Nagumo system: a transition from wave annihilation to classical wave behavior, *J. Theor. Biol.*, **205** (2000), 47–52.
2. L. Edelstein-Keshet, *Mathematical Models in Biology*, SIAM Classics in Applied Mathematics 46, 2004.
3. C.P. Fall, E.S. Marland, J.M.W. Wagner and J.J. Tyson, *Computational Cell Biology*, Springer, 2000.
4. E. Kaumann and U. Staude, Uniqueness and nonexistence of limit cycles for the FitzHugh equation, *Equadiff 82* (H.W. Knobloch and K. Schmitt, eds), Lecture Notes in math., vol 1017, Springer-Verlag, 1983, 313–321.
5. H. Korn and P. Faure, Is there chaos in the brain? II. Experimental evidence and related models, C. R. Biologies, **326** (2003), 787–840.
6. Yu. A. Kuznetsov, *Elements of Applied Bifurcation Theory*, Springer-Verlag, 1995.
7. S. Lefschetz, *Differential equations: geometric theory*, Interscience Publishers, 1957.
8. J. Murry, *Mathematical Biology* (2nd edition) Srpinger,1993.
9. T. Kostova, R. Ravindran and M. Schonbek, FitzHugh-Nagumo revisited: types of bifurcations, periodical forcing abd stability regions by a Lyapunov functional, submitted to *Int. J. Bifurcation and Chaos*.
10. J. Sugie, Nonexistence of periodic solutions for the FitzHugh nerve system, *Quart. Appl. Math.*, **49**(1991) 543–554.
11. J. Sugie and T. Hara, Nonexistence of periodic solutions of the Liénard system. *J. Math. Anal. Appl.* **159** (1991), 224–236.
12. S.A. Treskov and E.P. Volokitin, On existence of periodic orbits for the FitzHugh nerve system, *Quart. Appl. Math.*, **54** (1996) 601–107.

The Hermann-Martin Curve

Joachim Rosenthal[*]

Mathematics Institute, University of Zürich, Winterthurerstrasse 190, CH-8057 Zürich
`rosenthal@math.unizh.ch`

Dedicated to Clyde F. Martin on Occasion of his 60th Birthday

Summary. Every linear system can be naturally identified with a rational curve in a Grassmann variety. The associated curve is often referred to as Hermann-Martin curve of the system.

This article explains this crucial link between systems theory and geometry. The geometric translation also provides important tools when studying control design problems. In a second part of the article it is shown how it is possible to tackle some important control design problems by geometric means.

Keywords: Linear systems, Grassmann varieties, vector bundles over the projective line, Hermann-Martin curve, Grothendieck Quot scheme.

22.1 Introduction

In the late seventies Bob Hermann and Clyde Martin published a series of papers [14, 13, 19, 20] which showed a way how problems in linear systems theory can be translated into problems of algebraic geometry.

On the conceptual level this link provided a much deeper understanding for questions where topological properties of the class of linear systems played a role. The geometric understanding gave also tools at hand which helped to progress the research in several prominent problems like e.g. the static and dynamic pole placement problem.

In Section 22.2 we will explain the link between systems theory and geometry provided by the Hermann-Martin curve. For this purpose we will have to review some basic notions in systems theory as well as some basic notions in algebraic geometry.

---

[*]Supported in part by NSF grant DMS-00-72383

J. Rosenthal: *The Hermann-Martin Curve*, LNCIS **321**, 353–365 (2005)
`www.springerlink.com`

In Section 22.3 we will be concerned with topological properties of the set of linear systems having a fixed number of inputs, a fixed number of outputs and a fixed McMillan degree. An understanding of topological properties of this set has importance in the area of system identification and the area of robust controller design, to mention a few. The section will revisit two compactifications studied in the literature. Finally in Section 22.4 we will show how the geometric translation provided by the Hermann-Martin identification led to new results in the area of pole placement.

22.2 Linear Systems and Rational Curves in Grassmannians

Consider an m-inputs, p-outputs linear system Σ_n having McMillan degree n. In state space form this system is governed by the equations:

$$\Sigma_n \ : \ \begin{cases} \dot{x} = Ax + Bu \\ y = Cx + Du. \end{cases} \tag{22.1}$$

In terms of the frequency domain system (22.1) has an associated *transfer function*

$$G(s) := C(sI - A)^{-1}B + D. \tag{22.2}$$

By definition $G(s)$ is a $p \times m$ matrix with rational entries. The concept of a transfer function can be defined for an arbitrary field $\mathbb{K}$ and we will work to a large degree in this general setting. Whenever some additional properties on the field are required we will say so. The transfer function $G(s)$ captures the input-output behavior of the linear system. We say the matrices A, B, C, D form a realization of the transfer function $G(s)$. When A, B forms a controllable pair of matrices (i.e. the matrix pencil $[sI - A\ B]$ is left prime) and if A, C forms an observable pair (i.e. the matrix pencil $\begin{bmatrix} sI - A \\ C \end{bmatrix}$ is right prime), then we say that (22.1) forms a *minimal realization* of the transfer function (22.2). The size of the square matrix A in a minimal realization is called the McMillan degree of the transfer function $G(s)$.

Minimal realizations of proper transfer functions are unique in the following way: If $(\tilde{A}, \tilde{B})$ is a controllable pair and $(\tilde{A}, \tilde{C})$ is an observable pair with $G(s) = \tilde{C}(sI - \tilde{A})^{-1}\tilde{B} + \tilde{D}$ then there is a unique invertible matrix S of size $n \times n$ such that

$$(\tilde{A}, \tilde{B}, \tilde{C}, \tilde{D}) = (SAS^{-1}, SB, CS^{-1}, D). \tag{22.3}$$

We can identify the matrix S with an element of the general linear group Gl_n and we can view (22.3) as an orbit of a Gl_n action on a vector space. In this way we can view a linear system either through its transfer function $G(s)$ or as a Gl_n orbit in a vector space of dimension $n(n + m + p) + mp$.

For many applications it is very important to understand families of linear systems with a fixed number of inputs, fixed number of outputs and a fixed McMillan degree. Denote by $S_{p,m}^n$ the set of all proper $p \times m$ transfer functions having a fixed McMillan degree n. If the underlying base field are the real numbers Clark [3] showed in 1976 that $S_{p,m}^n$ has the structure of a smooth manifold of dimension $n(m+p) + mp$. Around the same time Hazewinkel [9] was able to show that $S_{p,m}^n$ has the structure of a quasi-affine variety, as soon as the base field $\mathbb{K}$ is algebraically closed. The basic proof techniques of Hazewinkel came from geometric invariant theory (GIT) and the interested reader is referred to [21, 36].

The work of Martin and Hermann [20] provided a new avenue to understand the algebraic and topological properties of the set $S_{p,m}^n$. For simplicity assume that the base field constitutes the complex numbers $\mathbb{C}$. Denote by

$$\mathbb{P}_{\mathbb{C}}^1 := \{\ell \subset \mathbb{C}^2 \mid \dim \ell = 1\} \hat{=} \{(x,1) \mid x \in \mathbb{C}\} \cup \{(1,0) =: \infty\}$$

the projective line over $\mathbb{C}$, i.e. the Riemann sphere and consider the Grassmann variety

$$\mathrm{Grass}(p, \mathbb{C}^{p+m}) := \{W \subset \mathbb{C}^n \mid \dim W = p\}$$

which parameterizes all p-dimensional linear subspace of the vector space $\mathbb{C}^n$. Martin and Hermann had the original idea to associate to each linear system a rational curve of genus zero inside $\mathrm{Grass}(p, \mathbb{C}^{p+m})$.

Definition 1 Let $G(s)$ be a $p \times m$ proper transfer function and consider the map

$$h : \ \mathbb{C} \longrightarrow \mathrm{Grass}(p, \mathbb{C}^{p+m}), \ s \mapsto \mathrm{rowspace}_{\mathbb{C}}[I_p \ G(s)]. \qquad (22.4)$$

Then h is called the *Hermann-Martin map* associated to the transfer function $G(s)$.

The Hermann-Martin map is a rational map. As the target space is compact all poles are removable and the map extends therefore to a holomorphic map:

$$\hat{h} : \ \mathbb{P}_{\mathbb{C}}^1 \longrightarrow \mathrm{Grass}(p, \mathbb{C}^{p+m}). \qquad (22.5)$$

The image of the map $\hat{h}$ defines a curve of genus zero and degree n inside the Grassmannian $\mathrm{Grass}(p, \mathbb{C}^{p+m})$, sometimes referred to as the *Hermann-Martin curve* associated to the linear system $G(s)$.

Note that every holomorphic map from the Riemann sphere $\mathbb{P}_{\mathbb{C}}^1$ to the Grassmannian is also rational. The following lemma is easily proved:

Lemma 2 *Let* $\hat{h} : \ \mathbb{P}_{\mathbb{C}}^1 \longrightarrow \mathrm{Grass}(p, \mathbb{C}^{p+m})$ *be a rational map with* $\hat{h}(\infty) = \mathrm{rowspace}_{\mathbb{C}}[M_1 \ M_2]$ *where* M_1 *is a* $p \times p$ *invertible matrix. Then* $\hat{h}$ *is the Hermann-Martin map of some proper transfer function* $G(s)$.

Given a system Σ_n described by the equations (22.1). Then one defines the *observability indices* of Σ_n as the left Kronecker indices of the pencil $\begin{bmatrix} sI - A \\ C \end{bmatrix}$. See [15, page 413] for details. Similarly one defines the controllability indices of Σ_n as the right Kronecker indices of the pencil $[sI - A \mid B]$. In [20] Martin and Hermann were able to connect the observability indices of Σ_n to the Grothendieck indices of an associated vector bundle. In order to make this precise we recall Grothendieck's theorem about the classification of vector bundles over the Riemann sphere:

Theorem 3 ([7]). *If ξ is a holomorphic vector bundle over $\mathbb{P}^1_{\mathbb{C}}$ then ξ decomposes as a sum of line bundles:*

$$\xi = O(\nu_1) \oplus \cdots \oplus O(\nu_p),$$

where $\nu_1, \ldots, \nu_p$ are the multiplicities of the line bundles. The nonnegative integers $\nu_1, \ldots, \nu_p$ depend up to order only on ξ.

The indices $\nu_1, \ldots, \nu_p$ are sometimes referred to as the Grothendieck indices of ξ. The integer $\nu = \sum_{j=1}^{p} \nu_j$ is called the degree of ξ.

Remark 4 Theorem 3 was derived by Grothendieck using general results from the theory of holomorphic vector bundles like Serre duality and splitting theorems for subbundles. At the time he was not aware that his result is also a straight forward consequence of some results by Dedekind and Weber [4]. The interested reader will find more details in [28].

Remark 5 Grothendieck's theorem is valid over any base field and a short elementary proof was given by Hazewinkel and Martin [11].

The Grassmann manifold is equipped with a natural vector bundle called the universal bundle U. Let U^* be its dual. The following theorem is due to Martin and Hermann [20]:

Theorem 6. *Let ξ be the pull back of the bundle U^* under the holomorphic map $\hat{h}$. Then the Grothendieck indices $\nu_1, \ldots, \nu_p$ of ξ are up to order equal to the observability indices of the system Σ_n. Moreover the degree $\nu = \sum_{j=1}^{p} \nu_j$ of ξ is equal to the McMillan degree of Σ_n.*

One way to gain more insight into the connection between the Grothendieck indices of a vector bundle and the observability indices of a system is via the concept of minimal bases as introduced by Forney [6]. For this assume that the transfer function $G(s) = C(sI - a)^{-1}B + D$ has a left coprime factorization $G(s) = D^{-1}(s)N(s)$. Without loss of generality we can assume that the rows of $[D(s) \mid N(s)]$ form a minimal polynomial basis of the rational vector space rowspace$_{\mathbb{C}(s)}[I_p \ G(s)]$ having ordered Forney indices $\nu_1 \geq \cdots \geq \nu_p$ and total degree $n = \sum_{j=1}^{p} \nu_j$. Then one has [6]:

Theorem 7. *1.* $\det D(s)$ *is equal to the characteristic polynomial of the transfer function* $G(s)$, *in particular* $n = \deg \det D(s)$ *is equal to the McMillan degree of* Σ_n.

2. The indices ν_j *are equal to the observability indices of* $G(s)$.

Remark 8 Since the observability indices are also equal to the Grothendieck indices it follows that under suitable translation the Forney indices of the polynomial matrix $[D(s) \mid N(s)]$ are also equal to the Grothendieck indices.

22.3 Compactification of the Set of Linear Systems having Fixed Mcmillan Degree

For many problems in linear systems theory such as questions of robustness and problems in identification theory it is very important to have an understanding of the topology of the set $S_{p,m}^n$ of systems having m inputs, p outputs and McMillan degree n. We already mentioned that $S_{p,m}^n$ has both the structure of a manifold and the structure of a quasi-affine variety [3, 9]. Further topological properties of $S_{p,m}^n$ have been derived over the years and we refer the reader e.g. to [12].

To understand degeneration phenomena of systems [10] or to understand the pole placement problem, a compactification of this space is desirable as well. The Hermann-Martin identification gives a natural way to achieve both these goals. We will explain this procedure for a general base field $\mathbb{K}$.

Denote by $\mathrm{Rat}_n(\mathbb{P}^1, \mathbb{P}^N)$ the set of rational maps from the projective line $\mathbb{P}^1_{\mathbb{K}}$ to the projective space $\mathbb{P}^N_{\mathbb{K}}$ having degree n. Every element of $\varphi \in \mathrm{Rat}_n(\mathbb{P}^1, \mathbb{P}^N)$ can be described through:

$$\varphi : \mathbb{P}^1 \longrightarrow \mathbb{P}^N, \ (s,t) \longmapsto (a_0(s,t), \ldots, a_N(s,t)), \tag{22.6}$$

where $a_i(x,y) \in \mathbb{K}[x,y], i = 0, \ldots, N$ are homogeneous polynomials of degree n. The description (22.6) is unique up to a nonzero constant factor $c \in \mathbb{K}^*$. In this way we can view an element of $\mathrm{Rat}_n(\mathbb{P}^1, \mathbb{P}^N)$ as a point in the projective space

$$\mathbb{P}(\mathbb{K}^{n+1} \otimes \mathbb{K}^{N+1}) = \mathbb{P}^{nN+n+N}_{\mathbb{K}}.$$

Since we can view a linear system as a rational map from the projective line to some Grassmannian and since a Grassmannian can be naturally seen as a subset of a projective space via the Plücker embedding we can view a linear system ultimately as a point of a projective space. This gives raise to an embedding of $S_{p,m}^n$ into a projective space as the following sequence of maps makes this precise:

$$
\begin{aligned}
S_{p,m}^n \ &\overset{\mathrm{Her.-Mar.}}{\longrightarrow} \ \mathrm{Rat}_n(\mathbb{P}^1, \mathrm{Grass}(p, m+p)) \\
&\overset{\mathrm{Pl\ddot{u}cker}}{\longrightarrow} \ \mathrm{Rat}_n(\mathbb{P}^1, \mathbb{P}^1(\wedge^p \mathbb{K}^{m+p})) \\
&\overset{\tau}{\longrightarrow} \ \mathbb{P}(\mathbb{K}^{n+1} \otimes \wedge^p \mathbb{K}^{m+p}).
\end{aligned}
\tag{22.7}
$$

Definition 9 Denote by $K^n_{p,m}$ the Zariski closure of the image of $S^n_{p,m}$ inside the projective space $\mathbb{P}(\mathbb{K}^{n+1} \otimes \wedge^p \mathbb{K}^{m+p})$.

In [26, 27] it has been shown:

Theorem 10. *For any base field* $\mathbb{K}$, $K^n_{p,m}$ *is a projective variety of dimension* $n(m+p) + mp$ *containing the set* $S^n_{p,m}$ *of* $p \times m$ *proper transfer functions of McMillan degree* n *as a Zariski dense subset.*

The space $K^n_{p,m}$ has also been studied in the area of conformal quantum field theory. For this reason Sottile [33, 34] calls the variety $K^n_{p,m}$ the *quantum Grassmannian.* In [1] this variety is also called the *Uhlenbeck compactification.* When $\min(m,p) = 1$ then $K^n_{p,m}$ represents simply a projective space. In general $K^n_{p,m}$ is however a singular variety [27].

On the side of $K^n_{p,m}$ there is a second well studied compactification of $S^n_{p,m}$ due to Grothendieck. In [8] Grothendieck showed that the set $\mathcal{Q}^n_{p,m}$ parameterizing all quotient sheaves $\mathcal{B}$ of $\mathbb{K}^{m+p} \otimes \mathcal{O}_{\mathbb{P}}$ having rank m, degree n and Hilbert polynomial $\chi(\mathcal{B}(x)) = px + p + n$ has naturally the structure of a scheme. In the algebraic geometry literature $\mathcal{Q}^n_{p,m}$ is usually referred as a *Quot scheme.* For the particular Quot scheme $\mathcal{Q}^n_{p,m}$ under consideration Strömme [35] showed that $\mathcal{Q}^n_{p,m}$ has the structure of a smooth projective variety of dimension $n(m+p)+mp$. Furthermore $\mathcal{Q}^n_{p,m}$ compactifies the space $\mathrm{Rat}_n(\mathbb{P}^1, \mathrm{Grass}(p, m+p))$ of all rational maps of degree n from the projective line to the Grassmannian $\mathrm{Grass}(p, m+p)$.

The fact that Grothendieck's Quot scheme $\mathcal{Q}^n_{p,m}$ has a relevance in linear systems theory was first recognized by Lomadze [18]. The author in collaboration with Ravi was able to give a direct systems theoretic interpretation of $\mathcal{Q}^n_{p,m}$ in terms of matrix pencils and polynomial matrices. We follow here the original description in [22, 23].

Let $\mathbb{K}$ be an arbitrary field and consider a $p \times (m+p)$ polynomial matrix

$$P(s,t) := \begin{pmatrix} f_{11}(s,t) & f_{12}(s,t) & \cdots & f_{1,m+p}(s,t) \\ f_{21}(s,t) & f_{22}(s,t) & \cdots & f_{2,m+p}(s,t) \\ \vdots & \vdots & & \vdots \\ f_{p1}(s,t) & f_{p2}(s,t) & \cdots & f_{p,m+p}(s,t) \end{pmatrix}. \tag{22.8}$$

We say $P(s,t)$ is homogeneous if each element $f_{ij}(s,t) \in \mathbb{K}[s,t]$ is a homogeneous polynomial of degree ν_i. We say two homogeneous matrices $P(s,t)$ and $\tilde{P}(s,t)$ are equivalent if after a possible permutation they have the same row-degrees and if there is a unimodular matrix $U(s,t)$ such that $P = U\tilde{P}$. Using this equivalence relation we define:

Definition 11 An equivalence class of full rank homogeneous polynomial matrices $P(s,t)$ will be called a *homogeneous autoregressive system.* The McMillan degree of a homogeneous autoregressive system is defined as the sum of the row degrees, i.e. through $n := \sum_{i=1}^{p} \nu_i$.

The main theorem of [22] states:

Theorem 12. *The set of $p \times (m + p)$ homogeneous autoregressive systems of degree n is in bijective correspondence to the points of the Grothendieck Quot scheme $\mathcal{Q}_{p,m}^n$. The set $S_{p,m}^n$ of proper transfer functions can be viewed as a Zariski open subset of $\mathcal{Q}_{p,m}^n$. In particular we can view $\mathcal{Q}_{p,m}^n$ as a smooth compactification of $S_{p,m}^n$.*

In the sequel we will elaborate on the connection of homogeneous autoregressive systems to algebraic geometry and to systems theory.

The connection to algebraic geometry can be seen in the following way: If the row degrees of the $p \times (m + p)$ matrix $P(s,t)$ are ν_i then there is a short exact sequence:

$$0 \longrightarrow \bigoplus_{i=1}^{p} \mathcal{O}_{\mathbb{P}}(-\nu_i) \xrightarrow{P(s,t)} \mathbb{K}^{m+p} \bigotimes \mathcal{O}_{\mathbb{P}} \xrightarrow{\Phi} \mathcal{B} \longrightarrow 0. \qquad (22.9)$$

In this way every homogeneous autoregressive system defines in a natural way a quotient sheaf $\mathcal{B}$.

There is also a direct connection to Hermann-Martin maps. When $m, p > 0$ then every $p \times (m + p)$ matrix $P(s,t)$ defines a rational map:

$$\hat{h} : \mathbb{P}_{\mathbb{K}}^1 \longrightarrow \mathrm{Grass}(p, \mathbb{K}^{p+m}), \quad (s,t) \longmapsto \mathrm{rowspace}_{\mathbb{K}} P(s,t). \qquad (22.10)$$

In case that $P(s,t)$ is left prime then the morphism $\hat{h}$ has no poles and therefore is a regular map.

Having introduced a smooth compactification of the set $S_{p,m}^n$ which parameterizes all m-input, p-output systems of McMillan degree n it is of course an interesting question if one can give a systems theoretic interpretation for the systems added in the compactification process. This can indeed be done.

We will need the notion of generalized state space systems as studied by Kuijper and Schumacher [17]. For this let G, F be matrices of size $n \times (m+n)$ and let H be a matrix of size $(m + p) \times (m + n)$. The matrices (G, F, H) describe over any field a discrete time linear system through:

$$Gz(t + 1) = Fz(t), \quad w(t) = Hz(t). \qquad (22.11)$$

In this representation $z(t) \in Z \simeq \mathbb{K}^{m+n}$ describes the set of "internal variables" and $w(t) \in \mathbb{K}^{m+p}$ describes the set of "external variables". The matrices G, F are linear maps from the space of internal variables $Z \simeq \mathbb{K}^{m+n}$ to the state space $X \simeq \mathbb{K}^n$. Corresponding to change of coordinates in X and Z one has a natural equivalence among pencil representations:

$$(G, F, H) \ \sim \ (SGT^{-1}, SFT^{-1}, HT^{-1}). \qquad (22.12)$$

In above equivalence it is assumed that $S \in Gl_n$ and $T \in Gl_{m+n}$. The realization (22.11) reduces to the familiar A, B, C, D representation (22.1) as soon as there is an invertible matrix T such that:

$$GT^{-1} = \begin{bmatrix} I & 0 \end{bmatrix}, \quad FT^{-1} = \begin{bmatrix} A & B \end{bmatrix}, \quad HT^{-1} = \begin{bmatrix} 0 & I \\ C & D \end{bmatrix}. \tag{22.13}$$

If there is no invertible matrix T to do such a transformation then (22.11) does not describe an input output system under the natural partitioning $w = \begin{bmatrix} u \\ y \end{bmatrix}$.

The main theorem of [23] states:

Theorem 13. *Let X be the set of all matrix triples (G, F, H) where G, F are matrices of size $n \times (m + n)$ and H is a matrix of size $(m + p) \times (m + n)$. Then the stable points in the sense of GIT [21] under the $Gl_n \times Gl_{m+n}$ action induced by (22.12) are given by the conditions:*

1. $[sG - tF]$ has full row rank n.

2. $\begin{bmatrix} sG - tF \\ H \end{bmatrix}$ has full column rank $m + n$ for all $(s, t) \in \mathbb{K}^2 \setminus \{(0, 0)\}$.

Finally the geometric quotient of the stable points modulo the group action is equal to the variety $\mathcal{Q}_{p,m}^n$.

We would like to conclude this section with the remark that generalized first order representations of the form (22.11) as well as Grothendieck's Quot scheme $\mathcal{Q}_{p,m}^n$ are well defined when $m = 0$. In systems theoretic terms we are then dealing with observable A, C systems having no inputs. When $m = 0$ the Hermann-Martin map (22.10) is however very degenerate and a sheaf theoretic interpretation is required to distinguish among the different points of $\mathcal{Q}_{p,0}^n$.

22.4 Results on Pole Placement by Geometric Methods

An area where algebraic geometric methods were very successfully applied in systems theory are the different questions of pole placement and stabilization of linear system. Crucial for the solution of these problems was the understanding of the manifold $S_{p,m}^n$ and its compactifications $K_{p,m}^n$ and $\mathcal{Q}_{p,m}^n$. In this section we will explain these results.

Consider a strictly proper linear system Σ_n of McMillan degree n:

$$\Sigma_n \; : \; \begin{cases} \dot{x} = Ax + Bu \\ y = Cx \end{cases} \tag{22.14}$$

A dynamic compensator of order q is a linear system of degree q, having the following state space representation:

$$\Sigma_q \; : \; \begin{cases} \dot{z} = Fz + Gy \\ u = Hz + Ky \end{cases} \tag{22.15}$$

In this representation, z is a q–vector, which describes the state of the compensator. The special case of $q = 0$ corresponds to the case of static feedback. The overall system is described by:

$$\begin{bmatrix} \dot{x} \\ \dot{z} \end{bmatrix} = \underbrace{\begin{bmatrix} A + BKC & BH \\ GC & F \end{bmatrix}}_{M} \begin{bmatrix} x \\ z \end{bmatrix} \tag{22.16}$$

$$y = Cx \tag{22.17}$$

which is a linear system of McMillan degree $n + q$. Stability of the closed loop system depends on the location of the eigenvalues of the matrix M.

We will parameterize the eigenvalues of the matrix M through its characteristic polynomial $\chi_M(x) = \det(xI - M) \in \mathbb{K}[x]$. The polynomial $\chi_M(x)$ is a monic polynomial of degree $n + q$ and we can identify this polynomial with a point in the vector space $\mathbb{K}^{n+q}$. Similarly we can identify a tuple of matrices F, G, H, K with a point in the vector space $\mathbb{K}^{q(m+p+q)+mp}$. With this identification we define the *affine pole placement map* through:

$$\varphi : \quad \mathbb{K}^{q(m+p+q)+mp} \longrightarrow \mathbb{K}^{n+q} \tag{22.18}$$

$$(F, G, H, K) \longmapsto \det(xI - M).$$

One says system (22.14) is arbitrary pole assignable with compensators of McMillan degree at most q as soon as the pole placement map (22.18) is surjective.

An important special case of the general question is the static pole placement problem. This is the situation when $q = 0$, i.e. the compensator (22.15) has the simple form $u = Ky$ and K is an $m \times p$ matrix.

In order to study this problem Hermann and Martin used the dominant morphism theorem of algebraic geometry to derive the result [14]:

Theorem 14. *Assume that the base field $\mathbb{K}$ is algebraically closed. Then for a generic set of systems Σ_n having the state space form (22.14) almost arbitrary pole placement by static compensators is possible if and only if $n \le mp$.*

To derive this theorem it was probably the first time that deeper methods from algebraic geometry were used to tackle a problem in control systems design. The dominant morphism theorem does a 'local computation' and therefore cannot achieve results of full surjectivity.

A couple of years later Brockett and Byrnes studied the static pole placement problem by considering also the effects 'at the boundary' of the parameter space. They realized that it is best to compactify the compensator space via the Grassmann variety $\mathrm{Grass}(m, \mathbb{F}^{m+p})$. Note that this compactification simply consists of all degree 0 Hermann-Martin maps from the projective line to $\mathrm{Grass}(m, \mathbb{F}^{m+p})$! The main result of [2] states:

Theorem 15. *Assume that the base field $\mathbb{K}$ is algebraically closed. Then for a generic set of systems Σ_n having the state space form (22.14) arbitrary pole placement by static compensators is possible if and only if $n \le mp$. Moreover when $n = mp$ the number of solutions (when counted with multiplicities) is exactly equal to the degree of the Grassmann variety:*

$$\deg \text{Grass}(m, m + p) = \frac{1!2! \cdots (p - 1)!(mp)!}{m!(m + 1)! \cdots (m + p - 1)!}. \tag{22.19}$$

In particular if $\deg \text{Grass}(m, m + p)$ *is odd, pole assignment by real static feedback is possible.*

This was quite a surprising result. The degree of the Grassmannian as described in formula (22.19) was computed in the 19th century by Schubert [32]. At the time Schubert's computations were not generally accepted and Hilbert devoted his 15th problem to the Schubert calculus. The modern way to see Schubert's number (22.19) as the degree of a Grassmann variety has its origin in the 20th century and the interested reader will enjoy the article of Kleiman [16] in this regard.

Both the results of Theorem 14 and Theorem 15 required that the base field is algebraically closed, e.g. the field of complex numbers. This is not surprising as some of the strongest results in algebraic geometry require that the base field is algebraically closed.

In 1992 Alex Wang adapted algebraic geometric methods for the study of the real Grassmannian to derive the following at the time very surprising result [38]:

Theorem 16. *For a generic set of real systems* Σ_n *having the state space form (22.14) arbitrary pole placement by static pole placement is possible as soon as* $n < mp$.

It was later realized that Wang's proof can be considerably simplified without requiring too deep results from algebraic geometry. The interested reader will find an elementary proof in [29].

In order to progress on the solution of the general pole placement problem with dynamic compensators it was necessary to come up with a suitable compactification of the space $S_{m,p}^q$, or equivalently the set of Hermann-Martin maps $\text{Rat}_q(\mathbb{P}^1, \text{Grass}(m, \mathbb{K}^{m+p}))$. As a generalization of the result by Brockett and Byrnes (Theorem 15), the author derived the following result [27]:

Theorem 17. *Assume that the base field* $\mathbb{K}$ *is algebraically closed. Then for a generic set of systems* Σ_n *having the state space form (22.14) arbitrary pole placement by dynamic compensators is possible if and only if*

$$n \leq q(m + p - 1) + mp.$$

Moreover when $n = q(m + p - 1) + mp$ *the number of solutions (when counted with multiplicities) is exactly equal to the degree of the quantum Grassmannian* $K_{m,p}^q$.

This theorem made it a challenge to compute the degree of the quantum Grassmannian and in this way to come up with a generalization of Schubert's famous formula (22.19). The result of this effort was:

Theorem 18 ([24, 25]). *The degree of the quantum Grassmannian $K_{m,p}^q$ is given by:*

$$(-1)^{q(m+1)}(mp + q(m + p))! \sum_{n_1+\cdots+n_m=q} \frac{\prod_{k<j}(j-k+(n_j-n_k)(m+p))}{\prod_{j=1}^{m}(p+j+n_j(m+p)-1)!} \tag{22.20}$$

The general pole placement problem over an arbitrary field $\mathbb{K}$ as described in this section is still not completely solved. Even for static compensators (when $q = 0$) and over the reals there is a gap of one degree of freedom in Wang's result. Eremenko and Gabrielov [5] have recently closed this gap for many cases but the gap still exists for infinite many cases. It would also be worthwhile to study the pole placement problem over other fields. E.g. convolutional codes can be viewed as linear systems over finite fields (see e.g. [30]) and the decoding problem seems to be closely connected to the problem of designing a linear observer.

22.5 Conclusion

In this paper we provided a survey about the Hermann-Martin curve, a crucial link between linear systems theory and algebraic geometry.

The Hermann-Martin curve provided a better understanding of the topology of the class of linear systems $S_{p,m}^n$ parameterizing the set of m-inputs, p-outputs system of McMillan degree n. The geometric point of view led to natural compactifications of the space $S_{p,m}^n$ and this was ultimately key in the progress on the static and dynamic pole placement problem.

It is our believe that a further investigation of the space $S_{p,m}^n$ would be very beneficial for many linear systems theory problems. E.g. the manifold $S_{p,m}^n$ comes with some natural metrics which allows one to compute distances between linear systems. A first attempt in this direction was done in [31]. Another area where a topological understanding of the space $S_{p,m}^n$ is important is the design of stable numerical algorithms while solving control problems. As an example we mention the recent paper by Verschelde and Wang [37] where this issue stands out.

As all these remarks make it clear the translation from systems theoretic questions to geometric questions has been very fruitful in the past and we expect that further results will come out from this. A crucial starting point to explore this connection is the paper by Martin and Hermann [20].

References

1. Bertram, A., G. Daskalopoulos, and R. Wentworth: 1996, 'Gromov Invariants for Holomorphic Maps from Riemann surfaces to Grassmannians'. *J. Amer. Math. Soc.* **9**(2), 529–571.

2. Brockett, R. W. and C. I. Byrnes: 1981, 'Multivariable Nyquist Criteria, Root Loci and Pole Placement: A Geometric Viewpoint'. *IEEE Trans. Automat. Control* **AC-26**, 271–284.

3. Clark, J. M. C.: 1976, 'The Consistent Selection of Local Coordinates in Linear System Identification'. In: *Proc. Joint Automatic Control Conference.* pp. 576–580.

4. Dedekind, R. and H. Weber: 1882, 'Theorie der algebraischen Functionen einer Veränderlichen'. *Journal für die reine und angewandte Mathematik* **92**, 181–291.

5. Eremenko, A. and A. Gabrielov: 2002, 'Pole Placement Static Output Feedback for Generic Linear Systems'. *SIAM J. Control Optim.* **41**(1), 303–312 (electronic).

6. Forney, Jr., G. D.: 1975, 'Minimal Bases of Rational Vector Spaces, with Applications to Multivariable Linear Systems'. *SIAM J. Control* **13**(3), 493–520.

7. Grothendieck, A.: 1957, 'Sur la classification des fibres holomorphes sur la sphere de Riemann'. *Amer J. Math.* **79**, 121–138.

8. Grothendieck, A.: 1966, 'Techniques de construction et théorèmes d'existence en géométrie algébrique IV: Les schèmas de Hilbert'. In: *Séminaire Bourbaki*, Vol. 1960/61, éxposé 221. New York: Benjamin.

9. Hazewinkel, M.: 1977, 'Moduli and Canonical Forms for Linear Dynamical Systems III: The Algebraic Geometric Case'. In: *Proc. of the 76 Ames Research Center (NASA) Conference on Geometric Control Theory.* pp. 291–336.

10. Hazewinkel, M.: 1980, 'On Families of Linear Systems: Degeneration Phenomena'. In: C. Byrnes and C. Martin (eds.): *Algebraic and Geometric Methods in Linear Systems Theory*, Vol. 18 of *Lectures in Applied Mathematics.* Amer. Math. Society, pp. 157–189.

11. Hazewinkel, M. and C. Martin: 1982, 'A Short Elementary Proof of Grothendieck's Theorem on Algebraic Vectorbundles over the Projective Line'. *J. Pure Appl. Algebra* **25**, 207–211.

12. Helmke, U.: 1985, 'The Topology of a Moduli Space for Linear Dynamical Systems'. *Comm. Math. Helv.* **60**, 630–655.

13. Hermann, R. and C. Martin: 1977a, 'Applications of Algebraic Geometry to Systems Theory. V. Ramifications of the Macmillan Degree'. In: *The 1976 Ames Research Center (NASA) Conference on Geometric Control Theory (Moffett Field, Calif., 1976)*. Brookline, Mass.: Math Sci Press, pp. 67–120. Lie Groups: History Frontiers and Appl., Vol. VII.

14. Hermann, R. and C. F. Martin: 1977b, 'Applications of Algebraic Geometry to System Theory Part I'. *IEEE Trans. Automat. Control* **AC-22**, 19–25.

15. Kailath, T.: 1980, *Linear Systems.* Englewood Cliffs, N.J.: Prentice-Hall.

16. Kleiman, S. L.: 1976, 'Problem 15: Rigorous Foundations of Schubert's Enumerative Calculus'. In: *Proceedings of Symposia in Pure Mathematics*, Vol. 28. Am. Math. Soc., pp. 445–482.

17. Kuijper, M. and J. M. Schumacher: 1990, 'Realization of Autoregressive Equations in Pencil and Descriptor Form'. *SIAM J. Control Optim.* **28**(5), 1162–1189.

18. Lomadze, V.: 1990, 'Finite-Dimensional Time-Invariant Linear Dynamical Systems: Algebraic Theory'. *Acta Appl. Math* **19**, 149–201.

19. Martin, C. F. and R. Hermann: 1977, 'Applications of Algebraic Geometry to Systems Theory. II. Feedback and Pole Placement for Linear Hamiltonian Systems'. *Proc. IEEE* **65**(6), 841–848.

20. Martin, C. F. and R. Hermann: 1978, 'Applications of Algebraic Geometry to System Theory: The McMillan Degree and Kronecker Indices as Topological and Holomorphic Invariants'. *SIAM J. Control Optim.* **16**, 743–755.

21. Mumford, D. and J. Fogarty: 1982, *Geometric Invariant Theory*, No. 34 in Ergebnisse. Springer Verlag, 2nd edition.

22. Ravi, M. S. and J. Rosenthal: 1994, 'A Smooth Compactification of the Space of Transfer Functions with Fixed McMillan Degree'. *Acta Appl. Math* **34**, 329–352.

23. Ravi, M. S. and J. Rosenthal: 1995, 'A General Realization Theory for Higher Order Linear Differential Equations'. *Systems & Control Letters* **25**(5), 351–360.

24. Ravi, M. S., J. Rosenthal, and X. Wang: 1996, 'Dynamic Pole Assignment and Schubert Calculus'. *SIAM J. Control Optim.* **34**(3), 813–832.

25. Ravi, M. S., J. Rosenthal, and X. Wang: 1998, 'Degree of the Generalized Plücker Embedding of a Quot Scheme and Quantum Cohomology'. *Math. Ann.* **311**(1), 11–26.

26. Rosenthal, J.: 1990, 'Geometric Methods for Feedback Stabilization of Multivariable Linear Systems'. Ph.D. thesis, Arizona State University.

27. Rosenthal, J.: 1994, 'On Dynamic Feedback Compensation and Compactification of Systems'. *SIAM J. Control Optim.* **32**(1), 279–296.

28. Rosenthal, J.: 2002, 'Minimal Bases of Rational Vector Spaces and their Importance in Algebraic Systems Theory'. In: R. E. Blahut and R. Koetter (eds.): *Codes, Graphs, and Systems*. Kluwer Academic Publishers, pp. 345–357.

29. Rosenthal, J., J. M. Schumacher, and J. C. Willems: 1995, 'Generic Eigenvalue Assignment by Memoryless Real Output Feedback'. *Systems & Control Letters* **26**, 253–260.

30. Rosenthal, J., J. M. Schumacher, and E. V. York: 1996, 'On Behaviors and Convolutional Codes'. *IEEE Trans. Inform. Theory* **42**(6, part 1), 1881–1891.

31. Rosenthal, J. and X. Wang: 1993, 'What is the distance Between Two Autoregressive Systems'. In: K. Bowers and J. Lund (eds.): *Computation and Control III*. Boston: Birkhäuser Verlag, pp. 333–340.

32. Schubert, H.: 1886, 'Anzahlbestimmung für lineare Räume beliebiger Dimension'. *Acta Math.* **8**, 97–118.

33. Sottile, F.: 2000, 'Real Rational Curves in Grassmannians'. *J. Amer. Math. Soc.* **13**(2), 333–341.

34. Sottile, F.: 2001, 'Rational curves on Grassmannians: systems theory, reality, and transversality'. In: *Advances in algebraic geometry motivated by physics (Lowell, MA, 2000)*, Vol. 276 of *Contemp. Math.* Providence, RI: Amer. Math. Soc., pp. 9–42.

35. Strömme, S. A.: 1987, 'On Parametrized Rational Curves in Grassmann Varieties'. In: F. Ghione, C. Peskine, and E. Sernesi (eds.): *Space Curves*, Lecture Notes in Mathematics # 1266. Springer Verlag, pp. 251–272.

36. Tannenbaum, A.: 1981, *Invariance and System Theory: Algebraic and Geometric Aspects*, Vol. 845 of *Lecture Notes in Mathematics*. Berlin: Springer Verlag.

37. Verschelde, J. and Y. Wang: 2004, 'Computing Dynamic Output Feedback Laws'. *IEEE Trans. Automat. Contr.* **49**(8), 1393–1397.

38. Wang, X.: 1992, 'Pole Placement by Static Output Feedback'. *Journal of Math. Systems, Estimation, and Control* **2**(2), 205–218.

Hamptonese and Hidden Markov Models

Mark Stamp and Ethan Le

Department of Computer Science, San Jose State University, San Jose, CA 95192
stamp@cs.sjsu.edu

Summary. James Hampton was a quiet and secretive man who left behind a monumental work of visionary art, along with a strange handwritten script. Hampton's script, or Hamptonese, bears no significant resemblance to any known written language. In this paper we analyze Hamptonese using hidden Markov models. This analysis shows that Hamptonese is not a simple substitution for English and provides some evidence that Hamptonese may be the written equivalent of "speaking in tongues".

23.1 James Hampton and Hamptonese

James Hampton was born in South Carolina in the year 1909. Hampton's father left his family to pursue his calling as an itinerant gospel singer and self-appointed Baptist minister. At the age of 19, Hampton moved to Washington, DC, where he struggled to find gainful employment during the Great Depression. Hampton was drafted into the Army in 1942 and served in Guam and Saipan. After his discharge from the Army in 1945, Hampton was employed as a janitor for the General Services Administration in Washington, DC, where he worked until his death in 1964.

Beyond the thumbnail sketch in the previous paragraph, very little is known about Hampton's life. In his post-war years, he lived in a small apartment in DC and rented a nearby garage. During this time period, he apparently had no family or close friends. Shortly after Hampton's death (due to stomach cancer), the owner of his rented garage discovered that it housed a collection of artwork that Hampton had dubbed "The Throne of the Third Heaven of the Nations' Millennium General Assembly".

Art critic Robert Hughes has written that Hampton's Throne "...may well be the finest work of visionary religious art produced by an American" [14]. The Throne is part of the permanent collection of the Smithsonian Museum

M. Stamp and E. Le: *Hamptonese and Hidden Markov Models*, LNCIS **321**, 367–378 (2005)
www.springerlink.com   © Springer-Verlag Berlin Heidelberg 2005

Fig. 23.1. Hampton and his Throne.

of American Art in Washington, DC. Hampton is pictured with his Throne in Figure 23.1.

Hampton's Throne is a remarkable collection of some 180 individual pieces. Each piece is constructed of discarded items—broken furniture, burnt out light bulbs, jars—carefully wrapped in gold and silver foil, with the foil itself taken from discarded items, such as cigarette packages. The artwork is extremely fragile, being held together with tacks, pins and tape.

The Throne is clearly a religious work as it includes several plaques with Biblical references. Hampton was a believer in the doctrine of "dispensationalism" [5], and the "7 dispensations" figure prominently in his work. For more information and speculation on Hampton's artwork and his life see [14, 15].

Along with the Throne, several notebooks were discovered in Hampton's garage. These notebooks contain some scattered English text, but consist primarily of a script of unknown origin that we refer to as Hamptonese. In total, there are 164 pages of Hamptonese. These 164 pages can be separated into two sets, one of which includes tombstone-shaped drawings, Roman numerals, Hamptonese characters and significant amounts of English text, including phrases such as "THE OLD AND NEW COVENANT RECORDED BY ST. JAMES" and "fear not". Many of these pages deal with the "7 dispensations" and some are similar to writing that appears on plaques included with the Throne.

The second set of Hamptonese text consists of 100 pages of essentially "pure" Hamptonese. This set has no drawings and only a handful of interspersed English words or letters, such as "Jesus", "Virgin Mary", "NRNR" and "Revelation". Each of these pages contains 25 or 26 lines of Hamptonese, with a highly variable number of symbols per line. There is no punctuation

or paragraphs and it is not clear that the pages should be read top-to-bottom
and left-to-right (although the ragged right margin might indicate that left-
to-right is appropriate).

Each page in the pure Hamptonese set is headed by "ST James" with
"REVELATiON" appearing at the bottom. Hampton also numbered the
pages, but his numbering is inconsistent with duplicates, gaps and some pages
having two number. Curiously, a few pages appear to have had certain lines
extended after the page was initially written, and in a few places it appears
that at least some lines may have been written from bottom to top, due to
the location of strategic gaps within the text. In addition, water damage significantly affects a few pages and some pages are marred by ink blotches.

It is not clear whether Hampton wrote his Hamptonese over a relatively
short period of time, or over a more extended timeframe. There is a reference
to Truman's inauguration on one page, which implies that Hampton could
have spent well over a decade writing his Hamptonese text. The writing style
does change substantially over the 100 pages of pure Hamptonese, becoming
far clearer—and easier to transcribe—as it progresses.

The 100 pages of "pure" Hamptonese is the basis for the analysis presented
in this paper. A partial page of Hamptonese appears in Figure 23.2. Scanned
images of all 100 pages are available online in JPEG and TIFF format at [12].

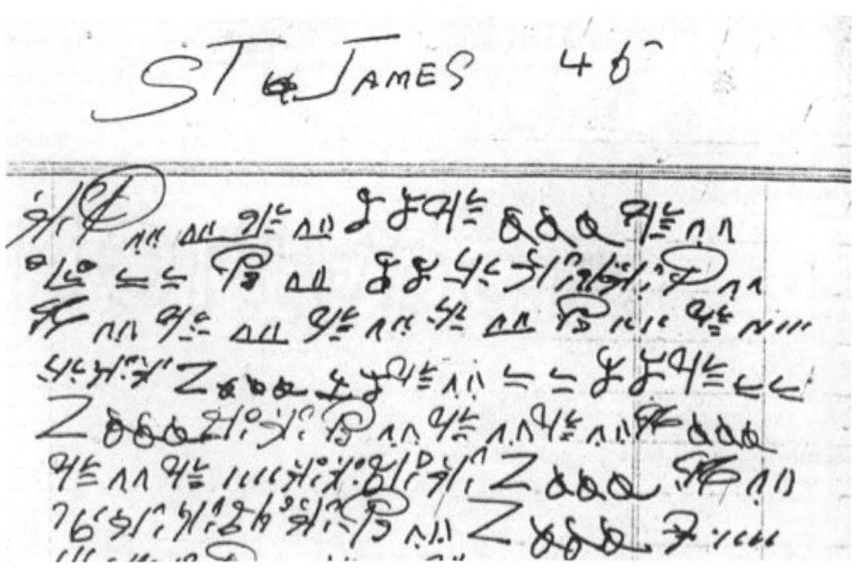

Fig. 23.2. Example of Hamptonese.

A comparison of Hamptonese with the encyclopedic collection of written languages at Ominglot [1] reveals no similar script. The only published analysis of Hamptonese that we are aware of is [9], which makes strong claims regarding the number of "consonants" and "vowels" in Hamptonese based on a small sample of the text. We believe that these claims are unsupported by the limited data and analysis provided.

23.2 Transcription

Before analyzing Hamptonese, we first transcribed it into a computer-friendly form. We believe that our transcription is reasonably accurate, but given the nature of Hamptonese, there are certain to be some errors.

We took a conservative approach, choosing to distinguishing symbols that might simply be variants or "typos". When doing analysis, it is always possible to group selected symbols together.

The Hamptonese symbols, the corresponding transcription keys and frequency counts appear in Table 23.1. A complete transcription of the 100 pages of Hamptonese is available at [12].

In the next section we present elementary entropy calculations which indicate that in some sense the information in Hamptonese is comparable to that in English. We then give a brief introduction to hidden Markov models followed by a discussion of the utility of this technique for analyzing English text. This is followed by a hidden Markov model analysis of Hamptonese and a discussion of our findings.

23.3 Entropy of Hamptonese

Shannon's entropy is the classic measure of information or uncertainty [8]. In terms of bits, entropy is computed as

$$H(X) = -\sum_{x \in X} P(x) \log(P(x)),$$

where the logarithm is to the base 2, and $0 \log(0)$ is taken to be 0. Another useful formula is that for conditional entropy [13],

$$H(X \mid Y) = -\sum_{x} \sum_{y} P_{X,Y}(x, y) \log(P_Y(y \mid x)). \tag{23.1}$$

It is well-known—and intuitively clear—that conditioning reduces entropy, i.e., $H(Y \mid X) \leq H(X)$, with equality holding if and only if X and Y are independent.

If English consisted of random selections from the 27 symbols (letters and space) then the uncertainty of English (at the level of individual symbols) would be

key	Hamptonese	count	relative frequency	key	Hamptonese	count	relative frequency
2		529	0.0180565	14		1313	0.0448169
ee		227	0.0077482	76		871	0.0297300
95		195	0.0066560	96		453	0.0154623
dc		709	0.0242004	EE		415	0.0141653
vv		6318	0.2156535	M		581	0.0198314
F		493	0.0168277	N		541	0.0184660
g		360	0.0122879	g7		46	0.0015701
GG		1365	0.0465918	Gi		889	0.0303444
Ki		2138	0.0729768	d3		510	0.0174079
d4		51	0.0017408	Y3		3578	0.1221285
Y4		644	0.0219818	qL3		1113	0.0379902
qL4		23	0.0007851	4L		272	0.0092842
uL		45	0.0015360	J1		186	0.0063487
JJ		587	0.0200362	LL		1014	0.0346111
nn		337	0.0115029	P		754	0.0257364
PL		113	0.0038571	P1		302	0.0103082
P2		665	0.0226986	o3		299	0.0102058
q3		558	0.0190463	S		329	0.0112298
3		89	0.0030379	T		67	0.0022869
A		21	0.0007168	A-		182	0.0062122
44		105	0.0035840	I		10	0.0003413
total						29297	1.0000000

Table 23.1. Hamptonese frequency counts

$$H(X) = -\sum 1/27 \log(1/27) = \log(27) \approx 4.75,$$

i.e., each letter would contain about 4.75 bits of information.

Of course, the distribution of English letters is not uniform, as is also true of the phonemes in phonetic English. Similarly, Hamptonese characters are not uniformly distributed. We have computed the empirical entropy for individual symbols for English letters and phonemes, as well as for Hamptonese characters. We have also computed the entropy for two and three consecutive symbols using the conditional entropy formula (23.1). For example, the term associated with the string abc in (23.1) is given by

$$P(abc)\log(P(c\,|\,ab)).$$

Our entropy results appear in Table 23.2.

	Entropy		
	1 symbol	2 symbols	3 symbols
English letters	4.11	3.54	2.68
English phonemes	4.76	3.87	3.00
Hamptonese	4.41	3.63	3.08

Table 23.2. Entropy comparison

The results in Table 23.2 indicate that for strings of three symbols or less, the information contained in Hamptonese is comparable to that of English. Of course, this does not prove that Hamptonese is a language, but it does suggest that it is reasonable to pursue the analysis further.

23.4 Hidden Markov Models

Hidden Markov models (HMMs) provide an efficient algorithmic solution to certain problems involving Markov processes. In a Markov process (of order one) the next state depends only on the current state and a fixed set of transition probabilities. HMMs generalize a Markov process to the case where the states cannot be directly observed. In an HMM, we observe some properties of the system that are related to the underlying states by fixed probability distributions.

A schematic view of an HMM is given in Figure 23.3, where $A = \{a_{ij}\}$ is the (row stochastic) matrix driving the underlying Markov process, the X_i are the states of the Markov process, the $\mathcal{O}_i$ are the observations, T is the length of the observed sequence, $B = \{b_{ij}\}$ is a (row stochastic) matrix that relates the states to the observations, and the dashed line is a "curtain" between the observer and the underlying Markov process. More precisely, a_{ij} is the probability of a transition to state j, given that the system is in state i, and b_{ij} is the conditional probability of observing "j" given that the Markov process is in state i. The matrices A and B, together with an initial state distribution π define the model, which is denoted $\lambda = (A, B, \pi)$.

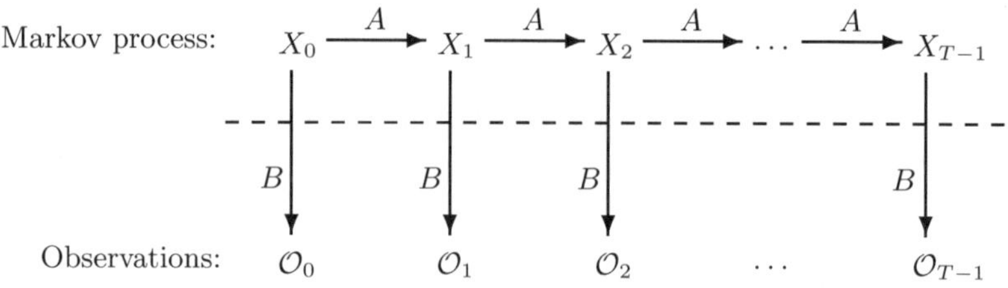

Fig. 23.3. Hidden Markov Model.

Efficient algorithms exist for solving the following three HMM problems.

Problem 1: Given the model $\lambda = (A, B, \pi)$ and a series of observations $\mathcal{O}$, find $P(\mathcal{O} \,|\, \lambda)$, that is, find the probability of the observed sequence given the (putative) model.

Problem 2: Given the model $\lambda = (A, B, \pi)$ and the observations $\mathcal{O}$, determine the most likely state sequence. In other words, we want to uncover the hidden part of the HMM.

Problem 3: Given the observations $\mathcal{O}$, "train" the model to best fit the observations. Note that the dimensions of the matrices are fixed, but the elements of A, B and π can vary, subject only to the row stochastic condition.

For example, consider speech recognition—an application where HMMs have been applied with great success. We could use the solution to Problem 3 to train a model to best match the word "no" and train another model to best match the word "yes". Then given a spoken word, we could use the solution to Problem 1 to determine whether it was more likely "yes", "no" or neither. In this scenario we do not need to solve Problem 2. However, a solution to Problem 2 (i.e., uncovering the hidden states) might provide additional insight into the underlying speech model.

In this paper we are interested in solving Problem 3. But first we must determine the sense in which the solution will "best fit" the observations. Perhaps the most intuitively appealing approach is to find the state sequence which yields the largest overall probability. In fact, this is precisely the solution found by dynamic programming. However, for HMMs, we instead maximize the expected number of correct states. For this reason, the HMM algorithm is sometimes referred to as the expectation maximization (EM) algorithm. This definition of "best" has a curious side-effect, namely, that invalid state transitions can occur in the optimal solution.

The HMM solution method (for Problem 3) is iterative, and can be viewed as a discrete hill climb on the parameter space defined by $\lambda = (A, B, \pi)$. As is typical of hill climbing techniques—discrete or not—we are only assured of finding a local maximum.

Further details on the HMM algorithms are beyond the scope of this paper. For more information, the standard introduction is [7], which also discusses the speech recognition application in some detail. The paper [11] gives a reasonably thorough introduction to HMMs with the emphasis on the algorithms and implementation.

23.5 HMMs and English Text

In this section we present two HMM experiments, one of which applies to letters in English text, and the other to phonetic English. In the following section we present analogous results for Hamptonese.

Cave and Neuwirth [3] apparently were the first to apply HMMs to English text. They selected the Brown Corpus [2] as a representative sample of English. This corpus of more than 1,000,000 words was carefully compiled (in the early 1960's) so as to contain a diverse selection of written English. Cave and Neuwirth eliminated all numbers, punctuation and special characters, and converted all letters to lower-case, leaving 27 distinct symbols—the letters plus inter-word space. They then assumed that there exists a Markov process with two hidden states, with the observations given by the symbols (i.e., letters) that appear in the Brown Corpus. This results in an A matrix that is 2×2 and a B matrix that is 2×27. They then solved HMM Problem 3 for the optimal matrices.

In repeating the Cave and Neuwirth experiment, using 10,000 observation and about 200 iterations, we obtain the results in Table 23.3, where the leftmost two columns give the initial B matrix, and the rightmost two columns give the final B. Note that the B matrix was initialized with approximately uniform random values. It is crucial that the matrix not be initialized to precisely uniform values, since that is a stationary point from which the algorithm cannot climb.

The results in Table 23.3 clearly show that hidden state 1 represents the "vowel state", while hidden state 0 is the "consonant state". Although it may not be surprising that consonants and vowels are distinguished, it is worth emphasizing that we made no *a priori* assumptions on the nature of the hidden states. This nicely illustrates the ability of HMMs to distill the most statistcally significant information from the data.

Cave and Neuwirth [3] obtain and interpret results for English text HMMs having up to 12 hidden states. For example, with three hidden states, the consonant state splits into a "pre-vowel" consonant state (i.e., consonants that tend to occur before vowels) and a "post-vowel" consonant state.

We have also computed HMM results for English phonemes instead of letters. For this phoneme HMM we first made a phonetic transcription of the Brown Corpus using the CMU pronouncing dictionary [4]. Neglecting accent marks, the pronouncing dictionary has 39 symbols. With two hidden states, using about 50,000 phonemes and running the HMM for about 200 iterations, we obtained the results in Table 23.4. Additional results are available at [12].

In the English phoneme HMM we again see a clear separation of the observation symbols into the two hidden states. Interestingly, in this phoneme HMM, one hidden state evidently represents consonant sounds and the other represents vowel sounds, analogous to the English text HMM.

23.6 HMMs for Hamptonese

We have computed HMM experiments on Hamptonese similar to those presented in the previous section for English. In our experiments, the number of distinct Hamptonese symbols was 42 and we computed results for 2, 3, 4 and

letter	Initial B		Final B	
	state 0	state 1	state 0	state 1
a	0.0372642	0.0366080	0.0044447	0.1306242
b	0.0386792	0.0389249	0.0241154	0.0000000
c	0.0358491	0.0338276	0.0522168	0.0000000
d	0.0353774	0.0370714	0.0714247	0.0003260
e	0.0349057	0.0352178	0.0000000	0.2105809
f	0.0344340	0.0370714	0.0374685	0.0000000
g	0.0400943	0.0370714	0.0296958	0.0000000
h	0.0344340	0.0347544	0.0670510	0.0085455
i	0.0349057	0.0370714	0.0000000	0.1216511
j	0.0391509	0.0366080	0.0065769	0.0000000
k	0.0363208	0.0356812	0.0067762	0.0000000
l	0.0353774	0.0403151	0.0717349	0.0000135
m	0.0344340	0.0366080	0.0382657	0.0000000
n	0.0410377	0.0370714	0.1088182	0.0000000
o	0.0396226	0.0398517	0.0000000	0.1282757
p	0.0377358	0.0338276	0.0388589	0.0000047
q	0.0377358	0.0398517	0.0011958	0.0000000
r	0.0344340	0.0403151	0.1084196	0.0000000
s	0.0358491	0.0366080	0.1034371	0.0000000
t	0.0377358	0.0352178	0.1492508	0.0134756
u	0.0349057	0.0361446	0.0000000	0.0489816
v	0.0405660	0.0370714	0.0169406	0.0000000
w	0.0377358	0.0384615	0.0286993	0.0000000
x	0.0382075	0.0370714	0.0035874	0.0000000
y	0.0382075	0.0389249	0.0269053	0.0000003
z	0.0382075	0.0338276	0.0005979	0.0000000
space	0.0367925	0.0389249	0.0035184	0.3375209

Table 23.3. English text HMM results

5 hidden states. The final B matrix for a typical case with two hidden states appears in Table 23.5.

Suppose that we require a threshold of a factor of 10 in order to assign a particular symbol to either state 0 or 1. Then the results in Table 23.5 show that the five symbols 14, M, q3, vv and o3 belong to state 0, while the 15 symbols 2, 96, F, g7, d3, d4, Y3, Y4, qL3, qL4, uL, P, P1, P2, and S, are in state 1. However, the remaining 22 symbols cannot be assigned to either state. This is in sharp contrast to the HMM results for English letters and phonemes, where—using the same criteria—virtually all symbols are assignable to one state or the other.

We can reasonably interpret an HMM solution as providing a "fingerprint" of the underlying data. Of course, the rows of the solution matrices can appear in any order and a relabeling of the observations will reorder the columns. But qualitative differences in the results—such as those mentioned in the previous

| | Final B | | | Final B | |
phoneme	state 0	state 1	phoneme	state 0	state 1
AH	0.0000000	0.2969759	EY	0.0000957	0.0460653
Z	0.0468621	0.0013497	F	0.0272705	0.0000000
AO	0.0000000	0.0353722	R	0.0772887	0.0000000
T	0.1115366	0.0164165	UW	0.0024273	0.0420961
W	0.0264391	0.0000000	N	0.1239146	0.0000000
AA	0.0000000	0.0443532	B	0.0270710	0.0000000
ER	0.0104758	0.0458586	G	0.0110797	0.0005440
K	0.0533438	0.0000000	M	0.0451959	0.0000000
D	0.0662517	0.0139417	EH	0.0000000	0.0763638
V	0.0367820	0.0000000	AE	0.0000000	0.0781199
IH	0.0000000	0.1464058	S	0.0804303	0.0101116
IY	0.0079505	0.0596027	L	0.0653828	0.0000000
OW	0.0000000	0.0273946	NG	0.0132029	0.0000000
SH	0.0161295	0.0000000	HH	0.0211180	0.0000000
AW	0.0001282	0.0131526	TH	0.0039602	0.0006984
AY	0.0001135	0.0282771	JH	0.0123050	0.0000000
P	0.0381176	0.0030524	CH	0.0094782	0.0000000
ZH	0.0007316	0.0000000	Y	0.0104094	0.0000000
DH	0.0545078	0.0000000	UH	0.0000000	0.0118911
OY	0.0000000	0.0019568			

Table 23.4. English phoneme HMM results

paragraph—indicate fundamentally different data. By comparing the results of the English HMMs with the Hamptonese HMM, we can conclude that Hamptonese characters do not represent English letters or phonemes.

23.7 Conclusions

There are several possible explanations for our Hamptonese HMM results. For example, it might be the case that we have insufficient Hamptonese data available. However, we can obtain HMM results on English with as few as 10,000 symbols and we have more than 29,000 Hamptonese characters available.

If the quantity of data is sufficient, then perhaps its quality is insufficient, i.e., the Hamptonese data is excessively noisy. This could be due to either poor transcription or errors inherent in Hampton's writings.

Another possible problem is incorrect interpretation of the Hamptonese data. It is conceivable that combinations of characters must be interpreted as the basic symbols. In this case the individual Hamptonese characters would have little or no semantic meaning and hence the HMM could not classify them. A potential example of this is given by the character "Ki", which occurs both individually and as "Ki Ki". The latter could certainly be considered as a distinct character.

symbol	Final B state 0	state 1	symbol	Final B state 0	state 1
2	0.0000000	0.0323915	14	0.0947011	0.0052195
ee	0.0059456	0.0091797	76	0.0504303	0.0132991
95	0.0013590	0.0108613	96	0.0071626	0.0220520
dc	0.0104561	0.0351127	EE	0.0141455	0.0141818
vv	0.4873287	0.0000000	M	0.0430382	0.0014101
F	0.0003230	0.0299307	N	0.0148755	0.0213175
g	0.0095264	0.0144809	g7	0.0005227	0.0024017
GG	0.0205943	0.0672325	Gi	0.0214469	0.0374094
Ki	0.0716056	0.0740697	d3	0.0000000	0.0312281
d4	0.0000000	0.0031228	Y3	0.0000000	0.2190866
Y4	0.0000000	0.0394331	qL3	0.0000000	0.0681507
qL4	0.0000241	0.0013892	4L	0.0068542	0.0112139
uL	0.0010389	0.0019307	J1	0.0055702	0.0069672
JJ	0.0089622	0.0288284	LL	0.0395888	0.0306004
nn	0.0108177	0.0120475	P	0.0000000	0.0461686
PL	0.0007135	0.0063528	P1	0.0000000	0.0184919
P2	0.0000000	0.0407190	o3	0.0230629	0.0000000
q3	0.0390770	0.0031463	S	0.0000000	0.0201452
3	0.0013213	0.0044007	T	0.0012893	0.0030790
A	0.0005299	0.0008652	A-	0.0044791	0.0075884
44	0.0027660	0.0042336	I	0.0004434	0.0002603

Table 23.5. Hamptonese HMM results

Perhaps Hamptonese is a cipher. At the extreme, Hamptonese could be the result of one-time pad encryption, in which case we have little hope of ever deciphering it. But given Hampton's background we might assume that he could only have developed a fairly weak encryption system—the non-randomness of Hamptonese could be taken as some limited evidence of this[1]. Our HMM experiment has ruled out that Hamptonese is a simple substitution cipher for English. After more cryptanalysis of Hamptonese, we might be able to make an argument to the effect that Hamptonese is probably not a good cipher, and we could probably break a bad cipher. Then if we cannot "break" Hamptonese, it is probably not a cipher.

Of course, there is a very real possibility that in spite of its language-like appearance, Hamptonese is simply the written equivalent of "speaking in tongues". But even if this is the case, it is not necessarily the end of the story. In the late 19th century, a French mystic, Hélène Smith, claimed that when in a trancelike state, spirits transported her to Mars, where she was able to communicate with Martians. To substantiate her claims, she produced written "Martian" messages in an unknown script. Her Martian writing appar-

---

[1]There is an obvious risk of underestimating James Hampton. His Throne clearly shows that he was capable of achieving the unexpected.

ently was consistent and had the appearance of a genuine language. Reportedly, Smith's Martian messages could be interpreted as a simple variation on French [6, 10].

References

1. S. Ager, Omniglot: A guide to writing systems,
   `http://www.omniglot.com/`
2. The Brown Corpus of Standard American English,
   `http://www.cs.sjsu.edu/faculty/stamp/Hampton/BrownCorpus.gz`
3. R. L. Cave and L. P. Neuwirth, Hidden Markov models for English, in *Hidden Markov Models for Speech*, J. D. Ferguson, editor, IDA-CRD, Princeton, NJ, October 1980.
4. The CMU Pronouncing Dictionary,
   `http://www.speech.cs.cmu.edu/cgi-bin/cmudict`
5. M. Esposito, L. Fleetwood and J. Brooks, Dispensationalism,
   `http://www.endtimes.org/dispens.html`
6. A. G. Hefner, Smith, Helene,
   `http://www.themystica.com/mystica/articles/s/smith_helene.html`
7. L. R. Rabiner, A tutorial on hidden Markov models and selected applications in speech recognition, *Proceedings of the IEEE*, Vol. 77, No. 2, February 1989,
   `http://www.cs.ucsb.edu/~cs281b/papers/HMMs%20-%20Rabiner.pdf`
8. C. E. Shannon, A mathematical theory of communication, *Bell System Technical Journal*, Vol. 27, pp. 379–423, 623–656, July, October 1948,
   `http://cm.bell-labs.com/cm/ms/what/shannonday/shannon1948.pdf`
9. D. J. Stallings, Hamptonese statistics,
   `http://www.geocities.com/ctesibos/hampton/hamptonese.html`
10. D. J. Stallings, Idiolect,
   `http://www.geocities.com/ctesibos/hampton/idiolect.html`
11. M. Stamp, A revealing introduction to hidden Markov models,
   `http://www.cs.sjsu.edu/faculty/stamp/Hampton/HMM.pdf`
12. M. Stamp and E. Le, Hamptonese,
   `http://www.cs.sjsu.edu/faculty/stamp/Hampton/hampton.html`
13. W. Trappe and L. C. Washington, *Introduction to Cryptography with Coding Theory*, Prentice Hall, 2002.
14. M. Walsh, The miracle of St. James Hampton,
   `http://www.missioncreep.com/tilt/hampton.html`
15. F. Weaver, James Hampton's Throne Of The Third Heaven Of The Nations Millenium General Assembly,
   `http://www.fredweaver.com/throne/thronebody.html`

Arithmetic Properties of Observable Time Dependent Systems

Dorothy Wallace

Dept. of Mathematics, Dartmouth College, Hanover, NH 03678
Dorothy.I.Wallace@Dartmouth.EDU

24.1 Introduction

Here we consider questions of discrete observability for linear systems of ordinary differential equations, extending work by Martin and Smith in [2]. The motivation for this paper is based on the observation that many of the questions that arise in observability of linear systems have been considered for the past half century by number theorists considering the problem of determining when a number is transcendental. In particular, Siegel's proof of Lindemann's theorem is very system theoretical in nature, [4], and in fact is reminiscent of the construction of controllability criteria for time varying linear systems developed by Silverman and Meadows in [5]. The primary contribution of this note is the use of the development of Gel'fond, [1], relating observability of linear systems with time varying coefficients to the transcendence of values of analytic functions.

24.2 Constant Coefficient Systems

We begin by reconsidering part of the work of Martin and Smith, [2], in order to develop the background necessary to consider the time varying case.

We consider the constant coefficient case:

$$\dot{x} = Ax(t)$$
$$y(t) = c'x(t)$$

The matrix A is an $n x n$ matrix of complex numbers, c' a row vector of length n, and $x(t)$ a column vector of length n. Let x_0 be a column vector of initial values. Then the solutions of

$$\dot{x} = Ax(t)$$
$$x(0) = x_0$$

D. Wallace: *Arithmetic Properties of Observable Time Dependent Systems*, LNCIS **321**, 379–388 (2005)
www.springerlink.com

is given by $x(t) = exp(tA)x_0$.

Definition 1. *Given an n - tuple of times $T = (t_1, ..., t_n), 0 \neq t_1 < t_2 < \cdots , < t_n$, and the system*

$$\dot{x} = Ax$$
$$y = c'x.$$

The triple (A, c, T) is discretely observable if for all k

$$y(t_k) = 0$$

if and only if

$$x_0 = \mathbf{0}.$$

Definition 2.1 is equivalent to:

Definition 2. *In the notation above, (A, c, T) is discretely observable if*

$$|M| = \left| \begin{pmatrix} c' expt_1 A \\ \vdots \\ c' expt_n A \end{pmatrix} \right| \neq 0.$$

Now we would like to translate the formation of the problem into an equivalent one where A is in Jordan normal form. We have

$$BAB^{-1} = \bar{A},$$
$$B \ \epsilon \ Sl(n, C)$$

and $\bar{A}$ in Jordan form, i.e.

$$\bar{A} = \begin{pmatrix} \bar{A}_1 & & & \\ & \bar{A}_2 & & \\ & & \cdots & \\ & & & \bar{A}_n \end{pmatrix}$$

and

$$\bar{A}_i = \begin{pmatrix} \lambda_i & a_{i2} & 0 & \cdots & \cdots & 0 \\ 0 & \lambda_i & a_{i3} & \cdots & \cdots & 0 \\ \vdots & & & & & \vdots \\ 0 & 0 & 0 & \cdots & \cdots & a_{ik} \\ 0 & 0 & 0 & \cdots & \cdots & \lambda_i \end{pmatrix}$$

where $a_{ik} = 0$ or 1 and $\lambda_i \neq \lambda_j$, if $i \neq j$. Likewise we have the following elementary fact:

Let A_i be a Jordan block with $a_{ij} = 1$ for all j. (Remember the a_{ij} are just the super diagonal entries.) Then

$$\exp tA_i = \begin{pmatrix} e^{\lambda_i t} & te^{\lambda_i t} & t^2 e^{\lambda_i t} & \cdots & t^{k-1} e^{\lambda_i t} \\ & e^{\lambda_i t} & te^{\lambda_i t} & \cdots & \\ & & e^{\lambda_i t} & & \\ & & & & e^{\lambda_i t} \end{pmatrix}.$$

Definition 3. *Call A admissible if for none of the Jordan block A_i in $\bar{A}$ are there any $a_{ij} = 0$. The following proposition is well known in control theory.*

Proposition 1. *If A is not admissible then the system (A, c, T) is not observable.*

Definition 4. *We call c admissible with respect to A if none of the $\bar{c}_i$ are zero, where i is the first entry of a Jordan block.*

Notice that for $\bar{c}_i \neq 0$ for such i means that c' is a linear combination of generalized left eigenvectors of $\bar{A}$ and the coefficients of actual eigenvectors are nonzero. That is $c' = \bar{d}_1 \bar{v}_1 + ... + \bar{d}_n \bar{v}_n$ is a set of generalized eigenvectors of $\bar{A}$ which span $\mathcal{R}^n$. Hence Definition 2.4 is equivalent to:

Definition 5. *The vector c is admissible with respect to A if it is a linear combination of a basis of $\mathcal{R}^n$ consisting of generalized eigenvectors of A and the coefficients of actual eigenvectors are nonzero.*

We now have the following elementary fact:
If c is not admissible with respect to A then the system (A, c, T) is not observable.

Definition 6. *The system (A, c, T) is admissible if A is admissible and c is admissible with respect to A.*

Theorem 1. *Let (A, c, T) be admissible. Then (A, c, T) is observable if and only if (A, c^*, T) is observable where*

$$c* = \sum_{i=1}^{k} v_i \quad k \leq n$$

and the v_i run through all the eigenvectors A.

Proof: First notice that c^* is a well defined sum because A admissible implies that no two v_i will correspond to the same eigenvalue, hence the choice of v_i is unique up to order.

We look at the part of $\bar{M}$ corresponding to a single Jordan block, with out loss of generality the first. The first k columns of $\bar{M}$ are

$$\begin{pmatrix} \bar{c}_1 e^{\lambda_1 t_1} & \bar{c}_1 t_1 e^{\lambda_1 t_1} + \bar{c}_2 e^{\lambda_1 t_1} & \cdots & \bar{c}_1 t_1^{k-1} e^{\lambda_1 t_1} + \cdots + \bar{c}_k e^{\lambda_1 t_1} \\ \vdots & \vdots & & \vdots \\ \bar{c}_1 e^{\lambda_1 t_n} & \bar{c}_1 t_n e^{\lambda_1 t_n} + \bar{c}_2 e^{\lambda_1 t_n} & \cdots & \bar{c}_1 t_n^{k-1} e^{\lambda_1 t_n} + \cdots + \bar{c}_k e^{\lambda_1 t_n} \end{pmatrix}.$$

382   D. Wallace

Note $\bar{c}_i \neq 0$ because c is admissible with respect to A. By column reduction the determinant will not change if these columns are replaced by

$$
\begin{pmatrix}
\bar{c}_1 e^{\lambda_1 t_1} & \bar{c}_1 t_1 e^{\lambda_1 t_1} & \cdots & \bar{c}_1 t_1^{k-1} e^{\lambda_1 t_1} \\
\vdots & \vdots & & \\
\bar{c}_1 e^{\lambda_1 t_n} & \bar{c}_1 t_n e^{\lambda_1 t_n} & \cdots & \bar{c}_1 t_n^{k-1} e^{\lambda_1 t_n}
\end{pmatrix}
$$

Hence we can take $\bar{c}_2, \cdots, \bar{c}_k = 0$. Furthermore the determinant of $\bar{M}$ will be a nonzero multiple of the one we get by taking $\bar{c}_1 = 1$. Thus we can take

$$
c' = (1, 0, \cdots, 0, 1, 0, \cdots, 0, \cdots)
$$

with 1's occurring at the beginning of Jordan blocks. That is,

$$
c' = \sum \bar{v}_i
$$

where the $\bar{v}_i$ are all the eigenvectors of $\bar{A}$. Therefore $c' = \sum \bar{v}_i$.

This theorem shows that if (A, c, T) is admissible and we want to decide if it is observable, we can replace it with $(\bar{A}, \bar{c}^*, T)$ and check it there. Thus we have reduced the problem to one which can be stated in a canonical fashion.

We now move to a special case where we can solve the problem of observability completely. For the time being let A have distinct eigenvalues $(\lambda_1, ..., \lambda_n)$ and let T be periodic, $T = (t_0 + t, t_0 + 2t, ..., t_0 + nt)$. Without loss of generality (by the above theorems), we can assume A is diagonal and $c = (1, ..., 1)$. M is then given by

$$
M =
\begin{pmatrix}
e^{\lambda_1(t_0+t)} & e^{\lambda_2(t_0+t)} & \cdots & e^{\lambda_n(t_0+t)} \\
\vdots & \vdots & & \vdots \\
e^{\lambda_1(t_0+nt)} & e^{\lambda_2(t_0+nt)} & \cdots & e^{\lambda_n(t_0+nt)}
\end{pmatrix}
$$

Now $e^{\lambda_1 t_0} \neq 0$ so we can factor these out without changing whether $det M = 0$. So, without loss of generality we can take

$$
M =
\begin{pmatrix}
e^{\lambda_1 t} & \cdots & e^{\lambda_n t} \\
\vdots & & \vdots \\
e^{\lambda_1 nt} & \cdots & e^{\lambda_n nt}
\end{pmatrix}.
$$

Setting $\alpha_i = e^{\lambda_i t}$, we have

$$
M =
\begin{pmatrix}
\alpha_1 & \alpha_2 & \cdots & \alpha_n \\
\vdots & \vdots & & \vdots \\
\alpha_1^n & \alpha_2^n & \cdots & \alpha_n^n
\end{pmatrix}
$$

and again we can factor a out of each column to give an equivalent problem where

$$M_0 = \begin{pmatrix} 1 & \cdots & 1 \\ \alpha_1 & & \alpha_n \\ \vdots & & \vdots \\ \alpha_1^{n-1} & \cdots & \alpha_n^{n-1} \end{pmatrix}$$

Also, we have that (A, c, T) is observable if and only if $|M_0| \neq 0$. Since M_0 is a Vandermonde matrix, it is invertible provided the α_i are distinct.

Setting

$$\alpha_i = e^{\lambda_i t} = e^{\lambda_j t} = \alpha_j$$

and letting $M_k = Re(\lambda_k), v_k = Im(\lambda_k)$ we have

$$e^{M_i t} e^{i v_i t} = e^{M_j t} e^{i v_j t}$$

which can happen only if

$$M_i = M_j$$

and

$$v_i t = v_j t + 2\pi m$$

so that

$$t = \frac{2\pi m}{v_i - v_j}$$

We immediately see that:

1. If the eigenvalues of A are real then (A, c, T) is observable. This was known before as a theorem about Vandermonde matrices.
2. If p is a fundamental period of the system then sampling at $t =$ integer multiples of $p/2$ results in a system which is not observable. This follows immediately from setting $v_j = -v_i$ since eigenvalues of A come in conjugate pairs.
3. Other than conjugate pairs, the zeros of M occur only when $Re\lambda_i = Re\lambda_j$ for two λ with different periods. For this the statement

$$t = \frac{2\pi m}{v_i - v_j}$$

characterizes those t for which (A, c, T) is not observable. We can now state a theorem.

Theorem 2. *Let (A, c, T) be an admissible system with periodic sampling of period t and let A have distinct eigenvalues. Then (A, c, T) is observable if and only if*

$$t \neq |\frac{2\pi m}{v_i \pm v_j}|$$

and

$$t \neq \frac{\pi m}{v_i}$$

where v_i and v_j are fundamental frequencies of the system $\dot{x} = Ax$.

24.3 Nonautonomous Systems

If the system $\dot{x} = Ax$ is given by a matrix A whose entries all lie in some finite algebraic extension of the rational numbers, then the eigenvalues obtained are also in a finite extension. Exponential functions whose coefficients are algebraic yield periodic solutions whose periods are transcendental numbers. The periodic sampling that destroys the observability of the system is given simply in terms of these transcendental numbers. Gel'fond exploits this relationship to give a more general statement that can apply to nonautonomous systems.

In this section of this paper we investigate the application to observability of a theorem of Gel'fond concerning the transcendence of values of an analytic function with specific properties. We follow closely the development in Gel'fond, [1].

Definition 7. *An entire analytic function $f(z)$,*

$$f(z) = \sum_0^\infty C_n \frac{z^n}{n!}$$

is called an E-function if

1. *all the coefficients C_n of the function $f(z)$ belong to a finite algebraic extension of the rationals,*
2. *$|\bar{C}_n| = 0(n^{n\epsilon})$ for all $E > 0$, where $|\bar{C}_n|$ is the largest absolute value of any conjugate of C_n in $\mathcal{K}$,*
3. *if $q_n > 0$ is the smallest rational integer for which the numbers $C_k q_n, k = 0, 1, ..., n_1$ are algebraic integers, then $q_n = 0(n^{n\epsilon})$ for arbitrary $\epsilon > 0$.*

The standard examples of E-functions are polynomials with algebraic coefficients, for which the estimates are satisfied trivially, and exponential functions, e^{wz}, where w is an algebraic number.

For the rest of this section we consider the system

$$\dot{x}(z) = A(z)x(z)$$

where the entries of $A(z)$ are rational functions of z. If the entries of $A(z)$ have coefficients in some algebraic field $\mathcal{K}$ and if we introduce initial conditions

$$x(0) = B$$

where the entries of B are in $\mathcal{K}$, then the recursion formula which gives the power series for $x(t)$ will yield coefficients in $\mathcal{K}$ also. Under some circumstances one can show these are E-functions.

Assume that we have interchanged the rows and columns of A to give a matrix in block diagonal form with the maximum number of blocks, $A_i, ..., A_r$. Let each A_t be of size m_t by m_t. Let

$$U_{s,t}(z) = \begin{pmatrix} U_{1,s,t}(z) \\ \vdots \\ U_{m_t,s,t}(z) \end{pmatrix} \quad 1 < s < m_t$$

be a system of solutions to

$$\dot{x} = A_t x,$$

not necessarily E-functions. Let

$$W_t = \begin{bmatrix} U_{1,1,t} & \cdots & U_{1,m_t,t} \\ \vdots & & \vdots \\ U_{m_t,1,t} & \cdots & U_{m_t,m_t,t} \end{bmatrix}$$

We denote the the entry of W_t in row k column s as $U_{k,s,t}$. Let W be the block diagonal matrix whose blocks are the W_t. Note the columns of W give a complete set of solutions to our original equation, $\dot{x} = A_t x$.

Let $p_{k,t}(z), C_{s,t}$ be an arbitrary collection of polynomials and constants respectively. Then we have the following definition:

Definition 8. *(Gel'fond) Let z be a complex variable. We shall say that the matrices $W_1, ..., W_r$, are linearly independent over the field of rational functions if the relation*

$$\sum_{t=1}^{r}\sum_{s=1}^{m_t}\sum_{k=1}^{m_t} p_{k,t}(z)C_{s,t}U_{k,s,t}(z) = 0$$

holds if and only if all of the products

$$p_{k,t}(z)C_{s,t}$$

vanish identically.

Let W be the block diagonal sum of the U_t. Let $W_{a,b}$ denote an entry of W. If we relabel the ordered pairs (k,t) and (s,t) as a and b respectively, then Definition 3.2 is equivalent to the following:

Definition 9. *(Gel'fond) Let z be a complex variable. We shall say that the matrices $W_1, ..., W_r$, are linearly independent over the field of rational functions if the relation*

$$\sum_{a=1}^{n}\sum_{b=1}^{n} p_a(z)C_b W_{a,b}(z) = 0$$

holds if and only if all of the products

$$p_a(z)C_b$$

vanish identically.

Notice that if there exists a basis for the solution space consisting of E-functions then we can replace the complex variable z by the real variable t, because E - functions are entire. The new definition will be equivalent to the old one. Under these circumstances the above definition is equivalent to an observability statement.

Theorem 3. *Let $\dot{x} = Ax$ be a system of equations where A is a matrix of rational functions given in block diagonal form with blocks A_t. Suppose there exists a basis for the solution set consisting of E-functions. Let $W_1, ..., W_r$ be the matrices of a basis for solutions of A_t, as above, given by these E-functions. For a given solution $x(z)$ denote by x_a the ath entry of x, written as a column vector. Let $P = (p_1, ..., p_n)$ be a vector of polynomials, none of which are identically zero. Then $W_1, ..., W_r$ are linearly independent over the field of rational functions if and only if the pair (A, P) is continuously observable.*

Proof: Let $x(t)$ be a solution of $\dot{x} = Ax$. We can express $x(t)$ as

$$x(z) = \sum_{t=1}^{r} \sum_{s=1}^{m_t} C_{s,t} U_{s,t}(z).$$

Using Definition 3.3, we can write row a of $x(z)$ as

$$x_a(z) = \sum_{b=1}^{n} C_b W_{a,b}(z)$$

Let $P = (p_1, ..., p_n)$ observe the system through output function

$$Q(z) = \sum_{a=1}^{n} p_a(z) x_a(z)$$

Then

$$Q(z) = \sum_{a=1}^{n} \sum_{b=1}^{n} p_a(z) C_b W_{a,b}(z) = \sum_{t=1}^{r} \sum_{s=1}^{m_t} \sum_{k=1}^{m_t} p_{k,t}(z) C_{s,t} U_{k,s,t}(z)$$

But $Q(z) = 0$ if and only if all of the products $p_{k,t}(z) C_{s,t}$ vanish. The polynomials are all nonzero, therefore $Q(z) = 0$ if and only if all of the $C_{s,t}$ equal zero, which implies that

$$x(z) = \sum_{t=1}^{r} \sum_{s=1}^{m_t} C_{s,t} U_{s,t}(z) = 0$$

Therefore the system, (A, P) is continuously observable.

Definition 10. *A system of mE − functions $f_1(z), \cdots, f_m(z)$ is said to be normal if*

1. *none of the $f_i(z)$ are identically zero,*
2. *the system of functions is a solution of a system of differential equations*

$$\dot{x} = Ax, \quad x' = (f_1, \cdots, f_m)$$

whose coefficients are rational functions with numerical coefficients in a finite algebraic extension of the rational numbers, whose primitive matrices are independent in the sense of Definition 3.3.

Theorem 4. *(Gel'fond) Suppose that the $mE - $ functions $f_1(z), ..., f_m(s)$ form a normal system and that $\alpha Q(\alpha) \neq 0$, where α is algebraic and Q is the least common multiple of the denominators of the entries of A in the system $\dot{x} = Ax$ whose solutions are $f_1, ..., f_m$. If, for arbitrary N, the $M = (m + n)!/m!n!$ functions*

$$f_1^{N_1}(z) \cdots f_m^{N_m}(z), \ N_i \geq 0, \ 1 \leq i \leq m, \sum_{i=1}^{m} N_i \leq N,$$

form a normal system of $E - $ functions, then the numbers $f_1(\alpha), ..., f_m(\alpha)$ cannot be interrelated by means of any algebraic relationship with algebraic coefficients which is not identically zero.

The above theorem has an application to discrete observability under some special conditions. First we need the following lemma.

Lemma 1. *Let $f_1(t), ..., f_n(t)$ be a normal family of $E - $ functions and let k be algebraic. Then the family*

$$f_{1,k}(t) = f_1(kt), f_{2,k}(t) = f_2(kt), ..., f_{n,k}(t) = f_n(kt)$$

is a normal family of $E - $ functions.

Proof: Suppose $f_{1,k}, ..., f_{n,k}$ are E - functions. Then they are a solutions of $\dot{x}_k A_k(t) x_k$ where $f_1, ..., f_n$, satisfy $\dot{x} = Ax, x_k = f_{1,k}, ..., f_{n,k}$ and $A_k = kA(kt)$. The primitive matrices for $f_1(t), ..., f_n(t)$ are independent, therefore the primitive matrices for translated families will also be independent. Clearly, none of the $f_{i,k}$ are identically zero. So it suffices to show that the $f_{i,k}$ are E-functions.

Let

$$x_i(t) = \sum_{n=0}^{\infty} \frac{C_n t^n}{n!}$$

and with the C_n, satisfying Definition 2.1. Then

$$f_{i,k}(t) = \sum_{n=0}^{\infty} \frac{c_n (kt)^n}{n!}$$

$$= \sum_{n=0}^{\infty} \frac{C_n k^n}{n!} e^n = \sum_{n=0}^{\infty} \frac{D_n t^n}{n!}$$

with $D_n = C_n k^n$.

We now have n normal families of n E - functions each. We can apply Gel'fond's theorem in the following situation.

Corollary 1. *Suppose we have a system $\dot{x} = Ax$ where A has entries which are rational functions of t over a finite algebraic extension field. Let $Q(t)$ be the least common multiple of these denominators. Suppose we have a complete set of solutions $x_1, ..., x_n$, whose entries are spanned over k by a collection $f_1, ..., f_m$ of nonzero $E - functions$. Suppose the family $\{f_{i,j}(t) = f_i(\alpha_j t), \alpha_j algebraic, distinct and nonzero\}$ are a normal family of $E - functions$ which satisfies the hypotheses of Theorem 2.2. Let C be a nontrivial row vector of n algebraic and $tQ(t) \neq 0$. Then the system (A, c, T) is observable.*

Proof: We need to show that the determinant of

$$M = \begin{bmatrix} cx(\alpha_1 t) & \cdots & cx_n(\alpha_1 t) \\ cx_1(\alpha_n t) & \cdots & cx_n(\alpha_n t) \end{bmatrix}.$$

is not zero, but $detM(t)$ is just a polynomial in the $\{f(t)\}$ with algebraic coefficients. Therefore $detM(t) \neq 0$ for all algebraic t. Thus the system (A, c, T) is observable.

The value of this corollary lies in setting t equal to some constant. The corollary then asserts that the original system, sampled at the α_i, is discretely observable. Notice that the conditions of independence of the primitive matrices associated to our observability problem may be difficult to verify. One can get around this to a limited extent using results of Shidlovskii [3] to replace these conditions with one that presupposes the algebraic independence of the $f_{i,k}(t)$, a condition that guarantees that the primitive matrices are algebraically independent, although it is easier to state. Shidlovskii also has intermediate results for subsystems of functions, guaranteeing similar results except at finitely many values.

References

1. A. 0. Gel'fond, *Transcendental and Algebraic Numbers,* Dover Publishing Company, New York: 1960.
2. Clyde Martin and Jennifer Smith *Approximation, Interpolation and Sampling,* Contemporary Mathematics, 68, 227-252, 1987.
3. A. B. Shidlovskii *On the Algebraic Independence of the Values of E-Functions,* Proc. Steklov Institute of Mathematics, 218, 439-443, 1997.
4. C. Siegel *Uber einige Anwendungen Diophantischer Approximationen,* Abh. Preuss. Akad, Wiss., 1929-1930, No. 1.
5. L. M. Silverman and H. E. Meadows, *Controllability and observability in time-variable linear systems,* SIAM Journal of Control, 5, 64-73, 1967.

A Fairness Index for Communication Networks

Wing Shing Wong and Fengjun Li*

Department of Information Engineering, Chinese University of Hong Kong
{wswong;fjli2}@ie.cuhk.edu.hk

Summary. This paper introduces a new fairness index in open architecture networks. The concept can be used to balance the load of self-authority servers and keep them operating in a fair manner. Properties, such as existence and uniqueness, of this index are investigated for some typical network structures. By connecting to von Neumann's equilibrium concept, the proposed fairness concept can be related via a pricing duality to an equilibrium index, which uniquely exists in general. We also investigate the problem of how distributed users can achieve a given set of target indices. A distributed, low data rate control algorithm is introduced and its convergence property is discussed.

25.1 Introduction

The altruistic spirit of routing sharing is a key factor that contributes to the rapid growth of Internet. Nodes in an open architecture network such as the Internet are assumed to participate in the routing of third party traffic whenever the demand is within their resource capacity limit. However, the hierarchical architecture of Internet dictates that it cannot be as flat and altruistic as one may imagine. Only nodes of similar caliber can peer with each other, sometimes through Bi-Lateral Peering Agreements (BLPA). Smaller nodes have to aggregate their traffic via backbone nodes. This kind of arrangement can be interpreted as a scheme to ensure fairness in routing contribution. However, multiple servers could participate in a fair peering without satisfying pair-wise symmetric conditions. The fairness issue is even more critical for wireless ad-hoc networks [1]. Unlike hierarchical networks, these networks are self-organizing and totally distributed in administration. Failure to address fairness issue could lead to congestions that may prove to be impossible to recover from. Moreover, unlike servers in a wired network, battery power is a critical resource for mobile servers. Routing for third party traffic demands

---

*The work reported in this paper was partially supported by a Direct Grant of the Chinese University of Hong Kong.

W.S. Wong and F. Li: *A Fairness Index for Communication Networks*, LNCIS **321**, 389–400 (2005)

not only communication channel resources but also consumes battery power. In [2]fairness issues in such a setting are investigated. In this paper, we report some of the key results in [2].

We present here a new fairness concept from [2]. An early version of this concept was first proposed in [3]. We begin by defining concepts of a *fairness index* and a *perfectly fair* solution in Sect. 25.2. The fairness index of a node is an indicator of its contribution to network routing. The ideal case where all the nodes have identical fairness indices is an important baseline reference point, and the corresponding network is called perfectly fair. One can connect the perfectly fair index with the von Neumann equilibrium by introducing a duality concept of pricing. The von Neumann equilibrium always exists for a network.

For implementation considerations, one needs to consider the problem how nodes in the network can achieve a perfectly fair state. Since each node can have only a local view of the network status, it is important that the desired state should be achieved through distributed algorithms. The problem is further complicated by the fact that in such a network, system parameters are basically unknown. In Sect. 25.3, a simple distributed controller is proposed to achieve a given set of fairness targets. A main result is to establish the convergence of such a class of distributed controllers. The algorithm can be extended to provide a heuristic solution to achieve a perfectly fair state without knowing the value of the perfectly fair index beforehand.

25.2 Fairness Index and Neumann Equilibrium

Nodes in a network tend to send as much traffic as possible into networks in order to achieve their maximum throughput unless being regulated. This behavior could lead to severe network congestion and unfairness in common resource usage. Many investigations use pricing and game theory to find efficient or fair operating states. By realizing the max-min fairness defined in [4] or the proportional fairness defined in [5] and [6], network link resources can be used efficiently. On the other hand, network traffic routing consumes resources at nodes as well as at links. For nodes peering in a network, it makes sense to demand that they are contributing to the routing function in a fair manner. We propose to use a *fairness index* for each node in a peer-to-peer network as a measure of whether routing load is shared fairly.

We conceptualize a communication network as a graph (V, E), with the nodes representing peering servers and an edge connects two nodes whenever there is a direct, duplex data link between two servers. Each node generates and consumes traffic data; it also routes traffic on behalf of other nodes.

Label the nodes in (V, E) from 1 to K. For each j, represent by r_j the rate of total traffic generated from node j to the network. Denote the vector of originating traffic rates, $(r_1, ..., r_K)^T$ by $\mathbf{r}$. A node, j, controls the network by means of r_j.

For any $i \neq j$, let $M_{ij}r_j$ represent the rate of the traffic generated from node j that is ultimately destined for node i. For all i, define

$$M_{ii} = 0. \tag{25.1}$$

Denote by $\mathbf{M}$ the K-by-K traffic distribution matrix with non-negative entries and column sums equal to 1,

$$\begin{pmatrix} 0 & M_{12} & \cdots & M_{1K} \\ M_{21} & 0 & \cdots & M_{2K} \\ \cdots & \cdots & \cdots & \cdots \\ M_{K1} & M_{K2} & \cdots & 0 \end{pmatrix} \tag{25.2}$$

The traffic from any source-destination pair can be routed over a variety of paths. We assume that the distributions of traffic into these alternative paths are arbitrary but known a priori and remain unchanged in a sufficiently long enough period for the consideration of this problem. Given a traffic distribution matrix and a set of routing schemes, the distribution of transitory traffic among intermediate nodes in the networks is fixed accordingly. Let $L_{ij}r_j$ denote the traffic rate of transitory traffic passing through node i that originates from node j, and $\mathbf{L}=(L_{ij})$ denote the corresponding K-by-K matrix. $\mathbf{L}$ is also a non-negative matrix with all entries bounded by 1. Moreover, $\mathbf{L}r$ is the column vector representing the total transitory data traffic rates passing through each node.

A *fairness index* for each node is, roughly speaking, the ratio of traffic directly attributed to the node as a source or destination to the amount of total traffic it handles. Depending on the economical model one adopts to account for the utility or the revenue from the traffic, three classes of fairness indices are defined below, all taking values between 0 and 1.

Definition 1: For a network, (V, E), suppose $r_i + \sum_{j=1}^{K}(M_{ij}+L_{ij})r_j > 0$, then the *source-weighted fairness index* for node i is defined to be the ratio

$$l_i = \frac{r_i}{r_i + \sum_{j=1}^{K}(M_{ij} + L_{ij})r_j}. \tag{25.3}$$

Otherwise, the index is defined to be 0.

Suppose $r_i + \sum_{j=1}^{K}(M_{ij}+L_{ij})r_j > 0$, then the *destination-weighted fairness index* for node i is defined to be the ratio

$$o_i = \frac{\sum_{j=1}^{K} M_{ij}r_j}{r_i + \sum_{j=1}^{K}(M_{ij} + L_{ij})r_j}. \tag{25.4}$$

Otherwise, the index is defined to be 0.

Suppose $r_i + \sum_{j=1}^{K}(M_{ij} + L_{ij})r_j > 0$, then the *combined fairness index* or simply the *fairness index* for node i is defined to be the ratio

$$\chi_i = \frac{r_i + \sum_{j=1}^{K} M_{ij}r_j}{r_i + \sum_{j=1}^{K}(M_{ij} + L_{ij})r_j}. \tag{25.5}$$

Otherwise, the index is defined to be 0.

The denominator of these indices accounts for the total traffic handled by node i. For the source-weighted fairness index, the numerator accounts for the total data rate generated by node i to the network; traffic received by a node as final destination is assumed to have no economical benefit to it. One can interpret the meaning of the numerator term for the other indices accordingly.

Definition 2: A *perfectly source-fair* solution exists if there is a non-negative, non-zero rate vector $\mathbf{r}=(r_1,...,r_K)^T$ so that the source-weighted fairness indices for all nodes are equal.

The concepts of a *perfectly destination-fair* and *perfectly fair* solution are defined similarly.

A perfectly fair solution exists if and only if there is a positive γ_C and a non-negative, non-zero rate vector so that

$$\gamma_C(\mathbf{I}+\mathbf{L}+\mathbf{M})\mathbf{r} = (\mathbf{I}+\mathbf{M})\mathbf{r}, \tag{25.6}$$

where $\mathbf{I}$ stands for the K-by-K identity matrix. One can establish similar equations for *perfectly source-fair* or *perfectly destination-fair* solutions:

$$\gamma_S(\mathbf{I}+\mathbf{L}+\mathbf{M})\mathbf{r} = \mathbf{r}, \tag{25.7}$$

$$\gamma_D(\mathbf{I}+\mathbf{L}+\mathbf{M})\mathbf{r} = \mathbf{M}\mathbf{r}. \tag{25.8}$$

Notice that perfectly fair solutions can be scaled uniformly without affecting their fairness properties. Hence with a proper scalar, a given set of link and node capacity constraints can always be satisfied.

The existence and uniqueness property of a perfectly fair solution is a natural question for investigation. For source-fair solution, the question can be settled by using the Perron-Frobenius Theorem on irreducible non-negative matrices [7]. The case for the other two types of indices is much more complicated. Mathematically, the issue hinges on finding generalized non-negative eigenvectors for a pair of nonnegative matrices. However, very little results have been reported in the literature on this subject. In [2], some existence and uniqueness properties of these solutions are reported. Furthermore, specific characterizations are found for networks with special topology.

Examples show that the existence of a perfectly fair solution depends on system parameters. The non-existence of a perfectly fair solution can be interpreted as an indication that nodes in the network are not suitable candidates for peering agreement. An alternative is to bring in a concept of pricing to compensate for the lack of perfect fairness. This latter approach makes contact with an equilibrium concept proposed by von Neumann [8].

A non-negative, non-zero price vector, $\mathbf{p}=(p_1,...,p_K)$, is introduced for a network as a dual to the rate vector. The interpretation of the price vector is that a node can charge according to the traffic it handles, including traffic it generates, it receives as destination, and traffic it routes for other nodes.

Denote $(\mathbf{I}+\mathbf{L}+\mathbf{M})$ by $\mathbf{B}$ and let $\mathbf{A}$ represent $\mathbf{I}$, $\mathbf{M}$, or $\mathbf{I}+\mathbf{M}$, depending on what type of fairness index is being considered. Introduce $1/\beta$ as the *minimum fairness ratio* so that for each node i,

$$\beta \sum_{j=1}^{K} a_{ij} r_j \geq \sum_{j=1}^{K} b_{ij} r_j. \tag{25.9}$$

Similarly, for each node, j, define a *maximum payment return ratio, $1/\alpha$* , so that for each node j,

$$\alpha \sum_{i=1}^{K} a_{ij} p_i \leq \sum_{i=1}^{K} b_{ij} p_i. \tag{25.10}$$

For a more concrete discussion, assume that both source and destination traffic have economic value, so that perfectly fair index is considered and $\mathbf{B}$ represents $\mathbf{I}+\mathbf{M}$. The discussion can be extended to the other two types of indices as well.

From a fairness consideration, each node, i, sends and receives data traffic of interest to it, amounting to $\sum_{j=1}^{K} a_{ij} r_j = r_i + \sum_{j=1}^{K} M_{ij} r_j$, while it handles the total amount $\sum_{j=1}^{K} b_{ij} r_j = r_i + \sum_{j=1}^{K} (M_{ij} + L_{ij}) r_j$ for the network. Normalized by the amount of useful traffic it receives, the contribution to a network by a node is maximal if equality holds in (25.9) for that node. A strict inequality implies the node enjoys more benefit from the network then nodes with equality.

For traffic originating from node j, (with rate r_j), a total charge of $\sum_{i=1}^{K} b_{ij} p_i = p_j + \sum_{i=1}^{K} (M_{ij} + L_{ij}) p_i$ per unit rate is incurred by the network for handling it. On the other hand $\sum_{i=1}^{K} a_{ij} p_i = p_j + \sum_{i=1}^{K} M_{ij} p_i$ can be viewed as the economic benefit generated by the traffic flow to the network. (One can assume the originating and destination nodes can charge end-users for traffic handling at a rate that is proportional to the shadow prices.) Hence, if equality holds in equation (25.10) for a node, traffic from that node can be viewed as enjoying the most efficient payment return ratio in the network; otherwise the payment return ratio is not optimal.

It is natural to define a rule requiring that nodes enjoying more routing benefit from the network than others should not require charges for traffic handling. Similarly, traffic rate from nodes that do not enjoy the optimal payment return ratio should be set to zero. Hence, one can adopt von Neumann's concept of an equilibrium solution here. That is an equilibrium solution to (25.9) and (25.10) is a pair of non-negative, non-zero vectors, $\mathbf{r}$ and $\mathbf{p}$, satisfying both set of equations for some positive constants, α and β, with the additional property that if an inequality holds for index i in (25.9) then $p_i = 0$ and if an inequality holds for index j in (25.10) then $r_j = 0$. That is:

$$\begin{cases} \alpha \sum_{i=1}^{K} a_{ij} p_i \leq \sum_{i=1}^{K} b_{ij} p_i, & \text{and} \quad r_j = 0 \quad \text{if} < \text{applies;} \\ \beta \sum_{j=1}^{K} a_{ij} r_j \geq \sum_{j=1}^{K} b_{ij} r_j, & \text{and} \quad p_i = 0 \quad \text{if} > \text{applies;} \end{cases} \tag{25.11}$$

From Neumann's result (see [8]), one can claim that there always exists a unique equilibrium value γ for the model discussed here so that $\alpha = \beta = \gamma$. If a perfectly fair index also exists for the system, then it is equal to or less than $1/\gamma$.

25.3 Distributed Controller and Its Convergence

Based on the approach first proposed in [9] and further extended in [10], we present here a tri-state distributed control algorithm that can achieve a pre-defined fairness index target. This algorithm assumes that each node only has a local view of the network: it knows the traffic rate it generates, the total rate of traffic ultimately destined to it, and the total rate of transitory traffic passing through it. Note that none of the nodes is assumed to know any global system-wide parameters. Each node computes its fairness index individually, and updates its originating traffic rate according to the following rule:

$$r_i(n+1) = \begin{cases} r_i(n)\delta & \text{if} \quad \gamma_i(n) > \varepsilon\lambda_i, \\ r_i(n)\delta^{-1} & \text{if} \quad \gamma_i(n) < \varepsilon^{-1}\lambda_i, \\ r_i(n) & \text{else.} \end{cases} \qquad (25.12)$$

In the above algorithm, δ and ε are scalar parameters which control the speed of convergence. They should satisfy the conditions $\delta > 1$ and $\delta^2 \leq \varepsilon$. Moreover, $\{\lambda_1, ...\lambda_K\}$ denotes a set of feasible performance targets and $\gamma_i(n)$ represents the fairness index for node i at iteration n. For example, if one considers the destination-weighted fairness index, then $\gamma_i(n)$ is defined as:

$$\gamma_i(n) = \frac{\sum_{j=1}^{K} M_{ij}r_j(n)}{r_i(n) + \sum_{j=1}^{K}(M_{ij} + L_{ij})r_j(n)}. \qquad (25.13)$$

The algorithm can start from any positive initial state $(r_1(0), ..., r_K(0))$. The traffic rates, $(r_1(n), ..., r_K(n))$, are adjusted according to 25.12 at each iteration. Therefore, for any server i the control levels are of the form $r_i(0)\delta^k$ for some integer k. It is shown in [2] that under suitable technical conditions, the simple algorithm defined by 25.12 always converges to $(r_1^*, ...r_K^*)$ and $\{\gamma_i^*\}$ such that

$$\gamma_i^* = \frac{\sum_{j=1}^{K} M_{ij}r_j^*}{r_i^* + \sum_{j=1}^{K}(M_{ij} + L_{ij})r_j^*}, \quad \text{and} \quad \varepsilon^{-1}\lambda_i \leq \gamma_i^* \leq \varepsilon\lambda_i. \qquad (25.14)$$

Numerical computations were carried out to examine the convergence property of the tri-state algorithm under different network topologies, system parameters, initial values, scalar parameters, etc. Some typical examples are shown below.

First consider a 6-node network with topology shown in Figure 25.1.

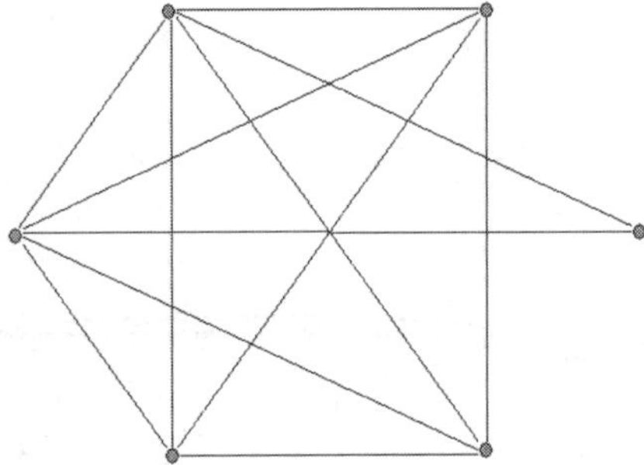

Fig. 25.1. Topology of a 6-node network.

Assume the traffic distribution, $\mathbf{M}$, and the transitory matrix, $\mathbf{L}$, are set as follows:

$$\mathbf{M} = \begin{bmatrix} 0 & 0.3433 & 0.1622 & 0.2240 & 0.2740 & 0.0724 \\ 0.1991 & 0 & 0.1615 & 0.2006 & 0.2915 & 0.1893 \\ 0.1772 & 0.2063 & 0 & 0.0626 & 0.2072 & 0.2564 \\ 0.1276 & 0.3604 & 0.4192 & 0 & 0.1923 & 0.1229 \\ 0.1489 & 0.0605 & 0.1479 & 0.1900 & 0 & 0.3589 \\ 0.3473 & 0.0295 & 0.1093 & 0.3229 & 0.0350 & 0 \end{bmatrix} \qquad (25.15)$$

$$\mathbf{L} = \begin{bmatrix} 0 & 0.4332 & 0.5532 & 0.5270 & 0.4876 & 0.6196 \\ 0.6417 & 0 & 0.6585 & 0.6401 & 0.5477 & 0.7089 \\ 0.6584 & 0.6100 & 0 & 0.7452 & 0.6119 & 0.6458 \\ 0.5912 & 0.4218 & 0.3820 & 0 & 0.5433 & 0.5849 \\ 0.5766 & 0.6217 & 0.5628 & 0.5502 & 0 & 0.4227 \\ 0.1650 & 0.1941 & 0.1831 & 0.1756 & 0.2459 & 0 \end{bmatrix} \qquad (25.16)$$

For these parameters, there exists a perfect destination-weighted index, λ, with value 0.2159. Set δ equal to 1.001, $\varepsilon = \delta^2$ and set the initial traffic rate to be $[1, 1, 1, 1, 1, 1]^T$. Under the tri-state distributed control algorithm, the fairness indices of the nodes converge to γ^* after 928 iterations, where $\gamma^* = [0.2162, 0.2163, 0.2155, 0.2155, 0.2162, 0.2155]^T$. The trajectories of fairness indices of the six nodes are shown in Figure 25.2.

As stated previously, the tri-state algorithm can also be used to achieve feasible performance targets other than perfectly fair solutions. As an example, consider again the previous network. Let a set of performance targets, $\{\lambda_i, 1 \le i \le K\}$, be specified by the vector,

$$\lambda = [0.2462, 0.1958, 0.2186, 0.2896, 0.2054, 0.1290]^T. \qquad (25.17)$$

This set of targets can be attained by setting the traffic rates to $\mathbf{r}_\lambda = [0.2, 0.5, 0.3, 0.1, 0.6, 0.8]^T$. Starting from an arbitrarily chosen initial state, the fairness index trajectories of all the nodes converge to this feasible target value under the tri-state distributed algorithm as shown in Figure 25.3.

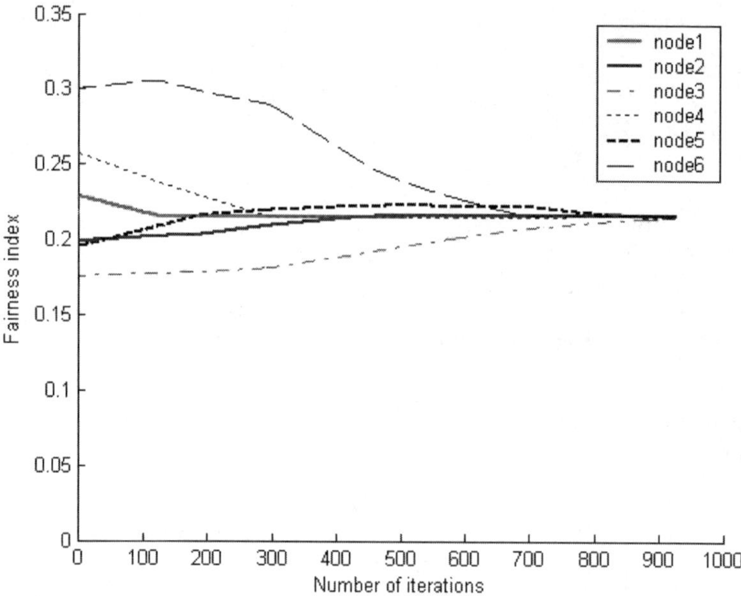

Fig. 25.2. Convergence to a perfectly destination-fair solution for a 6-node network.

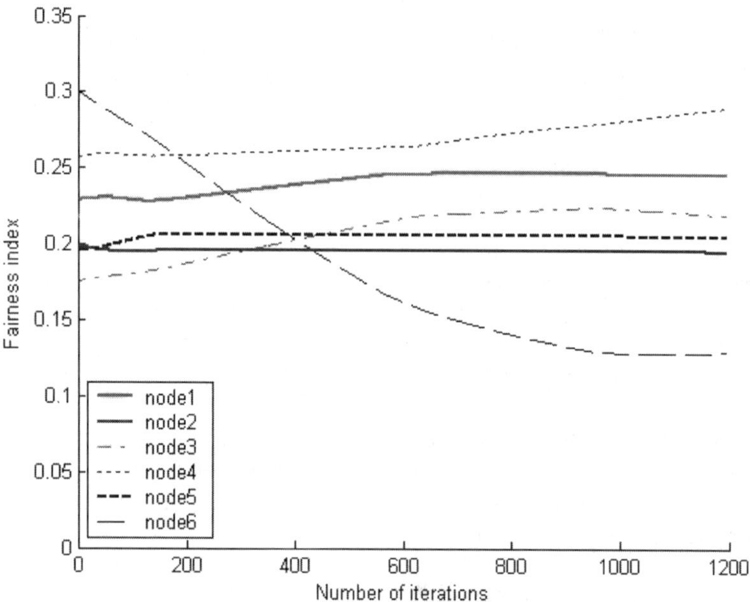

Fig. 25.3. Feasible performance target convergence for a 6-node network.

Consider now the 7-node network shown in Figure 25.4. By choosing the matrices $\mathbf{M}$ and $\mathbf{L}$ properly, one can find cases where the fairness index exists. Simulation results show that for a feasible set of performance targets, the tri-state algorithm always converges to the targeted performance band. Some simulation results for this network are listed in Table 25.1. (In these simulations, $\delta = 1.001$ and $\varepsilon = \delta^2$).

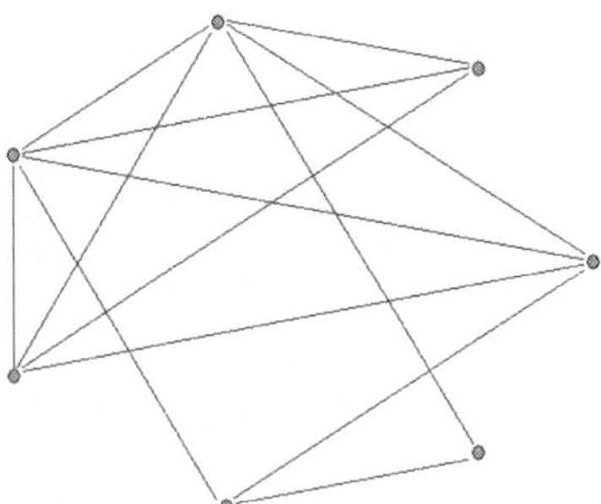

Fig. 25.4. Topology of a 7-node network.

Table 25.1. Convergence results of a 7-node network.

Target	0.1912	0.1986	0.1953	0.1979	0.1921
	0.1909	0.1983	0.1949	0.1982	0.1919
	0.1913	0.1984	0.1949	0.1979	0.1925
Converged	0.1916	0.1990	0.1956	0.1982	0.1925
	0.1908	0.1990	0.1956	0.1975	0.1919
Indexes	0.1909	0.1983	0.1949	0.1982	0.1917
	0.1916	0.1983	0.1949	0.1975	0.1922
	0.1916	0.1990	0.1956	0.1975	0.1917
No. of Iterations	1348	1305	2953	615	1223

The convergence process is affected by parameters such as the initial traffics rate and the scalars, δ and ε. Choosing these values correctly can speed up the convergence. Simulation runs were conducted for the previous 7-node network example. For one set of runs, the initial traffic rates were chosen independently from an identical distribution. For a second set of runs, the initial rates were chosen to be identically for all the nodes. Simulation results show that the algorithm converges faster for runs starting with identical initial rates, if other conditions are set identically. Effects of the adjusting the parameters, δ and ε, were also studied. In general, the lager these parameters are, the faster the convergence rate.

Table 25.2. Convergence rate under different conditions.

Target	δ	ε	Initial state chosen	No. of Iterations
0.1921	1.001	1.002001	Identically	1223
			Independently	1954
0.1937	1.001	1.002001	Identically	4949
			Independently	5602
0.1921	1.01	1.03	Identically	89
			Independently	243
0.1937	1.01	1.03	Identically	389
			Independently	536

From the table, the tri-state control algorithm appears to converge reasonably fast. Nevertheless, a weakness for the algorithm is that it requires the feasible performance target be known beforehand and any invalid target value will result in a process that never converges. It is difficult to determine a priori whether a set of performance targets is feasible or not in a large-scale network. However, a heuristic algorithm exists for reaching a perfectly fair solution, even if the index value is not known a priori. The basic idea is to apply the previous distributed algorithm using an estimated index value and updates the target value if the algorithm does not converge. The target update procedure implies that the algorithm is only partially distributed, since a central server is needed to exchange and updated target value to all the nodes. The heuristic algorithm is defined below:

1. Set proper initial values: Set δ and ε, $1 < \delta^2 \leq \varepsilon$, with relatively large values; set initial target value, γ_0, and initial rate vector, $\mathbf{r}_0$.
2. Perform the tri-state algorithm until it converges or remains unchanged at a value outside the target band after a maximal iteration number, κ.
3. If the maximum iteration number is reached, the nodes send their current index value to the central server. The mean of the current index values is set as the target value for the next turn.
4. If the tri-state algorithm stops due to convergence to currently set target zone, the error control parameters are decreased. (One approach is to set the new value to the square root of the current value.)
5. Re-do the target tracking process using the tri-state algorithm until δ and ε are decreased to predefined acceptable value δ_0 and ε_0 respectively.

Numerical computations were carried out to examine the convergence behavior of the heuristic algorithm. A 5-node network with a fairness index value of 0.2462 was studied. Key convergence stages of the heuristic algorithm are

Table 25.3. Convergence rate of the heuristic algorithm.

Target	δ	ε	Lower bound	Upper bound	Status	No. of Iterations
0.3	1.7320	3.0000	0.1	0.9	Converged	1
0.3	1.3160	1.7320	0.17321	0.5196	Converged	1
0.3	1.1471	1.3160	0.2280	0.3948	Converged	6
0.3	1.0710	1.1471	0.2615	0.3441	Not converged	281
0.2460	1.0710	1.1471	0.2145	0.2822	Converged	4
0.2460	1.0348	1.0710	0.2297	0.2635	Converged	11
..	...	...	...	...	...	...
0.2460	1.0005	1.0011	0.2457	0.2463	Not converged	761
0.2462	1.0002	1.0005	0.2459	0.2461	Converged	1499

shown in Table 25.3. The target value was initially set to be 0.3. It took 11 outer loops for algorithm to converge to the perfectly fair index value.

Conclusion

A new fairness index for communication networks is introduced in this paper. This index can be used as a reference point to measure the contribution of independent, self-authority servers to the networks. Some initial properties of this index are presented here. However, much remains to be done. Issues such as how the index can be applied to practical network control and traffic pricing are of great interest.

Acknowledgment

The authors would like to thank Professor Dah Ming Chiu for helpful discussions on this work. In particular, the discussions helped us reformulate the definition of fairness index from the form stated in [3] to the form stated here.

References

1. Andy Oram. *Peer-to-Peer: Harnessing the Power of Disruptive Technologies.* O'Reilly & Associates, 2001.
2. Wing Shing Wong and Fengjun Li. Fairness index and distributed control in communication networks. Preprint, 2004.
3. Wing Shing Wong. Fairness issues in peer-to-peer networks. In *Proceedings of the IEEE 2002 Conference on Decision and Control*, Las Vegas, Nevada, Dec 2002.
4. Jean-Yves Le Boudec. Rate adaptation, congestion control and fairness: a tutorial. Avaliable: http://ica1www.epfl.ch/PS_files/LEB3132.pdf.

5. Frank Kelly. Charging and rate control for elastic traffic. *European Transactions on Telecommunications*, 8:33–37, Jan 1997.
6. Richard J. Gibbens and Frank Kelly. Resource pricing and the evolution of congestion control. *Automatica*, 35:1969–1985, 1999.
7. E. Seneta. *Non-negative Matrices and Markov Chain*. Springer-Verlag, New York, second edition, 1981.
8. John von Neumann. A model of general economic equilibrium. *The Review of Economic Studies*, 13(1):1–9, 1945-1946.
9. Chi Wan Sung and Wing Shing Wong. A distributed fixed-step power control algorithm with quantization and active link quality protection. *IEEE Transactions on Vehicular Technology*, 48(2):553–562, 1999.
10. Wing Shing Wong and Chi Wan Sung. Robust convergence of low data rate distributed controllers. *IEEE Transactions on Automatic Control*, 49(1):82–87, 2004.

Lecture Notes in Control and Information Sciences

Edited by M. Thoma and M. Morari

Further volumes of this series can be found on our homepage:
springeronline.com

Vol. 257: Moallem, M.; Patel, R.V.; Khorasani, K.
Flexible-link Robot Manipulators
176 p. 2001 [1-85233-333-2]

Vol. 258: Isidori, A.; Lamnabhi-Lagarrigue, F.;
Respondek, W. (Eds.)
Nonlinear Control in the Year 2000 Volume 1
616 p. 2001 [1-85233-363-4]

Vol. 259: Isidori, A.; Lamnabhi-Lagarrigue, F.;
Respondek, W. (Eds.)
Nonlinear Control in the Year 2000 Volume 2
640 p. 2001 [1-85233-364-2]

Vol. 260: Kugi, A.
Non-linear Control Based on Physical Models
192 p. 2001 [1-85233-329-4]

Vol. 261: Talebi, H.A.; Patel, R.V.; Khorasani, K.
Control of Flexible-link Manipulators
Using Neural Networks
168 p. 2001 [1-85233-409-6]

Vol. 262: Dixon, W.; Dawson, D.M.; Zergeroglu, E.;
Behal, A.
Nonlinear Control of Wheeled Mobile Robots
216 p. 2001 [1-85233-414-2]

Vol. 263: Galkowski, K.
State-space Realization of Linear 2-D Systems with
Extensions to the General nD (n>2) Case
248 p. 2001 [1-85233-410-X]

Vol. 264: Baños, A.; Lamnabhi-Lagarrigue, F.;
Montoya, F.J
Advances in the Control of Nonlinear Systems
344 p. 2001 [1-85233-378-2]

Vol. 265: Ichikawa, A.; Katayama, H.
Linear Time Varying Systems
and Sampled-data Systems
376 p. 2001 [1-85233-439-8]

Vol. 266: Stramigioli, S.
Modeling and IPC Control of Interactive Mechanical
Systems – A Coordinate-free Approach
296 p. 2001 [1-85233-395-2]

Vol. 267: Bacciotti, A.; Rosier, L.
Liapunov Functions and Stability in Control Theory
224 p. 2001 [1-85233-419-3]

Vol. 268: Moheimani, S.O.R. (Ed)
Perspectives in Robust Control
390 p. 2001 [1-85233-452-5]

Vol. 269: Niculescu, S.-I.
Delay Effects on Stability
400 p. 2001 [1-85233-291-316]

Vol. 270: Nicosia, S. et al.
RAMSETE
294 p. 2001 [3-540-42090-8]

Vol. 271: Rus, D.; Singh, S.
Experimental Robotics VII
585 p. 2001 [3-540-42104-1]

Vol. 272: Yang, T.
Impulsive Control Theory
363 p. 2001 [3-540-42296-X]

Vol. 273: Colonius, F.; Grüne, L. (Eds.)
Dynamics, Bifurcations, and Control
312 p. 2002 [3-540-42560-9]

Vol. 274: Yu, X.; Xu, J.-X. (Eds.)
Variable Structure Systems:
Towards the 21st Century
420 p. 2002 [3-540-42965-4]

Vol. 275: Ishii, H.; Francis, B.A.
Limited Data Rate in Control Systems with Networks
171 p. 2002 [3-540-43237-X]

Vol. 276: Bubnicki, Z.
Uncertain Logics, Variables and Systems
142 p. 2002 [3-540-43235-3]

Vol. 277: Sasane, A.
Hankel Norm Approximation
for Infinte-Dimensional Systems
150 p. 2002 [3-540-43327-9]

Vol. 278: Chunling D. and Lihua X. (Eds.)
H_∞ Control and Filtering of
Two-dimensional Systems
161 p. 2002 [3-540-43329-5]

Vol. 279: Engell, S.; Frehse, G.; Schnieder, E. (Eds.)
Modelling, Analysis, and Design of Hybrid Systems
516 p. 2002 [3-540-43812-2]

Vol. 280: Pasik-Duncan, B. (Ed)
Stochastic Theory and Control
564 p. 2002 [3-540-43777-0]

Vol. 281: Zinober A.; Owens D. (Eds.)
Nonlinear and Adaptive Control
416 p. 2002 [3-540-43240-X]

Vol. 282: Schröder, J.
Modelling, State Observation and
Diagnosis of Quantised Systems
368 p. 2003 [3-540-44075-5]

Vol. 283: Fielding, Ch. et al. (Eds.)
Advanced Techniques for Clearance of
Flight Control Laws
480 p. 2003 [3-540-44054-2]

Vol. 284: Johansson, M.
Piecewise Linear Control Systems
216 p. 2003 [3-540-44124-7]

Vol. 285: Wang, Q.-G.
Decoupling Control
373 p. 2003 [3-540-44128-X]

Vol. 286: Rantzer, A. and Byrnes C.I. (Eds.)
Directions in Mathematical Systems
Theory and Optimization
399 p. 2003 [3-540-00065-8]

Vol. 287: Mahmoud, M.M.; Jiang, J.; Zhang, Y.
Active Fault Tolerant Control Systems
239 p. 2003 [3-540-00318-5]

Vol. 288: Taware, A. and Tao, G.
control of Sandwich Nonlinear Systems
393 p. 2003 [3-540-44115-8]

Vol. 289: Giarré, L. and Bamieh, B.
Multidisciplinary Research in Control
237 p. 2003 [3-540-00917-5]

Vol. 290: Borrelli, F.
Constrained Optimal Control
of Linear and Hybrid Systems
237 p. 2003 [3-540-00257-X]

Vol. 291: Xu, J.-X. and Tan, Y.
Linear and Nonlinear Iterative Learning Control
189 p. 2003 [3-540-40173-3]

Vol. 292: Chen, G. and Yu, X.
Chaos Control
380 p. 2003 [3-540-40405-8]

Vol. 293: Chen, G. and Hill, D.J.
Bifurcation Control
320 p. 2003 [3-540-40341-8]

Vol. 294: Benvenuti, L.; De Santis, A.; Farina, L. (Eds.)
Positive Systems: Theory
and Applications (POSTA 2003)
414 p. 2003 [3-540-40342-6]

Vol. 295: Kang, W.; Xiao, M.; Borges, C. (Eds.)
New Trends in Nonlinear Dynamics and Control,
and their Applications
365 p. 2003 [3-540-10474-0]

Vol. 296: Matsuo, T.; Hasegawa, Y.
Realization Theory of Discrete-Time
Dynamical Systems
235 p. 2003 [3-540-40675-1]

Vol. 297: Damm, T.
Rational Matrix Equations in Stochastic Control
219 p. 2004 [3-540-20516-0]

Vol. 298: Choi, Y.; Chung, W.K.
PID Trajectory Tracking Control
for Mechanical Systems
127 p. 2004 [3-540-20567-5]

Vol. 299: Tarn, T.-J.; Chen, S.-B.; Zhou, C. (Eds.)
Robotic Welding, Intelligence and Automation
214 p. 2004 [3-540-20804-6]

Vol. 300: Nakamura, M.; Goto, S.; Kyura, N.; Zhang, T.
Mechatronic Servo System Control
Problems in Industries and their Theoretical Solutions
212 p. 2004 [3-540-21096-2]

Vol. 301: de Queiroz, M.; Malisoff, M.; Wolenski, P.
(Eds.)
Optimal Control, Stabilization
and Nonsmooth Analysis
373 p. 2004 [3-540-21330-9]

Vol. 302: Filatov, N.M.; Unbehauen, H.
Adaptive Dual Control: Theory and Applications
237 p. 2004 [3-540-21373-2]

Vol. 303: Mahmoud, M.S.
Resilient Control of Uncertain Dynamical Systems
278 p. 2004 [3-540-21351-1]

Vol. 304: Margaris, N.I.
Theory of the Non-linear Analog Phase Locked Loop
303 p. 2004 [3-540-21339-2]

Vol. 305: Nebylov, A.
Ensuring Control Accuracy
256 p. 2004 [3-540-21876-9]

Vol. 306: Bien, Z.Z.; Stefanov, D. (Eds.)
Advances in Rehabilitation
472 p. 2004 [3-540-21986-2]

Vol. 307: Kwon, S.J.; Chung, W.K.
Perturbation Compensator based Robust Tracking
Control and State Estimation of Mechanical Systems
158 p. 2004 [3-540-22077-1]

Vol. 308: Tarbouriech, S.; Abdallah, C.T.; Chiasson, J.
(Eds.)
Advances in Communication Control Networks
358 p. 2005 [3-540-22819-5]

Vol. 309: Kumar, V.; Leonard, N.; Morse, A.S. (Eds.)
Cooperative Control
301 p. 2005 [3-540-22861-6]

Vol. 310: Janczak, A.
Identification of Nonlinear Systems Using Neural
Networks and Polynomial Models
197 p. 2005 [3-540-23185-4]

Vol. 311: Lamnabhi-Lagarrigue, F.; Loría, A.;
Panteley, V. (Eds.)
Advanced Topics in Control Systems Theory
294 p. 2005 [1-85233-923-3]

Vol. 312: Henrion, D.; Garulli, A. (Eds.)
Positive Polynomials in Control
313 p. 2005 [3-540-23948-0]

Vol. 313: Li, Z.; Soh, Y.; Wen, C.
Switched and Impulsive Systems
277 p. 2005 [3-540-23952-9]

Vol. 314: Gil', M.I.
Explicit Stability Conditions for Continuous Systems
193 p. 2005 [3-540-23984-7]

Vol. 315: Herbordt, W.
Sound Capture for Human/Machine Interfaces:
Practical Aspects of Microphone
Array Signal Processing
286 p. 2005 [3-540-23954-5]

Vol. 316: R.V. Patel; F. Shapdey
Control of Redundant Robot Manipulators
224 p. 2005 [3-540-25071-9]

Vol. 317: Chuan Ma; W. Murray Wonham
Nonblocking Supervisory Control
of State Tree Structures
208 p. 2005 [3-540-25069-7]

Vol. 318: Eli Gershon; Uri Shaked; Isaac Yaesh
H_∞ Control and Estimation
of State-multiplicative
Linear Systems 272 p. [1-85233-997-7]

Vol. 319: Hofbaur; W. Michael
Hybrid Estimation of Complex Systems
176 p. [3-540-25727-6]

Vol. 320: Thomas Steffen
Control Reconfiguration of Dynamical Systems
286 p. [3-540-25730-6]

Printing and Binding: Strauss GmbH, Mörlenbach